Statistical Method
in Biological Assay

Books on cognate subjects

Experiments: Design and Analysis	J.A. John & M.H. Quenouille
Experimental Design: Selected Papers	F. Yates
The Advanced Theory of Statistics	M.G. Kendall & A. Stuart
Multivariate Analysis	M.G. Kendall
Patterns and Configurations in Finite Spaces	S. Vajda
The Mathematics of Experimental Design: Incomplete Block Designs and Latin Squares	S. Vajda
Statistical Models and their Experimental Application	P. Ottestad
The Analysis of Variance: a Basic Course	A. Huitson
Maximum Likelihood Estimation in Small Samples	L.R. Shenton & K.O. Bowman
Families of Bivariate Distributions	K.V. Mardia
Families of Frequency Distributions	J.K. Ord
The Mathematical Theory of Infectious Diseases	N.T.J. Bailey

Complete list of statistical books available from the Publishers

PLACE IN RETURN BOX to remove this checkout from your record.
TO AVOID FINES return on or before date due.
MAY BE RECALLED with earlier due date if requested.

DATE DUE	DATE DUE	DATE DUE
JUL 2 0 2002	AUG 1 6 2004 09 17 06	
AUG 2 8 2003 DEC 2 1 2009 01 05 14		

1/98 c:/CIRC/DateDue.p65-p.14

Statistical Method in Biological Assay

David J. Finney, CBE, MA, ScD, FRS, FRSE

Professor of Statistics, University of Edinburgh, and
Director, Agricultural Research Council Unit of Statistics

THIRD EDITION

CHARLES GRIFFIN & COMPANY LTD
London and High Wycombe

CHARLES GRIFFIN & COMPANY LIMITED
Registered Office:
Charles Griffin House, Crendon Street
High Wycombe, Bucks HP13 6LE
England

Copyright © 1978

All rights reserved. No part of this publication may be reproduced or transmitted in any form or by any means, electronic or mechanical, including photocopying, recording, or by any information storage and retrieval system, without permission in writing from Charles Griffin & Company Limited.

First published	1952
Second edition	1964
Second impression	1971
Third edition	1978

ISBN 0 85264 252 0

Printed in Great Britain at the Alden Press
Oxford London and Northampton

Contents

Preface   *Page*   xi

1 Introduction

1.1	The purpose of biological assay	1
1.2	The history of biological assay	2
1.3	The structure of biological assay	3
1.4	International standards	5
1.5	Statistical science and biological assay	6
1.6	Scope of this book	7
1.7	Computation	8
1.8	Statistical tables	10
1.9	Notation and terminology	10
1.10	Glossary of symbols	12

2 Direct assays

2.1	Types of biological assay	17
2.2	The nature of direct assays	17
2.3	An assay of strophanthus	18
2.4	Precision of estimates	19
2.5	Dilution assays	22
2.6	Revised computations for the strophanthus assay	25
2.7	Adjustment for body weight	26
2.8	A direct assay with covariance	27
2.9	Efficiency and utility of concomitant measurements	36
2.10	The design of direct assays	37
2.11	A criticism of direct assays	38

3 Quantitative dose–response relations

3.1	Indirect assays	39
3.2	The dose–response regression	39
3.3	Similarity	40
3.4	Assay validity	41
3.5	Preliminary regression investigation	43
3.6	Monotony	43
3.7	Linearizing transformations	43
3.8	Essentially non-linear relations	46
3.9	A response curve for vitamin B_{12}	47
3.10	Heterogeneity of variance	54
3.11	Testing homoscedasticity	55
3.12	Estimation by maximum likelihood	57
3.13	Estimation for regression equations	59
3.14	Standard curve estimation	65
3.15	Standard slope estimation	66
3.16	Simultaneous trial estimation	67

4 Parallel line assays

4.1	Unsymmetric designs	69
4.2	Data for an unsymmetric assay	69
4.3	The dose–response diagram	70

4.4	Analysis of variance	72
4.5	Scedasticity	74
4.6	Linearity	75
4.7	The difference in preparations	75
4.8	Parallelism	76
4.9	Regression	76
4.10	Significance levels	76
4.11	Potency estimation	79
4.12	Fieller's theorem	80
4.13	Analogues of Fieller's theorem	82
4.14	Fiducial limits in the vitamin D_3 assay	86
4.15	Data for a symmetric assay	88
4.16	Analysis of variance	90
4.17	Validity tests	91
4.18	Potency estimation and precision	92
4.19	Constraints of design	93
4.20	Heterogeneous deviations from linearity	95
4.21	Missing values	96
4.22	Approximate analyses for missing values	98
4.23	Exact analysis for missing values	99

5   Symmetric dose-structure for parallel line assays

5.1	The general structure	105
5.2	An assay of vitamin B_{12}	105
5.3	The symmetric (k, k)-point design	109
5.4	The (2, 2) assay	112
5.5	The (3, 3) assay	115
5.6	The (4, 4) assay	116
5.7	Aids to arithmetic	117
5.8	Range estimation for standard deviations	119
5.9	Distribution-free methods	120
5.10	Control charts	121
5.11	Quadratic regression for parallel line assay	122
5.12	Computer programs	125

6   Principles of planning an assay

6.1	The purpose of an assay	133
6.2	Validity	133
6.3	The economics of design	134
6.4	Pilot investigations	135
6.5	Symmetry	136
6.6	The cost of statistical analysis	136
6.7	Randomization	137
6.8	Parallel line assays	139
6.9	Symmetric (k, k) designs	142
6.10	Choice of k	144
6.11	Comparison of assay techniques	146

7   Slope ratio assays

7.1	The power dose metameter	148
7.2	The multiple regression equation	149
7.3	An unsymmetric slope ratio assay	150
7.4	Analysis of variance	150
7.5	Validity tests	152
7.6	Potency estimation	155
7.7	General formulae	156

	7.8 The symmetric $(1, k, k)$-point design	157
	7.9 The $(1, 1, 1)$ assay	159
	7.10 The $(1, 2, 2)$ assay	159
	7.11 Other $(1, k, k)$ assays	162
	7.12 The symmetric $(0, k, k)$-point design	164
	7.13 Routine assays	165
	7.14 Other slope ratio problems	166
8	**Efficiency in slope ratio assays**	
	8.1 General principles	167
	8.2 Symmetric $(1, k, k)$ designs	169
	8.3 The asymmetric $(1, 1, 1)$ design	171
	8.4 The choice of k	173
	8.5 Symmetric $(0, k, k)$ designs	173
	8.6 Comparison of assay techniques	175
9	**Incomplete block designs**	
	9.1 Design in bioassay	177
	9.2 Confounding	178
	9.3 An incomplete block assay of vitamin A	179
	9.4 Criticism of the design for the assay of vitamin A	182
	9.5 A balanced incomplete block assay of vitamin A	183
	9.6 Statistical analysis of balanced incomplete block designs	191
	9.7 Catalogue of balanced incomplete block designs	192
	9.8 Unbalanced confounding in the assay of vitamin A	195
	9.9 Catalogue of confounded designs	197
	9.10 Blocks of unequal size	201
10	**Cross-over designs**	
	10.1 Intra-subject estimation	203
	10.2 The twin cross-over	204
	10.3 A cross-over assay of corticotrophin	204
	10.4 Efficiency of cross-over designs	208
	10.5 Catalogue of cross-over designs	209
	10.6 A cross-over assay of a parathyroid extract	215
	10.7 Single subject assays	219
	10.8 An exact analysis with missing subjects	223
11	**Multiple assays**	
	11.1 The economy of multiple assays	227
	11.2 An assay of two tuberculins	228
	11.3. A multiple assay of vitamin B_{12}	230
	11.4 Multiple slope ratio assays	234
	11.5 Designs for multiple assays	235
	11.6 The allocation of subjects to preparations	236
12	**The use of concomitant information**	
	12.1 The combination of measurements	240
	12.2 Adjustment by proportionality	241
	12.3 Adjustment by covariance analysis	242
	12.4 An example of covariance	243
	12.5 An assay of a parathyroid extract	247
	12.6 Validity tests from the concomitant	257
	12.7 The economics of covariance adjustment	258

13 Composite responses

- 13.1 Discriminants — 260
- 13.2 An assay analysis using a discriminant function — 261
- 13.3 Exact multivariate analysis — 265
- 13.4 Discriminants and concomitants — 268
- 13.5 The economics of discriminant analysis — 268

14 The combination of estimates

- 14.1 Weighted means — 269
- 14.2 Illustrative calculations — 271
- 14.3 The combination of parallel line assays — 273
- 14.4 An example of combining cross-over assays — 279
- 14.5 Replication within blocks — 281
- 14.6 Two assays of vitamin D_3 — 286
- 14.7 Combination of slope ratio estimates — 290
- 14.8 Antiserum activity — 290
- 14.9 Neutralization of antiserum — 294

15 Validity and the choice of metameters

- 15.1 Fundamental validity — 297
- 15.2 The validity of (2, 2) assays — 299
- 15.3 Statistical validity — 299
- 15.4 Comments on statistical validity — 301
- 15.5 The objectivity of statistical analysis — 305
- 15.6 Alternative metametric transformations — 307
- 15.7 Further notes on metameters — 313
- 15.8 Response metameters and covariance analysis — 314

16 General transformations and radioligand assays

- 16.1 Curvature and scedasticity — 316
- 16.2 An assay of *dl*-tryptophan — 316
- 16.3 Weighted analysis of the assay of *dl*-tryptophan — 320
- 16.4 Criticism of the design of the tryptophan assay — 326
- 16.5 Least squares by computer — 327
- 16.6 Radioligand assays — 328
- 16.7 The RLA sigmoids — 330
- 16.8 The variance function — 334
- 16.9 Estimation and computation for RLA — 336
- 16.10 Assay of oestradiol — 339
- 16.11 Alternative principles of estimation — 344
- 16.12 Design for RLA — 345
- 16.13 Programs for RLA — 346

17 Quantal responses and the tolerance distribution

- 17.1 The use of quantal responses — 349
- 17.2 Minimal effective dose — 350
- 17.3 Median effective dose — 350
- 17.4 The equivalent deviate transformation — 351
- 17.5 Estimation of the two parameters — 353
- 17.6 The test of homogeneity — 355
- 17.7 Variance of estimates — 356
- 17.8 Normal sigmoid — 357
- 17.9 Logistic sigmoid — 358
- 17.10 Wilson–Worcester sigmoid — 360

	17.11 Cauchy–Urban sigmoid	360
	17.12 Angle sigmoid	361
	17.13 'Linear' sigmoid	361
	17.14 Comparisons between transformations	362
	17.15 Choice of a transformation	366
	17.16 A generalized model	368
	17.17 Polytomous quantal responses	368
18	**Assays based on quantal responses**	
	18.1 The classical computational scheme	370
	18.2 An assay of insulin	375
	18.3 Speed of convergence	379
	18.4 Variants of the method	381
	18.5 Computer methods	381
	18.6 Approximate methods of analysis	390
	18.7 Spearman–Kärber and moving averages	394
	18.8 Comments and comparisons	398
	18.9 Minimum χ^2	401
19	**The design of assays based on quantal responses**	
	19.1 Principles of good design	404
	19.2 The choice of doses	406
	19.3 Block restrictions	410
	19.4 Cross-over designs	413
	19.5 Multiple assays	414
	19.6 Sequential sampling	414
	19.7 Quality control	416
	19.8 Comparison of assay techniques	416
20	**Special problems with quantal responses**	
	20.1 Concomitant variation	418
	20.2 Combination of estimates	418
	20.3 Natural response rates	419
	20.4 Assays with unknown numbers of subjects	423
	20.5 Dilution series	425
	20.6 Density estimation for rope bacillus	432
	20.7 The Fisher estimator	434
	20.8 Design for dilution assays	435
	20.9 Information from different types of assay	437
21	**Time as a response**	
	21.1 Time responses	440
	21.2 An assay of a virus	441
	21.3 Complete analysis	447

References   449

Appendix tables   465

    I The distribution of t 466
   II The distribution of the variance ratio   467
  III The Behrens distribution   468
  IV The distribution of χ^2 469
    V The probit transformation   469

VI Minimum and maximum working probits, ranges, and weighting coefficients	470
VII Working probits	471
VIII The logit transformation	481
IX Minimum and maximum working logits, ranges, and weighting coefficients	482
X Working logits	483
XI The angle transformation	485
XII Minimum and maximum working angles and ranges	486
XIII Working angles	488
XIV Transformation of angles to probits	490
XV Transformation of angles to logits	490
XVI The loglog transformation	491
XVII Minimum and maximum working loglog deviates, ranges, and weighting coefficients	492
XVIII Distribution of range of transition in dilution series	495

Index of Authors   497

Index of Subjects   501

Preface to the third edition

I intend this book to provide a comprehensive account of statistical methods and experimental designs for biological assays, and in doing so to be both a reference work for the practising assayist and a text-book for students. It assumes some familiarity with basic statistical techniques used in biology, notably those connected with analysis of variance and regression, but presents any theory specific to bioassay alongside discussion of applications.

In the Preface to the First Edition, I wrote:

'Many experimenters who are not themselves statisticians can, usefully and without undue difficulty, attain sufficient knowledge of statistical techniques to aid their own work. Nevertheless, I make no apology for emphasizing in this book the rôle of the professional statistician rather than that of a particular group of experimental scientists. The proper function of the statistician in scientific research and technology is no longer merely that of analyzing and summarizing large bodies of numerical data: he must also be prepared to advise on the plan and economy of each investigation, in the light of its operational efficiency for a particular purpose. The efficiency attainable will depend upon the nature and reliability of existing information. The statistician must therefore be a major contributor to the answering of the question: "In the present state of knowledge about the nature and behaviour of certain materials and subjects of experimentation (or of non-experimental observation), how should the next investigation be planned in order to obtain the most trustworthy information on specified points from limited resources of materials, subjects, or time?" He must later analyze the data from the investigation, report on the information they give for the immediate purpose, and be prepared to integrate this information with that previously existing, as a prelude to the planning of further investigations on the same or related topics. The cycle of design, analysis, report, and integration recurs frequently in the work of the statistician, and the intimate relationship between its parts is particularly well exemplified in biological assay. A familiarity with the details of assay design and analysis is to-day necessary to many statisticians working as consultants in scientific research; in addition, a study of the manner in which the parts of this cycle interlock may be enlightening even to those whose chief interests have no connexion with bioassay.'

After 27 years, I have not changed my belief that bioassay is a field of applied statistics with great educational value for the student. Within a relatively simple logical framework, he can see the close relations beteween estimation theory, significance testing, experimental design and planning, distribution theory, linear and non-linear models, and other facets of statistical science. In this time, however, the importance of the assayist himself having some statistical understanding has vastly increased. Involvement of professional statisticians is essential to the development of new types of assay and to large collaborative studies; they are less commonly engaged in the day-to-day handling of assay analyses, because computer packages now enable the assayist himself — or even a junior assistant — to use statistical techniques that once were seen as so complex and time-consuming as to demand attention from a statistician. Unless the assayist himself knows enough about the statistical logic and the structure of programs, he may fail to use methods suited to his experiments or may overlook indications that anomalies or unsuspected errors make a particular set of data misleading. The great expansion in use of some types of assay (notably the radioimmunoassays) that are essential to patient care surely renders neglect of quality in statistical analysis as blameworthy as neglect of proper maintenance of equipment or negligence in identifying and measuring samples.

Although this edition retains the same form as its predecessors, it has been completely rewritten. I have endeavoured to make clear how computers and general programs can remove much of the labour previously associated with bioassay, especially for quantal responses; I have introduced new illustrative examples, recomputed most analyses, and shown computer output. The simplicity and low cost of modern computation has justified me in omitting various approximate procedures and advising against the use of others. One of the greatest gains to statistics from computers is the ease of initiating and executing iterative calculations. Dose-response relations for quantitative and quantal responses often need non-linear regressions for their realistic representation; these can now be handled without undue labour, but careful study of data and equations is still needed in order to steer a safe course between inadequacy and overparametrization. Moreover, the classical probit and logit iterative regression calculations can be replaced by direct optimization techniques that lead to the same answers expeditiously and more accurately than before. New material introduced includes a recent developed general method for combining potency estimates from several assays, a discussion of the general principles of analysis for radioligand assay with emphasis on the four-parameter logistic curve, and ideas on antiserum assays. Many examples drew attention to the robustness of conclusions from good assay data, with emphasis on the way in which a computer permits robustness to be confirmed by repeated analyses of the same data.

I can never forget the debt I owe to Dr Eric C. Wood and to my late father for their help with the First Edition. Among many others to whom I am grateful are: Professor P.A.P. Moran who read the original typescript, Professor J. Berkson whose correspondence on quantal responses has always been stimulating, Professor S.Z. Cekan, Professor E. Diczfalusy and Dr D. Rodbard who have done much to help my understanding of radioimmunoassay, numerous friends who have drawn my attention to small errors, and not least my wife who has patiently read drafts and checked indexing. This Edition owes much to careful typing by various members of my staff and to thorough proof reading by Mrs E. Purves and Miss E.M. Heyburn. I acknowledge elsewhere my thanks for permission to quote data and to reproduce tables.

D. J. FINNEY

EDINBURGH
May 1978

1
Introduction

1.1 The purpose of biological assay

A biological assay (or bioassay) is an experiment for estimating the nature, constitution, or potency of a material (or of a process), by means of the reaction that follows its application to living matter. For example, a substance might be identified by means of a characteristic reaction produced in a particular species of organism. Such *qualitative assays* rarely present major statistical problems. *Quantitative assays*, with which this book is concerned, are similar to methods of physical measurement or of quantitative chemical analysis in that they lead to numerical assessment of some property of the material to be assayed; essential to this assessment is measurement of growth or change in animals, plants, animal tissue, micro-organisms, or some other form of living matter. A typical comparative experiment in biology involves applying known treatments to subjects, measuring the subjects, and so estimating the differences between the effects of treatments. A biological assay may employ the same experimental techniques, but its aim is to use the measurements as a foundation for comparing the potencies of the treatments. This new aim affects both the optimal experimental design and the statistical analysis. An investigation into the effects of different samples of corticotrophin on the ascorbic acid in rat adrenals is not necessarily a biological assay; it becomes one if the experimenter's interest lies in using the changes in ascorbic acid for estimating the potencies of the samples in standard units of corticotrophin. A field trial of the responses of potatoes to various phosphatic fertilizers would not generally be an assay; if the yields of potatoes are to be used in assessing the potency of one phosphatic compound relative to another, and perhaps even in estimating the nutritional availability of phosphorus, the experiment is an assay within the terms of the description here given.

Wood (1946a) wrote: 'If one attempts to classify analytical procedures with reference not to the substances analyzed nor to the ingredients determined, but to the fundamental principles involved – the strategy underlying the attack, as it were – one soon finds that many methods, which vary widely from the former view-point, can be broadly described in the same terms. First, some physical quantity must be found – it may be the weight of an animal, the volume of a reagent, the reading on the dial of an instrument – the magnitude of which depends on, and varies regularly with, the amount of the substance it is desired to estimate, and which I shall refer to as "Factor X". Second, the quantitative relation between the amount of "factor X" and the magnitude of the effect it causes is determined by performing parallel sets of operations with various known amounts or "doses" of the factor and measuring the result, which we may call the "response". The relation between the dose and the response may be concisely expressed either diagrammatically in the form of a graph, or algebraically in the form of an equation. Finally, a known amount of the material to be analyzed is put through an identical series of operations, the response is measured, and the amount of factor X present is deduced from the graph or the equation.' Wood's description includes purely physical procedures, such as absorptiometric determination of trace-elements in alloys, as well as true biological assays. The distinguishing feature of the latter is that the dependence upon living matter almost inevitably introduces considerable variability between measurements obtained by apparently identical operations. Within an experiment of practicable size, the estimation of the dose–response relation, and consequently of the potency of the material to be analyzed, is far from perfect; the methods of statistical science enable best use to be made of the available data.

Bacharach (1945) discussed the relation of biological assay to chemical analysis. He emphasized that, when a biological and a chemical method are available for the same purpose, the greater precision of the latter does not invariably make it preferable. If the result required is an estimate of the biological potency of a complex material, the specificity of a biological method may give it an advantage over a chemical analysis that must be followed by numerical synthesis of the potency of the original material from the potencies of its constituents.

1.2 The history of biological assay

Although biological assay could not become an instrument for precise measurement without adequate development of statistical science, the basic principle has been in use for a long time:

> And it came to pass at the end of forty days, that Noah opened the windows of the ark which he had made:
> And he sent forth a raven, which went to and fro, until the waters were dried up from off the earth.
> Also he sent forth a dove from him, to see if the waters were abated from off the face of the ground;
> But the dove found no rest for the sole of her foot, and she returned unto him into the ark, for the waters were on the face of the whole earth: then he put forth his hand, and took her, and pulled her in unto him into the ark.
> And he stayed yet other seven days; and again he sent forth the dove out of the ark;
> And the dove came in to him in the evening: and lo, in her mouth was an olive leaf pluckt off; so Noah knew that the waters were abated from off the earth. (Genesis 8: 6–11)

This is an excellent account of an assay that, although only qualitative in character, has features of modern quantal response techniques. The three essential constituents of an assay, stimulus and dose (depth of water), subject (the dove), and response (the plucking of an olive leaf), are well described. Knowledge of the response enabled Noah to estimate, or rather to place an upper limit on the size of the stimulus. His animal house could not provide the replication that would today be recommended, but in other respects his assay was admirable.

The serious scientific history of biological assay began late in the nineteenth century with Ehrlich's investigations into the standardization of diphtheria antitoxin. Since then, in pharmacology, endocrinology, plant pathology and other sciences, standardization of materials by means of the reactions of living matter has become a common practice. The development of pharmacological standardization has often been described, notably by Dale (1939), Gautier (1945), and Hartley (1935, 1945b). Not until early (1920–1925) attempts at standardizing insulin was any assessment of the trustworthiness of assay results attempted. Potency was at first measured in animal units, the unit being the amount required to produce a specified response in an animal of a particular species. In their responses to and in their tolerances of drugs, however, animals are as variable as they are in more easily measurable bodily characteristics. The cat unit of digitalis and the mouse unit of insulin were soon discovered not to be constants. The introduction of standard preparations of various drugs, against which others could be assayed, made possible the measurement of potency on a fixed scale, independently of the particular animals used: the difference is analogous to that between a spring balance and a balance using a standardized set of weights. To make this the general practice, said Burn (1930), could transform 'this whole subject from the plane of an insidious means of self-deception to that of a well-ordered and progressive science'.

Historical surveys by Irwin (1937, 1950), Bliss and Cattell (1943), and Gaddum (1953a) have given extensive bibliographies of publications on the theory and practice of assays. Finney (1947a) presented a systematic account of the statistical principles, and Jerne and Wood (1949) published an important detailed discussion of the assumptions common to most assay techniques. Papers by Miles (1948, 1950, 1951) and Lightbown (1961) covered similar ground, and reviewed the nature and function of bioassay in relation to modern pharmacology and medicine. In 1948, Emmens published the first book devoted purely to the statistical aspects of biological assay; Burn, Finney and Goodwin (1950) included a chapter on statistical technique. Bliss's long chapter for a book by György (1951), including much that complements the present book and many valuable numerical examples, was subsequently revised and published separately (1952a). Books by Coward (1947) and Gaddum (1948), though primarily concerned with the biological problems of assay, are historically important for having directed attention to statistical considerations.

1.3 The structure of biological assay

A typical assay involves a *stimulus* (for example, a vitamin, a hormone, a fungicide) applied to a *subject* (for example, an animal, a piece of animal tissue, a plant, a bacterial culture). The size of the stimulus may be varied, generally in accordance with a plan made by the investigator. This size, the *dose* given to the subject, can be measured (perhaps as a weight, a volume, or a concentration).

The *response* of the subject is a measurement of the final value of some characteristic of the subject (body weight, kidney weight, bone ash percentage) or of the change in a particular characteristic (increase in body weight, decrease in blood pressure); it may be a simple record of occurrence or non-occurrence of a phenomenon (recovery from disease, death, a muscular contraction). The magnitude or the frequency of the response depends upon the dose. The relation between dose and response will be obscured by random variations between replicate subjects, or, if the stimulus is one that can be applied more than once to the same subject, in the responsiveness of one subject on different occasions. Nevertheless, the relation enables the potency of a dose to be inferred from the responses it induces.

As subsequent chapters exemplify, biological assays are usually comparative, potency being estimated relative to a *standard preparation* of the stimulus. The standard preparation may be a sample of an internationally agreed standard, or of a more readily available working standard whose potency relative to the international standard has previously been carefully evaluated, or of a stock maintained as a provisional laboratory standard. The potency of the standard is expressed in ordinary units of weight or volume or in arbitrary units, such as the international units adopted for insulin and penicillin (Miles, 1949). The potency of any *test preparation* of the stimulus is assayed by finding the ratio between equivalent doses of it and the standard preparation, equivalence being interpreted as equality of the corresponding mean responses; experimentation with several different doses of one or both preparations is almost always needed in order to accomplish this satisfactorily.

This book presents methods in terms of an ideal situation in which the test and standard preparations are identical in their biologically active principle and differ only in the extents to which they are diluted by solvents or other inactive materials (§§2.5, 3.3, 3.4, 15.1–5). An assertion that all assays should be of this kind would be unrealistic. Indeed, certainty may imply so complete an understanding of the preparations that quantitative physical or chemical analysis can replace biological assay for potency estimation: such a change is not always desirable, but its advantages, in economy and in precision, need consideration (Lightbown, 1961). Unfortunately for the peace of mind of the statistician, many important assays will not conform to the ideal conditions. It then becomes necessary to discuss the nature of the departure from the ideal and the manner in which this should affect the interpretation of any statistical analysis. The primary considerations are not really statistical; they pertain to the use that is to be made of assay results and the question of whether a relative potency estimated from one response or one species of subject can be assumed to have even approximate validity for another response or species.

The British Pharmacopoeia (1973) and the United States Pharmacopeia (1970) were pioneers in including sections describing the computations required for the more important biological assays. The latter follows Bliss's style and notation, and has been supplemented by separate publication of a series of numerical examples (Bliss, 1956). The European Pharmacopoeia (1971) now contains a rewritten version of what used to appear in the British Pharmacopoeia, with many new examples. These publications do not discuss the statistical principles or the validity of the underlying biological and statistical assumptions.

1.4 International standards

Until 1914, standardized diphtheria antitoxin was supplied from Ehrlich's Institute in Frankfurt and from the Hygienic Laboratory in Washington. In 1921, a conference convened by the Health Committee of the League of Nations determined that these two standards agreed well, and recommended adoption of Ehrlich's unit as the international unit for the antitoxin. By 1924, the need for other international standards was apparent, and the League of Nations established the Permanent Commission on Biological Standardization. The State Serum Institute in Copenhagen and the National Institute for Medical Research in London acted as custodians for an increasing number of standards, 36 in 1939. The Expert Committee on Biological Standardization of the World Health Organization has continued this work; a later report (WHO, 1975) listed over 150 international standards for antigens, antibodies, antibiotics, hormones, vitamins, enzymes, and other pharmacological substances (including a few for veterinary use). At least 20 more were under consideration and 25 had been discontinued as no longer needed. These standards form an indispensable part of the system of measurement necessary to modern medicine.

The precise nature of internationally accepted standard preparations is not important to the presentation of this book, though it may be vitally important to the proper interpretation of an assay. Even the statistician, however, ought to be aware of the general character of the precautions taken in selecting, preparing and preserving a standard, and in distributing it for world-wide use. Hartley (1935) stated: 'The biological standard chosen for international adoption must fulfil certain conditions. In nearly every case the standard is a dry preparation of an arbitrarily chosen but representative sample of the substance for which it is to serve as a standard, or, when a pure substance is available, the sample must comply with accurately defined physical constants. It is obvious that, since the standard must satisfy world requirements for many years, the quantity set aside as a standard must be large. Secondly, the standard must be stable, a condition which is fulfilled by preparing it in the absolutely dry condition and preserving it constantly at temperatures at, or below, the freezing point, in sealed containers, and protected from the action of light, moisture and oxygen. Thirdly, the standard must be dispensed in such a form as to be readily accessible and capable of being brought into use by the laboratory worker with a minimum of trouble, additional manipulation or delay.' Gautier (1945) and Hartley (1945a) described the necessary careful control over all stages of the preparation and distribution. Dale (1939), Hartley (1945b), and Miles (1951) discussed the establishment of new standards. Jerne and Perry (1956) reported on the stability of some international standards. A standard preparation has much the same function as a set of standard weights for a balance: without a standard that has some permanence, at least in one laboratory and preferably more widely, biological assay is merely a scheme of comparisons between unknowns that has little absolute meaning.

1.5 Statistical science and biological assay

The statistician makes three contributions to the practice of bioassay. He must advise on the underlying statistical principles; he must construct experimental designs that, in the light of existing information, seem likely to give the most useful and reliable results; he must analyze, or instruct the investigator how to analyze, experimental data so as to make the best use of all evidence on potency. The word *design* (Fisher, 1966) implies specification of the number and magnitudes of doses of each preparation to be tested, the number of subjects to be used at each dose, the system of allocating subjects to doses, the order in which subjects shall be treated and measured, and related characteristics of the experiment. Advice on experimental design is often the most important function of the statistician. No ingenuity of statistical analysis can force a badly designed experiment to yield evidence on points neglected in the designing; even an imperfect or inefficient analysis of a well-designed experiment may give conclusions that are enlightening and substantially correct. The close connexion between the design of an experiment and the analysis of the results, however, ensures that questions of optimal design cannot be separated from those of statistical analysis. This book endeavours to show the relations between logical principles, design, and analysis, naturally with most to be said on the last. Unless the principles and design of an assay are sound, statistical analysis is at best a tentative numerical evaluation of the data, at worst a groping in the dark that may prove disastrous.

The experimental designs and methods of statistical analysis required in biological assay are essentially those familiar in many branches of scientific research. To understand fully the chapters that follow, the reader will need facility in the analysis of variance and covariance, in regression analysis, and in test and estimation procedures related to the normal distribution. Among the many standard texts written especially with the needs of biologists in mind are Colquhoun (1971), Dagnelie (1970), Fisher (1970), Pearce (1965), Snedecor and Cochran (1967), Sokal and Rohlf (1973), and Williams (1959). Especially concerned with the philosophy and the practice of experimental design are Cochran and Cox (1957), Cox (1958), Finney (1955, 1959a), Fisher (1966), Kempthorne (1952) and John and Quenouille (1977). Many more advanced books are available, but those mentioned are particularly useful for the bioassayist with limited mathematical skills. If discussion of the design and analysis of assays is to be relevant to assay practice, and not merely a series of exercises in statistical manipulation, the logical basis of the assay argument, especially as developed in Chapters 2 and 3, must be kept in mind and related to particular circumstances.

In some important respects, statistical methods for potency estimation differ from those needed for experiments in which the interest lies in the comparison of effects.

Study of bioassay benefits the outlook of statisticians on other fields of science, and should find a place in their professional training. It illustrates particularly well the union of a variety of techniques into a coherent discipline. As is still evident from many standard textbooks, the initial impetus to development in the design and analysis of scientific experiments came largely from agricultural research. One consequence has been a tendency to emphasize tests of significance rather than estimation. Because the part that each must play is clearly distinguishable, the analysis of biological assays is a valuable corrective to this attitude. The validity of an assay, without which the assay is worthless save in so far as the investigator is prepared to guarantee validity from information external to the experiment, needs to be checked by significance tests; nevertheless, the main purpose of every assay is to estimate the parameter representing the potency of a test preparation relative to a standard.

1.6 Scope of this book

This book is intended to present a connected account of statistical science for that class of bioassays known as dilution assays (§ 2.5). The amount of formal theory needed is small, the number of important variations on the main theme large. Most of the illustrative examples are taken from published work, although occasionally data have been modified in order to avoid an irrelevant complication. In one or two places, artificial 'data' have been used because no suitable records of experiments could be found.

The analysis of variance is central to most applications of statistical methods in the analysis of experiments. This is true of biological assay, but perhaps the fundamental importance of regression and related concepts is here particularly apparent. Techniques of covariance analysis and discriminant analysis have obvious relevance to bioassay, but have not yet been fully exploited; I illustrate how their value to the economy of assay practice may be assessed. I stress general problems of experimental planning, in relation to the choice of doses and of replication, and also to the relative merits of particular designs; the complex designs found valuable in other fields of research must sometimes be adopted if available materials are to be used to best advantage. Almost any proposal for the statistical analysis of biological data involves explicit or implicit assumptions about the inter-relations of the data that can best be expressed in terms of a mathematico-probabilistic model: the need for this model, and the importance of considering its effect upon conclusions, is especially evident in bioassay. Nowhere is this more clear than in the design and analysis of assays based upon quantal responses.

Chapter 2 is concerned with assays in which potency is assessed from direct measurement of the dose necessary to produce a specified response. Chapter 3 introduces the indirect assay, based upon the relation between dose and a quantitative response, the ramifications of which are followed until Chapter 16. Chapters 7 and 8 deal with slope ratio assays, the others primarily with parallel line assays, though many ideas are common to the two types. After general discussion of methods of analysis in Chapters 3, 4, 5, 7, and of efficiency of design in Chapters 6 and 8, the structure and analysis of special designs are considered in Chapters 9, 10, 11. Chapters 12 and 13 relate to the combination of measurements of different qualities of the subjects, Chapter 14 to the combination of potency estimates from different assays. In Chapter 15 the logic and validity of the assay method are re-examined in detail; special attention is given to statistical models, and to the objectivity of the conclusions drawn. Chapter 16 illustrates transformation procedures suggested in the general theory of Chapter 3, with special reference to radioligand assays. Quantal responses are considered in Chapters 17 to 20, not in such variety and detail as are quantitative responses, because many of the quantitative methods can easily be adapted to quantal assays. Although these chapters to some extent repeat methods and ideas presented in my *Probit Analysis* (1971), the amount of overlap of the two books is not great; the emphasis in one is on the study of the relation between quantal response and dose, and in the other is on the use of that relation in potency estimation. In particular, questions of experimental design are discussed much more thoroughly than in the earlier book. In Chapter 20 a method for estimating densities of bacterial suspensions from a dilution series is presented. Chapter 21 contains an account of special problems encountered when the response is a time measurement.

1.7 Computation

The first edition of this book was written for scientists whose only computational aids might be fairly simple desk calculators. Today any person who must analyze bioassay data will have access to an electronic calculator, with its merits of simple functions available on the keyboard, memories for intermediate steps, silence, and now commonly even programming facilities. Those who are responsible for analyzing large numbers of routine assays will almost certainly seek to use a larger electronic computer, with storage facilities and ready availability of good programs.

The many fully-worked numerical examples exhibit systematic patterns of computing suitable for desk or pocket calculators, though the ideal pattern will depend a little upon the particular machine. Early examples include considerable arithmetic detail; thereafter, the reader is assumed familiar with the standard steps and minor variations on them, so that attention can be concentrated on new features. Some standard computer programs for bioassay exist, but too often these are rigid in structure and do not allow for the variants of design and the special aspects of analysis with which this book is so largely concerned. Comments will be made later about computer use and the desiderata of a good program. For any standard assay, a program that will analyze data, take account of all special precautions and tests, and print a clear summary is easily written. Complications arise only when an attempt is made to provide for a wide range of designs, for accidents such as loss of some observations, for transformations, and so on; desirable as these features are, they call for programming expertise. Once a program exists, its routine use should require nothing more than the punching of a few cards to record for each assay an identification, a specification of design, the doses, and the responses. Rarely would more than twenty cards per assay be necessary, and computer processing time should be a very few seconds (even for quantal responses, on a computer of moderate size 30 seconds should suffice). The user's worries will commonly relate much more to the delays in queueing for access to the computer and waiting for distribution of output!

Whatever the methods of computation, careful checks that the data are reasonable and are correctly transcribed and that the highest standards of arithmetic accuracy are maintained should be considered mandatory. Ideally, when a desk calculator is used, a different operator should make check calculations; a good check is an independent sequence of calculations that steps across an analysis by an alternative route. I have not explicitly described checking routines; often nothing more complicated is needed than confirming that the sum of column totals agrees with the sum of row totals, or verifying that independently computed squares with single degrees of freedom add to the right component of an analysis of variance. The imposition of checks should always be regarded as a guarantee of freedom from mistakes in the main work, not as an excuse for carelessness at that stage.

In the earlier days of bioassay, much was written about rapid techniques of statistical analysis, especially for those who lacked the mechanical or human resources to undertake heavy arithmetic. Nomographs, use of ranges in place of sums of squares, graphical and simple arithmetical alternatives to maximum likelihood estimation, and so on all had their advocates. The very ingenious proposals that were made found little favour with the academic research worker who was not faced with large numbers of routine assays; they did much to help those in industrial laboratories who could not otherwise face the arithmetical labour of assays forming part of their daily task. Today arithmetic is so much cheaper relative to experimental costs that interest in rapid techniques should decline, except perhaps for interim assessments of data.

I have tried to adhere to reasonable standards in the number of decimal places carried on desk calculators. Bioassays are seldom sufficiently precise to warrant quotation of results to more than four significant digits: a statement that a test preparation is estimated to have a potency of 35·716 85 units per mg is both stupid and confusing. In intermediate stages of an analysis, more digits may be retained in order to maintain internal consistency. Although to waste effort by having too many digits is better than to err by having too few, retention of an excessive number may increase both the labour and the danger of mistakes. For routine work, the accuracy should never need to be greater than this book illustrates; for exposition of method, or discussion of statistical theory, additional digits may sometimes be desirable. Eight significant digits are enough for almost any set of calculations. Computer users should remember that the accuracy of calculation may be much less than is suggested by the numbers of digits output. A value shown in output to eight or ten digits may be numerically correct only to five or six, and such a simple change as entering data in a different order may alter the later digits because internal rounding of calculations affects them differently. This can affect comparisons between runs of the same program and data on different computers.

1.8 Statistical tables

The statistical analysis of scientific data requires frequent reference to numerical tables of special functions, such as values in the t, χ^2, and variance ratio distributions corresponding to selected probability levels, as well as of square roots, logarithms, and trigonometric functions. A good volume of statistical tables is as essential to a bioassay laboratory as to a professional statistician. To aid the reader who is studying statistical method rather than using the text for reference in practical assay work, simplified versions of the more important tables appear as Appendices I–IV; for numerical examples, however, values have usually been obtained, directly or by interpolation, from Fisher and Yates (1963). Certain tables that relate to assays based on quantal responses are not so readily available elsewhere, and are included as Appendices V–XVII.

1.9 Notation and terminology

The literature of biological assay is confused by differences in notation. Many papers on short-cut schemes of computation, in particular, assigned meanings to large numbers of symbols specifically for single designs; the more popular letters of the alphabet were used repeatedly, with no coherent plan and no hope of fitting them into a general notation.

When a new statistical problem in bioassay is encountered, whether it be a minor modification in design or a method of analyzing data of a totally unfamiliar type, the only sure way of avoiding gross blunders is to begin from the dose–response relation and its roots in regression theory and practice. If the general form of the appropriate statistical analysis is understood, anyone with a moderate facility in elementary algebra can streamline the details of computation so that superfluous steps are avoided. This general approach requires a uniform system of notation and terminology; the policy adopted here, therefore, is a more parsimonious use of the Roman and Greek alphabets with an attempt to establish usages that are readily extended and adapted to new circumstances. Uniqueness and perfect consistency are impracticable without the use of unfamiliar type-founts and the adornment of symbols with many subscripts. Only the pedant would insist that 'r' or 'S' should be used always with the same meaning, when in fact these symbols can conveniently play two or more roles without ambiguity.

The main conventions adopted, some of which inevitably conflict with the usages of other writers on bioassay, are listed below and in the glossary that forms § 1.10.

(i) The words 'stimulus', 'subject', 'dose', and 'response' indicate the entities of an assay in general form (§ 1.3). A particular example may relate to a dose of oestrogen given to rats, but the same statistical technique may be appropriate to a completely different stimulus and subject, such as vitamin B_{12} applied to tubes inoculated with *Lactobacillus leichmannii*.

(ii) The letters S and T (italics) refer to the standard and test preparations respectively. With numerical subscripts, they indicate doses, or groups of subjects at a dose, of either preparation; when themselves used as subscripts, they indicate totals, means, or other quantities relating to all the data on either preparation.

(iii) So far as is reasonably convenient, Greek letters are used for the parameters of a population, Roman for values determined from observational or experimental data.

(iv) The natural logarithm of a number (logarithm to base e) is indicated by ln; log indicates a logarithm to a specified base (e.g. $\log_3$) or, in the absence of any specification, to base 10. Correspondingly, exp and antilog indicate the inverse functions, though where typographical convenience permits $e^{2 \cdot 2}$ or 10^M may be used. In the transformation of doses, logarithms to base 10 or to a base determined by a fixed dose ratio are customary, though with modern computation there is no special advantage in this. For some theoretical purposes (e.g. in § 20.5) natural logarithms are especially appropriate.

(v) The true relative potency is denoted by ρ, its estimate from an assay by R. The symbols μ, M represent the logarithms of ρ, R (to any convenient base), or the logarithms reduced by a constant dependent upon the scales of measurement of the doses.

(vi) The symbols z, u are used for the dose and response actually measured, but x, y are the quantities used in statistical calculations. In general, x and y are obtained as metametric transformations (§ 3.7) of z and u, but either can be identical with the quantity measured.

(vii) Summations over observations relating to one preparation are indicated by S, summations over different preparations or different assays by Σ. This is unambiguous because S referring to the standard preparation seldom occurs *in formulae* except as a subscript. A condensed symbolism represents sums of squares and products of deviations. For example, S_{xy} is the sum of products of the deviations of corresponding values of the variates x, y from their means for one preparation:

$$S_{xy} = S\{(x - \bar{x})(y - \bar{y})\};$$

where data are differentially weighted, a weighted sum of products is implied. Summation over preparations is indicated by Σ; ΣS_{xx} is the sum of squares of deviations of a variate x, summed for the two or more preparations in an assay or over several assays according to the context.

(viii) Percentages are a frequent source of confusion, especially when used by mistake in formulae and computations planned for proportions. The general practice will be to express probabilities as decimals (0·98, not 98 percent) and quantal response rates as decimal proportions (5 subjects responding out of 21 is a response of 0·238, not 23·8 percent), but occasionally percentages are more convenient.

(ix) Except where the contrary is stated, all significance tests are made at probability 0·05, and all fiducial limits are for probability 0·95. This convention is adopted to simplify presentation of analyses, and should be regarded as without prejudice to the choice of probability for particular purposes (cf. § 4.10); certainly it carries no implication that the 0·05 level is best, or even that an investigator ought always to use the same level.

(x) In the first edition of this book, the word *sensitivity* referred to the change of response as dose increases, the dose–response regression coefficient. Pharmacologists apply the word to describe the minimal dose of a drug that produces an observable response. In this edition, pharmacological usage is respected, and the rate of change of response is termed *responsiveness*.

1.10 Glossary of symbols

The following are the more important usages adopted throughout the book, except where the text makes clear that an alternative meaning has been given to a symbol:

Symbol		*Meaning*
a		Constant term in linear regression equation, estimating α
antilog		Function inverse to $\log_{10}$
b		Regression coefficient of y on x, estimating β
c		In slope ratio assay, ratio of lowest to highest dose of either preparation (other meanings elsewhere, e.g. Chapters 11, 19, 20)
d		Sukhatme's d-statistic (§ 4.13); on the null hypothesis, distributed as shown in Appendix Table III
	OR	Interval between successive log doses
e		Base of natural logarithms, 2·718 28 ...
exp		Function inverse to ln
f		A number of degrees of freedom
$f(\ )$		A regression, a metametric, or a probability density function
g		Index of regression significance, equal to $t^2 \operatorname{Var}(b)/b^2$ (§ 4.12)
h		In parallel line assay, $(M - \bar{x}_S + \bar{x}_T)/X$; in slope ratio assay, RX_T/X_S (other meanings elsewhere, e.g. § 10.7, Chapter 13)
i, j		Index of particular dose, as x_i ($i = 1, 2, \ldots, k$); alternatively an index for a variate, a parameter, etc.
k		Number of doses per preparation
ln		Natural logarithm (base e)
log		Logarithm to any stated base, or to base 10 if no other stated
m		Estimate of μ
n		Number of subjects per dose
p		r/n in assays with quantal responses (other meanings elsewhere, e.g. Chapters 6, 13, 14)
q		$(1 - p)$
r		Number of subjects manifesting a quantal response in a batch of n at one dose (other meanings elsewhere, e.g. Chapters 14, 19)
s		Estimated standard deviation for the distribution of log tolerances or of response metameters at a fixed dose, according to

§ 1.10　　　　　　　　　　INTRODUCTION　　　　　　　　　　13

Symbol　　　　　　　　　　　　*Meaning*

	context. Most usually occurs as s^2, the estimated variance obtained from a sum of squares of deviations
OR	Number of survivors (§ 20.4)
t	The familiar t-statistic; on the null hypothesis, distributed as shown in Appendix Table I
u	Response as measured (other meaning in § 13.3)
v_{ii}	The variance multiplier for the estimate of parameter i, whose variance is therefore $\sigma^2 v_{ii}$ or $s^2 v_{ii}$; a diagonal element of V
v_{ij}	The covariance multiplier for the estimates of parameters i and j, whose covariance is therefore $\sigma^2 v_{ij}$ or $s^2 v_{ij}$; a non-diagonal element of V
w	A weighting coefficient, especially for quantal responses (other meaning in § 13.3)
x	Dose metameter, usually equal to z or $\log z$ (other meanings in §§ 2.8, 3.9, 20.3)
y	Response metameter, usually but not always equal to u; in quantal response assays, the working equivalent deviate at dose x
z	Dose as measured
OR	A concomitant variate (Chapter 12)
A	The range, used in calculating working responses, especially for quantal response assays where it is $1/Z$ (other meanings in Chapters 6, 8)
B	With appropriate subscript, a block total
OR	A regression coefficient for a concomitant (§ 12.4)
OR	Bound count (Chapter 16)
OR	Bias variance (§ 18.7)
C	The blanks, or the sum of the responses for the blanks, in a slope ratio assay (other meaning in applications of Bartlett's test)
Cov (,)	Covariance of quantities in parentheses
D	The ratio between successive doses in a symmetric assay
C, D	Also used as additional parameters in Chapters 3, 16, 17, 18, 20
E( )	Statistical expectation of quantity in parentheses
F	Free count (Chapter 16)
F or $F(,)$	A variance ratio which, on the null hypothesis, is distributed as shown in Appendix Table II
$F( )$	Function of z or x used as regression function
I	Square of semi-fiducial interval for M in parallel line assays, or for R in slope ratio assays
J	Parameter of a variance function (Chapter 16)
$J(\mu)$	A function used in generalized calculation of fiducial ranges (Chapter 14)
K	Used with different meanings in Chapters 6, 8, 9, 10, 14, 15
L	Logarithm of likelihood, always to base e (Chapters 3, 14, 16, 17)
OR	A typical contrast between responses

Symbol	Meaning
L_p	In parallel line assay, preparation contrast
L_1	In parallel line assay, regression contrast
$L_2, L_2', L_3, L_3', \ldots$	In parallel line assay, various curvature contrasts
L_B	In slope ratio assay, blanks contrast
L_I	In slope ratio assay, intersection contrast
$L_{2S}, L_{2T}, L_{3S}, L_{3T}, \ldots$	In slope ratio assay, various curvature contrasts
M	Estimated log potency, usually obtained from equation (4.11.5)
M_L, M_U	Lower and upper fiducial limits for M, at probability 0·95 unless otherwise stated
N	Total number of subjects in an assay
OR	Non-specific count (Chapter 16)
N_S, N_T	Total number of subjects for standard and test preparation respectively
P	Probability of occurrence of a quantal response
Q	$(1-P)$
R	Estimate of ρ
R_L, R_U	Lower and upper fiducial limits for R, at probability 0·95 unless otherwise stated
R_0	Guessed value of relative potency used in planning an assay
S	Indicator of the standard preparation
$S_1, S_2, S_3, \ldots$	Doses, or sums of responses to doses, of the standard preparation
$S(\ )$	Summation of quantity in parentheses over data for a particular preparation; for a slope ratio assay it usually represents summation over all data
S_{xx}	$S\{(x-\bar{x})^2\}$ or $S\{nw(x-\bar{x})^2\}$
S_{xy}	$S\{(x-\bar{x})(y-\bar{y})\}$ or $S\{nw(x-\bar{x})(y-\bar{y})\}$
S_{yy}	$S\{(y-\bar{y})^2\}$ or $S\{nw(y-\bar{y})^2\}$
T	Indicator of the test preparation
OR	Estimate of θ (§ 14.8)
OR	Total count (Chapter 16)
$T_1, T_2, T_3, \ldots$	Doses, or sums of responses to doses, of the test preparation
U	$E(u)$
$\mathbf{V}$	Matrix of variance and covariance multipliers (v_{ij}), often calculated as an inverse matrix in the solution of linear equations
V, V_1, V_2	Parameters of a variance function (Chapter 16)
Var$(\ )$	Variance of quantity in parentheses
$V_E(\ )$	Effective variance (Chapter 19)
W	Total weight of observations (§ 20.4)
X	In parallel line assay, interval between highest and lowest log dose

Symbol		Meaning
X_S, X_T		In slope ratio assay, highest doses of standard and test preparations
Y		Response metameter corresponding to U, usually expressed as linear regression function of empirical or working metameter on x
Y_0, Y_1		Minimum and maximum working equivalent deviates
Z		Standardized ordinate of tolerance distribution
Z_S, Z_T		In § 3.3, equipotent doses of standard and test preparations
α		Constant term in true regression equation $Y = \alpha + \beta x$
β		True regression coefficient of y on x
γ		Coefficient of x^2 in quadratic regression equation (§§ 3.8, 5.11)
	OR	A block parameter (§ 10.7)
	OR	Unneutralized fraction of a hormone (§ 14.8)
δ		Used only in combination with other symbols, to denote increments to estimates during iteration, as $\delta\alpha$, $\delta\beta$ (§ 3.13)
ϵ		Used as a typical base of logarithms (§ 4.2), as a random error (§ 14.5), or as ln 10 (§ 16.3)
η		Working loglog deviate
η_0, η_1		Minimum and maximum working loglog deviates
θ		Angle used as ancillary statistic in association with Sukhatme's d (§ 4.13)
	OR	Neutralizing potency (§ 14.8)
	OR	Dummy variate (Chapter 17)
λ		Index of z in definition of power dose metameter (§§ 3.7, 7.1)
	AND NEVER	s/b or $1/b$, as often in the literature of bioassay
μ		Log ρ
	OR	Mean of log tolerance distribution (Chapter 17)
	OR	Density of bacterial suspension (§ 20.4)
ν		Total number of doses (§§ 17.5, 17.6)
	OR	Unknown total number of subjects (§ 20.4)
π		Ratio of circumference to diameter of a circle, 3·141 59 ...
ρ		True potency of test preparation relative to standard
σ		Standard deviation for the distribution of log tolerances or of response metameters at a fixed dose, according to context
ϕ		Special meaning in § 10.7
$\phi(U)$		Expression of σ^2 as a function of U (§ 3.10, Chapter 16)
χ^2		Statistic which, on the null hypothesis, is distributed as shown in Appendix Table IV
ψ^2		Inter-litter variance of β (§ 14.5)
ω		Weight used in combination of estimates (§ 14.1)
Σ		Summation over data for different preparations in one assay, or over different assays

Some symbols not listed above, and several of those that are listed, are occasionally used with different meanings where there is no fear of confusion.

A bar over the top of a symbol ($\bar{x}, \bar{s}^2, \ldots$) indicates a mean value, unweighted or weighted according to the context; a subscript distinguishes values corresponding to different preparations ($b_S, b_T, \ldots$) or to different variates, different parts of an assay, or different assays ($s_1, s_2, b_1, b_2, \ldots$); an asterisk denotes values that have been adjusted for regression on a concomitant variate or a transformed, compounded or modified variate ($y^*, b^*, \ldots$). A 'hat' ($\hat{\alpha}, \hat{C}, \ldots$) denotes an estimator, usually maximum likelihood, of the hatted symbol.

Differences in content and purpose cause Bliss's book (1952a), which in so many respects parallels and complements this, to have a completely different use of symbols. In particular Bliss and others (including the *European Pharmacopoeia*) write in terms of C instead of g (§ 4.13), where

$$C = 1/(1-g).$$

Thus C ranges from slightly more than 1·0 for good assays to very large values for poor assays.

2
Direct assays

2.1 Types of biological assay

Two types of quantitative biological assay must be distinguished, the *direct* and the *indirect*. Although direct assays are of limited applicability and their use is declining, they are of fundamental importance to the logic of assay. The problems of design and statistical analysis are similar to those familiar to biometricians for other investigations involving planned experimentation. Nevertheless, a chapter is assigned to these assays because of their intrinsic interest and because an understanding is essential to proper appreciation of indirect assays.

Indirect assays may be divided according as they are based on quantitative or on quantal ('all-or-nothing') responses. All make use of dose–response regression relations, though the full calculations are usually more complex for quantal responses (Chapter 17). In logical structure, assays with quantal responses are more closely related to direct assays than to those with quantitative responses.

2.2 The nature of direct assays

The principle of a direct assay is to measure the doses of the standard and test preparations that produce a specified response. By definition, the potency of the test preparation relative to the standard is the amount of the standard equivalent in effect to one unit of the test. Hence the ratio of these critical doses estimates potency.

A direct technique is practicable only for certain stimuli and subjects, for it depends upon the possibility of measuring just the dose needed, not merely one that is at least large enough. The response must be unambiguous and easily recognized, and the dose must be administered in such a manner that the exact amount needed to produce the response can be recorded. The critical dose will generally vary considerably from subject to subject, and even from one occasion to another with the same subject; hence only an estimate of potency is obtained, and calculation from averages over a number of trials is desirable. Ideally, trials of both preparations would be made on each subject used, so that estimation is independent of differences between subjects and, in consequence, more precise. Often this is impossible, however, because once a subject has responded it may not be usable again, or may be so changed as to make a second trial far from comparable with the first; often, indeed, the response is death of the subject.

A typical example of a direct assay is the 'cat' method for the assay of digitalis (Burn, Finney and Goodwin, 1950; Hatcher and Brody, 1910). The standard or the test preparation is infused, at a fixed rate, into the bloodstream of a cat until the heart stops beating. The total time of infusion multiplied by the rate measures the dose. This is repeated on several cats for each preparation, and the mean doses are compared. The most obvious form of calculation is illustrated in § 2.3; criticisms of the method and of the statistical analysis follow.

In the early days of bioassay, attempts were made to measure digitalis potency in cat units, the unit being the average of the just-lethal doses. Investigators soon realized that the source and nature of the animals as well as uncontrolled, and perhaps uncontrollable, conditions in the laboratory affected this quantity. The practice of comparative assay therefore developed. Bliss (1944a) summarized several sets of records for which the general level of the lethal dose differed substantially between laboratories, and even between repeated trials within the same laboratory at intervals of a few weeks. The same records showed very satisfactory stability of the ratio of lethal doses for a pair of preparations, both between laboratories and within laboratories over time.

2.3 An assay of strophanthus

Table 2.3.1 shows the fatal doses, or *tolerances*, of three groups of cats for two tinctures of strophanthus and a preparation of ouabain (Burn *et al.*, 1950). The experimental procedure was essentially as described in § 2.2 for digitalis. The doses were recorded as quantities per kg body weight of cat. Unfortunately the absolute doses are not known: the tolerance was assumed to vary in proportion to rather than independently of body weight. Provided that the cats were assigned at random to the different preparations, either form of expression of dose is valid for estimating potency, but neither necessarily makes the best possible use of information on body weight (§ 2.7).

TABLE 2.3.1 Tolerances of cats for tinctures of strophanthus and ouabain

Preparation	Strophanthus 1 (μl/kg)	Strophanthus 2 (μl/kg)	Ouabain (μg/kg)
Tolerances	15·5	24·2	52·3
	15·8	18·5	99·1
	17·1	20·0	47·6
	14·4	22·7	65·1
	12·4	17·0	66·8
	18·9	14·7	57·6
	23·4	22·0	49·3
	—	—	45·8
	—	—	66·9
Total	117·5	139·1	550·5
Mean	16·8	19·9	61·2

Suppose that tincture 2 is to be regarded as the standard preparation, and tincture 1 is to be compared with it as a test preparation. From the means at the foot of Table 2.3.1, $16 \cdot 8 \, \mu l$ of 1 is estimated to produce the same results as $19 \cdot 9 \, \mu l$ of 2, either being just sufficient, on an average, to kill a cat. Hence the relative potency is estimated to be

§ 2.4 DIRECT ASSAYS

$$R = \frac{19\cdot 9}{16\cdot 8} \qquad (2.3.1)$$

$$= 1\cdot 18;$$

1 μl of tincture 1 is estimated to be equivalent to 1·18 μl of tincture 2. As a statement of the findings in an experiment of this kind, the conclusion is satisfactory, except that, of course, allowance must be made for statistical errors of estimation (§ 2.4). An investigator, however, often wishes to make a more general inference, here that 1 μl of tincture 1 contains the same amount of some effective constitutent as does 1·18 μl of tincture 2. The ensuing complications for the logic of the argument are discussed again in § 2.5, and more fully in Chapters 3 and 15.

The two tinctures of strophanthus may be regarded as containing the same effective constituent, so that the relative potency will estimate the ratio between the concentrations of this constituent in the tinctures. Ouabain, on the other hand, though closely related, is not an identical substance. Although a numerical relative potency may be useful in comparing it with strophanthus, that figure must not be interpreted as having the same analytical significance. The calculation is made in the same manner:

$$R = \frac{19\cdot 9}{61\cdot 2} \qquad (2.3.2)$$

$$= 0\cdot 325;$$

thus 1 μg ouabain is estimated to be equivalent to 0·325 μl of tincture 2.

2.4 Precision of estimates

An estimate of potency, based upon individual observations that vary widely, is of little use unless supported by some indication of its limits of error. Standard errors of the mean tolerances can be calculated by familiar processes, and from these must be derived a measure of precision.

Consider first the two strophanthus tinctures alone. Writing z as a typical dose, sums of squares of deviations are used to estimate the variances of $\bar{z}_1, \bar{z}_2$, the two mean doses. For tincture 1,

$$S(z - \bar{z})^2 = (15\cdot 5)^2 + (15\cdot 8)^2 + \ldots + (23\cdot 4)^2 - (117\cdot 5)^2/7$$

$$= 2048\cdot 19 - 1972\cdot 32$$

$$= 75\cdot 87.$$

Similarly for tincture 2,

$$S(z - \bar{z})^2 = 2832\cdot 27 - 2764\cdot 12$$

$$= 68\cdot 15.$$

Since both sums of squares have 6 degrees of freedom, the tinctures have about the same variability in their tolerances. Hence the usual practice of pooling the sums of squares, so as to obtain a single estimate of the variance per cat, seems reasonable: the logical flaw is discussed in § 2.5. This involves writing

$$s^2 = \frac{75 \cdot 87 + 68 \cdot 15}{6 + 6}$$

$$= 12 \cdot 00$$

with 12 degrees of freedom (d.f.). Both $\bar{z}_1$ and $\bar{z}_2$ are means of 7 observations, and therefore have variances

$$\text{Var}(\bar{z}_1) = \text{Var}(\bar{z}_2) = \frac{s^2}{7} = 1 \cdot 71.$$

The square root of this, $1 \cdot 31$, is the standard error (SE) of $\bar{x}_1$ and $\bar{x}_2$.

Equation (2.3.1) can be written

$$R = \frac{\bar{z}_2}{\bar{z}_1}. \qquad (2.4.1)$$

A standard error for R can be calculated with the aid of a commonly used approximate formula for the coefficient of variation of a ratio of two independent quantities. The formula is

$$\left[\text{CV}\left(\frac{a}{b}\right)\right]^2 = [\text{CV}(a)]^2 + [\text{CV}(b)]^2, \qquad (2.4.2)$$

where CV, the coefficient of variation,[*] is defined as

$$\text{CV}(a) = (\text{SE of } a)/a. \qquad (2.4.3)$$

Application of (2.4.2) to R leads to

$$\text{Var}(R) = [\text{V}(\bar{z}_1) + R^2 \text{V}(\bar{z}_2)]/\bar{x}_2^2 \qquad (2.4.4)$$

or

$$\text{Var}(R) = \frac{s^2}{\bar{z}_2^2}\left(\frac{1}{N_1} + \frac{R^2}{N_2}\right), \qquad (2.4.5)$$

where N_1, N_2 are the numbers of subjects for the two preparations. Numerically,

$$\text{Var}(R) = \frac{12 \cdot 00}{16 \cdot 8^2}\left(\frac{1}{7} + \frac{1 \cdot 18^2}{7}\right)$$

$$= 0 \cdot 0145,$$

and $\quad R = 1 \cdot 18 \pm 0 \cdot 120.$

Throughout this book, the results of an assay are summarized by limits between which the unknown true potency is almost certain to lie. Although

---

[*] Coefficients of variation are often expressed as percentages, but the usage here is less confusing.

§ 2.4   DIRECT ASSAYS   21

certainty is unachievable, rules of calculation can ensure that the statement has a high probability of being correct. Two systems of inference, those of *confidence intervals* and of *fiducial intervals* (Kendall and Stuart, 1973, Chapters 19 and 20), are in common use in all branches of statistical science. The logical difference between them is important to the theorist, but this is not the place for studying it. Often, but not always even for simple problems, the two give identical intervals. In this book, fiducial inference will be used, though (as in the next paragraph) limits quoted are sometimes approximations based upon statistical theory that must later be discussed critically and modified.

In the present assay, s^2 is based on 12 d.f., and the corresponding t-deviate for a probability of 0.05 is 2.18 (Appendix Table I; Fisher and Yates, 1963, Table III). The fiducial range will therefore extend $2 \cdot 18 \times 0 \cdot 120$, or 0.26 on either side of R: the fiducial limits may be written

$$R_L = 0 \cdot 92,$$
$$R_U = 1 \cdot 44.$$

With a degree of confidence expressed by the probability 0.95, $1 \, \mu l$ of 1 may be asserted to be not less potent that $0 \cdot 92 \, \mu l$ of 2 and not more potent than $1 \cdot 44 \, \mu l$ of 2.

This is simple enough. In fact, R_L is the smallest value for the true potency that would not be rejected by a significance test based upon the assay results (since $(1 \cdot 18 - 0 \cdot 92)/0 \cdot 120 = 2 \cdot 18$), and R_U is the greatest value that would not be rejected by such a test. The limits closely approximate those from the preferred system of calculation (§ 2.6).

A complication enters when the potency of ouabain is considered. For the 9 cats in this group

$$S(z - \bar{z})^2 = 35\,843 \cdot 61 - 33\,672 \cdot 25$$
$$= 2171 \cdot 36.$$

In order to avoid additional problems arising in the simultaneous testing of the three preparations, tincture 1 will now be ignored. An assumption of equal variance per cat for tincture 2 and ouabain would be quite contrary to the evidence of much greater variability for ouabain. Separate variances for the two groups must therefore be taken as

$$s_2^2 = \frac{68 \cdot 15}{6} = 11 \cdot 36,$$

$$s_0^2 = \frac{2171 \cdot 36}{8} = 271 \cdot 4$$

respectively. In § 2.3, the value 0.325 was obtained for R, and equation (2.4.4) gives

$$\mathrm{Var}(R) = \frac{1}{61 \cdot 2^2} \times \left[\frac{11 \cdot 36}{7} + \frac{271 \cdot 4 \times (0 \cdot 325)^2}{9} \right]$$
$$= 0 \cdot 001\,284,$$

whence

$$R = 0{\cdot}325 \pm 0{\cdot}036.$$

This standard error cannot be used so simply for the assessment of fiducial limits, for the t-distribution is not applicable to a standard error compounded of two independent variance estimates. Use of a t-deviate with 6 d.f. would provide a cautious assessment of limits, but in reality a radically modified approach is needed. The difficulty is best removed by a logarithmic transformation (§§ 2.5, 2.6). An alternative is to use Fieller's theorem in one of its generalized forms (§ 4.13). The calculations are not shown in detail as they are not recommended here. The reader may verify that equation (4.13.12) leads to fiducial limits for the potency of $1\,\mu l$ of tincture 1 at $0{\cdot}95\,\mu l$ and $1{\cdot}48\,\mu l$ of tincture 2. With a little more arithmetic, the fiducial limits for ouabain are obtained from equation (4.13.12) as $0{\cdot}252\,\mu l$ per μg and $0{\cdot}427\,\mu l$ per μg.

2.5 Dilution assays

The process of estimating potency and assessing fiducial limits should take account of the nature of the frequency distribution of tolerances. This distribution is fully discussed in Chapter 17, but needs comment here. The analyses in § 2.4 assumed the tolerances of individuals for a particular preparation, in respect of the stimulus and response used, to be normally distributed. In statistical analyses of experimental data, the assumption of normality is often made without considering possible consequences of its failure to correspond to reality. By virtue of the central limit theorem (Cramér, 1946, § 17.4; Kendall and Stuart, 1977, § 7.26), unless the basic distribution of individual tolerances is very different from the normal, the distribution of means of N will be satisfactorily approximated by the normal. Although techniques developed for the normal distribution will seldom be seriously misleading on this score, further examination of the principles implicit in earlier sections is desirable and helpful.

In many assays, the test preparation, T, behaves as though it were simply a dilution (or a concentration) of the standard preparation S, in a diluent that is inert in respect of the response used. Thompson (1948) pointed out that not all valid assays need be of this character. Assays in current use, however, are almost all either strictly or approximately *dilution assays*, to which category this book is restricted. If there are strong reasons for believing that all constituents of T other than one are without effect on the response of the subjects, an assay against a standard preparation of the effective constituent is logically equivalent to a chemical analysis for that constituent in T: this is termed an *analytic dilution assay*. In other circumstances, the two preparations may for assay purposes behave as though qualitatively the same in effective constituent, though in fact they are known not to be the same; such a *comparative dilution assay* is of more limited value as a basis for inference. For example, two insecticides (of different but related chemical composition) may behave in their effects on one species of insect as though one was a dilution of the other, yet the apparent 'dilution' may depend upon the species or upon the conditions of experiment. The statistical methods appropriate to analytical assays may still be applied to estimate a relative potency figure that may then be useful as a concise expression of the results of a particular experiment.

Suppose that any pair of equivalent doses of two preparations, A and B, is represented by z_A, z_B, and that B behaves exactly as a dilution of A by a factor ρ, the relative potency. This implies that

$$\rho z_A = z_B, \qquad (2.5.1)$$

for all possible pairs of doses, z_A, z_B. Consequently, whatever the distributions of tolerances, the variance of the B distribution must be ρ^2 times that of the A distribution. When ρ is near to unity, as for the two strophanthus tinctures in Table 2.3.1, little harm will result from assuming equality of variance. When ρ is very different from unity, as in the comparison of the ouabain preparation with strophanthus, the inconsistency of this assumption may have more serious consequences and lead to wrong conclusions.

Equation (2.5.1) may be written

$$\log z_A + \log \rho = \log z_B, \qquad (2.5.2)$$

which shows that the logarithms of equivalent doses of A and B must have distributions identical except for a shift of $\log \rho$ in the mean. In particular, the variances of the log tolerances in the two distributions are necessarily the same. An analysis in terms of log doses may therefore be preferable to one in absolute units. Indeed a little consideration suggests that normality of distribution is likely to be better approximated by log doses. A tolerance distribution must have all doses positive, yet need not have a sharply defined upper limit; hence it is likely to be positively skewed. Logarithmic doses are measured on a scale that can extend from large negative to large positive values, and their distribution may be more nearly symmetric. The logarithmic transformation also removes the possibility that the lower fiducial limit to a potency estimate is negative, an absurdity that the type of calculation shown in § 2.4 can produce.

In direct assays, assumption of a normal distribution for log tolerances has many advantages. All variance estimates may be pooled, so allowing a more precise estimation of the population variance than if each preparation must provide its own. Also, since the estimate of relative potency is obtained as the antilogarithm of the difference of two means, instead of as a ratio of two means, fiducial limits are calculated from simple standard error formulae (without appeal to Fieller's theorem). There can scarcely be any demonstration that normality represents absolute truth, but use of the normal distribution is as reasonable here as in the many other branches of statistical science in which it is regularly applied. The argument may seem specious, but empirical and practical support can be given.

Bliss (1944a; see also Bliss and Hanson, 1939) presented experimental evidence that the lognormal distribution fits well the frequency distribution of lethal doses of digitalis in cats. This is perhaps one of the best-documented situations, and more information on other drugs and subjects would be welcome. Wherever the facts do not violently conflict with lognormality, the central limit theorem is a fair insurance that the approach to normality of distributions of mean log doses will improve as the number of observations is increased. Undoubtedly there are philosophical objections to the assumption, but in general the alternatives are less attractive:

(i) No other parametric formulation of the distribution has even as strong a theoretical justification as the normal. Unless experimental evidence strongly indicates a specific alternative as desirable, to adopt one is analytically more complicated without any compensating advantage.

(ii) Distribution-free methods for the interpretation of many types of statistical data have been developed. These make use of various properties of the observations but employ only such distributions as arise solely from the randomness and independence of the observations. The methods are ingenious but usually too complicated for estimation problems in complex experiments. Moroever, except for the smallest sets of data, they seldom give conclusions differing to any important extent from those of parametric analyses. For example, non-parametric estimation from Table 2.3.1 would be practicable, but could scarcely be considered for Table 2.8.1. Colquhoun (1971) has given an excellent account of non-parametric methods for biometric problems, though he did not discuss their application to direct assays. When the form of the tolerance distribution is unknown, the exact worth of numerical observations may be in doubt, but to deny all value to the data except for the limited processes of non-parametric techniques seems deplorably wasteful. However, techniques have been proposed that seem remarkably successful in obtaining precision comparable with that of parametric analyses (Sen, 1963, 1964, 1965; Shorack, 1966); they are still relatively laborious to handle, and perhaps scarcely justify themselves as an alternative route to numerical results almost identical with those in § 2.6.

(iii) A third possibility is to deny the legitimacy of any quantitative assessment of relative potency from experimental data of the kind here described. Those who adopt this point of view will be spared the need of reading a book that accepts the opposite as axiomatic. They must not imagine that they are instead at liberty to reject sophisticated statistical techniques but to retain the more primitive. To reject the methods of analysis illustrated in § 2.6 or later in this book, on the grounds that the conditions of the statistical theory are not exactly fulfilled, and then to adopt the methods of §§ 2.3, 2.4 or the standard curve method of § 3.14 because they appear simple common sense, would be as irrational as to refute Newton's cosmology by appeal to Einstein and then to accept Ptolemaic views because 'obviously the sun goes round the earth'. Old and new statistical methods rest upon similar foundations, but the new at least have a self-consistency that the old often lack.

2.6 Revised computations for the strophanthus assay

The computational advantages of the logarithmic transformation are readily seen by application to the data of Table 2.3.1. The logarithms of the tolerances in that table are shown in Table 2.6.1, together with calculations leading to the

TABLE 2.6.1 Logarithms of the tolerances in Table 2.3.1

Preparation	Strophanthus 1	Strophanthus 2	Ouabain
Tolerances[*]	0·190	0·384	0·718
	0·199	0·267	0·996
	0·233	0·301	0·678
	0·158	0·356	0·814
	0·093	0·230	0·825
	0·276	0·167	0·760
	0·369	0·342	0·693
	–	–	0·661
	–	–	0·825
Total: $S(x)$	1·518	2·047	6·970
Mean: $S(x)/N$	0·217	0·292	0·774
$S(x^2)$	0·3759	0·6338	5·4858
$[S(x)]^2/N$	0·3292	0·5986	5·3979
S_{xx}	0·0467	0·0352	0·0879

[*] Each entry has been reduced by 1·0 to simplify the arithmetic. This is equivalent to dividing every dose by 10·0, a change of scale that has no effect on calculations and statements of relative potency.

sums of squares of deviations; the symbol x is used for log dose, in accordance with the convention of § 3.8. As equation (2.5.2) shows, homogeneity of variance is a necessary condition for fundamental validity, the basic hypothesis of a dilution assay (§ 15.1). If a proper test of this were needed, Bartlett's test for the homogeneity of a set of variances (§ 3.11) might be used. Here, inspection of the sums of squares is sufficient to show that the mean squares for the three preparations do not differ significantly: the ratio of the largest mean square to the smallest,

$$F = \frac{0·0879}{8} \bigg/ \frac{0·0352}{6}$$

$$= 1·87,$$

is far from being judged significant even by a variance ratio test that makes no allowance for the selection of extremes. The variance per cat is therefore estimated as

$$s^2 = \frac{0·0467 + 0·0352 + 0·0879}{6 + 6 + 8}$$

$$= 0·008\ 49,$$

with 20 d.f.

The symbol M, as an abbreviation for $\log R$, will be used for the estimate of log potency. From equation (2.5.2), its value for tincture 1 relative to tincture 2 is

$$M = \bar{x}_2 - \bar{x}_1 \qquad (2.6.1)$$

$$= 0 \cdot 292 - 0 \cdot 217$$

$$= 0 \cdot 075.$$

Moreover, the variance of M is simply

$$\mathrm{Var}(M) = s^2 \left(\frac{1}{N_1} + \frac{1}{N_2} \right) \qquad (2.6.2)$$

$$= 0 \cdot 008\,49\, (\tfrac{1}{7} + \tfrac{1}{7})$$

$$= (0 \cdot 0493)^2.$$

On the assumption of normal distribution of the log tolerances, fiducial limits to M are found exactly by direct application of the t distribution. For 20 d.f., the $0 \cdot 05$ t-deviate is $2 \cdot 086$, and therefore

$$M_L = 0 \cdot 075 - 2 \cdot 086 \times 0 \cdot 0493 = \bar{1} \cdot 972$$

$$M_U = 0 \cdot 075 + 2 \cdot 086 \times 0 \cdot 0493 = 0 \cdot 178.$$

The antilogarithms of M and its limits give the relative potency and its limits: these are $1 \cdot 19 \,\mu\mathrm{l}$ per $\mu\mathrm{l}$ with limits at $0 \cdot 94 \,\mu\mathrm{l}$ per $\mu\mathrm{l}$ and $1 \cdot 51 \,\mu\mathrm{l}$ per $\mu\mathrm{l}$, very similar to the results in § 2.4. Similarly, for the potency of the ouabain preparation relative to tincture 2,

$$M = 0 \cdot 292 - 0 \cdot 774$$

$$= \bar{1} \cdot 518;$$

also

$$\mathrm{Var}(M) = s^2(\tfrac{1}{7} + \tfrac{1}{9})$$

$$= (0 \cdot 0464)^2,$$

and therefore

$$M_L, M_U = \bar{1} \cdot 421, \bar{1} \cdot 615.$$

Hence the potency of ouabain is estimated as $0 \cdot 330 \,\mu\mathrm{l}$ per $\mu\mathrm{g}$, with limits at $0 \cdot 264 \,\mu\mathrm{l}$ per $\mu\mathrm{g}$, $0 \cdot 412 \,\mu\mathrm{l}$ per $\mu\mathrm{g}$. The fiducial range is appreciably narrower than that reported in § 2.4, primarily because of the greater number of degrees of freedom available for calculation of the variance.

2.7 Adjustment for body weight

The tolerances shown in Table 2.3.1 were expressed as amounts per kg body weight of cat, a procedure common with pharmacologists, at least where large subjects are expected to have greater tolerances than small. Others would choose to state the absolute dose that produces a specified response, irrespective of the size of the subject. Provided that subjects are assigned to preparations entirely at random, or in accordance with an experimental design for which the appropriate statistical analysis is made, either procedure is legitimate in the sense that it should lead to a valid estimate of potency. Each ensures that relative potency is calculated from comparable measures of dose, and the choice between them may be made on the basis of precision of estimation.

When information on body weight is available, to ignore its possible correlation with tolerance is wasteful. Equally undesirable, however, is the too ready assumption that adjustment of doses by proportionality is ideal (Braun and Siegfried, 1947). Where the range of body weights is not great, the proportional adjustment may prove fairly satisfactory, though perhaps no great improvement on the unadjusted doses. For a wider range, proportional adjustment may sometimes represent so serious an oversimplification of the weight–tolerance relation as to do more harm than good.

The analysis of covariance, a standard statistical technique, allows data to determine their optimal adjustment. The regression coefficient of log dose on log weight measures the average effect of change in the latter quantity on the former, and an estimate of this coefficient can be used to adjust mean log doses to a standard weight. A zero regression coefficient corresponds to absence of any need for adjustment, and a coefficient of unity corresponds to the proportional adjustment. Intermediate values may occur, as also may values that exceed unity or are negative; for such data, neglect of any adjustment or a dogmatic assumption that doses should be expressed per unit of body weight will lead to estimates of potency less precise than the best that can be formed. The covariance technique permits simultaneous adjustment for two or more concomitant observations, so resolving the dilemma of the investigator who is unsure whether to express doses as per unit of body weight or per unit weight of some particular organ.

At one time, many workers were discouraged from using covariance analysis by the heavy computations. With a computer, even an analysis using two or more independent variates simultaneously requires only slight extension of a standard program for a direct assay. The important question today is whether the labour of making the concomitant measurements for each subject (body weights, heart weights, or any others) increases the precision by at least as much as would an equally costly use of additional subjects (§ 2.9); if not, the practice of adjustment should be abandoned for that class of assays.

2.8 A direct assay with covariance

Unfortunately the body weights of the cats for Table 2.3.1 are no longer known. A more extensive experiment (Chen *et al.*, 1942), relating to the assay of digitalis-like principles in ouabain and other cardiac substances, well illustrates the covariance technique in direct assays. A suitable dilution of a drug was slowly infused into an anaesthetized cat until death occurred (Hatcher and Brody, 1910), at which point the dose was measured and recorded. Three observers collaborated in tests of twelve drugs:

A: α-Antiarin   G: Coumingine HCl
B: β-Antiarin   H: Cymarin
C: Bufotalin   I: Emicymarin
D: Calotoxin   J: Ouabain (taken as standard)
E: Calotropin   K: Periplocymarin
F: Convallotoxin   L: Uscharin

Each drug was to be tested on twelve cats, but an observer could test only four cats per day. Because the general level of tolerance might change from day to day, a 12 × 12 Latin square was used to balance observer differences and day differences (Table 2.8.1). The columns of the square contain the twelve cats tested on any one day; the rows represent the first and second cat (a, b) tested by each of the observers (I, II, III) in the morning and afternoon of each day. The square was derived from a standard 12 × 12 Latin square by randomization of the order of rows, scarcely as complete a randomization as most statisticians would think desirable.

For each cat, the body weight, the heart weight, and the fatal dose of drug were measured. Table 2.8.1 shows these expressed as logarithms:

$$x_1 = 1000 \times \log(\text{body weight in kg}),$$
$$x_2 = 100 \times \log(\text{heart weight in g}),$$
$$x = 1000 \times \log(\text{dose in } \mu g).$$

The doses had to be reconstructed from the published records of dose per kg body weight and dose per g heart weight.

The heart weight might be expected to be more valuable than the body weight in the adjustment of tolerances; both might be useful, and therefore the multiple covariance analysis of x_1, x_2, x needs to be computed (Table 2.8.2). Details are not discussed here, as the calculation of the analysis of variance for each variate separately should be a familiar process; those to whom covariance analysis is new should need no more instruction than that they must use products of pairs of corresponding observations or totals in exactly the same manner as they used squares for the analyses of variance. Some of the books mentioned in § 1.5 describe the method. The 11 degrees of freedom for 'rows' in the analysis could be subdivided into the main effects of time of day (1 d.f.), observers (2 d.f.), first versus second cat (1 d.f.), and the interactions of these. In a comparative experiment these would be interesting, but they are not relevant to the assay: the experimental design has deliberately been balanced in such a way as to eliminate their average effects. Although large effects might complicate the assay by making the error variance different for different observers, days, or times of day, inspection of the data makes clear that no such worries arise here. Mean squares for the analyses of variance of x_1 and x_2 are not shown in Table 2.8.2, but inspection of the table shows the absence of any significant differences between drugs in respect of body or heart weight; had differences occurred, they would presumably have indicated a non-random allocation of drugs to cats, and the data then might have had to be entirely rejected.

§ 2.8 DIRECT ASSAYS

The error line of Table 2.8.2 shows x to be correlated with both x_1 and x_2 separately. The partial regression coefficients $b_{1.2}$ and $b_{2.1}$ therefore require study; these are the solutions of

$$\left. \begin{array}{l} b_{1.2} S_{x_1 x_1} + b_{2.1} S_{x_1 x_2} = S_{xx_1}, \\ b_{1.2} S_{x_1 x_2} + b_{2.1} S_{x_2 x_2} = S_{xx_2}, \end{array} \right\} \quad (2.8.1)$$

where $S_{x_1 x_1}$ represents the error sum of squares for x_1, $S_{x_1 x_2}$ the error sum of products for x_1, x_2, and so on. Numerically,

$$\left. \begin{array}{l} 341\,700 b_{1.2} + 30\,995 b_{2.1} = 250\,743, \\ 30\,995 b_{1.2} + 5\,185 b_{2.1} = 35\,056. \end{array} \right\}$$

Hence

$$\left. \begin{array}{l} b_{1.2} = 0\cdot263\,298\,75, \\ b_{2.1} = 5\cdot187\,088\,74. \end{array} \right\}$$

The large number of decimal places is desirable for calculating the sum of squares (2 d.f.) accounted for by the regression, namely

$$b_{1.2} S_{xx_1} + b_{2.1} S_{xx_2} = 250\,743 b_{1.2} + 35\,056 b_{2.1} \quad (2.8.2)$$

$$= 247\,859.$$

Of this, a regression on x_1 alone would account for

$$b_1 S_{xx_1} = (S_{xx_1})^2 / S_{x_1 x_1} \quad (2.8.3)$$

$$= 250\,743^2 / 341\,700$$

$$= 183\,998,$$

where b_1, the simple regression coefficient on x_1, is given by

$$b_1 S_{x_1 x_1} = S_{xx_1}. \quad (2.8.4)$$

A similar calculation may be made for x_2 alone. The analysis in Table 2.8.3 then follows, and standard variance ratio tests may be applied.

Table 2.8.3 shows that inclusion of x_2 in a regression equation after x_1 has been used gives a significant reduction in the residual mean square, but that inclusion of x_1 in an equation that already contains x_2 gives scarcely any advantage. Since x_1 and x_2 are themselves fairly closely correlated, it is scarcely surprising that a regression on either accounts for much the same variation as a regression on the other. Allowance for the regression on x_2 appears to be sufficient here. The significance tests given by Table 2.8.3 are in fact tests of the partial regression coefficients of x on x_1, x_2. The regression coefficient on x_2 alone is

$$b_2 = S_{xx_2} / S_{x_2 x_2} \quad (2.8.5)$$

TABLE 2.8.1 Tolerances of cats for various cardiac substances

(Table shows values of x_1, x_2, x_3, in that order, for 144 cats: for explanation, see text)

DATE	6/3	7/3	8/3	9/3	13/3	14/3	16/3	21/3	24/3	27/3	30/3	3/4	TOTAL
a.m. Ia	I 359/95/525	J 410/104/273	B 323/89/315	L 363/91/557	H 208/85/189	G 290/92/228	F 375/88/54	K 323/87/473	D 273/83/165	E 270/86/254	A 336/88/193	C 271/90/358	3801/1078/3584
a.m. Ib	K 424/108/737	G 304/97/345	J 262/83/195	H 425/100/425	I 346/100/444	B 439/112/350	L 344/99/557	C 312/86/209	E 280/97/335	F 247/76/22	D 300/98/605	A 371/93/237	4054/1149/4461
a.m. IIa	B 314/89/293	L 336/92/427	G 329/92/371	C 384/103/413	D 359/103/515	J 450/104/307	K 355/99/446	E 261/79/301	H 352/93/266	A 286/93/368	F 274/86/198	I 357/96/515	4057/1129/4420
a.m. IIb	E 309/94/299	D 424/102/437	F 363/100/411	G 359/92/400	J 225/85/173	K 353/100/661	A 324/91/250	L 410/100/449	C 402/91/573	I 327/89/316	B 395/97/347	H 253/81/228	4144/1122/4544
a.m. IIIa	C 289/91/601	K 306/93/398	A 381/98/400	B 422/113/502	F 426/106/385	L 294/89/443	I 337/95/329	D 260/86/394	G 435/92/444	H 361/87/377	J 336/85/307	E 353/95/247	4200/1130/4827
a.m. IIIb	F 231/73/35	H 313/94/355	K 263/87/384	E 375/95/451	G 247/82/378	C 230/85/444	D 372/98/394	B 291/83/211	A 288/87/218	L 336/81/442	I 258/81/473	J 257/89/239	3461/1035/4024

§2.8 DIRECT ASSAYS

p.m. Ia	J 323/92/376	C 341/87/540	E 406/97/350	K 364/86/512	A 356/98/501	I 383/107/674	H 316/89/256	F 239/81/126	B 284/96/253	G 401/98/336	L 366/95/373	D 260/88/132	4039/1114/4429
p.m. Ib	D 329/93/313	F 440/93/284	I 358/91/348	A 287/90/326	L 280/90/537	E 416/105/501	C 330/86/523	G 446/108/402	J 387/98/199	B 277/90/270	H 388/88/387	K 347/96/473	4285/1128/4563
p.m. IIa	A 294/81/309	B 320/85/294	C 357/94/446	D 391/94/336	E 393/100/349	F 399/102/283	G 356/99/322	H 308/88/377	I 312/91/477	J 365/96/305	K 249/86/650	L 256/85/580	4000/1101/4728
p.m. IIb	H 342/97/261	E 384/96/419	L 357/94/625	J 342/97/368	C 407/100/426	F 450/100/460	B 281/86/211	I 294/91/348	K 417/98/716	D 291/82/289	G 248/83/402	F 340/100/181	4153/1124/4706
p.m. IIIa	G 297/89/363	I 367/101/606	D 372/94/651	F 405/106/360	K 257/93/453	H 376/93/336	J 321/95/185	A 287/91/205	L 347/93/437	C 381/99/632	E 253/85/337	B 248/76/167	3911/1115/4732
p.m. IIIb	L 340/94/521	A 408/104/387	H 386/96/326	I 384/103/692	B 423/100/461	D 406/100/369	E 350/98/348	J 311/93/313	F 279/87/139	K 391/98/439	C 301/88/447	G 265/85/398	4244/1146/4840
Total	3851/1096/4633	4353/1148/4765	4157/1115/4822	4501/1170/5342	3927/1142/4811	4486/1189/5056	4061/1123/3875	3742/1073/3808	4056/1106/4222	3933/1075/4050	3704/1060/4719	3578/1074/3755	48349/13371/53858
Drug totals:-	A 4068/1114/3854	B 4017/1116/3674	C 4005/1100/5612	D 4037/1121/4600	E 4050/1127/4191	F 4018/1098/2478	G 3977/1109/4389	H 4028/1091/3783	I 4082/1140/5747	J 3989/1121/3240	K 4049/1131/6342	L 4029/1103/5948	

TABLE 2.8.2 Analysis of variance and covariance for the data of Table 2.8.1

Adjustments for means		16 233 513	4 489 406	1 241 553	18 083 198	5 000 940	20 143 640
Nature of variation	d.f.	$S_{x_1 x_1}$	$S_{x_1 x_2}$	$S_{x_2 x_2}$	$S_{x_1 x}$	$S_{x_2 x}$	S_{xx}
Rows (observers, times)	11	46 707	5 833	900	51 422	7 060	121 133
Days (columns)	11	83 120	10 281	1 574	100 193	13 752	254 750
Drugs	11	865	212	197	11 201	5 657	1 307 259
Error	110	341 700	30 995	5 185	250 743	35 056	1 113 282
Total	143	472 392	47 321	7 856	413 559	61 525	2 796 424

§ 2.8 DIRECT ASSAYS 33

TABLE 2.8.3 Error regression analysis for Table 2.8.2

Nature of variation	d.f.	Sum of squares	Mean square
Regression on x_1 alone	1	183 998	
Additional for x_2	1	63 861	63 861
Regression on x_1, x_2	2	247 859	
Regression on x_2 alone	1	237 015	
Additional for x_1	1	10 844	10 844
Regression on x_1, x_2	2	247 859	
Residual	108	865 423	8 013
Total	110	1 113 282	

Note that the "additional" mean square for x_2 is large relative to the residual mean square, but that the "additional" mean square for x_1 is almost equal to the residual.

$$= \frac{35\,056}{5185}$$

$$= 6{\cdot}761.$$

Moreover, the residual variance is now

$$s^2 = \frac{1\,113\,282 - 237\,015}{109}$$

$$= 8039$$

with 109 d.f. Hence, by the usual regression formula,

$$\text{Var}(b_2) = s^2/S_{x_2 x_2} \tag{2.8.6}$$

$$= \frac{8039}{5185}$$

$$= (1{\cdot}245)^2.$$

If variations in tolerance were proportional to variations in heart weight, in the logarithmic units used here the regression coefficient ought to be 10·0. Chen *et al.* suggested that tolerance was proportional to (heart weight)$^{2/3}$, to which rule would correspond a regression coefficient of 6·67. Clearly b_2 is significantly less than 10·0, but differs little from 6·67.

The second column in Table 2.8.4 contains the mean value of x for each of the twelve drugs, *unadjusted* for variation in x_2. The standard error of these means is obtained from

$$\text{Var}(\bar{x}) = \frac{1\,113\,282}{110 \times 12}$$

$$= (29{\cdot}0)^2.$$

TABLE 2.8.4 Unadjusted and adjusted mean log tolerances, from the data of Table 2.3

Drug	$\bar{x}$	$\bar{x}_2$	$b_2(\bar{x}_2 - \bar{\bar{x}}_2)$	$\bar{x}'$
A	321·2	92·83	− 0·1	321·3
B	306·2	93·00	1·0	305·2
C	467·7	91·67	− 8·0	475·7
D	383·3	93·42	3·9	379·4
E	349·2	93·92	7·2	342·0
F	206·5	91·50	− 9·1	215·6
G	365·8	92·42	− 2·9	368·7
H	315·2	90·92	− 13·0	328·2
I	478·9	95·00	14·5	464·4
J	270·0	93·42	3·9	266·1
K	528·5	94·25	9·5	519·0
L	495·7	91·92	− 6·3	502·0
Mean	374·0	92·85	0·0	374·0
Standard error	± 29·0	−	−	± 25·9

In order to adjust for x_2, a column of means for this variate must be added. The regression calculations estimate that the correlation of x with x_2 has increased each observation on x by (6·761 × deviation of the corresponding x_2 from its general mean). Hence each $\bar{x}$ may be adjusted to an $\bar{x}'$, an estimate of the value it would have had if all cats had had the same heart weight:

$$\bar{x}' = \bar{x} - b_2(\bar{x}_2 - \bar{\bar{x}}_2), \qquad (2.8.7)$$

where $\bar{\bar{x}}_2$ is the general mean. The variance of $\bar{x}'$ is

$$\mathrm{Var}(\bar{x}') = s^2 \left[\frac{1}{N} + \frac{(\bar{x}_2 - \bar{\bar{x}}_2)^2}{S_{x_2 x_2}} \right], \qquad (2.8.8)$$

where N is the number of subjects tested for a particular drug. This variance depends upon which drug is in question, and the variance of a difference between two values of $\bar{x}'$ cannot be found simply by addition of variances. Unless the concomitant variate (here x_2) varies widely in its means for different treatments (drugs), a variance averaged over all comparisons may reasonably be used; this merely requires that s^2, the variance per response, be multiplied by

$$1 + \frac{\text{Mean square of } x_2 \text{ for treatments}}{\text{Error sum of squares for } x_2} \qquad (2.8.9)$$

in every comparison made (Finney, 1946a). Here the mean square for drugs in the x_2 analysis has already been observed to be small compared with the error, and the factor (2.8.9) is

$$1 + \frac{197}{5185 \times 11} = 1 \cdot 0035.$$

Hence

§ 2.8 DIRECT ASSAYS

$$\mathrm{Var}(\bar{x}') = \frac{8039}{12} \times 1{\cdot}0035$$

$$= (25{\cdot}9)^2.$$

With J, ouabain, as the standard, the remaining steps are just as in § 2.6. For drug A, and without adjustment for heart weight, equation (2.6.1) gives

$$M = \bar{x}_J - \bar{x}_A$$
$$= -51{\cdot}2,$$

with a standard error from equation (2.6.2)

$$\mathrm{SE}(M) = 29{\cdot}0\sqrt{2}$$
$$= 41{\cdot}0.$$

Since x was $1000 \times$ log dose,

$$\log R = M/1000$$
$$= \bar{1}{\cdot}9488$$

and therefore $R = 0{\cdot}889$.

The standard error is based upon 110 d.f., for which $t = 1{\cdot}982$; hence

$$\log R_L = (-51{\cdot}2 - 1{\cdot}982 \times 41{\cdot}0)/1000 = \bar{1}{\cdot}8675,$$
$$\log R_U = (-51{\cdot}2 + 1{\cdot}982 \times 41{\cdot}0)/1000 = 0{\cdot}0301,$$

and the fiducial limits to the estimate of relative potency are

$$R_L = 0{\cdot}737,$$
$$R_U = 1{\cdot}072.$$

The same process applied to the $\bar{x}'$ column in Table 2.8.4 gives for A

$$M' = \bar{x}'_J - \bar{x}'_A$$
$$= -55{\cdot}2 \pm 36{\cdot}6,$$

whence

$$\log R' = \bar{1}{\cdot}9448,$$
$$\log R'_L = \bar{1}{\cdot}8723,$$
$$\log R'_U = 0{\cdot}0173,$$

and therefore

$$R' = 0{\cdot}881,$$
$$R'_L = 0{\cdot}745,$$
$$R'_U = 1{\cdot}041.$$

Table 2.8.5 collates these results. In every case the potency figures agree well, but fiducial ranges are narrowed a little by adjustment for heart weight. The table also shows values of R obtained by Chen *et al.*, from calculations based upon dose/(heart weight)$^{2/3}$ in absolute (not logarithmic) terms, an analysis that did not give fiducial limits.

TABLE 2.8.5   Estimates of relative potency and their fiducial limits, for the data of Table 2.8.1

Drug	Unadjusted		Adjusted for heart weight		Chen *et al.*
	R	Limits	R	Limits	R
A	0·889	0·737–1·072	0·881	0·745–1·041	0·873
B	0·920	0·763–1·109	0·914	0·773–1·080	0·903
C	0·634	0·526–0·765	0·617	0·522–0·729	0·611
D	0·770	0·639–0·929	0·770	0·652–0·910	0·763
E	0·833	0·691–1·005	0·840	0·711–0·992	0·830
F	1·157	0·960–1·396	1·123	0·951–1·327	1·115
G	0·802	0·665–0·967	0·790	0·668–0·933	0·781
H	0·901	0·747–1·086	0·867	0·734–1·024	0·858
I	0·618	0·513–0·745	0·633	0·536–0·749	0·628
J	*1·000*	–	*1·000*	–	*1·000*
K	0·551	0·457–0·665	0·559	0·473–0·660	0·553
L	0·595	0·493–0·717	0·581	0·492–0·686	0·575

In practice, only the adjusted potencies would be reported; detailed calculation of the unadjusted would not be required after a statistically significant regression of x on x_2 has been demonstrated. The arithmetic can be simplified by forming Table 2.8.4 in terms of totals of 12 subjects instead of means. Exactly the same procedure would lead to adjusted totals of x', and a number of divisions would thereby be saved. The method used here illustrates better the structure of the calculations, and might be required in a less symmetric experiment.

2.9 Efficiency and utility of concomitant measurements

After the discovery of a highly significant regression, the gain in precision, shown by the narrowing of the fiducial range, perhaps appears surprisingly small. The gain may best be expressed in terms of the ratio of the variances of unadjusted and adjusted dose means in Table 2.8.4:

$$\frac{(29 \cdot 0)^2}{(25 \cdot 9)^2} = 1 \cdot 25;$$

to obtain the same gain without the covariance adjustment, 25 percent more cats would be needed, a total of 180 instead of 144. This ignores the fact that the Latin square balance could not be achieved with a larger number of cats (except for multiples of 144), and precision might decline further if a different design had to be adopted. Any computational saving resulting from avoidance of covariance analysis should be negligible, but the saving of experimental resources through no longer requiring dissection and weighing of hearts might more than compensate for the cost of a larger number of animals. An important consideration for the practical economics of the assay is that a covariance adjustment based upon body weight would have been almost as good as that using heart weight. Although Table 2.8.3 showed a significantly greater reduction in error variance from use of heart weight, body weight as a covariate gives 19 percent increase in precision. Presumably the body weight can be determined at much less cost, and the better policy for future assays might therefore be to use x_1 as the concomitant variate. The decision is important to the economic deployment of resources; it is not just a matter of statistical theory, but depends upon relative costs, in time, labour, or money, of various experimental operations.

In this example, much the same precision would have been obtained if an assumption of proportionality of tolerance and heart weight (or body weight) had been made, either by direct analysis of their ratio and formation of R from ratios of means (as in § 2.3) or — theoretically preferable (§ 2.5) and computationally simpler in the fiducial limit calculations — by a logarithmic analysis of $(x - 10x_2)$. Better still would have been an assumption of proportionality of the tolerances to (heart weight)$^{2/3}$, preferably by analysis of $(x - 6\cdot 67x_2)$. All are open to the objection that the relation is either guessed or taken from other experiments. Unless evidence from past experiments indicates strongly that a particular adjustment is suitable, estimation of the relation from the data of the current assay is desirable. Results based on a guessed relation between tolerance and a concomitant variate may occasionally be much less precise than if the concomitant were ignored.

Exploration of the possibility that a simply determined concomitant measurement, such as body weight, will improve the precision of assay results is often worth while. Complicated though the calculations may appear, they should take much less time than an equivalent increase in the number of subjects. Of course, if several similar assays showed no increase in precision from covariance analysis, thereafter covariance would be abandoned for this class of assay. A concomitant that is more trouble to measure, such as heart weight, may give better results, but the improvement may not always compensate for the labour of measuring it. Each type of assay needs to be considered on its merits.

2.10 The design of direct assays

The primary object of a direct dilution assay is to obtain the most precise estimate of a difference between two mean log doses that is possible with the resources available. If normality of the distribution of individual log tolerances is to be assumed, the aim is exactly that underlying most comparative experiments: measurements are to be made on two groups of differently treated subjects with a view to forming an estimate of the difference in means attributable to the contrast of treatments (Cox, 1958; Finney, 1955, 1959a). Out of the need for planning experiments so as to give precise estimates of treatment differences has grown the whole subject of experimental design. Any account of the design of direct assays would be much the same as a general account of experimental design, and would duplicate such excellent texts as Fisher (1966) and Cochran and Cox (1957). When only a single test preparation is to be assayed against a standard, little complexity will be needed. In *multiple assays* such as that of § 2.8, however, greater finesse is possible and often desirable. The user of direct assay techniques needs also to be familiar with complete and incomplete randomized block designs and the simpler factorial designs.

Indirect assays, the subject of the remaining chapters of this book, have problems of experimental design distinct from those general to all experimentation, and these are discussed in Chapters 6, 8, 9, 10, 11, and 19.

2.11 A criticism of direct assays

One grave objection to the technique of direct assay seems to have received insufficient attention (Bliss and Allmark, 1944; Gaddum, 1953a), though Miles and Perry (1950) have noted a related point. If two stimuli of unequal potency are applied at equal rates (e.g. two drugs infused at equal speeds), subjects receiving the less potent stimulus will have longer average times under treatment than those receiving the more potent. Any time-lag in the production of effects on the subject, or any cumulative effect not of a simply additive nature, will bias comparison of the stimuli. This difficulty would be overcome by applying equipotent doses of each stimulus per unit of time, but that presupposes knowledge of the relative potency; to say how much even quite small deviations from the ideal affect the validity of the potency estimate may be impossible. The question is one for the physiologist and pharmacologist to consider, in the light of their knowledge of the mode of action of the drugs or other stimuli studied. The decision as to whether the method is valid may vary from one type of assay to another; the statistician has fulfilled his duty by warning of a danger which has been ignored throughout this chapter.

3
Quantitative dose-response relations

3.1 Indirect assays

Biological assay seeks to estimate equally effective doses of the standard and test preparations, that is to say doses whose inverse ratio will estimate the potency of the test preparation relative to the standard. One objection to direct techniques (Chapter 2), namely bias produced by time lag, has been mentioned in § 2.11. Even without this danger, ensuring that subjects receive exactly the right dose to produce the characteristic response may be difficult: to measure individual tolerances of cats for digitalis may require only reasonable skill and care, but to measure individual tolerances of aphids for an insecticide is a greater problem.

In an indirect assay, specified doses are given, each to several subjects. The record for each administration of a dose may state merely that a characteristic response, such as death, is or is not produced: this is a *quantal* or 'all-or-nothing' response. Alternatively, some property of the subject (weight, weight of a particular organ, blood calcium, time of survival) may be measured: this is a *quantitative* response. Quantal response assays are closely related to direct assays, but analytically they are more complicated and discussion is deferred until Chapters 17 and 18. One could envisage analyzing quantitative response assays as though they were direct assays, after rejection of all records except those for responses in a defined narrow range. Even if each subject can be tested repeatedly with different doses until an acceptable response is found, such a procedure is wasteful of effort and information. Far more valuable is a method that takes account of the relation between dose and the magnitude of a response. If the form of this relation is known for the two preparations, equally effective doses may be estimated and, by appropriate statistical procedures, the precision of the corresponding estimate of potency may be assessed (Bliss, 1940b).

3.2 The dose–response regression

Suppose that a subject receives a dose z of a particular stimulus, and that the response subsequently measured is u. In random sampling from a population of subjects (or repeated trials with the same subject, where these can be made without affecting the independence of successive responses), the average or expected response to the dose may be written

$$E(u|z) = U. \quad \text{expected response} \tag{3.2.1}$$

A response which is to be of value for assay purposes must depend in some manner upon the dose; for assays at present in use,(*) this involves dependence of the expected response upon z:

$$U = F(z), \quad (\text{response is a function of dose}) \qquad (3.2.2)$$

Equation (3.2.2) is a regression equation of u on z. At present, no restriction need be placed on the form of the function $F(z)$, except that it must be a single-valued real function of z for all doses in the range that will be used in assays.

3.3 Similarity

Suppose that regression functions for the two preparations are

$$\begin{aligned} U_S &= F_S(z), \\ U_T &= F_T(z), \end{aligned} \qquad (3.3.1)$$

respectively. For a selected U, doses Z_S, Z_T of the preparations can be defined such that each has U as its expected response; at this level of response, Z_S and Z_T are equally effective doses. Nothing yet stated excludes the possibility of response curves (e.g. Fig. 3.6.2) for which the value of z corresponding to certain values of U is not uniquely determined, and a conventional rule for associating equivalent doses of the two preparations would be needed; other values of U might have no corresponding z (cf. Finney, 1971, § 11.10). The potency of the test preparation relative to the standard, *at the level of response U*, is Z_S/Z_T. For two completely unrelated stimuli, the relation between this potency ratio and U could be of any form, and a request that a single numerical value be assigned to the relative potency cannot be met.

For an analytic dilution assay (§ 2.5), these difficulties disappear. Wood (1946a) stated the basic requirements:

'(a) that the response supposed to be produced by the known amounts of "factor X" (i.e. the effective constitutent of the standard preparation) is actually due to the factor itself and not to some other substance associated with it, e.g., an impurity; and (b) that the response produced by the material to be analysed is also due solely to the presence in it of "factor X", without augmentation, diminution, or modification by any other substance also present. In other words, if we use the terms "Standard Preparation" and "Test Preparation" to denote respectively the solution of allegedly pure "factor X" and the solution prepared from the material to be analysed, we assume that the Std. Prep. contains no substance, other than factor X itself, contributing to the response we measure, and that the Test Prep. behaves for the purpose of the analysis so similarly to the Std. Prep. that it may be regarded simply as a dilution of the Std. Prep. in a completely inert diluent'.

With the convention that the term 'dilution' includes 'concentration', so as to

---

(*) In theory, an assay might be based upon a system in which U was constant, but changes in z affected, say, the variance of u. This is unlikely to be of practical importance.

cover assays in which the test preparation is more potent than the standard, this fully describes *analytic dilution assays*.

If two preparations contain the same effective constituent, or the same effective constituents in fixed proportions, and all other constituents are without effect on U, so that one behaves as a dilution of the other in an inert diluent, the potency ratio, Z_S/Z_T, must be independent of U;

$$\rho = Z_S/Z_T. \tag{3.3.2}$$

Consequently, the two regression functions must be related by

$$F_T(z) = F_S(\rho z) \tag{3.3.3}$$

for all z; ρ is a constant, the potency of T relative to S. This is the algebraic statement of the *condition of similarity*, a prerequisite of all dilution assays.

The fundamental assay problem is: Under specified conditions of testing, a standard preparation, whose content of the effective constituent either is known or is defined in arbitrary units, has a dose–response regression represented by equation (3.2.2), where the function $F(\ )$ is unknown. A test preparation is known to have the same effective constituent, and therefore its dose–response regression function must be

$$U = F(\rho z), \tag{3.3.4}$$

where ρ is the relative potency, the number of units of effective constituent contained in or equivalent to unit dose of the test preparation. An estimate, R, of ρ must be formed from observations on dose–response tests, usually after simultaneous estimation of parameters of the regression function. An understanding of this argument ensures avoidance of such gross mistakes as using a ratio of mean responses as an estimate of ρ (Kolb *et al.*, 1961).

3.4 Assay validity

If the responses cannot adequately be represented by the same form of regression function for both preparations, either the conditions of testing have differed for the two preparations or the basic assumption of similarity is false. Judged as a dilution assay, the assay is invalid, either because of insufficient care in the control of the experiment or because of the inherent incommensurability of the preparations. When the two regression functions are markedly different, no other conclusions can be drawn. Sometimes the functions are nearly the same over a wide range of responses, so that the ratio of equally effective doses varies little in that range, even though at extremes it is far from constant. Many experiments may then fail to show clear evidence against similarity, or may give a ratio of equally effective doses whose variations are not practically important; an average value of the ratio may still be a convenient expression of the relative potency of the preparations in the production of a particular type of response. The result of such a *comparative dilution assay* need not have any validity under conditions other than those of its estimation.

In an analytic dilution assay, on the other hand, the choice of assay technique, subject, response, and experimental conditions should be irrelevant to the potency estimate, although these may affect precision of estimation (Gaddum, 1950). This requirement is of great practical, as well as theoretical, importance. In such an assay, the biological system plays a part analogous to that of a balance in weighing an object: it is an instrument, not a factor influencing the magnitude of the result. The practice of the WHO Expert Committee on Biological Standardization is of interest in this connexion. By international agreement, the Committee has established standard preparations of many substances, the potency of each being defined in arbitrary international units (§ 1.4). The potency of a test preparation may be assayed by comparison with a sample of the appropriate international standard. In reviewing the work of the League of Nations Commission that preceded the Committee, Hartley (1935) wrote:

'As a rule a method of assay is described; while this method is one which has proved practicable and satisfactory, it is emphasized that its use is not in any way compulsory, individual workers being expected to use those methods which they have in regular use and in which they have confidence. The Commission takes the view that, while standards and units should be fixed and stable, and determinations of potency should always be carried out in strict comparison with the standard preparation (or its exact equivalent) and expressed in international units, no attempt should be made to fix or impose any particular method by which these comparative tests should be carried out. Improvement in existing methods of assay and the devising of new ones, and the progress of research, are more likely to be advanced by leaving to individual workers freedom of choice as to the method by which assays are carried out, rather than by insistence upon the details of a particular method which, on the one hand, may be difficult to describe adequately, and, on the other, may appear to give an air of finality in a field of biological standardisation in which every encouragement should be given for improvement and advance.'

The fact that two preparations give responses whose measurements agree with, or do not significantly deviate from, equation (3.3.3) is no demonstration that they have a common effective constituent: the condition is necessary, but not sufficient. No statistical process can distinguish between Wood's 'factor X' and a chemically distinct substance that affects the measured response in the same manner. Under particular conditions, even chemically very different preparations may act in conformity with equation (3.3.3). Lightbown (1961) comments on the existence of at least two distinct neomycins, so that some different neomycin preparations behave as though similarity were applicable but others may not allow valid assays. Miles (1951) describes how the existence of a family of penicillins was discovered because the potencies of different preparations varied according to the subject used in assay. He states:

'A biological standard, in fact, serves two purposes: it is primarily the stable vehicle of a certain specific biological activity; but it is also a reference point for the detection of heterogeneity among the substances that possess the specific activity.'

Bliss and Cattell (1943) mention materials whose relative potencies in man are completely different from estimates obtained with laboratory mammals. Such assays are comparative, not analytic. The statistical analysis may be the same, but the important logical difference may affect the choice of an optimal experimental design (Wood, 1944b).

3.5 Preliminary regression investigation

Before any relation between dose and response can be the basis of an assay, something must be known about the form of the regression function. Data from each single assay will not be sufficient for a study of this *ab initio*, though they may provide useful confirmation that an assumed form is not seriously wrong. Often knowledge of the function has slowly accumulated, without a special study. Bliss (1946c) described the kind of investigation of $F(z)$ that should precede use of a dose–response relation in assays, and the topic is discussed further in § 3.9. Wood (1946a) has emphasized the need for separating this study from regular assay work, especially because the requirements in respect of experimental design are different. In particular, many different doses must be tested, but this is seldom the most economic procedure in an assay (§§ 6.10, 8.4).

3.6 Monotony

For almost all useful assays, $F(z)$ is a *strictly monotonic function* of z within the dose range of normal experimentation. For simplicity of presentation, attention is here restricted to increasing functions; decreasing functions require only the obvious sign changes. The general condition is that, for every pair of doses z_1, z_2, if

then
$$z_2 > z_1,$$
$$F(z_2) > F(z_1);$$

equality is explicitly excluded. Fig. 3.6.1 illustrates some strictly increasing regression functions.

Non-monotonic dose–response relations can occur and are important for the information they give on the modes of action of stimuli. For a relation such as is shown in Fig. 3.6.2, the dose of the standard preparation equipotent with a stated dose of the test preparation may not be uniquely determined. If equation (3.3.3) applies, however, the one curve must still be a magnification of the other; conceptually, estimation is still possible. Obviously $F(z)$ is mathematically complicated and a procedure for estimating ρ would be difficult. Such a regression is unlikely to be a good basis for a practical assay, unless attention can be restricted to a monotonic segment.

3.7 Linearizing transformations

Graphical simplicity and arithmetic convenience suggest the advantages of a linear dose–response regression, if this can be achieved without violating the facts. Often a transformation of the measured dose helps. A family of transformations introduced by Box and Cox (1964) is useful:

$$x = (z^\lambda - 1)/\lambda; \tag{3.7.1}$$

this is meaningful for all values of λ, positive and negative, with the understanding that it includes the limiting form as λ tends to zero:

$$x = \ln z. \tag{3.7.2}$$

A more general form has $(z - z_0)$ in place of z, where z_0 is an arbitrary adjustment to the dose, but only $z_0 = 0$ will be considered here.

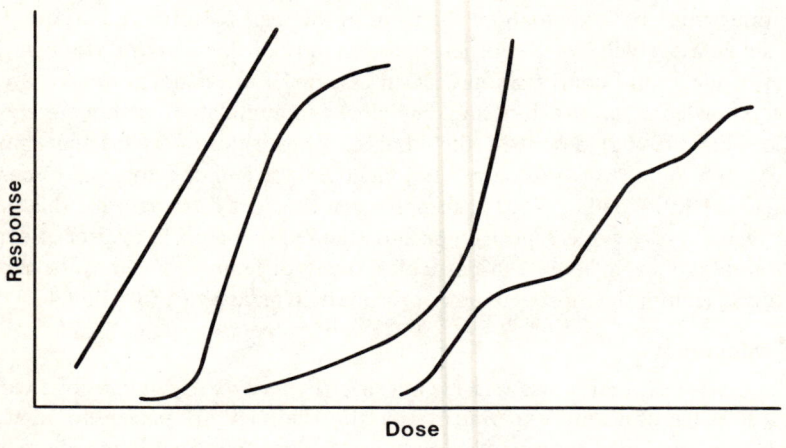

Fig. 3.6.1   Examples of strictly monotonic (increasing) dose–response regression curves

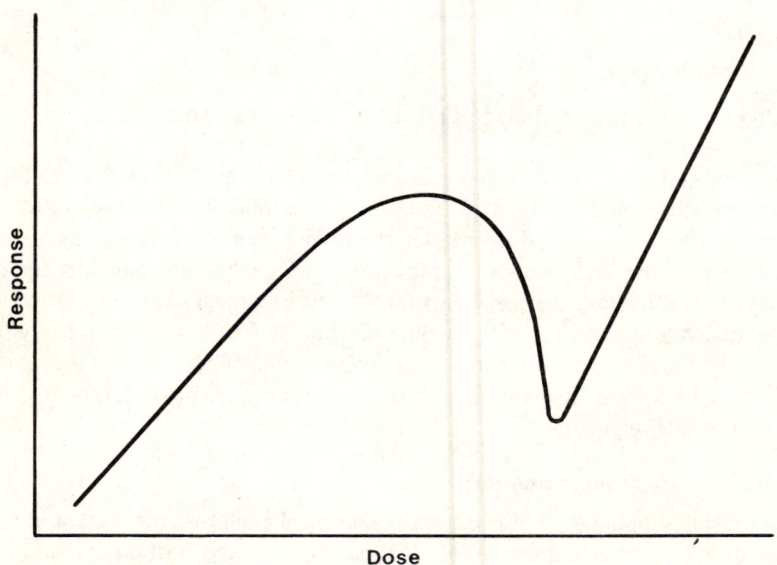

Fig. 3.6.2   An example of a non-monotonic dose–response curve (cf. Finney, 1971, § 11.10)

§ 3.7 QUANTITATIVE DOSE–RESPONSE RELATIONS

Equations (3.7.1) and (3.7.2) are in mathematically consistent form. Except where details of this approach to a limit are under discussion, in practice the transformation can be taken as

$$\left.\begin{array}{l} x = z^\lambda, \quad \lambda \neq 0; \\ x = \log z, \quad \lambda = 0; \end{array}\right\} \quad (3.7.3)$$

any convenient base of logarithms is now permissible, not only e as required in equation (3.7.2). For many dose–response relations, equation (3.2.2) appears to be well represented by

$$U = f(\alpha + \beta x), \quad (3.7.4)$$

in which $f(\ )$ is a completely specified function that satisfies the condition of monotony, so that α, β, λ are the only unknown parameters. Although conclusions about relative potency must eventually be expressed on the scale of z, use of x in the statistical analysis is a legitimate intermediate step; x is termed the *dose metameter* (Bacharach et al., 1942), though it may be called the dose when there is no fear of confusion. A *response metameter*, Y, may now be defined by the inverse function $f^{-1}(\ )$, so that

$$Y = f^{-1}(U). \quad (3.7.5)$$

Consequently for S, equation (3.7.4) can be rewritten

$$Y_S = \alpha + \beta x, \quad (3.7.6)$$

and statistical analysis of the dose–response relation can be put into the form of a process for estimating a linear regression between the metameters.

Similarity ensures that the same metameters linearize both the S and T regression functions. Moreover, from equation (3.3.3), T must have

$$\left.\begin{array}{l} Y_T = \alpha + \beta \rho^\lambda x, \quad \lambda \neq 0; \\ Y_T = \alpha + \beta \log \rho + \beta x, \quad \lambda = 0. \end{array}\right\} \quad (3.7.7)$$

If $\lambda \neq 0$, the S and T lines intersect at $x = 0$; writing β_S, β_T for their slopes,

$$\rho = (\beta_T/\beta_S)^{1/\lambda}. \quad (3.7.8)$$

If $\lambda = 0$, the lines are parallel; writing α_S, α_T for the constants and β for the common slope,

$$\log \rho = (\alpha_T - \alpha_S)/\beta. \quad (3.7.9)$$

Data from a single bioassay are seldom extensive enough to discriminate satisfactorily between different values of λ. Fortunately potency estimates themselves seem remarkably insensitive to moderate changes in λ (§ 15.5). The best practice, therefore, seems to be to regard the preliminary investigation (§ 3.5) as determining both the form of $f(\ )$ and the value of λ. The most common values of λ are 0 and 1, and probably most data can be adequately represented by these or other simple ones such as $\frac{1}{2}, \frac{1}{3}, -1, -\frac{1}{2}$. The response metameter also is usually taken as a simple function; examples are

$$Y = U, \qquad (3.7.10)$$

$$Y = \log U, \qquad (3.7.11)$$

$$Y = 1/U, \qquad (3.7.12)$$

$$Y = U^{1/2}. \qquad (3.7.13)$$

If no such simple choice is adequate, a linearizing transformation for a class of assays might be constructed empirically. Adapting ideas from Ipsen (1941), mean responses in a preliminary investigation of the standard preparation might be plotted against a logarithmic dose metameter. A line could be fitted by eye, by polynomial regression calculations, or by any other procedure that gives a reasonable approximation to the statistical dependence of u, an observed response, on x (cf. Kapteyn, 1903; Kapteyn and Van Uven, 1916). Expressed graphically or as a table, this could be written

$$U = F_0(x), \qquad (3.7.14)$$

and the inverse function

$$Y = F_0^{-1}(U) \qquad (3.7.15)$$

may then be tried as a response metameter in subsequent assays. If the dose–response relation is stable, an assay should be well fitted by

$$Y_S = x,$$

$$Y_T = \log \rho + x;$$

even if conditions have changed since the preliminary investigation, two more general parallel lines (as for $\lambda = 0$ above) should be satisfactory. Northam and Norris (1952) illustrated a procedure essentially of this type.

When λ is known and $f(\ )$ is a known function, the natural method of estimating potency from data on two preparations is to express doses and responses metametrically, and to fit two straight lines subject to the constraint either of intersection at $x = 0$ or of parallelism. The lines might be fitted by eye on a diagram and an estimate, R, of ρ obtained by measurement, or equations to the lines might be obtained by calculation. Much of this book is concerned with methods of calculation for various types of data. These can involve special cases of the general theory in § 3.13, but in practice the simpler possibilities usually suffice.

3.8 Essentially non-linear relations

Not every monotonic dose–response regression conforms to equation (3.7.4). If the range of doses is wide, additional complexity and the need for more parameters are to be expected. Emmens (1940) found a logistic function

$$U = \frac{D}{1 + \exp\{-2(\alpha + \beta x)\}}, \qquad (3.8.1)$$

where D is an additional parameter denoting an upper limit to U as x increases, to be a good representation of several series of data. In view of the importance of the logistic function as a growth curve, its relevance to assays in which the responses are growth measurements is not surprising. If D were known or reliably estimated from previous investigations, the metametric transformation

$$Y = \tfrac{1}{2} \ln \{U/(D-U)\} \tag{3.8.2}$$

would reduce equation (3.8.1) to the form of equation (3.7.5), and the method of § 3.13 would apply. More commonly, D will be unknown, and the equations for maximum likelihood estimation must be generalized so as to estimate it also. An analogous model using

$$U = D \int_{-\infty}^{\alpha+\beta x} \frac{1}{\sqrt{2\pi}} \exp(-\tfrac{1}{2}t^2) \, dt \tag{3.8.3}$$

(Butler, Finney and Schiele, 1943; Finney, 1971, § 10.4) is practically indistinguishable from the logistic on empirical evidence.

These have been little used in bioassay, though radioimmunoassay techniques now alter the situation (§§ 16.6–16.8). For large negative x, each curve flattens and approaches a limiting value zero; for large positive x it again flattens and approaches a limit D. The flat portions of the curve will be of little use in assays, because small changes in response correspond with large dose differences; without much loss, interest may be confined to a middle portion of the curve that is approximately linear, even though the variance of individual responses may be less at extremes of x.

If the dose–response regression cannot be put into the form of equation (3.7.4), or of some simple modification involving perhaps an extra parameter as in equation (3.8.1), the assay problem is more complicated. Even if S has a quadratic regression

$$U_S = \alpha + \beta x + \gamma x^2, \tag{3.8.4}$$

that for T takes the troublesome form

$$\left. \begin{array}{ll} U_T = \alpha + \beta \rho^\lambda x + \gamma \rho^{2\lambda} x^2, & (\lambda \neq 0), \\ U_T = \alpha + \beta (\log \rho + x) + \gamma (\log \rho + x)^2, & (\lambda = 0). \end{array} \right\} \tag{3.8.5}$$

No assay dependent upon such a specification, with simultaneous estimation of $\rho, \alpha, \beta, \gamma$, seems to be in use at present (cf. § 5.11).

3.9 A response curve for vitamin B_{12}

Data from turbidimetric measurements on the growth response of *Lactobacillus leichmannii* to vitamin B_{12} (Emery, Lees and Tootill, 1951) provide a good illustration of a preliminary investigation of a dose–response relation. Table 3.9.1 shows the responses to eight different doses of vitamin B_{12}, as measured in six independent tubes at each dose. For discussion here, the fact that the tubes were arranged in three racks, each with two tubes from each dose, is neglected.

TABLE 3.9.1 Turbidimetric measures of response of *L. leichmannii* to vitamin B_{12}

Dose (ng/tube)	0·23	0·35	0·53	0·79	1·19	1·78	2·67	4·00
Responses	0·15	0·28	0·36	0·51	0·68	0·85	1·06	1·21
	0·14	0·20	0·36	0·53	0·63	0·80	0·91	1·22
	0·19	0·23	0·34	0·54	0·64	0·71	1·09	1·29
	0·19	0·25	0·37	0·45	0·61	0·85	0·93	1·24
	0·17	0·23	0·33	0·57	0·65	0·94	1·09	1·18
	0·16	0·23	0·38	0·49	0·68	0·83	1·12	1·24
Totals	1·00	1·42	2·14	3·09	3·89	4·98	6·20	7·38
Means	0·167	0·237	0·357	0·515	0·648	0·830	1·033	1·230

Although the appropriate modification of the analysis is easy, it makes the example less typical and diverts attention from more important issues; for these particular records, it makes no important difference to the conclusions.

The total and mean for each dose are shown at the foot of the table. The constant ratio, 1·5, between successive doses is an indication that a simple relation with log dose was expected; Fig. 3.9.1 therefore shows mean responses plotted against log dose. The obvious transformation

$$x = \log z \tag{3.9.1}$$

could be used, but the alternative

$$x = \log_{1.5}(z/4 \cdot 00) \tag{3.9.2}$$

has the advantage that the values of x are $-7, -6, -5, \ldots, 0$. Better still is

$$x = 2\log_{1.5}(z/4 \cdot 00) + 7 \tag{3.9.3}$$

for which the values are $-7, -5, -3, \ldots, 7$, symmetric about the middle of the range. This kind of trick in modifying the dose metameter is useful in reducing the arithmetic. Since the three definitions of x are linearly related, any one of them can be used to explore the form of the response curve, and finally a conversion to another can be made if required.

Inspection of the figure suggests a marked curvature, but an analysis of variance based on $Y = U$ is needed for a full examination. The standard regression calculations are as follows:

$$Sx = 6 \times (-7) + 6 \times (-5) + \ldots + 6 \times 7$$
$$= 0,$$
$$Sy = 1 \cdot 00 + 1 \cdot 42 + \ldots + 7 \cdot 38$$
$$= 30 \cdot 10,$$
$$\bar{x} = 0,$$
$$\bar{y} = 30 \cdot 10/48$$
$$= 0 \cdot 6271.$$

§ 3.9   QUANTITATIVE DOSE–RESPONSE RELATIONS   49

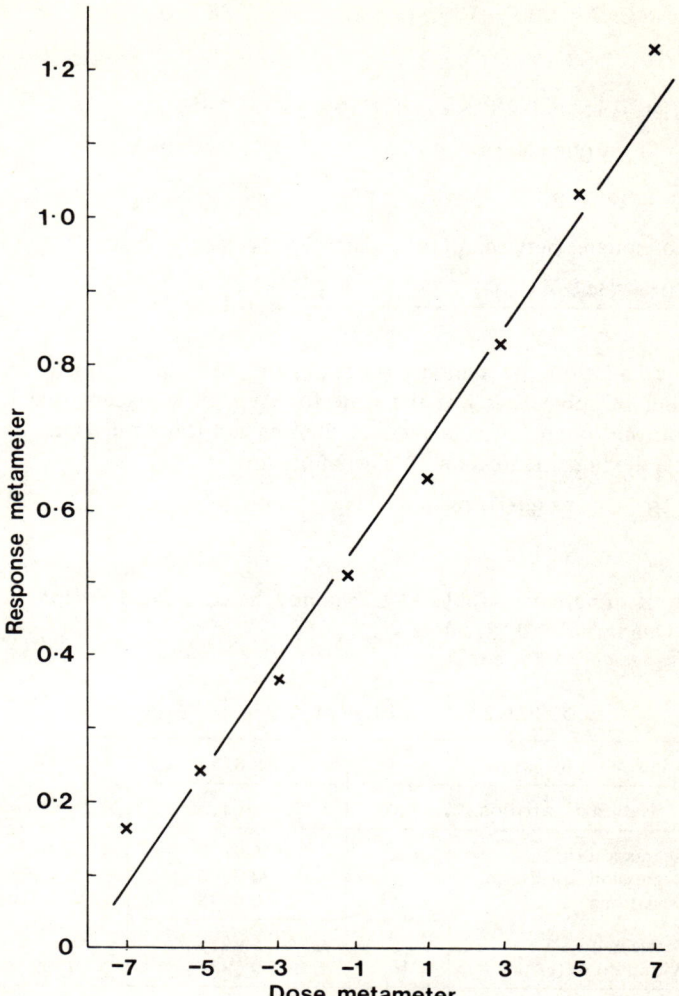

Fig. 3.9.1 Dose–response relation for the data of Table 3.9.1, growth of *L. leichmannii* at different doses of vitamin B_{12}

×: Mean responses

The straight line has been fitted to the data as described in the early part of § 3.9

Using the formula

$$S_{xy} = Sxy - (Sx)(Sy)/N \qquad (3.9.4)$$

and corresponding formulae for S_{xx}, S_{yy},

$$S_{xx} = 6 \times (-7)^2 + 6 \times (-5)^2 + \ldots + 6 \times (7)^2 - 0^2/48$$
$$= 1008,$$

$$S_{xy} = -7 \times 1\cdot00 - 5 \times 1\cdot42 - \ldots + 7 \times 7\cdot38 - 0 \times 30\cdot10/48$$

$$= 77\cdot88$$

$$S_{yy} = 0\cdot15^2 + 0\cdot28^2 + \ldots + 1\cdot24^2 - 30\cdot10^2/48$$

$$= 25\cdot0924 - 18\cdot8752$$

$$= 6\cdot2172.$$

The sum of squares between doses, with 7 d.f., is calculated as

$$\frac{1\cdot00^2 + 1\cdot42^2 + \ldots + 7\cdot38^2}{6} - \frac{30\cdot10^2}{48} = 6\cdot1216.$$

All these calculations are standard for regression; they apply equally well even if the number of subjects is not the same for every dose, except that in the last calculation the squares of totals would then have different divisors. Of this sum of squares, the linear regression on x accounts for

$$S_{xy}^2/S_{xx} = 77\cdot88^2/1008 \tag{3.9.5}$$

$$= 6\cdot0172$$

The analysis of variance, Table 3.9.2, can now be completed except for the lines labelled 'Quadratic' and 'Residual'.

TABLE 3.9.2 Analysis of variance for Table 3.9.1

Adjustment for mean		18·8752	
Nature of variation	d.f.	Sum of squares	Mean square
Regression (linear)	1	6·0172	6·0172
Regression (quadratic)	1	0·0996	0·0996
Deviations	5	0·0048	0·0010
Between doses	7	6·1216	
Within doses (error)	40	0·0956	0·00239
Total	47	6·2172	

The mean square for linear regression is very significantly greater than the error within doses, a matter scarcely requiring comment because no one would advocate this response as a basis for bioassay unless it were known to be greatly affected by vitamin B_{12}. The residual variation between doses gives a mean square

$$(6\cdot1216 - 6\cdot0172)/6 = 0\cdot0174,$$

also significantly greater than error, indicating that a straight line does not adequately represent the regression. Table 3.9.2 goes farther by focusing attention on a major component of non-linearity, expressed by a regression on x^2. Write

$$x' = x^2$$

and compute the partial regression coefficients on x, x'. The additional arithmetic is

$$S_{xx'} = 0,$$
$$S_{x'x'} = 16\,128,$$
$$S_{x'y} = 40 \cdot 08.$$

Hence the regression coefficients are given by the equations

$$1008b + 0b' = 77 \cdot 88,$$
$$0b + 16\,128b' = 40 \cdot 08,$$

and the regression accounts for an amount

$$bS_{xy} + b'S_{x'y} = 6 \cdot 1168.$$

The additional amount attributable to the quadratic component is therefore

$$6 \cdot 1168 - 6 \cdot 0172 = 0 \cdot 0996.$$

In this instance, the result could have been obtained more easily by the use of orthogonal polynomial coefficients (Fisher and Yates, 1963, Table XXIII); the general procedure illustrated here would still be applicable to a less symmetric experiment that did not have $S_{xx'} = 0$.

The significance of the quadratic regression and lack of significance of the residual deviations are in accordance with the smooth curve that Fig. 3.9.1 suggests. For any type of response, at extremes of dose non-linearity of regression is almost inevitable. As will soon be apparent, bioassay techniques are much more easily applied with linear regressions, and often an assayist employs a linear relation as an approximation over a range of doses without any assumption that the true regression is perfectly linear. One natural course of action, therefore, is to omit the lowest dose, the greatest indicator of curvature, and to repeat the whole scheme of computation on the remaining 42 responses. Although the quadratic component is then relatively less important, it is still statistically significant. Indeed, not until the three lowest doses are omitted does the evidence against linearity become non-significant, suggesting that to assume linearity between doses of about 0·8 ng/tube to about 4·0 ng/tube could be satisfactory.

An alternative is to look for a response metameter that gives a simpler regression. To write a computer program that would make similar analyses for several different metameters, involving a sequence of values of λ in equation (3.7.1) or some other definition of a set of metameters, is easy. Here however the form of Fig. 3.9.1 suggests one simple possibility, that of equation (3.7.13),

$$Y = U^{\frac{1}{2}}. \tag{3.9.6}$$

Table 3.9.3 contains the metametric transforms of the original data, to only 2

TABLE 3.9.3 Response metameters for data of Table 3.9.1, according to the relation $Y = U^{\frac{1}{2}}$

Dose (ng/tube)	0·23	0·35	0·53	0·79	1·19	1·78	2·67	4·00
Response metameters	0·39	0·53	0·60	0·71	0·82	0·92	1·03	1·10
	0·37	0·45	0·60	0·73	0·79	0·89	0·95	1·10
	0·44	0·48	0·58	0·73	0·80	0·84	1·04	1·14
	0·44	0·50	0·61	0·67	0·78	0·92	0·96	1·11
	0·41	0·48	0·57	0·75	0·81	0·97	1·04	1·09
	0·40	0·48	0·62	0·70	0·82	0·91	1·06	1·11
Totals	2·45	2·92	3·58	4·29	4·82	5·45	6·08	6·65
Means	0·408	0·487	0·597	0·715	0·803	0·908	1·013	1·108

TABLE 3.9.4 Analysis of variance for Table 3.9.3

Adjustment for mean		27·3612	
Nature of variation	d.f.	Sum of squares	Mean square
Regression (linear)	1	2·6149	2·614 9
Regression (quadratic)	1	0·0000	0·000 0
Deviations	5	0·0028	0·000 56
Between doses	7	2·6177	
Within doses (error)	40	0·0357	0·000 892
Total	47	2·6534	

places of decimals because the accuracy of the data does not justify more, and Fig. 3.9.2 is the response diagram. Inspection alone will convince that a linear regression is now exceedingly good for the whole range; Table 3.9.4, constructed exactly as was Table 3.9.2, removes any doubts that remain. Though use of a linear regression of dose metameter on x beyond the range of doses tested cannot be justified, it appears to fit exceedingly well over this 17-fold range of doses.

The regression coefficient of y on x, for this response metameter, is

$$b = S_{xy}/S_{xx} \tag{3.9.7}$$

$$= 51·34/1008$$

$$= 0·050\,93.$$

From the analysis of variance in Table 3.9.4, the variance per response is estimated as

$$s^2 = 0·000\,892.$$

Therefore, for b,

$$\text{Var}(b) = s^2/S_{xx} \tag{3.9.8}$$

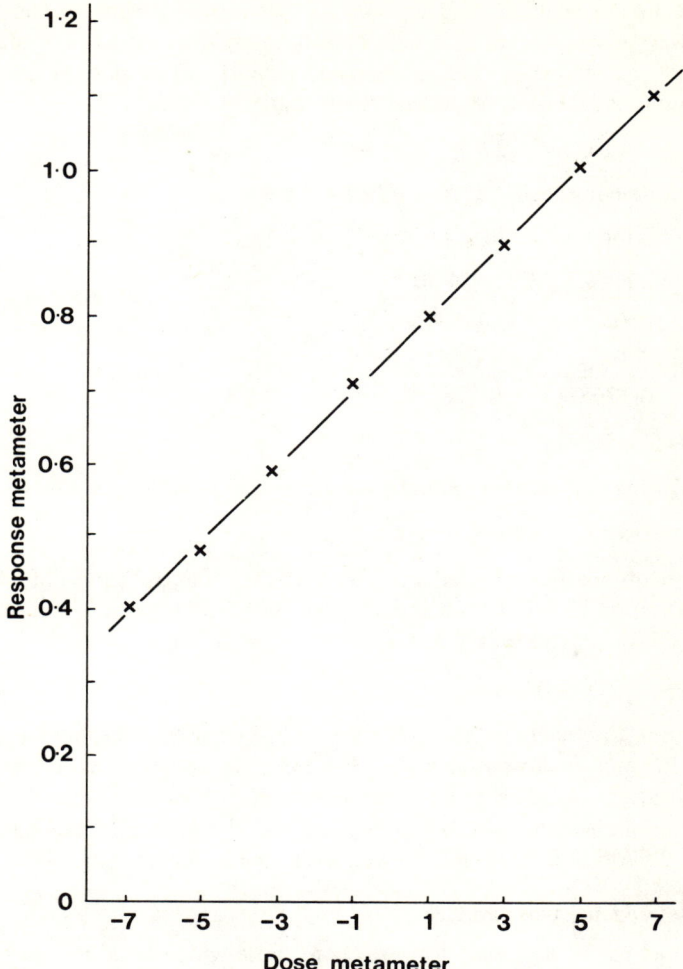

Fig. 3.9.2 Dose–response relation for Table 3.9.1 after transformation of responses according to equation (3.9.6)
×: Mean response metameters
The straight line is that for equation (3.9.10)

$$= 0.000\ 892/1008$$
$$= 0.000\ 000\ 885 = (0.000\ 941)^2.$$

The linear regression equation is, in the usual form,

$$Y = \bar{y} + b(x - \bar{x}), \tag{3.9.9}$$

or

$$Y = 0.7550 + 0.050\ 93x. \tag{3.9.10}$$

The dose metameter in equation (3.9.3) was chosen for its computational simplicity in the analysis. Had a computer been used, the metameter could as easily have been $\log_{10} z$ or $\ln z$, and indeed one of these is to be preferred for any use of the regression that goes beyond analysis of the data in Table 3.9.1. Conversion is easy. From the general result that

$$\log_f g = \log_h g / \log_h f \qquad (3.9.11)$$

for any positive numbers f, g, h, it follows that

$$\begin{aligned}
x &= 2\log_{1.5} z - 2\log_{1.5} 4{\cdot}00 + 7 \\
&= \frac{2\log_{10} z}{\log_{10} 1{\cdot}5} - \frac{2\log_{10} 4{\cdot}00}{\log_{10} 1{\cdot}5} + 7 \\
&= \frac{2\log_{10} z}{0{\cdot}176\,09} - \frac{2 \times 0{\cdot}602\,06}{0{\cdot}176\,09} + 7 \\
&= 11{\cdot}36 \log_{10} z + 0{\cdot}1619.
\end{aligned} \qquad (3.9.12)$$

By substituting in equations (3.9.10), the estimated regression can be written

$$Y = 0{\cdot}7632 + 0{\cdot}5786 \log_{10} z. \qquad (3.9.13)$$

The regression coefficient in equation (3.9.10) was multiplied by 11·36 to give that in equation (3.9.13); its standard error, 0·000 941, must be multiplied by the same factor, to give on the new scale

$$b = 0{\cdot}5786 \pm 0{\cdot}0107. \qquad (3.9.14)$$

A regression coefficient is an estimate of the increase in response metameter per unit increase in dose metameter. For example, over the range of the data, a multiplication of dose by 1·5 (the step from one experimental dose to the next) is an increase of 0·1761 in $\log_{10} z$, and the estimated increase in y is 0·1761 × 0·5786 or 0·102, agreeing well with the last line of Table 3.9.3.

3.10 Heterogeneity of variance

The most useful responses for assays are those which are direct measures of size or growth. For them, the variance of individual values of u at a specified dose is often found to be practically independent of U, at least over such range of doses as is of interest to the experimenter. The regression of u on z is then said to be *homoscedastic*. In the method of estimating the regression equation to be described in § 3.13, homoscedasticity is assumed.

Even for responses of this type, and still more for scores and other arbitrary measures, the possibility of heteroscedasticity must be borne in mind. If the preliminary investigation shows that $\sigma^2(u)$, the variance of u for a specified z, is markedly dependent upon U, a *scedasticity transformation* is required. Slight heterogeneity of variance will not much affect the statistical analysis, and a transformation that is approximately correct is adequate. Sometimes theoretical considerations suggest

$$\sigma^2(u) = \phi(U), \qquad (3.10.1)$$

where $\phi(\ )$ is a known function; sometimes the preliminary investigation enables estimates of $\sigma^2(u)$ to be formed at different doses, and the plotting of these against corresponding mean responses points to such a function. Write

$$u^* = \int \frac{du}{\sqrt{(\phi(u))}}. \tag{3.10.2}$$

Then, to a first-order approximation (Bartlett, 1947),

$$\sigma^2(u^*) = 1. \tag{3.10.3}$$

Hence the transformation of responses from u to u^* may be expected to give a homoscedastic regression.

Important cases for bioassay (and other branches of biometry) are

$$\sigma^2(u) = \text{constant} \times U, \tag{3.10.4}$$

which gives

$$u^* = \text{constant} \times u^{\frac{1}{2}}, \tag{3.10.5}$$

and

$$\sigma^2(u) = \text{constant} \times U^2, \tag{3.10.6}$$

which gives

$$u^* = \text{constant} \times \log u. \tag{3.10.7}$$

If responses must lie between two fixed limits, as did the line test scores once commonly used in vitamin D assay (Burn *et al.*, 1950), the variance is likely to be reduced in the neighbourhood of the limits. When the limits are known, the scale of measurement can be chosen to make them 0, 1. For such a response,

$$\sigma^2(u) = \text{constant} \times U(1 - U) \tag{3.10.8}$$

is often a satisfactory approximation; it leads to

$$u^* = \text{constant} \times \sin^{-1}(u^{\frac{1}{2}}), \tag{3.10.9}$$

the well-known angle transform of the proportion u (Fisher and Yates, 1963, Table XII).

3.11 Testing homoscedasticity

Analyses such as Tables 3.9.2, 3.9.4 assume the variance of y about its expectation for a fixed dose to be constant. Inspection of Table 3.9.1 throws doubt on this assumption for $Y = U$, the dispersion of the responses at doses of 1·79 and 2·67 ng/tube looking particularly large. Table 3.11.1 summarizes variance estimates, each with 5 d.f., calculated from the eight columns of Table 3.9.1, and the corresponding quantities from Table 3.9.3; the variations are large and further examination is desirable.

Bartlett's test (Bartlett, 1937; Emmens, 1948, § 9.6; Snedecor and Cochran, 1967, § 10.21) is often used to assess the evidence for the heterogeneity of several independent variance estimates. If the estimate within dose group i is s_i^2

TABLE 3.11.1 Variance estimates for the data of Tables 3.9.1 and 3.9.2

Dose (ng/tube)	d.f.	Response $= u$		Response $= u^{\frac{1}{2}}$	
		$10^4 s^2$	$\log_{10}(10^4 s^2)$	$10^4 s^2$	$\log(10^4 s^2)$
0·23	5	4·27	0·630	7·77	0·890
0·35	5	7·07	0·849	7·07	0·849
0·53	5	3·47	0·540	3·47	0·540
0·79	5	17·50	1·243	7·90	0·898
1·19	5	7·77	0·890	2·67	0·427
1·78	5	56·40	1·751	18·17	1·259
2·67	5	81·07	1·909	21·47	1·332
4·00	5	13·60	1·134	2·97	0·473
All	40	23·89	1·378	8·94	0·951
$f \log_{10} s^2 - S f_i \log_{10} s_i^2$		–	10·390	–	4·700

with f_i degrees of freedom, the test criterion is

$$\chi^2_{[k-1]} = [f \ln s^2 - S(f_i \ln s_i^2)]/C; \qquad (3.11.1)$$

here k is the number of variance estimates, s^2 and f relate to a pooled variance estimate given by

$$f = S(f_i), \qquad (3.11.2)$$

$$fs^2 = S(f_i s_i^2), \qquad (3.11.3)$$

and the adjustment factor C is defined by

$$C = 1 + \frac{1}{3(k-1)}[S(1/f_i) - 1/f]. \qquad (3.11.4)$$

Table 3.11.1 also contains logarithms of the variance estimates; a multiplicative factor (here 10^4) can be introduced into the variances without affecting the value of χ^2, and the use of logarithms to base 10 merely requires that a factor $\log_e 10 \,(= 2\cdot 3026)$ be introduced into the χ^2 formula. In this example, all f_i are equal, and therefore

$$C = 1 + \left(\frac{8}{5} - \frac{1}{40}\right) \Big/ (3 \times 7)$$

$$= 1\cdot 075.$$

Hence for the original responses

$$\chi^2_{[7]} = 2\cdot 3026 \times 10\cdot 390/1\cdot 075$$

$$= 22\cdot 25,$$

substantially greater than the 0·05 probability level, 14·07 (Appendix Table IV; Fisher and Yates, 1963, Table IV). On the other hand, for the recommended response metameter,

§ 3.12

$$\chi^2_{[7]} = 2\cdot3026 \times 4\cdot700/1\cdot075$$
$$= 10\cdot07,$$

which is not statistically significant.

Box (1953) criticized reliance on Bartlett's test as an indicator of whether inference from an analysis of variance can be trusted. He found the test sensitive to non-normality, whereas analysis of variance tests are 'robust' in the face of non-normality. Consequently, such a preliminary test on variances is, he says, 'like putting to sea in a rowing boat to find out whether conditions are sufficiently calm for an ocean liner to leave port!' Experience shows that only great heterogeneity of variance will seriously affect conclusions drawn from a good assay (Chapters 15, 16). Despite the evidence of heterogeneity for Table 3.9.1, little harm need have been feared from ignoring it if for other reasons (notably linearity) the regression for this response had seemed satisfactory, especially as no pattern of dependence of variance on response appears. Nevertheless, the disappearance of heteroscedasticity is a further merit of the metameter in Table 3.9.3. Very commonly, indeed, a simple metametric transformation that performs well as a linearizer also stabilizes variance, though the empirical finding has no theoretical basis (Fieller, 1947).

In practice, no one should rely on a single experiment to determine the form of a dose–response regression and the appropriate metameters. Emery *et al.* regarded repetition of an experiment such as that recorded in Table 3.9.1 as part of the regular routine of their vitamin B_{12} assays. All that can be said here is that the one experiment strongly supports the idea that the regression of $u^{\frac{1}{2}}$ on $\log z$ is linear and homoscedastic over a wide range of doses, and that this regression might be a better basis for assay than the original authors' choice of u on $\log z$ over a more restricted range.

3.12 Estimation by maximum likelihood

The problem of how to estimate an unknown parameter or set of parameters is central to statistical inference. Certain experimental or observational data are known or assumed to be related according to a mathematical law involving one or more parameters: what can be said about the values of these parameters? Perhaps the simplest relevant example is that of values of a response metameter, y, believed to be related to the dose metameter, x, according to a homoscedastic linear regression; that is to say

$$E(y) = Y = \alpha + \beta x, \tag{3.12.1}$$

and
$$E[(y - Y)^2] = \sigma^2. \tag{3.12.2}$$

The parameters are then α, β, σ^2. In order to make the problem amenable to theoretical discussion, some assumption about the frequency distribution of y is necessary, such as: 'For any x, the distribution of y is normal with mean Y and variance σ^2'. Verbal description of the theory of statistical estimation, adopted here for the benefit of the non-mathematical reader, cannot be wholly satisfactory. This section lacks logical and mathematical rigour; anyone wishing for more precise statements about maximum likelihood and other principles of estimation should consult a textbook of statistical theory (Rao, 1965).

Many general principles of estimation proposed by theoretical statisticians are discussed in standard textbooks. That most favoured today is *maximum likelihood*, originated by R.A. Fisher (1912, 1922). The *likelihood* that a set of parameters should have any particular values is defined to be a quantity proportional to the probability that, if these be the parameters, the totality of observations should be the data recorded. Fisher (1925) fixed the factor of proportionality by the condition that the maximum value of the likelihood is unity, but this is seldom important. The method of maximum likelihood then states that the parameters are to be estimated so as to make the likelihood take its maximum value. In the most usual circumstances, the maximum likelihood estimator of a parameter has a number of important properties, chief of which are the following:

(i) The estimate is *consistent*. Essentially this means that if the estimator is applied to the 'true' distribution instead of to the sample of values produced by an experiment it will be exactly equal to the parameter estimated. Formal expression of this idea is not easy (Rao, 1965). In practice, it can usually be regarded as equivalent to convergence in probability: as the number of observations is increased without limit, the probability that the estimate differs from the parameter by less than an amount δ approaches $1 \cdot 0$, however small δ may be. In order to avoid paradoxes (Fisher, 1956), careful specification of the limiting process is needed and the definition of the estimator in relation to the number of observations must avoid certain absurdities. For bioassay, one consequence is that the number of observations *at each dose* shall increase without limit and the number of doses shall be enough to provide information on the parameter. The first type of definition, known as *Fisher consistency*, is preferable.

The distinction between consistency and freedom from bias must be kept clear. For unbiasedness, the expectation of the estimator is required to equal the parameter estimated. An unbiased estimator need not be consistent and a consistent estimator need not be unbiased. Although many important estimators have both properties, one of the difficulties with maximum likelihood estimation is that it commonly involves a bias of unknown size.

(ii) The frequency distribution of the estimator is *asymptotically normal*. For any fixed number of observations, repeated samplings from the system under study would give different sets of data, from each of which a maximum likelihood estimate could be formed, and thus would generate a frequency distribution for the estimator. As the number of observations per dose is increased without limit, this distribution approaches the normal form.

(iii) The estimator is *asymptotically efficient*. A consistent estimator is said to be asymptotically efficient if, as the number of observations increases without limit, the information on the parameter contained in the estimator tends to equality with the total information on the parameter contained in the observations. Here 'information' is used in a specialized sense of statistical theory, closely related to the reciprocal of variance. Less exactly, therefore, this property means that in large samples the variance of the estimator approaches the minimum that could be achieved by any conceivable consistent estimation from the same data.

(iv) For finite numbers of observations, the specification of a particular problem permits a minimal variance for any unbiased estimator to be stated. In many circumstances, maximum likelihood estimators, or simple adaptations of these that remove bias, are *efficient* in the sense of achieving this minimum variance. Consequently there is some reason to hope for good performance of this estimation principle even with relatively small amounts of data.

(v) If a *sufficient* estimator of the parameter exists, the maximum likelihood estimator will be a function of it (or even equal to it). Such an estimator has the remarkable property of utilizing all the relevant information contained in a set of observations.

Although restrictions on the mathematical specification of distributions are necessary for the truth of (i)–(v) above, these are not so severe as to be often important in general practice.

Maximum likelihood is not above criticism, nor is it the only general principle that can be proposed for estimating parameters; it is the most serviceable general principle yet enunciated. It has certain obvious flaws, notably its possible bias and the uncertainty as to its efficiency in small samples when no fully efficient estimate is known (Shenton and Bowman, 1977). At least for problems of the type considered in this book, however, no other general principle is any more widely applicable or escapes the disadvantages and difficulties that have been noted. In the absence of evidence to its discredit in any particular situation, it may reasonably form a standard of comparison for other methods. In biological assay, it is often equivalent to the less versatile *principle of least squares*, and for quantal responses it is closely related to the *method of minimum* χ^2 (§ 19.9). As will be seen from subsequent chapters, it leads to computational processes that are easily arranged for routine work, and it often reduces to the common-sense calculations that would be adopted by a reader who ignored § 3.13.

3.13 Estimation for regression equations

Although a response metameter that produces homoscedasticity often also gives effective linearity of regression over a useful range of doses, this is not invariably so (Fieller, 1947). For example, a situation envisaged as corresponding to equation (3.10.8) might involve a sigmoidal response curve that is not linearized by equation (3.10.9). Even so, when $\sigma^2(u)$ is clearly not independent of U, a transformation of u to an appropriate u^* in order to equalize the variances seems very desirable, though perhaps only as a preliminary to a further transformation (Finney, 1947a).

If it be supposed that u now represents a response *after* this scedasticity transformation, the remaining problem is that of estimating α, β in the regression equation (3.7.4), where individual responses for various values of x have been determined by experiment, and the variance of any such u about U is σ^2, a constant. By argument analogous to that of §2.5, a case may be advanced for assuming the frequency distribution of u in repeated sampling for a fixed x to be normal with mean U and variance σ^2. Corresponding results might be derived for distributions other than normal, provided that they were exactly specified in terms of U and σ^2, but there seems no good reason for adopting any alternative here. Most assays will use means of several observations at each dose. Conclusions based upon an assumption of normality are therefore likely to be protected by the central limit theorem: unless the distribution of individual responses is very markedly non-normal, the distribution of means will tend to normality as the number of observations is increased (see §15.4).

A scedasticity transformation might be followed by a linearization, this facilitating graphical estimation of potency. If individual response metameters, or means for doses, are plotted against dose metameters, two straight lines can be drawn by eye, subject to the constraint either of intersection at $x = 0$ or of parallelism. The ratio of slopes or the distance between the lines (parallel to the x-axis) then leads to an estimate R. If no assessment of precision is wanted, this procedure may be sufficiently exact; indeed, experience enables lines drawn after brief inspection of the diagram to give a value of R close to that obtainable by calculation. The method lacks objectivity, and might bias results for a user who, despite honest intention, had an interest in showing a particular conclusion. Moreover, a measure of the reliability of a value of R as an indicator of ρ is usually needed, to which end computation of R and of its precision is essential.

The standard calculations for bioassay with quantitative responses (Chapters 4–15) are adaptations of unweighted linear regression techniques. The general theory presented below is mathematically more complicated than is necessary for most purposes. The iterative process includes the familiar methods of analysis as particular cases based upon simplifying assumptions. The reader anxious to reach the more practical aspects of assay analysis should omit the remainder of this section, perhaps returning to it after Chapter 17.

Consider first the standard preparation alone, for which N pairs of values of z, u have been observed (some doses may have several independent measures of responses, as in Table 3.9.1). The likelihood function for the three parameters is a function of α, β, σ^2 proportional to the product of the probabilities of all the observations; it is therefore proportional to the product of expressions like

$$\frac{1}{\sigma} \exp\left\{-\frac{(u-U)^2}{2\sigma^2}\right\}, \tag{3.13.1}$$

where U is as in equation (3.7.4). Simultaneous estimation of the parameters can be effected by maximizing the likelihood, or its logarithm:

$$L = \text{constant} - N \log \sigma - \frac{S(u-U)^2}{2\sigma^2}, \tag{3.13.2}$$

§ 3.13   QUANTITATIVE DOSE–RESPONSE RELATIONS

where S denotes summation over all observations. Since U does not involve σ, the maximization of L requires minimization of $S(u - U)^2$. As far as α, β are concerned, least squares estimation is equivalent to maximum likelihood, and this will be first considered.

General computer procedures can be used to explore the behaviour of the *objective function*, $S(u - U)^2$, and to determine what values of the parameters minimize it. This approach is examined further in § 18.5. The classical method is to consider the estimates, a and b, as solutions of

$$\left.\begin{aligned}\frac{\partial L}{\partial \alpha} &= 0, \\ \frac{\partial L}{\partial \beta} &= 0.\end{aligned}\right\} \qquad (3.13.3)$$

Suppose that a_1, b_1 are any approximations to a, b; they are readily obtained by forming response metameters for each observed u from equation (3.7.5) and then using either eye estimates from a diagram or a rough calculation of an unweighted linear regression equation. By Taylor–Maclaurin expansion to the first order of small quantities, improved values for the estimates will be $a_2 = a_1 + \delta a_1$, $b_2 = b_1 + \delta b_1$, where the increments $\delta a_1, \delta b_1$ are the solutions of

$$\left.\begin{aligned}\frac{\partial L}{\partial \alpha_1} + \delta a_1 \frac{\partial^2 L}{\partial \alpha_1^2} + \delta b_1 \frac{\partial^2 L}{\partial \alpha_1 \partial \beta_1} &= 0, \\ \frac{\partial L}{\partial \beta_1} + \delta a_1 \frac{\partial^2 L}{\partial \alpha_1 \partial \beta_1} + \delta b_1 \frac{\partial^2 L}{\partial \beta_1^2} &= 0.\end{aligned}\right\} \qquad (3.13.4)$$

Addition of a subscript 1 to α, β indicates replacement by a_1, b_1 after differentiation. Repetition with a_2, b_2 in equation (3.13.4) in place of a_1, b_1 and solving for increments $\delta a_2, \delta b_2$ establishes an iterative process that can be continued until a satisfactorily close approach to a, b is achieved. The equations must next be made more suitable for computation. The limits of the iteration are unaffected if in equations (3.13.4) the second differential coefficients of L are replaced by their expectations, a convenient change because expected values can be tabulated, whereas observed values need special calculation. Write

$$\frac{\partial U}{\partial Y} = f'(Y), \qquad (3.13.5)$$

and define the *weighting coefficient*, w, by

$$w = \{f'(Y)\}^2. \qquad (3.13.6)$$

From equation (3.13.2)

$$\frac{\partial L}{\partial \alpha} = \frac{1}{\sigma^2} S\left\{(u - U)\frac{\partial U}{\partial \alpha}\right\}. \qquad (3.13.7)$$

A second differentiation, followed by substitution of $E(u)$ for u, gives

$$E\left(\frac{\partial^2 L}{\partial \alpha^2}\right) = -\frac{1}{\sigma^2} S\left(\frac{\partial U}{\partial \alpha}\right)^2$$

$$= -\frac{1}{\sigma^2} Sw\left(\frac{\partial Y}{\partial \alpha}\right)^2$$

$$= -Sw/\sigma^2 \qquad (3.13.8)$$

from equation (3.7.6).

Similarly

$$E\left(\frac{\partial^2 L}{\partial \alpha \partial \beta}\right) = -Swx/\sigma^2, \qquad (3.13.9)$$

and

$$E\left(\frac{\partial^2 L}{\partial \beta^2}\right) = -Swx^2/\sigma^2. \qquad (3.13.10)$$

Corresponding to a_1, b_1 is an expected value of the response metameter:

$$Y_1 = a_1 + b_1 x. \qquad (3.13.11)$$

Substitution into equations (3.13.4) from equations (3.13.5)–(3.13.10) gives

$$\left.\begin{aligned}\delta a_1 Sw_1 + \delta b_1 Sw_1 x &= S\left\{\frac{w_1(u - U_1)}{f'(Y_1)}\right\}. \\ \delta a_1 Sw_1 x + \delta b_1 Sw_1 x^2 &= S\left\{\frac{w_1 x(u - U_1)}{f'(Y_1)}\right\}.\end{aligned}\right\} \qquad (3.13.12)$$

Next define the *working response*, y, as

$$y = Y + \frac{u - U}{f'(Y)}. \qquad (3.13.13)$$

In general, y differs from the empirical response metameter, $f^{-1}(u)$. In the many assays for which the 'transformation' $Y = U$ is used, the two are identical and a special symbol to represent the empirical response metameter is then unnecessary. Expected values Y and U are always related by equation (3.7.5); u represents the original observation, or the value obtained after a scedasticity transformation, y is used only for the working response. Now add $Sw_1 Y_1$ to each side of the first of equations (3.13.12) and $Sw_1 x Y_1$ to each side of the second. The result may be written

$$\left.\begin{aligned}a_2 Sw_1 + b_2 Sw_1 x &= Sw_1 y_1, \\ a_2 Sw_1 x + b_2 Sw_1 x^2 &= Sw_1 x y_1.\end{aligned}\right\} \qquad (3.13.14)$$

Hence from simple regression theory

$$Y = a_2 + b_2 x \qquad (3.13.15)$$

is obtainable as the weighted linear regression equation of y_1 on x. If the weighted

§ 3.13 QUANTITATIVE DOSE–RESPONSE RELATIONS

means of x and y_1 are written

$$\bar{x}_1 = Sw_1 x / Sw_1, \quad (3.13.16)$$

and

$$\bar{y}_1 = Sw_1 y_1 / Sw_1, \quad (3.13.17)$$

and the notation for sums of squares and products of deviations is now used for weighted sums, so that

$$S_{xy} = Swxy - (Swx)(Swy)/Sw, \quad (3.13.18)$$

then the revised estimates are

$$b_2 = S_{x_1 y_1} / S_{x_1 x_1} \quad (3.13.19)$$

and

$$a_2 = \bar{y}_1 - b_2 \bar{x}_1. \quad (3.13.20)$$

Iteration involves replacing a_1, b_1 by a_2, b_2 and calculating a_3, b_3 from a new set of weighting coefficients and working responses; further cycles follow in the same manner, until the smallness of the differences between successive approximations indicates close approach to a, b. Thus estimation of α, β from experimental data requires only repeated weighted linear regression calculations. Finney (1947a) developed this process on the analogy of Fisher's (1935) method for quantal responses and extensions to it (Garwood, 1941; Finney 1949b, 1951c).

The variances of the estimates may also be obtained from the usual expressions for linear regression:

$$\text{Var}(\bar{y}) = \sigma^2 / Sw, \quad (3.13.21)$$

$$\text{Var}(b) = \sigma^2 / S_{xx}, \quad (3.13.22)$$

where all quantities now relate to the limit of iteration. From the maximum likelihood point of view, the variances take these forms asymptotically in accordance with general theory as in § 17.7. In practice, Sw and S_{xx} taken from the last iteration computed will be close enough to their limits, and the true variance per response, σ^2, will be replaced by s^2, an estimate from a residual sum of squares of deviations of u.

For a metametric transformation that must be used often, special tables should be prepared. First a table of the transformation (3.7.5) is required, from which may be read the empirical response metameters. Secondly, a table of w, equation (3.13.6), the *minimum working response*,

$$Y_0 = Y - \frac{U}{f'(Y)}, \quad (3.13.23)$$

and the *range*

$$A = \frac{1}{f'(Y)}, \quad (3.13.24)$$

as functions of Y, is formed. The names are given on the analogy of quantal responses (Chapter 17), though Y_0 is not necessarily an absolute minimum and A not a true range. The first step in calculation is to read from the first table the empirical response corresponding to each observed u. This is plotted against x and a straight line is drawn, by eye, through the points; with experience, some allowance for unequal weights can be made. To correspond with each x in the observations, an *expected response*, Y, is read from the line. From the second table, the weighting coefficient for each Y is read, and the corresponding working response is formed as

$$y = Y_0 + uA. \tag{3.13.25}$$

The weighted linear regression of y on x gives a new set of values for expected responses, with which a second cycle of iteration is initiated. Further cycles may be computed until a_{r-1}, b_{r-1} are found to be almost the same as a_r, b_r; the latter are then regarded as the maximum likelihood estimates of α, β.

This process rarely requires many cycles: often the first or second cycle will give results sufficiently good for practical purposes. The iteration is easily programmed for a computer. Although for quantitative responses the general form is seldom required, its adaptation to quantal responses is important.

Next the estimation of σ^2 must be considered. Reference to equation (3.13.2) and differentiation with respect to σ shows that, whatever the values of α, β, maximization of L gives

$$s^2 = S(u - U)^2/N \tag{3.13.26}$$

as an estimator for σ^2. As with maximum likelihood variance estimators in much simpler problems, this will be biased, and division by $(N - 2)$ instead of N would be preferred as an allowance for the estimation of the other two parameters. However, if some doses have several independent responses, the sum of squares $S(u - U)^2$ can be analyzed into a component between doses and a component within doses. The latter is independent of any estimation of α, β, and can be used for estimating σ^2 without reference to the maximum likelihood iteration. The component between doses is a measure of deviation of the data from the model represented by equation (3.7.4); its mean square can be compared with the mean square within doses as a test of goodness of fit. This procedure is essentially that illustrated in discussion of Tables 3.9.2, 3.9.4. Formulae need not be displayed here, but henceforth s^2 will be taken as the mean square within doses wherever this is available.

If the scedasticity transformation also effectively linearizes the regression on a simple dose metameter such as $\log z$, equation (3.7.5) is

$$Y = U. \tag{3.13.27}$$

Hence

$$f'(Y) = 1, \tag{3.13.28}$$

and equations (3.13.6), (3.13.13) become

$$w = 1, \tag{3.13.29}$$

$$y = u. \tag{3.13.30}$$

The iterative process then reduces to calculation of an unweighted linear regression, and one cycle completes it. Subsequent discussion of quantitative responses will be concerned mostly with this model; no claim is made that σ^2 is absolutely constant or that the regression is perfectly linear, but the accumulated evidence of many experiments often fails to show consistent or large departures from these conditions. An application of the full technique is presented in §§ 16.6–16.8.

3.14 Standard curve estimation

What use can be made of equation (3.9.13),

$$Y = 0{\cdot}7632 + 0{\cdot}5786x,$$

the regression of the square root of a growth measurement on log dose of vitamin B_{12}? It could facilitate assay in any of three ways; despite their interest as illustrating the argument, two of these are untrustworthy in practice.

Standard curve estimation need not be restricted to a linear regression. Suppose that 8 responses measured at unit dose per tube of a test preparation are $0{\cdot}83, 0{\cdot}87, 0{\cdot}96, 0{\cdot}92, 0{\cdot}80, 1{\cdot}03, 0{\cdot}96, 0{\cdot}82$. For the response metameter

$$\bar{y}_T = (0{\cdot}91 + 0{\cdot}93 + 0{\cdot}98 + 0{\cdot}96 + 0{\cdot}89 + 1{\cdot}01 + 0{\cdot}98 + 0{\cdot}91)/8$$

$$= 0{\cdot}946.$$

From equation (3.9.13), the equivalent dose of the standard preparation is

$$x = (0{\cdot}946 - 0{\cdot}763)/0{\cdot}5786$$

$$= 0{\cdot}316.$$

The antilogarithm of this then gives $2{\cdot}07$ ng as the estimated vitamin B_{12} content of unit dose of the test preparation.

If the tubes of test preparation for the assay were tested under conditions identical with those that obtained for Table 3.9.1, this method would give a valid estimate of potency. Unfortunately it is practically useless, as even slight variations in experimental conditions within a laboratory are liable to alter the true dose–response relation in an unpredictable manner. Any general assumption that a response curve once determined can be used in future assays is unacceptable. An example from a quantal response assay is discussed in § 18.6. Bliss and Packard (1941) reported one of the few known exceptions: they found the curve relating percentage survival of eggs of *Drosophila melanogaster* to X-ray dosage to remain constant for several years, so that it might be used as a basis for the standardization of dosage.

Even less permissible is the use of a standard curve from one laboratory as a standard of comparison for test preparations in another. The curve is then likely to be quite irrelevant because of differences in strains of animals or other subjects and in experimental technique. Except where the contrary is demonstrated by careful experiment, it must be regarded as axiomatic that tests of the standard preparation, S, shall be run simultaneously with, and as an integral part of, the assay of any test preparation, T.

3.15 Standard slope estimation

Though the response regression is likely to shift from day to day, for some types of response that shift may represent a change in position only. In other words, the response to each dose may change, but with the increase in response per unit increase in dose metameter remaining constant. The standard curve method, if it were trustworthy, could be applied whatever the form of the regression relation for the standard preparation; the standard slope method is limited to assays for which the regression of response on log dose is linear. If experience has shown the assumption of constant slope for a linear regression on log dose to be justified for S, a test preparation T can be assayed by simultaneous experimentation with one dose of S and one of T under comparable conditions. The two doses should be chosen well within the range of linearity and as nearly equivalent as existing information makes possible.

Suppose that, at the same time as the eight tubes recorded in § 3.14, a further six tubes given the known dose 1·2 ng per tube under otherwise identical conditions had the responses 1·12, 1·15, 0·98, 0·74, 1·12, 1·05. The mean of the response metameter is

$$\bar{y}_S = (1{\cdot}06 + 1{\cdot}07 + 0{\cdot}99 + 0{\cdot}86 + 1{\cdot}06 + 1{\cdot}02)/6$$
$$= 1{\cdot}010.$$

The difference in mean responses, 0·064, corresponds to a difference

$$-0{\cdot}064/0{\cdot}5786 = -0{\cdot}111$$

in log dose, on the assumption that the regression coefficient in equation (3.9.13) can be used. Since

$$\text{antilog } \bar{1}{\cdot}889 = 0{\cdot}774,$$

the test dose is estimated to contain 0·774 times the vitamin B_{12} of the standard dose: the estimated vitamin content is therefore 0·93 ng.

The general result can be written

$$M = x_S - x_T - (\bar{y}_S - \bar{y}_T)/b, \tag{3.15.1}$$

where x_S, x_T are the dose metameters and M is the logarithm of the estimated relative potency. Thus the required potency is

$$R = \text{antilog } M. \tag{3.15.2}$$

For the data,

$$M = 0{\cdot}079 - 0{\cdot}000 - (1{\cdot}010 - 0{\cdot}946)/0{\cdot}5786$$
$$= \bar{1}{\cdot}968$$

and $R = 0{\cdot}93$ as before. Note that the scale of measurement used for the test dose can be entirely arbitrary, but R will always correspond to the unit on that scale. For example, if the test dose had been 0·4 in terms of some arbitrary scale on which this material was being measured, x_T would be $\bar{1}\cdot 602$ (or $-0\cdot 398$) instead of 0·000, and the same steps would give $M = 0\cdot 366$; the test preparation would be estimated to contain 2·32 ng of vitamin B_{12} per unit of this scale.

For the standard slope method, the precision of the estimate can be expressed in terms of fiducial limits to the potency. The procedure, essentially as in § 4.14, is justifiable only if the regression coefficient has remained constant. This assumption is not recommended: when the only data available are responses to one dose of each preparation, no statistical theory or test can demonstrate that the standard slope method is valid. Herein lies danger, for the simplicity of the method is a temptation. The investigator must recognize that unexpected changes can make his results entirely misleading. If $\bar{y}_S$ and $\bar{y}_T$ are nearly equal, the effect of errors in b upon M is practically negligible, and choice of x_S, x_T as equivalent dosages is therefore desirable. Even this cannot ensure that fiducial limits will be correctly assessed. The regression coefficient measures the responsiveness of subjects to changes in dose; it might vary with season, it might show steady increase over a period in which subjects were being bred selectively for responsiveness and homogeneity, or it might vary erratically because of lack of control of experimental conditions.

3.16 Simultaneous trial estimation

The only way of overcoming the objections to the standard curve and standard slope methods of estimating potency is to perform simultaneous trials of preparations, under strictly comparable conditions, using two or more doses of each. All quantities required in the computation of potency are then obtainable from evidence internal to the current assay, and if the design has been well chosen, any validity tests needed can be made on the same data. The preliminary investigation is used to indicate metameters for which the regression is nearly linear, and to locate a suitable range of doses. The statistical analysis and estimation are otherwise entirely self-contained, in accordance with widely accepted principles of good experimental design. The apparent precision of estimation of potency is likely to be less than if the standard slope method were used, because the regression coefficient is less precise. In a good assay, however, the variance of the regression coefficient is not a major factor in the assessment of the precision of the potency estimate. A regression coefficient determined for the assay itself, even though its variance be relatively large, is preferable to one with a small variance that may be inappropriate to the current assay.

Simultaneous trial is the basis of subsequent chapters. The general iterative procedure (§ 3.13) requires little modification for assays, and is seldom needed in its full complexity. Empirical responses for both S and T must be plotted against x, and two provisional regression lines drawn, subject to the constraint either of intersection at $x = 0$ or of parallelism. Weighted regression calculations are applied to improve the approximations to the estimates of all parameters involved, the intersection or parallelism constraint being maintained. Eventually

ρ is estimated, as in the graphical method, either from the ratio of slopes or from the horizontal distance between parallel lines.

The three forms of estimation may be summarized as follows:

(i) Standard curve: Regression equation, not necessarily linear, assumed to remain fixed in position;

(ii) Standard slope: Regression equation on log dose assumed linear, with constant regression coefficient;

(iii) Simultaneous trial, two doses of each preparation: Regression on dose metameter assumed linear, but each assay provides its own estimate of all parameters;

(iv) Simultaneous trial, three or more doses of each preparation: As (iii), but each assay also gives a test of deviations from linearity.

The standard curve and standard slope methods will not be discussed further.

4
Parallel line assays

4.1 Unsymmetric designs

The most widely used type of simultaneous trial assay is that for which a simple response metameter has a homoscedastic linear regression on log dose. For such an assay, the condition of similarity requires the lines for the standard and test preparations to be parallel (§3.7). As will become apparent in Chapters 5 and 6, symmetry in the number and spacing of doses and in the allocation of subjects to doses usually improves precision and eases computation. Nevertheless, an accident may convert a symmetric design into an unsymmetric, or shortage of material may force adoption of an unsymmetric design. A general unsymmetric assay, such as is discussed below, is also the best illustration of the whole structure of the computations.

4.2 Data for an unsymmetric assay

Table 4.2.1 (British Standards Institution, 1940) relates to an assay of vitamin D_3 in cod-liver oil by means of its antirachitic activity in chickens, using percentage bone ash as the response. When the measured responses are percentages, both non-linearity and heteroscedasticity of the regression are likely, at least at extreme doses (§3.9). In this assay, almost all the responses lay between 30 percent and 45 percent, and no difficulties of statistical invalidity (§§4.5, 4.6) were encountered in an analysis based on a linear regression of response on log dose. Although no response metameter was needed, a linear transformation (commonly termed a *coding*) was applied. Each bone ash percentage, u, was transformed by

$$y = 10(u - 30), \tag{4.2.1}$$

so that a response 33·5 was coded as 35, etc. Such coding can reduce the magnitudes of quantities used in calculations, remove decimal digits, and make most values positive, all of which are conveniences in desk calculation though seldom worth while on a computer. Responses recorded to halves or quarters of arbitrary units can be coded by multiplication by 2 or 4. A linear metametric transformation does not affect scedasticity, linearity, tests of validity, or potency estimates. Table 4.2.1 contains the coded data for the vitamin D_3 assay; by inversion of equation (4.2.1), the original percentages are recoverable:

$$u = 30 + 0 \cdot 1y. \tag{4.2.2}$$

One good feature of the assay is that the doses were at equal logarithmic spacing: for both preparations, the ratio of successive doses was 5/3. With a logarithmic dose metameter, such a choice can ensure that the dose range is adequately

TABLE 4.2.1 Responses in an assay of cod-liver oil for vitamin D_3

	Dose of standard preparation, S (BSI units per 100 g food)			Dose of test preparation, T (mg oil per 100 g food)			
	5·76	9·6	16	32·4	54	90	150
	35	62	116	20	26	57	140
	30	67	105	39	60	89	133
	24	95	91	16	48	103	142
	37	62	94	27	−8	129	118
	28	54	130	−12	46	139	137
	73	56	79	2	77	128	84
	31	48	120	31		89	101
	21	70	124			86	
	−5	94					
		42					
n	9	10	8	7	6	8	7
Sy	274	650	859	123	249	820	855
$\bar{y}$	30·4	65·0	107·4	17·6	41·5	102·5	122·1
x	−2	0	2	−3	−1	1	3

covered and simplify the arithmetic. The calculations could be executed with $\log_{10} z$ as the metameter. If instead

$$x = \log_\epsilon z - \log_\epsilon 9 \cdot 6 \tag{4.2.3}$$

for S, and

$$x = \log_\epsilon z - \tfrac{1}{2}(\log_\epsilon 54 + \log_\epsilon 90) \tag{4.2.4}$$

for T, where

$$\epsilon = (5/3)^{1/2}, \tag{4.2.5}$$

the doses are represented by the simple integers shown in the last line of Table 4.2.1. Alternatively, in order to avoid negative values, metameters might have been chosen so that the values of x were 0, 1, 2, for S and 0, 1, 2, 3 for T; equations (4.2.3), (4.2.4) have the advantage of making $\bar{x}_S$, $\bar{x}_T$ almost zero (not exactly, because of unequal numbers of subjects per dose). The essential feature of the metameter scale is that the same base of logarithms is used for both preparations. The choice of base and the addition of different quantities to the logarithms for the two preparations affect only the intermediate arithmetic; a simple final adjustment removes their effects.

4.3 The dose–response diagram

A diagram (Fig. 4.3.1) showing mean responses plotted against x leads to a rapid estimation of potency; it also protects against gross errors or misinterpretations of the statistical analysis. The experienced user of assay techniques may often dispense with the diagram, at least for symmetric designs, because he can visualize its form without drawing it, but to others a sketch is practically essential.

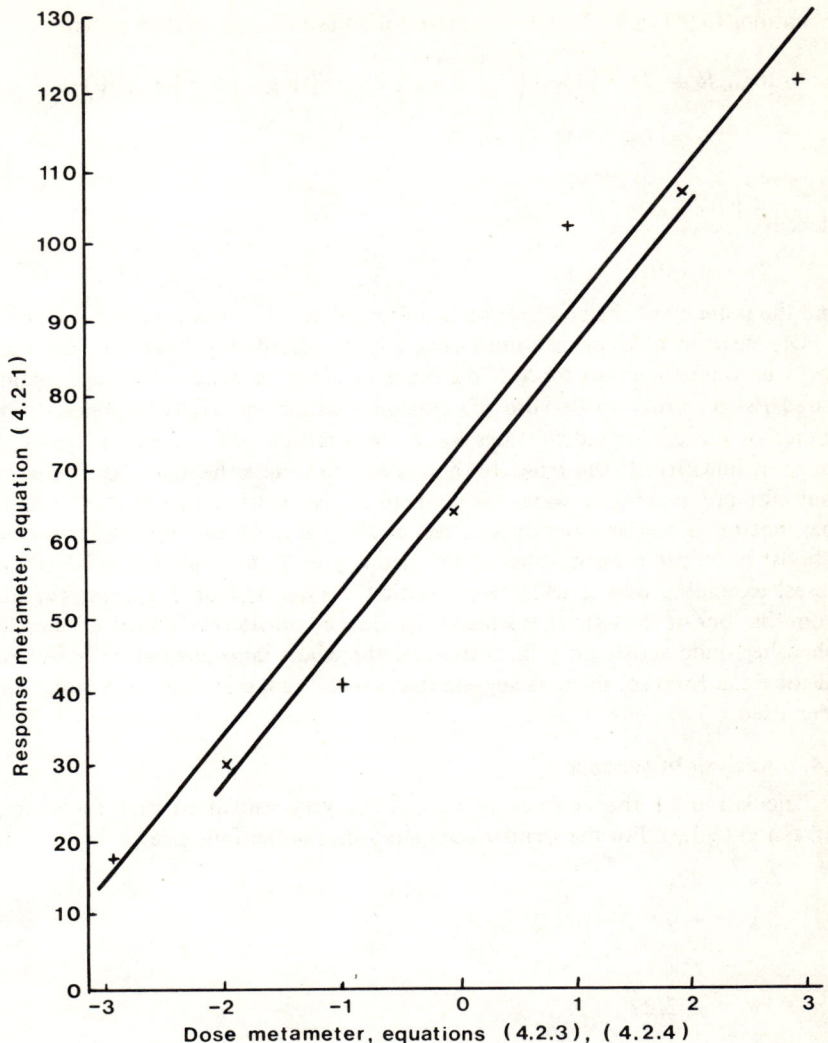

Fig. 4.3.1 Linear dose–response regressions for the assay of vitamin D_3, Table 4.2.1

×: Mean responses to standard preparation
+: Mean responses to test preparation

The straight lines are those drawn by eye (§4.3), but the calculated equations (§4.11) are almost identical with them.

In Fig. 4.3.1, two parallel lines have been drawn by eye so as to fit the points approximately. The horizontal distance between these lines roughly estimates the difference in x between doses giving equal responses:

$$M = 0\cdot 3. \tag{4.3.1}$$

Equations (3.9.11), (4.2.3), (4.2.4) give for an estimate of relative potency:

$$\log_{10} R = M \times \tfrac{1}{2}\log_{10}\left(\frac{5}{3}\right) + \log_{10} 9\cdot 6 - \tfrac{1}{2}(\log_{10} 54 + \log_{10} 90)$$

$$= 0\cdot 1109 M - 0\cdot 8610 \qquad (4.3.2)$$

$$= \bar{1}\cdot 1723.$$

Hence

$$R = 0\cdot 149, \qquad (4.3.3)$$

and the potency of the cod-liver oil is estimated as $0\cdot 149$ units per mg.

One feature of this assay must generally be regarded as bad: the dose range for T is wider than that for S. If the range for S was as wide as the experimenter dared risk in order to be sure of remaining within the region of linearity, his choice of a wider spread of doses for T was almost sure to take him outside the range of linearity. If the range for S was not as wide as he dared risk, he was at fault for not making it wider, at least up to the width used for T. The design may not be as bad as strict application of the principles of §6.8 suggests. If the assayist had little previous idea of the potency of T, he might have chosen four doses, extending over a wide range, with the intention of discarding the data from that one of the extremes which proved to be outside the region of linearity; when he found satisfactory linearity over the whole range, he naturally retained all the data. Even so, the data suggest that a wider range of doses of S could have been used.

4.4 Analysis of variance

Calculation of the analysis of variance is very similar to that for a single regression (§3.9). For the standard preparation, summations give:

$$N_S = 27,$$

$$S x = -9 \times 2 + 8 \times 2$$

$$= -2,$$

$$\bar{x}_S = -2/27$$

$$= -0\cdot 0741,$$

$$S y = 1783,$$

$$\bar{y}_S = 66\cdot 04,$$

$$S_{xx} = 9 \times 4 + 8 \times 4 - \frac{(-2)^2}{27}$$

$$= 67\cdot 8519,$$

$$S_{xy} = -2 \times 274 + 2 \times 859 - \frac{(-2) \times 1783}{27}$$

$$= 1302\cdot 07.$$

§ 4.4 PARALLEL LINE ASSAYS

Similarly, for the test preparation,

$N_T = 28,$

$Sx = 2,$

$\bar{x}_T = 0{\cdot}0714,$

$Sy = 2047,$

$\bar{y}_T = 73{\cdot}11,$

$S_{xx} = 7 \times 9 + 6 \times 1 + 8 \times 1 + 7 \times 9 - \dfrac{2^2}{28}$

$\phantom{S_{xx}} = 139{\cdot}8571,$

$S_{xy} = -3 \times 123 - 1 \times 249 + 1 \times 820 + 3 \times 855 - \dfrac{2 \times 2047}{28}$

$\phantom{S_{xy}} = 2620{\cdot}79.$

The analysis of variance, Table 4.4.1, may then be completed. Notes on each component are given below the table, in the order in which they are most conveniently computed.

TABLE 4.4.1 Analysis of variance for Table 4.2.1

1	Adjustment for mean		266 707	Mean square
	Nature of variation	d.f.	Sum of squares	
4	Preparations	1	687	687
5	Regression	1	74 088	74 088
6	Parallelism	1	10	10
7	Linearity	3	2 312	771
3	Between doses	6	77 097	
8	Error (within doses)	48	22 928	477·67
2	Total	54	100 025	

Notes on formation of sums of squares:—

(1) $\dfrac{(1783 + 2047)^2}{55} = 266\,707$

(2) $35^2 + 30^2 + 24^2 + \ldots + 84^2 + 101^2 - 266\,707 = 100\,025$

(3) $\dfrac{274^2}{9} + \dfrac{650^2}{10} + \dfrac{859^2}{8} + \dfrac{123^2}{7} + \dfrac{249^2}{6} + \dfrac{820^2}{8} + \dfrac{855^2}{7} - 266\,707 = 77\,097$

(4) $\dfrac{1783^2}{27} + \dfrac{2047^2}{28} - 266\,707 = 687$

(5) Pooled regression component, given by

$\dfrac{(\Sigma S_{xy})^2}{\Sigma S_{xx}} = \dfrac{(1302{\cdot}07 + 2620{\cdot}79)^2}{67{\cdot}8519 + 139{\cdot}8571}$

$\phantom{\dfrac{(\Sigma S_{xy})^2}{\Sigma S_{xx}}} = \dfrac{(3922{\cdot}86)^2}{207{\cdot}7090}$

$\phantom{\dfrac{(\Sigma S_{xy})^2}{\Sigma S_{xx}}} = 74\,088$

(6) Difference between fitting two independent regression coefficients and one pooled value is

$$\Sigma\left\{\frac{(S_{xy})^2}{S_{xx}}\right\} - \frac{(\Sigma S_{xy})^2}{\Sigma S_{xx}} = \frac{1302 \cdot 07^2}{67 \cdot 8519} + \frac{2620 \cdot 79^2}{139 \cdot 8571} - 74\,088$$
$$= 10$$

(7) $77\,097 - (687 + 74\,088 + 10) = 2312$

(8) $100\,025 - 77\,097 = 22\,928.$

The components of the analysis must next be examined. Unless evidence of its unsuitability is found, the mean square from the error line,

$$s^2 = 477 \cdot 67, \qquad (4.4.1)$$

will be used as the basic variance estimate.

4.5 Scedasticity

The error sum of squares comprises contributions from each of the seven doses. These may be examined separately in a study of evidence for heteroscedasticity. Tests of significance for other validity tests (§§4.6–4.8), and assessment of precision of the potency estimate (§4.14), in theory require that $\sigma^2(y)$ be the same for all doses (§3.10), though experience shows that even quite large departures from homoscedasticity do not matter much. In more complex designs, to test the homogeneity of variance may be impracticable, but care in the preliminary investigation and in the choice of doses will generally remove the risk of heteroscedasticity so severe as to disturb the estimation seriously.

TABLE 4.5.1 Test of variance heterogeneity for Table 4.2.1

	f_i	Sum of squares	s_i^2	$\log s_i^2$
S	8	3268	408·5	2·611
	9	2808	312·0	2·494
	7	2280	325·7	2·513
T	6	1854	309·0	2·490
	5	4356	871·2	2·940
	7	5392	770·3	2·887
	6	2971	495·2	2·695
	48	22929	477·7	2·679

Table 4.5.1 shows the seven mean squares, s_i^2, with their degrees of freedom, f_i; the last line of the table contains the total number of degrees of freedom, f, and the pooled mean square, s^2. A final column shows the logarithms of the mean squares. Bartlett's test (§3.11) then gives, by equations (3.11.4) and (3.11.1)

$$C = 1 + \left(\frac{1}{5} + \frac{2}{6} + \frac{2}{7} + \frac{1}{8} + \frac{1}{9} - \frac{1}{48}\right) \bigg/ 18 = 1 \cdot 0575$$

§ 4.7   PARALLEL LINE ASSAYS   75

$$\chi^2_{[6]} = 2 \cdot 3026 \times 1 \cdot 648 / 1 \cdot 0575$$
$$= 3 \cdot 59;$$

χ^2 is so well below the 0·05 significance level (12·6) that adjustment by the factor C could have been omitted. The data cause no worry about heteroscedasticity.

4.6 Linearity

A preliminary investigation will be presumed to have established that, over a range of responses such as occurs here, the regression of bone ash percentage on log dose is practically linear. The mean square for 'Linearity', or, more fully, 'Deviations from linearity' should still be examined, as a check that nothing has seriously disturbed this linearity. Unless accompanied by other danger signals, a mean square that is large relative to the error would most probably indicate *statistical invalidity*, that is to say inappropriateness of the form of analysis adopted. For example, a bad choice of doses for either preparation might take most of the observations off the linear portion of the response curve; the conditions of similarity and monotony might be fulfilled, so that in theory the data would still be suitable for an assay, but the assumption of a linear regression would no longer be justified. This need not be evidence against the inherent comparability of the two preparations. The assay might be rejected, however, because a satisfactory linearizing transformation could not be found without more extensive data; it might be rejected because changes in dose had so little effect on response as to make any estimate of ρ hopelessly imprecise; or it might still be usable (see also §4.22).

The mean square for linearity in Table 4.4.1 is greater than that for error, but not significantly so; the ratio is not great enough to occasion any alarm.

4.7 The difference in preparations

In an ordinary experiment for comparing treatments, major interest attaches to differences between treatment means. Here the difference between S and T in their mean responses is not of intrinsic interest. A large difference, however, will seldom arise unless the responses to either the lowest or the highest doses of T lie far outside the range of responses to S, though the converse is not necessarily true. As already implied (§4.3), this should not happen: if it does, either the range of doses for S ought to have been wider, or that for T was too wide and extended beyond the region of linearity. Moreover, as will be apparent from §4.14, a large difference in mean responses will decrease the precision of potency estimation (§6.8).

In this assay, at both extremes of dose the responses to T lie outside the range for S, and in the mean response these extremes compensate for one another. Though the mean square for preparations is only a little greater than the error mean square, the assay is certainly open to criticism on account of the choice of doses of T. A large mean square for the difference in preparations is always a danger signal, but a small one is no assurance that all is well. Results of an assay like the present should be treated with some reserve, and the response diagram should be inspected for any indications of non-linearity at the extremes of the test preparation regression.

4.8 Parallelism

If other tests had disclosed no significantly large mean squares for linearity or preparations, a large mean square for the component based on deviations from parallelism would indicate *fundamental invalidity* of the assay. When log dose is used as a metameter, an essential condition for an analytic dilution assay is that the regression curves are parallel (§3.3). If these curves are linear (§3.7), non-parallelism would violate the condition of similarity. Whether the initial assumption that T behaved as a dilution of S was inherently false or whether it had been obscured by an impurity in one preparation, the data would have to be discarded. This is without prejudice to the possibility of deliberately using non-parallel regressions for other types of assay, as suggested by Thompson (1948). Table 4.4.1 shows no evidence of deviations from parallelism.

If danger signals appear simultaneously in several validity tests, assignment of the cause to one explanation may be impossible. The whole assay is then suspect, and should be discarded; whether it is fundamentally invalid or statistically intractable matters little, except in so far as the planning and experimental technique for the next assay may be affected.

4.9 Regression

No assay should be undertaken without strong prior belief in the existence of a regression, without which the dose–response relation is useless for estimating potency. In a good assay, therefore, the variance ratio for the regression component will generally be highly significant. Here it is

$$F = \frac{74\,088}{477 \cdot 67}$$
$$= 155 \cdot 1.$$

Only when F is large are fiducial limits to the potency narrow enough to be useful (§4.14).

4.10 Significance levels

In the validity tests described in §§4.5–4.8, though a test of statistical significance at a probability of 0·05 was implied, each conclusion was so clear that any reasonable probability would have given the same answer. The assayist need not use the same probability here as for the fiducial limits. What level is ideal?

Some might think that stringent tests should be applied, especially for parallelism, because of the importance of rejecting invalid analyses: perhaps a probability of 0·10 should be used instead of 0·05. Others might think this extravagant, because many good sets of data would be rejected on account of the mischances of random sampling. Experience suggests that, in a well planned assay, even fairly large deviations from the strict theoretical requirements of

§ 4.10   PARALLEL LINE ASSAYS   77

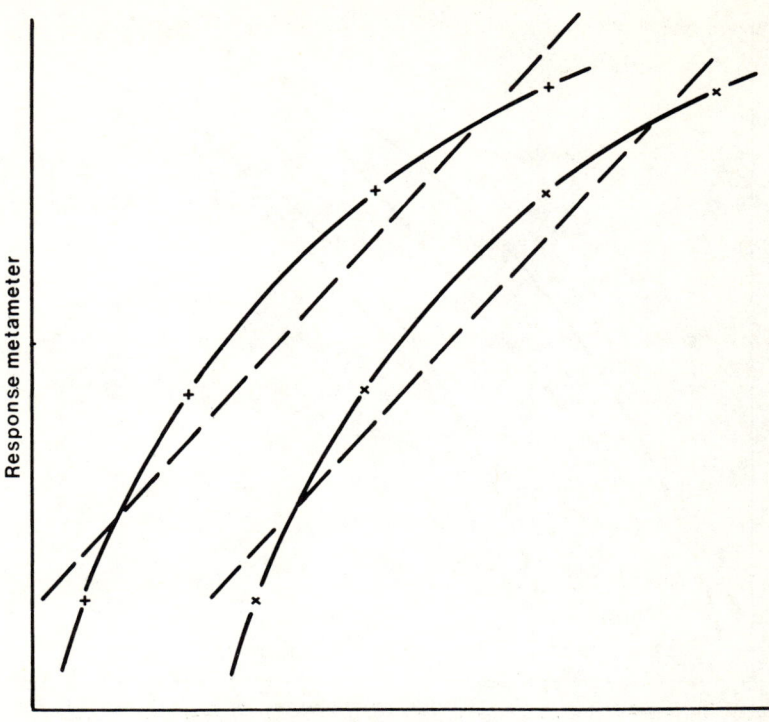

Fig. 4.10.1   A successful choice of doses permits a valid estimation of potency from parallel linear regressions, even though the true regressions are curved

×: Mean responses to standard preparation
+: Mean responses to test preparation

Full lines indicate regression curves, broken lines are hypothetical estimated linear regressions.

statistical validity will have little effect on the estimate of ρ and not much on the assessment of its precision. If the design is symmetric, with the same number of doses for both preparations and equal numbers of subjects at every dose (Chapter 5), and if the assayist guesses his doses of T so successfully that they are almost exactly equivalent to those of S, a determination of R from the horizontal distance between linear regressions will be valid, even though the true regression is curved (Fig. 4.10.1). In such an assay, the assessment of error may be seriously upset (but see §5.11). With the same true regression, a bad choice of doses may give apparent parallelism but a hopelessly biased estimate (Fig. 4.10.2), or complete non-parallelism (Fig. 4.10.3), in spite of the fundamental validity of the assay.

Perhaps the chief danger is that of fundamental invalidity, and a reasonably stringent parallelism test therefore seems desirable. A deviation from parallelism significant at a probability of 0·05 should be regarded as sufficient cause for

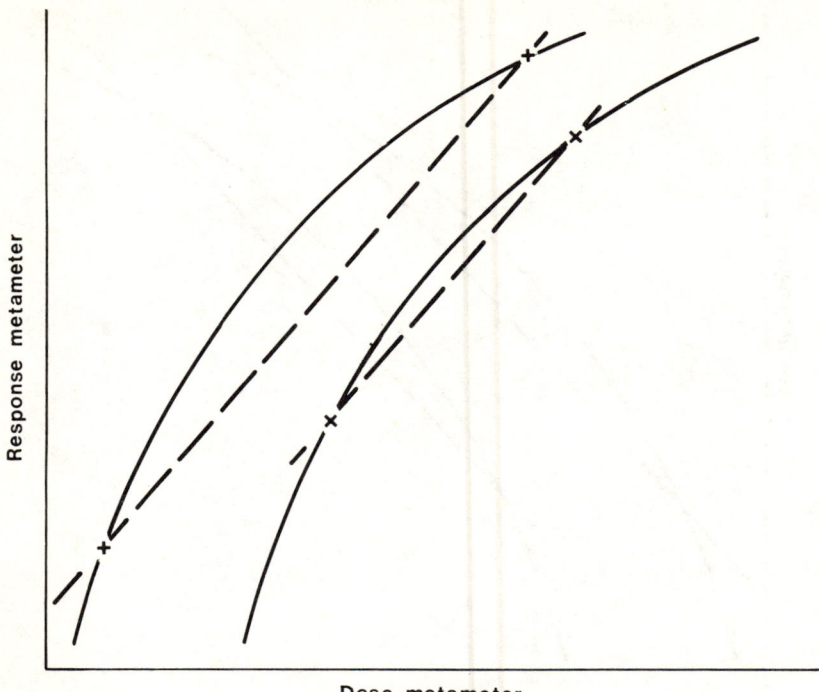

Fig. 4.10.2 Hypothetical results of a (2, 2) assay showing parallelism of linear regressions but biased estimation of potency; regression curves conform to the condition of similarity but dose intervals are unequal

×: Mean responses to standard preparation
+: Mean responses to test preparation

Full lines indicate regression curves, broken lines are hypothetical estimated linear regressions.

rejection of an assay, unless extenuating circumstances not only explain the situation but ensure a statistically valid analysis. Usually the history of an assay technique for estimating the potency of test preparations with respect to a particular effective constituent, by an accepted experimental procedure and on a known stock of subjects, provides a strong presumption of parallelism; without this, even a deviation significant at a probability of 0·10 should be a little suspect.

When experience of a technique has given grounds for belief in similarity and in the statistical validity of the method of evaluation of the data, and also when the current assay gives no evidence against parallelism, less stringent tests for heteroscedasticity, deviations from linearity, and the difference between preparations might be allowed. As a working rule to be applied and interpreted intelligently, not uncritically, acceptance of the analysis as statistically valid unless one or more of these criteria are significant at a probability of 0·01 seems reasonable. The question is not entirely one of statistics; the knowledge of the chemist and

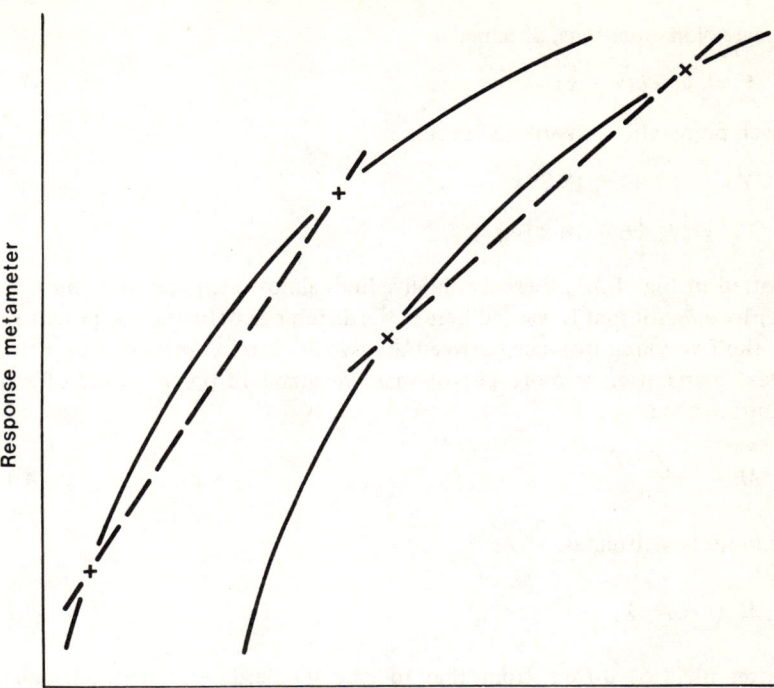

Fig. 4.10.3 Hypothetical results of a (2, 2) assay showing non-parallelism of linear regressions; regression curves conform to the condition of similarity but doses fail to correspond

×: Mean responses to standard preparation
+: Mean responses to test preparation

Full lines indicate regression curves, broken lines are hypothetical estimated linear regressions.

biologist about the materials of the assay must also be taken into account. For routine assays, quality control techniques may help interpretation of the criteria of statistical validity (§5.10).

4.11 Potency estimation

The cod-liver oil assay is free from evidence of invalidity, and estimation can proceed. The regression coefficient is estimated as

$$b = \frac{\Sigma S_{xy}}{\Sigma S_{xx}} \tag{4.11.1}$$

$$= \frac{3922 \cdot 86}{207 \cdot 7090}$$

$$= 18 \cdot 89. \tag{4.11.2}$$

The regression equations, obtained as

$$Y = \bar{y} + b(x - \bar{x}) \tag{4.11.3}$$

for each preparation, therefore become

$$Y_S = 67 \cdot 44 + 18 \cdot 89 x,$$

$$Y_T = 71 \cdot 76 + 18 \cdot 89 x.$$

If plotted in Fig. 4.3.1, these would give lines almost identical with those drawn earlier by eye. All that is wanted here is the difference between equipotent values of x, the horizontal distance between the two lines, to be reckoned as positive if the test preparation is more potent than the standard on the scales of x used. This difference is

$$M = \frac{Y_T - Y_S}{b}, \tag{4.11.4}$$

more usefully written as

$$M = \bar{x}_S - \bar{x}_T - \frac{\bar{y}_S - \bar{y}_T}{b}; \tag{4.11.5}$$

equation (4.11.5) differs from that for the standard slope method, equation (3.15.1), by having $\bar{x}_S, \bar{x}_T$ for x_S, x_T. Here

$$M = -0 \cdot 0741 - 0 \cdot 0714 - \frac{66 \cdot 04 - 73 \cdot 11}{18 \cdot 89}$$

$$= -0 \cdot 1455 + \frac{7 \cdot 07}{18 \cdot 89}$$

$$= 0 \cdot 2288. \tag{4.11.6}$$

This must be transformed by equation (4.3.2) to give

$$\log_{10} R = \bar{1} \cdot 1644,$$

whence

$$R = 0 \cdot 1460,$$

as compared with the graphical estimate, $0 \cdot 149$, in equation (4.3.3).

4.12 Fieller's theorem

In equation (4.11.5), $(\bar{x}_S - \bar{x}_T)$ is a constant set by the choice of doses and numbers of subjects. Consequently, fiducial limits for M depend upon the second term, $(\bar{y}_S - \bar{y}_T)/b$. They are obtained from an important theorem first fully enunciated by Fieller (1940), though others (Bliss, 1935b) had earlier stated the particular case for zero covariance.

§ 4.12 PARALLEL LINE ASSAYS

The conventions of symbolism are suspended for this and the next section. Suppose that α, β are two parameters, and write

$$\mu = \frac{\alpha}{\beta}. \tag{4.12.1}$$

Suppose further that a, b are unbiased estimates of α, β, each a linear function of a set of observations with normally distributed errors; the typical situation is that in which a, b are means, differences between means, or regression coefficients calculated from experimental data. Freedom from bias implies that

$$\mathrm{E}\,(a) = \alpha,$$
$$\mathrm{E}\,(b) = \beta,$$

where $\mathrm{E}\,(a)$ is the *expectation* of a. An analysis of variance of the data will give an error mean square, s^2, with f degrees of freedom. The estimated variances of a, b, and their covariance may be expressed as $s^2 v_{11}, s^2 v_{22}, s^2 v_{12}$ respectively, where v_{11}, v_{22}, and v_{12} depend only on the coefficients of the observations in the definitions of a, b. For example, if a is the arithmetic mean of certain observations, v_{11} is the reciprocal of the number of these observations; if b is a regression coefficient of y on x, v_{22} is the reciprocal of the sum of squares of deviations of the values of x about their mean. Now the natural estimator of μ is

$$m = \frac{a}{b}. \tag{4.12.2}$$

Fieller's theorem states that upper and lower fiducial limits to μ are[*]

$$m_L, m_U = \left[m - \frac{g v_{12}}{v_{22}} \pm \frac{ts}{b} \left\{ v_{11} - 2m v_{12} + m^2 v_{22} - g \left(v_{11} - \frac{v_{12}^2}{v_{22}} \right) \right\}^{1/2} \right] \bigg/ (1 - g), \tag{4.12.3}$$

where

$$g = \frac{t^2 s^2 v_{22}}{b^2} \tag{4.12.4}$$

and t is the t-deviate with f degrees of freedom (Appendix Table I). The proof follows from consideration of the expression $(a - \mu b)$. For any μ, this also is a linear function of the observations; it has expectation

$$\mathrm{E}\,(a - \mu b) = \alpha - \mu \beta = 0, \tag{4.12.5}$$

and an estimated variance

$$\mathrm{Var}(a - \mu b) = s^2 (v_{11} - 2\mu v_{12} + \mu^2 v_{22}), \tag{4.12.6}$$

---

[*] The symbol '±' generally introduces a standard error. In Fieller's theorem or one of its analogues, the symbol indicates the alternative operations of subtraction and addition needed for the lower and upper limits.

with f degrees of freedom. Hence (assuming normality), with probability appropriate to the t-deviate.

$$(a - \mu b)^2 \leqslant t^2 s^2 (v_{11} - 2\mu v_{12} + \mu^2 v_{22}); \tag{4.12.7}$$

the equality sign gives a quadratic equation in μ, whose solution is (4.12.3).

When b is large relative to its standard error, g will be small; if g can be neglected, equation (4.12.3) becomes

$$m_L, m_U = m \pm ts(v_{11} - 2mv_{12} + m^2 v_{22})^{1/2}/b, \tag{4.12.8}$$

a formula equivalent to using the expression

$$\text{Var}(m) = s^2 (v_{11} - 2mv_{12} + m^2 v_{22})/b^2 \tag{4.12.9}$$

as though it were the variance of m. This variance formula, obtained in other ways, is often used as a way of attaching a standard error to a ratio, especially in the form applicable when $v_{12} = 0$:

$$\frac{\text{Var}(m)}{m^2} = s^2 \left(\frac{v_{11}}{a^2} + \frac{v_{22}}{b^2} \right). \tag{4.12.10}$$

The approximation (4.12.8) is adequate if g is less than 0·05, which for limits at probability 0·95 requires b to be at least nine times its standard error. For large g, equation (4.12.3) is essential: the approximation much underestimates the width of the fiducial interval when g exceeds 0·2. New complications arise if g exceeds 1·0, as b then does not differ significantly from zero (Fieller, 1954). Detailed study of the inequality (4.12.7) shows that when $g = 1·0$ one of m_L, m_U becomes infinite. When $g > 1·0$, the range of values of μ that satisfy the inequality remains infinite, and the limits set by (4.12.3) become exclusive instead of inclusive: the assertion made with the chosen probability is that μ lies outside the interval m_L, m_U. The logic is sound, but the practical value of an assay with $g \geqslant 1·0$ is small.

Biological assay often requires that fiducial limits be assigned to a ratio of two means, a ratio of two regression coefficients, or a horizontal distance between two parallel regression lines. Although many publications on bioassay have used equation (4.12.9) or some equivalent formula, the extra labour of calculating from (4.12.3) is so small that routine employment of the full formula is preferable to argument about the adequacy of the approximation. Even for desk calculation, this should be standard practice; when a computer is programmed for assay analyses, there is no excuse for not incorporating equation (4.12.3).

4.13 Analogues of Fieller's theorem

Valuable though Fieller's theorem is, it is not applicable to every bioassay in which estimation involves a ratio. Complications arise when more than one error mean square must be used; this situation will not be encountered until Chapter 9, but generalized theorems based on the Behrens distribution are for convenience presented here. First the nature of this distribution itself, a generalization of the t distribution, must be considered.

Again suppose that a, b are unbiased estimates of α, β, defined as linear functions of observations with normally distributed errors, but now with estimated variances

$$\left.\begin{array}{l} \text{Var}(a) = s_1^2 v_{11}, \\ \text{Var}(b) = s_2^2 v_{22}, \end{array}\right\} \quad (4.13.1)$$

and zero covariance, where s_1^2, s_2^2 are independent mean squares with f_1, f_2 degrees of freedom respectively. The variance of $(a - b)$ is estimated as

$$\text{Var}(a - b) = s_1^2 v_{11} + s_2^2 v_{22}. \quad (4.13.2)$$

If s_1^2, s_2^2 were the same mean square, the deviation of $(a - b)$ from its expectation divided by the estimated standard error,

$$\frac{(a - b) - (\alpha - \beta)}{(s_1^2 v_{11} + s_2^2 v_{22})^{1/2}}, \quad (4.13.3)$$

would follow the t distribution with f degrees of freedom. When s_1^2, s_2^2 are independent mean squares, however, this is no longer true, and the distribution of the ratio (4.13.3) is the Behrens distribution (Appendix Table III; Fisher and Yates, 1963, Table V_1). The ratio, the Sukhatme d-statistic, has a distribution defined in terms of the degrees of freedom f_1, f_2, and the angle θ such that

$$\tan^2 \theta = \frac{s_1^2 v_{11}}{s_2^2 v_{22}}. \quad (4.13.4)$$

When θ is $0°$, d is distributed as t with f_2 degrees of freedom; when θ is $90°$, it is distributed as t with f_1 degrees of freedom. For other angles, the value of d for any probability is generally (but not always) intermediate between $t_{[f_1]}$ and $t_{[f_2]}$. When $f_1 = f_2$, the value of d for any probability is about equal to, but a little less than, the corresponding t, irrespective of the size of θ. The d-test is appropriate for testing a difference between two means or two regression coefficients whose variances are based on independent mean squares that cannot be assumed estimates of the same population variance and therefore must not be pooled. To attempt to refer the ratio (4.13.3) to the t distribution may mislead seriously if f_1 and f_2 are small.

Suppose now that a and b are known to be estimates of the same quantity $(\alpha = \beta)$. They might be two estimates of the same mean, or two estimates of a regression coefficient, from different sets of observations with different variances. The most precise estimate that can be compounded of a and b is a weighted mean having the reciprocals of the variances as weights; this is $\bar{a}$, where

$$\bar{a}\left(\frac{1}{s_1^2 v_{11}} + \frac{1}{s_2^2 v_{22}}\right) = \frac{a}{s_1^2 v_{11}} + \frac{b}{s_2^2 v_{22}}, \quad (4.13.5)$$

and

$$\text{Var}(\bar{a}) = \left(\frac{1}{s_1^2 v_{11}} + \frac{1}{s_2^2 v_{22}}\right)^{-1}. \quad (4.13.6)$$

The deviation of $\bar{a}$ from α, divided by its standard error,

$$(\bar{a} - \alpha)\left(\frac{1}{s_1^2 v_{11}} + \frac{1}{s_2^2 v_{22}}\right)^{1/2}, \qquad (4.13.7)$$

follows the Behrens distribution with degrees of freedom f_1, f_2, and an angle θ defined by

$$\tan^2 \theta = \frac{s_2^2 v_{22}}{s_1^2 v_{11}} \qquad (4.13.8)$$

(Yates, 1939; Finney, 1951b). Hence the same distribution may be used in testing the significance of the deviation of a weighted mean, $\bar{a}$, from a theoretical value; alternatively, with d taken as the tabular value for the chosen probability, it allows fiducial limits to α to be placed at

$$d\left(\frac{1}{s^2 v_{11}} + \frac{1}{s^2 v_{22}}\right)^{-1/2}$$

on either side of $\bar{a}$. This neglects information on the distribution of $\bar{a}$ given by the magnitude of $(a - b)$; Fisher (1961a, b) has shown how to take account of the information, which can be important, but no tables exist.

This theory leads to two analogues of Fieller's theorem. For now suppose again that

$$\mu = \frac{\alpha}{\beta}$$

is to be estimated. The ratio of $(a - \mu b)$ to its standard error estimated from

$$\text{Var}(a - \mu b) = s_1^2 v_{11} + \mu^2 s_2^2 v_{22} \qquad (4.13.9)$$

follows the Behrens distribution, with f_1, f_2 degrees of freedom and an angle θ given by

$$\tan^2 \theta = s_1^2 v_{11} / \mu^2 s_2^2 v_{22}. \qquad (4.13.10)$$

As in §4.12, the fiducial limits are the roots of the quadratic equation

$$(a - \mu b)^2 = d^2(s_1^2 v_{11} + \mu^2 s_2^2 v_{22}). \qquad (4.13.11)$$

Since d is dependent upon θ, which is in turn a function of μ, no explicit solution of equation (4.13.11) can be given. The solution may be written as

$$m_L, m_U = \left[m \pm \frac{d}{b}\left\{s_1^2 v_{11}(1-g) + m^2 s_2^2 v_{22}\right\}^{1/2}\right]\bigg/(1-g), \qquad (4.13.12)$$

where

$$g = \frac{d^2 s_2^2 v_{22}}{b^2}; \qquad (4.13.13)$$

numerical evaluation of m_L, m_U, requires interpolation or iteration, since each must have its value of d corresponding to the θ given by equation (4.13.10) when the fiducial limit itself is substituted for μ.

§ 4.13 PARALLEL LINE ASSAYS

The second analogue of Fieller's theorem requires no iterative calculations, but its statement is more complicated. Suppose that a_1, b_1 and a_2, b_2 are independent pairs of unbiased estimates of α, β as were a, b in §4.12. The variances and covariance for the first pair, $s_1^2 v_{11}$, $s_1^2 v_{22}$, and $s_1^2 v_{12}$, are based upon a single mean square, s_1^2 with f_1 degrees of freedom. The corresponding quantities for a_2, b_2 are based upon an independent mean square s_2^2 with f_2 degrees of freedom, but are in the same ratios as those for a_1, b_1; they may therefore be written $s_2^2 k v_{11}$, $s_2^2 k v_{22}$, and $s_2^2 k v_{12}$, where k is known. The second adaptation of the Behrens distribution can give fiducial limits for a ratio of weighted means of a_1, a_2 and b_1, b_2.

The theorem in §4.12 applies directly to the determination of fiducial limits for a_1/b_1 or a_2/b_2 as estimates of μ. If mean values $\bar{a}, \bar{b}$ are determined by weighting inversely as the variances, so that

$$\bar{a}\left(\frac{1}{s_1^2 v_{11}} + \frac{1}{s_2^2 k v_{11}}\right) = \frac{a_1}{s_1^2 v_{11}} + \frac{a_2}{s_2^2 k v_{11}} \tag{4.13.14}$$

and

$$\bar{b}\left(\frac{1}{s_1^2 v_{22}} + \frac{1}{s_2^2 k v_{22}}\right) = \frac{b_1}{s_1^2 v_{22}} + \frac{b_2}{s_2^2 k v_{22}} \tag{4.13.15}$$

the ratio

$$\bar{m} = \bar{a}/\bar{b} \tag{4.13.16}$$

seems likely to be a more precise estimate of μ than either m_1 or m_2. Now

$$\mathrm{Var}(\bar{a}) = v_{11}\left(\frac{1}{s_1^2} + \frac{1}{ks_2^2}\right)^{-1}, \tag{4.13.17}$$

$$\mathrm{Var}(\bar{b}) = v_{22}\left(\frac{1}{s_1^2} + \frac{1}{ks_2^2}\right)^{-1}, \tag{4.13.18}$$

and the covariance of $\bar{a}, \bar{b}$ is

$$\mathrm{Cov}(\bar{a}, \bar{b}) = v_{12}\left(\frac{1}{s_1^2} + \frac{1}{ks_2^2}\right)^{-1}. \tag{4.13.19}$$

The weights used in forming $\bar{a}, \bar{b}$ are proportional; the variances and covariance of the weighted means therefore preserve the same ratios as those of a_1, b_1 and a_2, b_2. Consideration of $(\bar{a} - \mu\bar{b})$ then gives a quadratic equation for the fiducial limits of $\bar{m}$:

$$(\bar{a} - \mu\bar{b})^2 = d^2(v_{11} - 2\mu v_{12} + \mu^2 v_{22})\bigg/\left(\frac{1}{s_1^2} + \frac{1}{ks_2^2}\right), \tag{4.13.20}$$

where d is a tabular value for f_1, f_2 degrees of freedom and

$$\tan^2 \theta = ks_2^2/s_1^2. \tag{4.13.21}$$

If s^2 be defined by

$$s^2 = \left(\frac{1}{s_1^2} + \frac{1}{ks_2^2}\right)^{-1}, \tag{4.13.22}$$

equation (4.13.20) becomes identical with the equality in (4.12.7), except that d replaces t. Hence the solution of equation (4.13.20) may be written in the same form as equation (4.12.3):

$$\bar{m}_L, \bar{m}_U = \left[\bar{m} - \frac{gv_{12}}{v_{22}} \pm \frac{ds}{b}\left\{v_{11} - 2\bar{m}v_{12} + \bar{m}^2 v_{22} - g\left(v_{11} - \frac{v_{12}^2}{v_{22}}\right)\right\}^{1/2}\right] \bigg/ (1-g),$$

(4.13.23)

where

$$g = \frac{d^2 s^2 v_{22}}{b^2}.$$

(4.13.24)

The condition that the variances and covariance of a_2, b_2 should be in the same ratio as those of a_1, b_1 might seem so restrictive as to make the theorem useless. It will be fulfilled, however, by two experiments of the same basic design but possibly different replication – for example, two randomized block experiments for the same treatments, but with different numbers of blocks. The result is therefore useful in the combination of evidence from two assays. If ks_2^2 is large relative to s_1^2, the information on μ provided by a_2, b_2 may be negligible, $\bar{m}$ becomes the same as m_1, d becomes t for f_1 degrees of freedom, and in the limit Fieller's theorem applies.

Generalization of the Behrens distribution to include more than two component variances is conceptually possible, but neither the theory nor tables for a generalized d have been developed. This could form the basis of a further extension to equations (4.13.5)–(4.13.8), and hence to an analogue of Fieller's theorem appropriate to a combination of several assays of similar design. The principle is obvious, and the problem arises again in Chapter 14. A different approach to the combination of estimates from two or more assays is presented in §14.3.

4.14 Fiducial limits in the vitamin D_3 assay

Equation (4.11.5) can be written

$$M - \bar{x}_S + \bar{x}_T = \frac{\bar{y}_T - \bar{y}_S}{b}.$$

(4.14.1)

Since $(\bar{x}_S - \bar{x}_T)$ is a constant imposed by the choice of doses, fiducial limits to M may be found by applying Fieller's theorem to the ratio $(\bar{y}_T - \bar{y}_S)/b$ and adding $(\bar{x}_S - \bar{x}_T)$ to the results. From equation (4.12.3), the limits to $(M - \bar{x}_S + \bar{x}_T)$ are

$$\left[M - \bar{x}_S + \bar{x}_T \pm \frac{st}{b}\left\{(1-g)\left(\frac{1}{N_S} + \frac{1}{N_T}\right) + \frac{(M - \bar{x}_S + \bar{x}_T)^2}{\Sigma S_{xx}}\right\}^{1/2}\right] \bigg/ (1-g),$$

(4.14.2)

where

$$g = \frac{t^2 s^2}{b^2 \Sigma S_{xx}}.$$

(4.14.3)

With s^2 taken from equation (4.4.1),

$$g = \frac{(2 \cdot 010)^2 \times 477 \cdot 67}{(18 \cdot 89)^2 \times 207 \cdot 7090}$$

$$= 0 \cdot 0260.$$

§ 4.14 PARALLEL LINE ASSAYS

An alternative form of calculation is sometimes more convenient; g is the ratio of the tabulated significance point for the variance ratio for 'regression' to the value calculated from the analysis of variance. The variance ratio here has $(1, 48)$ degrees of freedom, whence (Appendix Table II)

$$g = \frac{4 \cdot 04 \times 477 \cdot 67}{74\,088}$$

$$= 0 \cdot 0260.$$

From the form of equation (4.11.6), the limits to $(M + 0 \cdot 1455)$ are

$$\left[0 \cdot 3743 \pm \frac{2 \cdot 010}{18 \cdot 89} \sqrt{\left\{ 0 \cdot 9740 \left(\frac{1}{27} + \frac{1}{28} \right) + \frac{(0 \cdot 3743)^2}{207 \cdot 709} \right\} \times 477 \cdot 67} \right] \bigg/ 0 \cdot 9740$$

$$= [0 \cdot 3743 \pm 0 \cdot 6220]/0 \cdot 9740$$

$$= -0 \cdot 2543,\ 1 \cdot 0229.$$

Therefore

$$M_L = -0 \cdot 3998,$$

$$M_U = 0 \cdot 8774.$$

Equation (4.3.2) gives

$$\log_{10} R_L = \bar{1} \cdot 0947,$$

$$\log_{10} R_U = \bar{1} \cdot 2363,$$

whence

$$R_L = 0 \cdot 1244,$$

$$R_U = 0 \cdot 1723.$$

The expression

$$\operatorname{Var}(M) = \frac{s^2}{b^2} \left\{ \frac{1}{N_S} + \frac{1}{N_T} + \frac{(M - \bar{x}_S + \bar{x}_T)^2}{\Sigma S_{xx}} \right\}, \tag{4.14.4}$$

frequently quoted as the variance of M, is equation (4.12.9) in the present notation. For these data, approximate fiducial limits to M are obtained by subtracting and adding $2 \cdot 010$ times the standard error (here $0 \cdot 3135$) as in equation (4.12.8). Table 4.14.1 summarizes the results. Evidently g is so small that it could be ignored without harm. Nevertheless, the safer practice of always using Fieller's theorem is little more trouble. Both pairs of limits are presented here for the sake of comparison. Where the concern is with conclusions from assays and not with statistical methodology, quotation of both 'exact' and 'approximate' fiducial limits is to be deprecated as a waste of space on irrelevant and misleading quantities. When g is large, Fieller's theorem should always be used, and values based on any formula for $\operatorname{Var}(M)$ are wrong.

Thus the vitamin content of the cod-liver oil is estimated as $0 \cdot 1460$ units per mg, and the assertion is made that the true potency lies between 85 percent

TABLE 4.14.1 Estimated potency of cod-liver oil (units vitamin D_3 per mg)

	Graphical	Calculated	
		Ignoring g	Fieller's theorem
Potency	0·149	0·1460	0·1460
Lower limit	—	0·1243	0·1244
Upper limit	—	0·1715	0·1723

and 118 percent of this. The fiducial limits are calculated from the internal evidence of a single assay, yet their subsequent use is likely to assume that they measure the agreement to be expected between results of repeated assays of the same test preparation (cf. Finney, 1971, §9.6). Provided that the condition of similarity is fulfilled and that assumptions implicit in the statistical analysis (linearity, homoscedasticity, normality, etc.) are substantially correct, this is justifiable. Potency ought then to be independent of assay technique (§3.4), and the assessment of sampling variation expressed by the fiducial limits ought to have universal validity. The assayist must guard against a too-ready belief that all conditions are satisfied, and that repeated assays will agree within the limits indicated by intra-assay variances. Published experimental verifications are few. Young and Romans (1948) reported satisfactorily consistent potency estimates for 21 insulin samples when each was assayed several times within a few days. Jones (1945) found X-ray and line test assays of vitamin D to agree well during a period of more than three years. Sheps and Munson (1957) proposed a method for taking account of inter-assay variance of M as well as of intra-assay; in a series of androgen assays, they found an important inter-assay component, but did not reconcile this with the general theory of bioassay. The *European Pharmacopoeia* (1971) explicitly counselled that, where possible, precision should be assessed in terms of a simple error mean square calculated from potency estimates from independent assays.

4.15 Data for a symmetric assay

The analysis of the vitamin D_3 assay was complicated because of the unsymmetric design. The coming discussion of the principles of design (especially Chapter 6) may be anticipated by the statement that symmetry is one desirable feature. The simplest symmetric parallel line assays have only two doses of each preparation: the high and low doses of the two preparations have the same difference on the logarithmic scale, and the total number of subjects is divided equally between doses.

Table 4.15.1 contains data from an assay of oestrone using 7 litters of 4 ovariectomized female rats each. Each rat was injected daily with one of the four experimental doses, 0·2 μg and 0·4 μg of the standard oestrone and 0·0075 ml and 0·015 ml of the test preparation. The response was the weight of the uterus, expressed as mg per 100 g body weight and measured at a fixed number of days after treatment of the animals (Bülbring and Burn, 1935). This method of

§ 4.15   PARALLEL LINE ASSAYS   89

TABLE 4.15.1   Weights of uteri of ovariectomized rats, in mg per 100 g body weight

Litter	Daily dose				Totals
	Oestrone		Test preparation		
	0·2 µg	0·4 µg	0·0075 ml	0·015 ml	
I	54	152	61	92	359
II	49	71	74	63	257
III	51	112	51	(87)	301
IV	(50)	58	60	102	270
V	81	102	(82)	120	385
VI	63	(111)	83	105	362
VII	126	(133)	83	108	450
Totals	474	739	494	677	2384
Means	67·7	105·6	70·6	96·7	

Five responses are shown in parentheses, for reasons explained in §4.21.

adjusting for the sizes of the subjects will be criticized in §12.6; here records of body weights are no longer available and therefore covariance analysis could not be tried. If litters were to differ in mean uterine weight, as might be expected, the precision of the assay could have been adversely affected by inter-litter variation. This was avoided by adopting a common device, a randomized block design; one animal from each litter, selected at random, was assigned to each of the four doses. For simplicity, some liberties have been taken with the data, as explained in §4.21.

The reader should draw a dose–response diagram for the mean responses in Table 4.15.1. A convenient dose scale, using logarithms to base $\sqrt{2}$, makes $(x - \bar{x})$ equal to -1 for the lower, $+1$ for the upper dose of either preparation. The two preparations may have different origins on the x scale, and a simple choice is that which makes the two lower doses have the same scale point. Parallel regression lines drawn by eye in this diagram give a rough estimate of potency. One version of the diagram had lines 0·18 apart in a direction parallel to the axis of x. Hence

$$M = \bar{x}_S - \bar{x}_T - 0.18,$$

and, using a formula similar to equation (4.3.2),

$$\log_{10} R = \log_{10} 0.2 - \log_{10} 0.0075 - 0.18 \times \tfrac{1}{2} \times \log_{10} 2, \qquad (4.15.1)$$

since symmetry makes $(\bar{x}_S - \bar{x}_T)$ equal to the difference in x values for corresponding doses. Therefore

$$R = \frac{0.2}{0.0075} \text{ antilog } \bar{1}\cdot 973 \qquad (4.15.2)$$

$$= 25,$$

and 1 ml of the test preparation is estimated to contain 25 µg oestrone.

4.16 Analysis of variance

Table 4.16.1 shows the analysis of variance of the 28 entries in the body of Table 4.15.1. The subdivision of the total sum of squares into components 'between doses', 'between litters', and 'error' follows the usual procedure for a randomized block design: for example, the sum of squares between doses is

$$(474^2 + 739^2 + 494^2 + 677^2 - 7 \times 202\,981)/7.$$

Subdivision of the dose component into squares for three separate degrees of freedom can be effected by the same steps as were used to give lines 4, 5, 6 in Table 4.4.1. For a symmetric design, orthogonal contrasts as illustrated in Table 4.16.2 are more convenient. The dose totals are multiplied in turn by each row of coefficients and summed to give the sums of products in the last column, each of which will be denoted by the letter L with a distinguishing subscript. These sums are squared and divided by the divisors shown; the quotients are the components for Table 4.16.1.

TABLE 4.16.1 Analysis of variance for Table 4.15.1

Adjustment for mean		202 981	
Nature of variation	d.f.	Sum of squares	Mean square
Preparations	1	63	63
Regression	1	7 168	7168
Parallelism	1	240	240
Between doses	3	7 471	
Between litters	6	7 069	
Error	18	7 165	398·1
Total	27	21 705	

TABLE 4.16.2 Coefficients of orthogonal contrasts for the (2, 2) design, applied to Table 4.15.1

Dose	S_1	S_2	T_1	T_2	Divisor	Sum
Response total	474	739	494	677	28	2384
Preparations (L_p)	−1	−1	1	1	28	−42
Regression (L_1)	−1	1	−1	1	28	448
Parallelism (L_1')	1	−1	−1	1	28	−82

Coefficients such as those in Table 4.16.2 lead to a subdivision of the sum of squares for doses *only* if:

(i) Each row of coefficients represents a *contrast* amongst the individual responses; that is to say, if each sum in Table 4.16.2 is written at length in terms of individual responses, the set of coefficients for these responses adds to zero;

(ii) Every pair of contrasts is *orthogonal*; that is to say, if each contrast is expressed in terms of individual responses, the products of corresponding coefficients for any pair add to zero.

§ 4.17   PARALLEL LINE ASSAYS   91

Because each dose total in a symmetric design contains the same number of individual responses, these conditions also apply to the coefficients of dose totals. Since

$$-1+1-1+1 = 0$$

the line for 'Regression' represents a contrast, and since

$$(-1) \times 1 + 1 \times (-1) + (-1) \times (-1) + 1 \times 1 = 0$$

the contrasts for 'Regression' and 'Parallelism' are orthogonal. The divisors are calculated as the sums of the squares of the coefficients of individual responses; for 'Regression',

$$7 \times (-1)^2 + 7 \times 1^2 + 7 \times (-1)^2 + 7 \times 1^2 = 28.$$

If the variance per response is σ^2, the variance of a contrast value is σ^2 multiplied by this 'divisor'. A consequence of these conditions is that the squares for the separate contrasts must add to the sum of squares for doses; here,

$$\frac{(-42)^2}{28} + \frac{448^2}{28} + \frac{(-82)^2}{28} = 7471.$$

Any sum of squares of deviations can be subdivided into single squares for mutually orthogonal contrasts, in number equal to the degrees of freedom, in an unlimited number of ways. The set in Table 4.16.2 is chosen as peculiarly relevant to the object of the analysis of variance. The three squares are exactly the same as would have been obtained from the general method of §4.4. The first contrast gives the difference between totals for the two preparations; the second has $(x - \bar{x})$ as its coefficients, so that the sum is ΣS_{xy}; and the third gives the difference between values of S_{xy} for the preparations. Bliss and Marks (1939a, b) showed the advantages of such coefficients in the analysis of assays (cf. Chapter 5).

4.17   Validity tests

As in §§ 4.7 and 4.8, the mean squares for preparations and parallelism must be compared with the error mean square. Neither is significantly large, so that on these counts the validity of the assay need not be doubted. The great flaw in the design is that it gives no test for linearity, since only two points on each response curve are studied. A genuinely non-linear regression might manifest itself as non-parallelism, at least if the doses had been so unsuccessfully chosen as to give also a large difference between preparations (Fig. 4.10.3). On the other hand, if doses of T were so chosen that they were almost equal in effect to corresponding doses of S, the fitted lines would appear satisfactorily parallel, even though the true relation was far from linear (Fig. 4.10.1). As pointed out in §4.10, this will not bias the estimation of potency appreciably, but may upset the assessment of precision. Assays should usually be planned to include at least three doses of each preparation (§6.10), unless the material under assay is so well understood as to remove all fear of non-linearity. No assay designed in randomized blocks allows a test of scedasticity, but this is usually less important once an assay technique is well established.

4.18 Potency estimation and precision

The construction of Table 4.16.2 makes clear that the contrast labelled L_p is the difference in total responses for the two preparations. Hence

$$\bar{y}_T - \bar{y}_S = L_p/14 \tag{4.18.1}$$
$$= -42/14$$
$$= -3{\cdot}00.$$

Moreover, the divisor and sum for the regression contrast, L_1, are ΣS_{xx} and ΣS_{xy} respectively, and therefore, by equation (4.11.1),

$$b = L_1/28 \tag{4.18.2}$$
$$= 448/28$$
$$= 16{\cdot}00.$$

Equation (4.11.5) now gives

$$M = \bar{x}_S - \bar{x}_T - 0{\cdot}1875,$$

and, by equation (4.15.1),

$$R = \frac{0{\cdot}2}{0{\cdot}0075} \text{ antilog}(-0{\cdot}1875 \times 0{\cdot}1505)$$
$$= 25{\cdot}0.$$

Fiducial limits may be found by Fieller's theorem, using the error mean square in Table 4.16.1:

$$s^2 = 398{\cdot}1$$

as the variance per response. By the well-known elementary property of a linear function of independent observations, the variance of any contrast can be written as the variance per response multiplied by the sum of the squares of the coefficients of individual responses in the contrast. The rule by which the divisors in Table 4.16.2 are calculated shows this to be expressible as

$$\text{Var}(L) = s^2 \times \text{Divisor} \tag{4.18.3}$$

for any contrast, or

$$\text{Var}\left(\frac{L}{\text{Divisor}}\right) = \frac{s^2}{\text{Divisor}}. \tag{4.18.4}$$

Consequently, from equations (4.18.1) and (4.18.2),

$$\text{Var}(\bar{y}_T - \bar{y}_S) = s^2/7 \tag{4.18.5}$$

and
$$\text{Var}(b) = s^2/28. \tag{4.18.6}$$

By equation (4.14.3)
$$g = \frac{(2\cdot101)^2 \times 398\cdot1}{(16\cdot00)^2 \times 28}$$
$$= 0\cdot2452,$$

or, from Table 4.16.1 and the alternative method in §4.14,
$$g = \frac{4\cdot41 \times 398\cdot1}{7168}$$
$$= 0\cdot2449,$$

a value that is arithmetically slightly less accurate. From equation (4.14.2), the fiducial limits to $(M - \bar{x}_S + \bar{x}_T)$ are

$$\left[-0\cdot1875 \pm \frac{2\cdot101}{16\cdot00} \left\{ \left(\frac{0\cdot7548}{7} + \frac{0\cdot1875^2}{28} \right) \times 398\cdot1 \right\}^{1/2} \right] \Big/ 0\cdot7548$$
$$= [-0\cdot1875 \pm 0\cdot8653]/0\cdot7548$$
$$= -1\cdot3948, 0\cdot8980.$$

Therefore
$$R_L = \frac{0\cdot2}{0\cdot0075} \text{ antilog } \bar{1}\cdot7901 = 16\cdot5,$$
$$R_U = \frac{0\cdot2}{0\cdot0075} \text{ antilog } 0\cdot1351 = 36\cdot4.$$

Thus the potency is estimated to be $25\cdot0\,\mu g$ per ml, with fiducial limits at $16\cdot5\,\mu g$ and $36\cdot4\,\mu g$ per ml.

4.19 Constraints of design

The design of any experiment determines the character of the proper statistical analysis. In the oestrone assay, litter-mate control was adopted so that differences between litters would not affect the estimate of potency or the assessment of its precision; to analyze the data of Table 4.15.1 ignoring the litter classification would be logically wrong and possibly very misleading.

An assayist might find it convenient to put all identically treated animals into one cage (or, in a microbiological assay, to put all tubes of the same dose in adjacent positions in the incubator). Rarely is this advisable, for it *confounds* (§9.2) differences between doses with differences between cages: the several animals in one cage are not true replicates of the treatment for comparison with differently treated animals in another cage. Interaction between the animals in one cage, such as competition for food, may make individual responses different from what they would have been had all animals been caged separately, so producing a variance between cages different from that within cages. Anyone who analyzes the responses without regard to cage differences is in effect asserting that these differences are negligible, and that he may legitimately assess the precision of potency estimate from variation *within* cages in spite of the fact that the dose contrasts used are made *between* cages; the experiment itself can provide no test of the validity of this assumption, unless each dose group is spread over two or more cages. Even though the animals were caged individually, the same difficulty would arise if all cages for one dose were placed close together in the animal house. From extensive experience, Emmens (1948, §13.5) wrote: 'There has been in biological work a considerable tendency to ignore the possibility of differences in reaction due to animals being caged in distinct groups and it seems to have been tacitly assumed that variation between cages must be negligible. It must be a rarely designed animal house in which conditions are so uniform that this assumption can be justified, and in the light of our knowledge that a variety of responses are influenced by health, temperature, light, feeding and many other factors, it would always seem worth while so to arrange our preliminary trials that the contributions of these factors to differences in the location of test objects may be examined'.

The statistical analysis of an experiment is a small part of the total labour, and its costs should not influence the choice of design (§6.6). The arrangement of subjects in cages, the randomization of order in an animal house, or the randomization of order of testing, may bring the convenience or even the practicability of an experiment into conflict with the ideal statistical conditions. Complex designs (Chapters 9 and 10) sometimes enable the statistician to overcome these difficulties, but the very complexity of a design can also make it impossible of application. Individual caging of subjects, or a complicated arrangement of tests, may so much increase the risk of gross mistakes or the costliness of an experiment as to make it completely impracticable. The statistician must recognize that these situations do arise, and must be prepared for some compromise with the exigencies of experimentation. He will need to make clear to the assayist the price that must be paid for the use of a statistically inferior design: loss of precision or, more serious, conclusions whose validity rests upon an untestable assertion about the unimportance of certain sources of variation. If the assayist is satisfied that these disadvantages do not outweigh the advantages of the design, the statistician's responsibility is ended.

Ideally, animals in an assay are caged individually, caged in groups corresponding to one of the uninformative classifications of the experiment (such as litters), or caged in some new groupings, orthogonal with all others, for which a sum of squares can be isolated in the analysis of variance. If individual caging is impracticable, and the assayist is reluctant to cage together animals that are being differently treated (possibly for the good reason that they would affect one another), he should at least aim at dividing each dose group between two or more cages. The analysis of variance will then show separate residual mean squares between and within cages; if the first is significantly larger than the second, it must be used as the basic s^2 for subsequent calculations.

When faced with the results of an assay for the detailed design of which he was not responsible, the statistician must discover exactly how the experiment was arranged and conducted. Bitter experience will teach him how easily an experimenter may fail to mention the existence of a constraint, because of failure to realize its relevance to the statistical analysis: in some instances, the appearance of the data may arouse the suspicions of an alert statistician, but in others only the most careful discussion of the experiment with the person responsible for its execution will elicit information that vitally affects the statistical analysis.

4.20 Heterogeneous deviations from linearity

Deviations from linearity of regression were discussed in §4.6. If an assay in which several doses of each preparation were included shows significant non-linearity, inspection of the dose–response diagram may show either of two situations. Systematic deviation of the points from the calculated straight lines may indicate that the true regressions are curved (cf. Fig. 4.10.1); the linear regression model is wrong, and must be rejected in favour of a different metametric transformation or a different method of analysis. Alternatively, the points may show considerable scatter about the lines yet appear completely erratic in their deviations; this may be a manifestation of an unusually complicated dose–response relationship, but often a more plausible explanation is heterogeneity of the batches of subjects at different doses. Even though the true regression be linear, if subjects were not assigned at random to doses (or if indeed the batches were knowingly made up from different sources), the deviations of mean responses from the regression lines will be greater than is predicted from variations within batches. Similar trouble may arise if different doses have to be tested on different occasions, and experimental conditions change between occasions.

If the assayist is prepared to accept heterogeneity between the dose groups, rather than a very complex regression curve, as the explanation of erratic deviations from linearity, the mean square for deviations from linearity may be used as the estimated variance per response in all subsequent tests and assessments of fiducial limits. Randomized allocation of groups, occasions, or other classifications, to doses is essential. For example, if only one dose per day can be tested, and the doses are used in systematic order on successive days, any secular trend in experimental conditions will bias estimation of the regression coefficient; a random order of doses will ensure that deviations about regression lines give a valid estimate of the random errors of experimentation. In an assay of an insecticide, successive batches of insects from a single culture might show a steady trend in sex-ratio; random allocation of batches to doses ensures that the estimation of potency and the assessment of precision are unbiased by any correlation of sex-ratio with response (Bliss, 1939; Murray, 1937). When the only alternative is to reject data as worthless, the temptation to accept the mean square for deviations from linearity as s^2 is strong. The critical reader will appreciate the need for restraint. To assume randomness when no random element has been incorporated into the design is a great risk. For validity of such a variance estimate, theory requires the number of subjects to be the same at each dose, but in practice slight inequalities do not matter.

When only two or three doses of each preparation are tested, discrimination between the two types of significant deviation from linearity is impossible. This might appear opposed to the recommendations of §6.10 on the number of dose levels. In reality, the right course is almost always to choose an assay design that makes proper randomness consistent with inevitable restrictions on experimental technique, thus avoiding any need for an estimate of variance based upon deviations from linearity. If this is not practicable, the case for four or more doses of each preparation is strengthened. The data of Table 4.2.1 give only 3 d.f. for the linearity mean square; if fiducial limits had to be based on this, they would have suffered from the imprecision in the estimate of variance. An illustration of the use of the linearity mean square as s^2 occurs in §16.2. In assays using quantal responses, the same problem may arise: the heterogeneity factor (§18.1) has the same function, and the same unsatisfactory basis, as the variance estimate just described.

It is important to avoid any automatic rule of rejecting assays on account of non-linearity or other aspects of statistical invalidity. As Humphrey *et al.* (1953) have emphasized, a rule based solely on individual significance tests would merely result in the most precise assays being rejected! A truly linear regression is a rarity, and to penalize all assays in which high precision detects non-linearity is folly. To formulate an ideal policy is difficult, as significant non-linearity at least indicates that precision is less good than the error mean square suggests. Humphrey's practice seems somewhat less desirable than the use of a different mean square for s^2, but the question deserves closer study by those concerned with large numbers of related routine assays.

4.21 Missing values

Even in the most carefully conducted experiment, an accident or unforeseeable circumstance may cause the loss of a subject and so destroy the symmetry of a design. Restrictions on the subjects available for use may even prevent adoption of a symmetric design. In the assay to which Table 4.15.1 refers, the responses recorded for five of the litters relate only to three rats; whether this was because only three female litter-mates were available or because animals were lost during the experiment is not now known. The missing records correspond to the positions marked by parentheses in Table 4.15.1. If the reason was that five litters had only three females, the design adopted was about the best that could be contrived; certainly it was preferable to omitting one dose entirely or from each of the five litters.

When observations are missing from a randomized block or more complicated design, special procedures are required to prevent the gaps inducing biased comparisons of dose means. This is because the orthogonality is destroyed: if the coefficients of Table 4.16.2 are applied to a new version of Table 4.15.1 that contains only the 23 genuine responses, neither the contrast nor the orthogonality condition is satisfied. Inspection of the data indicates large differences between litters, so that simple averaging of columns would give an unfair representation. Litter IV, for example, gave low results, and therefore the average of the

§ 4.21 PARALLEL LINE ASSAYS

responses of the other six litters for the lower dose of S would give too high a value relative to other doses.

One way of overcoming the difficulty is to calculate from the genuine records values that, when inserted in the empty spaces of the table, will remove any distortion in the means. The general procedure is to use symbols $y_1, y_2, y_3, \ldots$ to represent the missing values, to perform an analysis of variance of all the data in terms of these symbols, to express the error sum of squares as a quadratic function of the unknowns, and then to determine $y_1, y_2, y_3, \ldots$ by the condition that the error sum of squares shall be a minimum. This is a standard adaptation of the analysis of variance. When a single value must be calculated in a randomized block design, it leads to the formula

$$y = \frac{rR + cC - G}{(r-1)(c-1)}; \qquad (4.21.1)$$

r is the number of rows (here litters) in the table of results and R the total of all genuine data in the same row as the missing entry,[*] c is the number of columns (here doses) and C the total of all genuine data in the same column as the missing entry, and G is the total of all the data. Equation (4.21.1) represents a compromise between the average of all other entries in the same row and the average of all other entries in the same column. When more than one entry is missing, the formula may be applied iteratively. Values are guessed for all except the first, and the formula is used to calculate the first; the result, together with all guessed values except that for the second missing entry, is then used in a calculation of the second, and so on until all have been calculated. The process is repeated so as to revise the first, second, ... values with the aid of the results of the first set of calculations, and the iteration is continued until two successive cycles agree closely. The final values are independent of the initial guesses.

In the oestrone assay, y_1, y_2, y_3, y_4, y_5 represent the missing responses in Litters III, IV, V, VI, VII, and the formula for iteration is

$$y = (7R + 4C - G)/18.$$

The process may start with any values, for example 100 for each of y_2, y_3, y_4, y_5. Remembering that R, C, G must include every response except the one currently being recalculated, the formula gives

$$y_1 = (7 \times 214 + 4 \times 590 - 2321)/18 = 85$$

and then

$$y_2 = (7 \times 220 + 4 \times 424 - 2306)/18 = 52$$

and so on. The complete iteration is

---

[*] A momentary use of R in a sense different from relative potency should not confuse the reader; it will occur only when missing data have to be discussed.

	y_1	y_2	y_3	y_4	y_5
Start	—	100	100	100	100
Cycle 1	85	52	84	105	132
Cycle 2	87	50	82	111	133
Cycle 3	87	50	82	111	133

Any alternative starting values would lead to the same results. Although iteration could be continued to establish one or more decimal digits, in practice it is pointless to evaluate to greater accuracy than that of the recorded responses.

The possibility of calculating 'missing values' in this manner is no justification for careless experimentation, either in choice of design or in failure to make complete records. Though the calculated values are estimates of the responses that would have been found in a complete experiment, they do not create information by some statistical trick. Their primary function is to eliminate bias in the comparison of means. Table 4.15.1 was in fact constructed by insertion of the values just calculated, and in §§4.16–4.18 the assay was analyzed without comment. However, in an analysis of variance using calculated entries as though they were genuine, the error mean square is an unbiased estimate of the variance per response only if the number of degrees of freedom for error is reduced by the number of missing entries inserted. The loss of information is felt when mean responses for different doses are compared. In particular, though the functions of mean responses used in the formation of $(\bar{y}_T - \bar{y}_S)$ and b are unbiased, their variances are increased.

4.22 Approximate analyses for missing values

The simplest adjustment that can be made to take account of the missing values in the oestrone assay is merely to use the correct degrees of freedom for error, $(18 - 5)$. From Table 4.16.1, the variance should be estimated as

$$s^2 = 7165/13$$
$$= 551 \cdot 2.$$

Repetition of the calculations in §4.18 with this value for s^2 and 13 d.f. gives fiducial limits 14·3 μg and 40·6 μg per ml.

Had only one response been missing, this would have been fairly satisfactory. With this high proportion of 5 missing out of 28, the approximation is not very good. A further modification will commonly give conclusions near enough to those from the exact analysis in §4.23. This consists in obtaining L_p, L_1 from the estimated values in §4.21, using equations (4.18.1) and (4.18.2), but writing

$$\text{Var}(L_p/14) = s^2 \left(\frac{1}{6} + \frac{1}{5} + \frac{1}{6} + \frac{1}{6} \right) \bigg/ 4 \qquad (4.22.1)$$
$$= 0 \cdot 175 \, s^2$$

$$\text{Var}(L_1/14) = s^2 \left(\frac{1}{6} + \frac{1}{5} + \frac{1}{6} + \frac{1}{6} \right) \bigg/ 16 \qquad (4.22.2)$$
$$= 0 \cdot 043 \, 75 s^2;$$

§ 4.23   PARALLEL LINE ASSAYS

these are based upon assigning to the mean for each dose an effective variance $s^2/6$ or $s^2/5$, instead of $s^2/7$, because only 6 or 5 genuine response measurements are available. Recalculation from this point on exactly as in §4.18 gives

$$g = 0.4395,$$

and then a potency estimate of $25.0\,\mu g$ per ml but with fiducial limits, obtained by using equations (4.22.1) and (4.22.2) in Fieller's theorem (assuming zero covariance), at $12.8\,\mu g$ and $44.1\,\mu g$ per ml.

The limits found by this approximation are a little narrower than the correct ones in §4.23, but are close enough for practical purposes. Had only one or two entries been missing, the approximation would have been still more satisfactory. On the other hand, had the missing entries left the design even more unbalanced (e.g. three missing from the S_1 dose and two from T_2), it might not have been good enough. Yates's method (1933) for calculating the variances of contrasts between means after adjustment for missing entries could be adapted for use here. In any assay for which neither method in the present section is good enough, however, the full analysis illustrated in §4.23 is today so little more laborious that it should be used.

4.23   Exact analysis for missing values

The contrasts expressed by the coefficients in Table 4.16.2 are mutually orthogonal only when true responses have been recorded for all subjects, as was originally assumed for Table 4.15.1. The calculations in §4.21 provide for the five empty spaces values that are functions of the other 23. If these were written in full, it would become apparent that L_p, L_1, L_1' are still contrasts but are no longer orthogonal and no longer have variances given by equation (4.18.3). Exact analysis could proceed on these lines, with examination of the contrasts for L_p, L_1, determination of their variances and covariance as linear functions of the 23 responses, and application of Fieller's theorem to L_p/L_1.

This would be tedious, and would require new examination of contrasts for each new configuration of missing values encountered. General methods of fitting constants or solving least squares problems for linear models can be applied; unless care is exercised in formulation of the problem and the parameters, the arithmetic can become heavy, involving inversion of a fairly large matrix. An easier procedure to adopt as standard for bioassay is that based upon multiple linear regression. The great advantage is that the computing routine is widely known, standard programs are available on almost all scientific computers, and even on desk calculators an operator with a little experience of statistical calculation will need minimal special instruction. Although described here in terms of a very simple assay, the method readily adapts to other situations.

The key lies in the realization that all the information required for validity tests and for potency estimation is contained in a set of contrasts between doses. In the oestrone assay, these are three, and they can be formally identified with a set of three partial regression coefficients. Introduce three *dummy variates*, x_1, x_2, x_3, defined to take the following values at the four dose levels:

	S_1	S_2	T_1	T_2
$x_1 =$	-1	-1	1	1
$x_2 =$	-1	1	-1	1
$x_3 =$	1	-1	-1	1

The variates have the same values as the coefficients in Table 4.16.2, and x_2 is the $(x - \bar{x})$ of the earlier analysis. For complete data with no missing entries, the regressions on x_1, x_2, x_3 would give the sums in the last column of Table 4.16.2 as the various S_{xy}. Because of orthogonality of the dummy variates, the three squares for preparations, regression and linearity in Table 4.16.1 would then be computed as the squares attributable to the separate regressions.

The method may be applied to the genuine data of Table 4.15.1, after omission of bracketed entries, and is numerically equivalent to other methods mentioned above. An analysis of variance and covariance for x_1, x_2, x_3, y is made, all distinctions between columns (doses) being dropped since these are taken account of by the independent variates. Sums of squares and products are calculated for the 22 d.f. between the 23 responses, and components with 6 d.f. for litter differences are subtracted. Some litters contain four responses and some only three. For y, for example, the sum of squares for litters is

$$\tfrac{1}{4}(359^2 + 257^2) + \tfrac{1}{3}(214^2 + 220^2 + 303^2 + 251^2 + 317^2) - \frac{1921^2}{23} = 4786;$$

similarly, the sum of products of x_2, y for litters is

$$\tfrac{1}{3}(-1 \times 214 + 1 \times 220 + 1 \times 303 - 1 \times 251 - 1 \times 317 - \frac{(-1) \times 1921}{23} = 6 \cdot 81.$$

The reader should have no difficulty in checking the details of the analysis in Table 4.23.1.

The within-litter regression coefficients, b_1, b_2, b_3, are the solutions of

$$\left. \begin{aligned} \frac{64}{3}b_1 + \frac{4}{3}b_2 - \frac{4}{3}b_3 &= -\frac{22}{3}, \\ \frac{4}{3}b_1 + \frac{64}{3}b_2 + \frac{4}{3}b_3 &= \frac{1006}{3}, \\ -\frac{4}{3}b_1 + \frac{4}{3}b_2 + \frac{64}{3}b_3 &= -\frac{116}{3}. \end{aligned} \right\} \quad (4.23.1)$$

In these equations, exact numerical values have been inserted (instead of their decimal expressions in Table 4.23.1) because they are here so simple in form. Solution requires inversion of the 3 × 3 matrix of coefficients:

$$\frac{1}{3} \begin{pmatrix} 64 & 4 & -4 \\ 4 & 64 & 4 \\ -4 & 4 & 64 \end{pmatrix}. \quad (4.23.2)$$

§ 4.23 PARALLEL LINE ASSAYS 101

TABLE 4.23.1 Analysis of variance and covariance for exact analysis of Table 4.15.1

Adjustments for means		0·0435	0·0435	0·0435	−0·0435	0·0435	−0·0435	83·52	−83·52	83·52	160 445
Nature of variation	d.f.	$S_{x_1 x_1}$	$S_{x_2 x_2}$	$S_{x_3 x_3}$	$S_{x_1 x_2}$	$S_{x_1 x_3}$	$S_{x_2 x_3}$	$S_{x_1 y}$	$S_{x_2 y}$	$S_{x_3 y}$	S_{yy}
Between litters Within litters	6 16	1·6232 21·3333	1·6232 21·3333	1·6232 21·3333	−0·2898 1·3333	0·2898 −1·3333	−0·2898 1·3333	6·81 −7·33	−2·81 335·33	62·15 −38·67	4 786 12 652
Total	22	22·9565	22·9565	22·9565	1·0435	−1·0435	1·0435	−0·52	332·52	23·48	17 438

By standard procedures, such as are well described by Searle (1966), the inverse matrix is found to be

$$\mathbf{V} = \frac{3}{952} \begin{pmatrix} 15 & -1 & 1 \\ -1 & 15 & -1 \\ 1 & -1 & 15 \end{pmatrix}. \qquad (4.23.3)$$

Those unfamiliar with matrices may note that, for an initial matrix that is symmetric, the sum of products of corresponding rows is unity; for example, from the second rows in (4.23.2) and (4.23.3)

$$(-1 \times 4 + 15 \times 64 - 1 \times 4)/952 = 1.$$

Sums of products of non-corresponding rows are zero; from the second row of (4.23.2) and the third of (4.23.3)

$$(1 \times 4 - 1 \times 64 + 15 \times 4)/952 = 0.$$

To be able to take advantage of simple expressions in fractions is unusual. In many practical situations, the inversion would have to be done decimally; many digits would need to be retained in $\mathbf{V}$ as an aid to checking and as a guard against arithmetical inaccuracy arising because differences between nearly equal quantities are evaluated. Matrix inversion is an arithmetical operation especially well suited to high-speed computers, and appropriate routines are always included in standard programs for regression and covariance analyses; even these can be inaccurate for large matrices. A desk calculator can be used, but the labour of inverting a 6 × 6 or larger matrix is then liable to be very heavy.

The element in row i, column j of $\mathbf{V}$ may be denoted v_{ij}. The regression coefficients are then obtained as sums of products of the quantities on the right of equations (4.23.1) with each row of $\mathbf{V}$ in succession. Thus

$$b_i = (-22v_{i1} + 1006v_{i2} - 116v_{i3})/3, \qquad (4.23.4)$$

using the obvious exact fractional values. Hence

$$b_1 = (-22 \times 15 - 1006 - 116)/952$$
$$= -1.5252,$$
$$b_2 = 15.9958,$$
$$b_3 = -2.9076.$$

The regression accounts for a sum of squares

$$(-22b_1 + 1006b_2 - 116b_3)/3 = 5488. \qquad (4.23.5)$$

Regression on log dose (x_2) alone would account for a component of this obtained by omitting x_1, x_3 and repeating the calculations, or more directly from the familiar formula

$$(335.3333)^2/21.3333 = 5271.$$

§ 4.23 PARALLEL LINE ASSAYS 103

The difference, with 2 d.f., is a composite test of preparations and parallelism, which cannot be separated completely because of non-orthogonality. As Table 4.23.2 shows, the two together are too small to occasion any concern; a test of either could be made by omitting x_2 or x_3 and finding out how much of the 5488 is left when a regression on x_1 and the other is formed, but the portions so obtained would not be independent and additive (cf. Table 2.8.3). Note that the error sum of squares is equal to that in Table 4.16.1; the procedures that gave the two are algebraically identical, though a small discrepancy appears because the arithmetic of estimating missing values in §4.21 was carried only to the nearest integers. Table 4.23.2, however, gives unbiased validity tests.

TABLE 4.23.2 Final analysis of variance for exact analysis of Table 4.15.1

Adjustment for mean		160 445	
Nature of variation	d.f.	Sum of squares	Mean square
Regression on log dose	1	5 271	
Preparations and Parallelism	2	217	108
Doses	3	5 488	
Litters, ignoring doses	6	4 786	
Error	13	7 164	551·1
Total	22	17 438	

The definitions show that the regression coefficient on x_2 estimates the regression of response on log dose, the quantity usually called b. Similarly, the regression coefficient on x_1 is an estimate of one-half the difference in mean responses for the two preparations, the quantity usually called $(\bar{y}_T - \bar{y}_S)$. The method has ensured that these estimates are adjusted for non-orthogonality; they differ from those in §4.21 only because of the rounding of decimal digits in that section. Hence M is taken as

$$M = \bar{x}_S - \bar{x}_T - \frac{2 \times 1 \cdot 5252}{15 \cdot 9958}$$

$$= \bar{x}_S - \bar{x}_T - 0 \cdot 1907. \tag{4.23.6}$$

In order to construct the fiducial limits for M, the variances and covariance of $2b_1$ and b_2 are needed. They are obtained from the matrix $\mathbf{V}$ as

$$\left.\begin{aligned} \text{Var}(2b_1) &= 4s^2 v_{11} = 45s^2/238, \\ \text{Var}(b_2) &= s^2 v_{22} = 45s^2/952, \\ \text{Cov}(2b_1, b_2) &= 2s^2 v_{12} = -3s^2/476. \end{aligned}\right\} \tag{4.23.7}$$

Hence

$$g = \frac{(2 \cdot 160)^2 \times 45 \times 551 \cdot 1}{952 \times (15 \cdot 9958)^2}$$

$$= 0 \cdot 4750. \tag{4.23.8}$$

Fieller's theorem gives fiducial limits for $(M - \bar{x}_S + \bar{x}_T)$; from equations (4.23.7) they are

$$\left[-0 \cdot 1907 + \frac{0 \cdot 4750 \times 6}{45} \pm \frac{2 \cdot 160}{15 \cdot 9958} \{(180 - 2 \times 0 \cdot 1907 \times 6 + 0 \cdot 1907^2 \times 45 - 0 \cdot 4750 \times 179 \cdot 2) \times 551 \cdot 1/952\}^{1/2} \right] \Big/ 0 \cdot 5250$$

$$= (-0 \cdot 1907 + 0 \cdot 0633 \pm 0 \cdot 9974)/0 \cdot 5250$$

$$= -2 \cdot 1425, \; 1 \cdot 6571.$$

Once again from equation (4.15.1),

$$R = \frac{0 \cdot 2}{0 \cdot 0075} \text{ antilog } \bar{1} \cdot 9713 = 25 \cdot 0,$$

and similarly

$$R_L = \frac{0 \cdot 2}{0 \cdot 0075} \text{ antilog } \bar{1} \cdot 6776 = 12 \cdot 7,$$

$$R_U = \frac{0 \cdot 2}{0 \cdot 0075} \text{ antilog } 0 \cdot 2494 = 47 \cdot 4.$$

The potency estimate could differ from that in §4.22 only because more digits were retained in the present calculations; to the accuracy that may reasonably be reported, it is identical with that obtained earlier. The widening of the limits to $12 \cdot 7 \, \mu g$ and $47 \cdot 4 \, \mu g$ per ml represents the effect of failure to adjust the first analysis adequately on account of the missing entries in Table 4.15.1.

5
Symmetric dose - structure for parallel line assays

5.1 The general structure

Chapter 4 has illustrated both the advantages of symmetry in the design of biological assays and the inadequacy of the simplest symmetric assay in respect of validity tests. A more general symmetric dose-structure, termed a (k, k)-point assay, has k doses of each preparation such that successive doses of either preparation bear a ratio D to one another $(D > 1)$, with n subjects at each dose. The total number of subjects,

$$N = 2nk, \tag{5.1.1}$$

includes nk for each preparation. Before discussion of the general analysis, an example of a $(3, 3)$ assay is presented.

5.2 An assay of vitamin B_{12}

Emery *et al.* (1951) reported an assay conducted according to the principles of the dose–response relation studied in §3.9. This assay has $k = 6$, $N = 36$, and $D = 1.5$. By taking the dose metameter

$$x = \log_{1.5}(z/z_c), \tag{5.2.1}$$

where z_c is the central dose (1·2 ng of S, 6 units of T), the values of x are made -1, 0, 1 for each preparation. Table 5.2.1 contains the response metameters obtained from equation (3.9.6). Fig. 5.2.1, showing mean responses plotted against x, suggests both linearity and parallelism.

TABLE 5.2.1 Response metameters in an assay of vitamin B_{12}

Standard preparation (ng/tube)			Test preparation (units/tube)		
0·8	1·2	1·8	4	6	9
S_1	S_2	S_3	T_1	T_2	T_3
0·96	1·06	1·17	0·91	1·09	1·15
0·91	1·07	1·14	0·93	1·04	1·15
0·92	0·99	1·14	0·98	0·97	1·14
0·76	0·86	1·13	0·96	1·06	1·16
1·03	1·06	1·13	0·89	1·04	1·10
0·93	1·02	1·15	1·01	1·02	1·15
5·51	6·06	6·86	5·68	6·22	6·85

The analysis of variance, Table 5.2.2, subdivides the total sum of squares into components between doses and within doses (error); the assay has no classification analogous to litters in the example of §4.15. The 5 d.f. between doses can

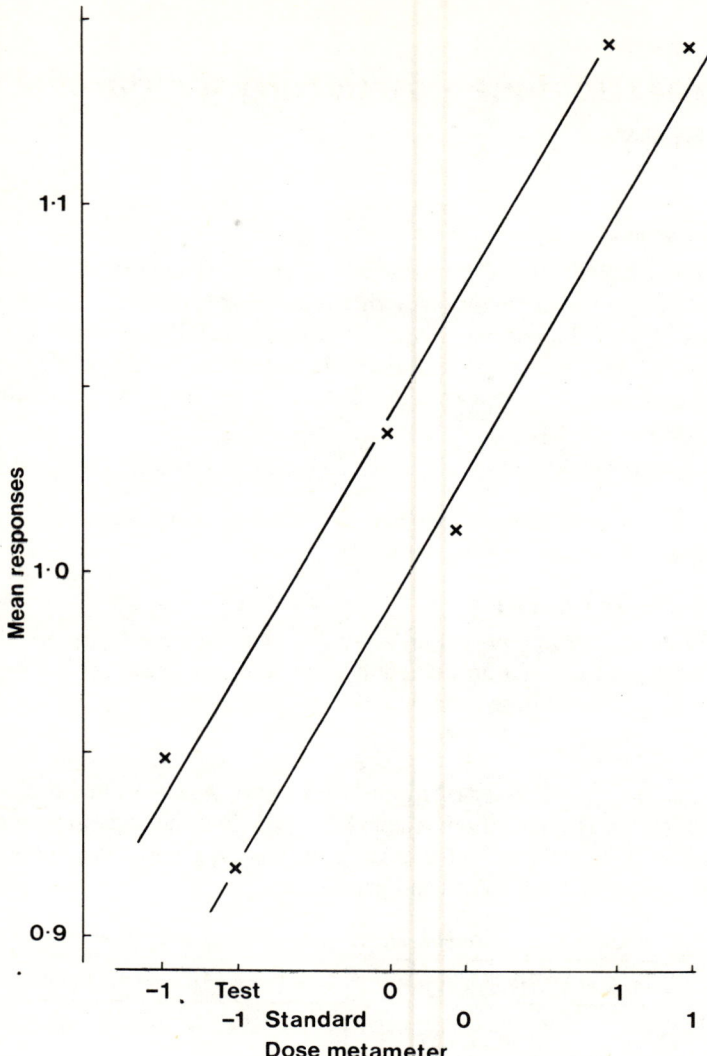

Fig. 5.2.1 Linear dose–response regressions for the assay of vitamin B_{12}, Table 5.2.1
×: Mean responses
The straight lines are those obtainable from the calculations of § 5.2.

be analyzed into five components, each of which aids interpretation of the assay. Again orthogonal coefficients may be used for obtaining the appropriate squares. One contrast must be defined as the difference between preparations; a second, leading to the average linear regression of response on x, must have as coefficients the values of $(x - \bar{x})$. By changing the signs of the coefficients for S in the regression component but leaving those for T unaltered, a contrast for the

§ 5.2   SYMMETRIC DOSE-STRUCTURE   107

difference in regressions, leading to a test of parallelism, is obtained. The remaining 2 d.f. measure deviations from linearity of regression. Tables of orthogonal polynomials (Fisher and Yates, 1963, Table XXIII) give $1, -2, 1$ as the coefficients for the quadratic component in a set of three equally spaced levels, and assignment of these values to both preparations leads to a component corresponding to the average quadratic curvature. Reversal of signs for S gives the fifth component, the difference between quadratic curvatures for the two preparations. Details are shown in Table 5.2.3, where every pair of contrasts satisfies the condition for orthogonality in §4.16; hence the rules of that section subdivide the sum of squares between doses into squares for the five contrasts. The contrasts for a (3, 3) design may be referred to as L_p (preparations), L_1 (regression), L_1' (parallelism), L_2 (quadratic curvature), and L_2' (difference of quadratics).

TABLE 5.2.2   Analysis of variance for Table 5.2.1

Adjustment for mean		38·398 678	Mean square
Nature of variation	d.f.	Sum of squares	
Preparations	1	0·002 844	0·002 844
Regression	1	0·264 600	0·264 600
Parallelism	1	0·001 350	0·001 350
Quadratic	1	0·001 606	0·001 606
Difference of quadratics	1	0·000 356	0·000 356
Between doses	5	0·270 756	
Error	30	0·092 766	0·003 092
Total	35	0·363 522	

TABLE 5.2.3   Coefficients of orthogonal contrasts for the (3, 3) design, applied to Table 5.2.1

Dose	S_1	S_2	S_3	T_1	T_2	T_3	Divisor	Sum
Response totals	5·51	6·06	6·86	5·68	6·22	6·85	36	37·18
L_p	−1	−1	−1	1	1	1	36	0·32
L_1	−1	0	1	−1	0	1	24	2·52
L_1'	1	0	−1	−1	0	1	24	−0·18
L_2	1	−2	1	1	−2	1	72	0·34
L_2'	−1	2	−1	1	−2	1	72	−0·16

Before estimating potency, assay validity must be examined. Inspection of Table 5.2.1 suggests freedom from serious heterogeneity of variance; calculation of sums of squares within the six dose groups shows some decrease of variance as dose increases, amounting to a ratio of about 15 between the highest and lowest doses, but the importance of this will not be discussed until §15.4. The mean square for regression in Table 5.2.2 is obviously very highly significant. The mean squares for preparations and parallelism are smaller than the error mean square. The design permits an additional validity test, that for linearity of regression, provided by the mean square for the quadratic component; this is so close to the error mean square as to occasion no fear of deviation from linearity. Except by chance, a large mean square for the component L_2' is unlikely to occur unless L_2 is large; if it does occur, it certainly raises grave doubts about validity. Here the square for L_2' also is small.

As ordinarily applied, the variance ratio test is a one-tail test; that is to say, it serves only to detect mean squares significantly larger than the error mean square, and the probability levels relate to this use. If it is to be used for examining mean squares that are smaller than the error, the reciprocal ratio of the mean squares must be taken (so as to keep the larger value on top) and compared with a tabular value in which the usual order of the degrees of freedom is reversed. Thus the mean square for difference of quadratics in Table 5.2.2 gives

$$F = \frac{0.003\ 092}{0.000\ 356}$$
$$= 8.4, \text{ with } (30, 1) \text{ d.f.}$$

The appropriate tabular entry in tables of the 0.05 significance level[*] (Appendix Table II; Fisher and Yates, 1963, Table V) is 250, so that the value of F found here causes no alarm. Moreover, special attention has been directed to L_2' because it is the most extreme criterion encountered in validity tests for this assay. If allowance were made for this selection, the probability assigned to the variance ratio would be even less extreme (Finney, 1941). Significantly small mean squares for important contrasts undoubtedly occur sometimes. When asked to explain one, the experienced statistician will first suspect a mistake (arithmetical or theoretical) in the statistical analysis; if this has been adequately checked, the possibility of non-independence of observations or lack of proper randomization must be investigated (cf. §§6.7, 15.3). The explanation of a significantly small mean square for, say, parallelism, will be entirely different from that of a large mean square; but the same conclusion, rejection of the assay as invalid, may follow.

The data of Table 5.2.1 have shown no suspicion of invalidity for the assay. From Table 5.2.3, since the coefficients of L_1 are values of $(x - \bar{x})$, the regression coefficient is

$$b = 2.52/24$$
$$= 0.1050.$$

Since x has been defined so as to have the same mean for both preparations, equation (4.11.5) gives

---

[*] The significance test now used allows extremes on either side of the error mean square to be judged significant, and is a two-tail test at the 0.10 probability level. In order to provide a test of this kind at the 0.05 level, a variance ratio table like Appendix Table II but computed for a one-tail probability of 0.025 is needed; Pearson and Hartley (1954, Table 16) give 1001 as the significance level for (30, 1) d.f.

§ 5.3 SYMMETRIC DOSE-STRUCTURE

$$M = -(\bar{y}_S - \bar{y}_T)/b$$
$$= -(1 \cdot 0239 - 1 \cdot 0417)/0 \cdot 1050$$
$$= 0 \cdot 170;$$

the numerator may be found directly from Table 5.2.1, or by dividing L_p by half the divisor in Table 5.2.3. From equation (4.14.3),

$$g = \frac{(2 \cdot 042)^2 \times 0 \cdot 003\ 092}{(0 \cdot 1050)^2 \times 24}$$
$$= 0 \cdot 0487.$$

By equation (4.14.2), therefore,

$$M_L, M_U = \left[0 \cdot 170 \pm \frac{2 \cdot 042}{0 \cdot 1050} \left\{ \left(0 \cdot 9513 \times \frac{2}{18} + \frac{0 \cdot 170^2}{24} \right) \times 0 \cdot 003092 \right\}^{1/2} \right] \bigg/ 0 \cdot 9513$$
$$= -0 \cdot 193, 0 \cdot 551.$$

Multiplication by $\log_{10} 1 \cdot 5$ converts M, M_L and M_U into logarithms to base 10; hence, with allowance for the scaling introduced by equation (5.2.1):

$$R = \frac{1 \cdot 2}{6} \text{ antilog } 0 \cdot 0299 = 0 \cdot 214,$$

$$R_L = 0 \cdot 2 \text{ antilog } \bar{1} \cdot 966 = 0 \cdot 185,$$

$$R_U = 0 \cdot 2 \text{ antilog } 0 \cdot 097 = 0 \cdot 250.$$

Thus the arbitrary unit used for the test preparation is estimated to contain 0·214 ng vitamin B_{12}, and the true potency is likely to lie between 0·185 ng and 0·250 ng.

5.3 The symmetric (k, k)-point design

Examples have been given of (2, 2) and (3, 3) symmetric assays, and the theory will now be generalized to the (k, k)-point introduced in §5.1. Let Z_S, Z_T be a pair of corresponding doses, for convenience the largest. The doses of the test preparation will usually be chosen in such a way as to make Z_S/Z_T equal to R_0, a value for the relative potency guessed from any available evidence (§6.8). Write S_i, T_i for the totals of responses to the n subjects on dose i of the preparations. As in §5.2, the contrast between preparations is defined to be

$$L_p = (-S_1 - S_2 \ldots - S_k) + (T_1 + T_2 + \ldots + T_k), \tag{5.3.1}$$

and consequently

$$\bar{y}_S - \bar{y}_T = -2L_p/N. \tag{5.3.2}$$

Formulae for k even and k odd must be distinguished. When k is even, the most convenient dose metameter is the logarithm to base $D^{1/2}$, giving as values of $(x - \bar{x})$

$$-(k-1), -(k-3), \ldots, -3, -1, 1, 3, \ldots, (k-3), (k-1).$$

The regression contrast is

$$L_1 = -(k-1)(S_1 + T_1) - (k-3)(S_2 + T_2) - \ldots + (k-1)(S_k + T_k), \quad (5.3.3)$$

with divisor

$$4n\{1^2 + 3^2 + 5^2 \ldots (k-1)^2\} = N(k^2 - 1)/3. \quad (5.3.4)$$

Hence

$$b = \frac{3L_1}{N(k^2 - 1)}. \quad (5.3.5)$$

When k is odd, the more natural metameter is log dose to base D, giving for $(x - \bar{x})$

$$-\tfrac{1}{2}(k-1), -\tfrac{1}{2}(k-3), \ldots -2, -1, 0, 1, 2, \ldots \tfrac{1}{2}(k-3), \tfrac{1}{2}(k-1).$$

The regression contrast is now

$$L_1 = -\tfrac{1}{2}(k-1)(S_1 + T_1) - \tfrac{1}{2}(k-3)(S_2 + T_2) - \ldots + \tfrac{1}{2}(k-1)(S_k + T_k), \quad (5.3.6)$$

with divisor

$$4n\left\{1^2 + 2^2 + 3^2 + \ldots \left(\frac{k-1}{2}\right)^2\right\} = N(k^2 - 1)/12. \quad (5.3.7)$$

Hence

$$b = \frac{12L_1}{N(k^2 - 1)}. \quad (5.3.8)$$

By equation (4.11.5),

and

$$\left.\begin{aligned} R &= \frac{Z_S}{Z_T} \text{ antilog} \left\{\frac{d(k^2-1)L_p}{3L_1}\right\} \quad (k \text{ even}), \\ R &= \frac{Z_S}{Z_T} \text{ antilog} \left\{\frac{d(k^2-1)L_p}{6L_1}\right\} \quad (k \text{ odd}), \end{aligned}\right\} \quad (5.3.9)$$

where

$$d = \log_{10} D. \quad (5.3.10)$$

Commonly $D = 2$ or $D = 4$ is taken, but a value such as $D = 10^{0.5}$ has advantages since $d = 0.5$ simplifies the arithmetic (Bliss, 1944b). Simplification is dearly bought, however, if it requires a much wider or a much narrower range of doses than would otherwise be chosen. Extra arithmetic is a small price to pay for avoiding the evils of non-linearity because the range is too wide or of reduced precision because it is too narrow (Chapter 6).

The attraction of equations (5.3.9) is that they facilitate rapid calculation of R from two easily constructed contrasts, L_p and L_1, without an analysis of variance or other intermediate steps. This can encourage neglect of validity tests and uncritical acceptance of estimates based on faulty premises. The formulae give an answer, but whether or not the answer has any meaning is uncertain without

§ 5.3 SYMMETRIC DOSE-STRUCTURE

validity tests. Moreover, evaluation of fiducial limits for the estimate requires further calculations.

For validity tests and fiducial limits, the variance per response is needed, but this does not always require the analysis of variance. The value of s^2 for the vitamin B_{12} assay may be found by adding sums of squares, each with 5 d.f. from the six dose groups.

$$30s^2 = \left(0{\cdot}96^2 + 0{\cdot}91^2 + \ldots + 0{\cdot}93^2 - \frac{5{\cdot}51^2}{6} + \ldots \ldots \right)$$

$$= 0{\cdot}092\,766,$$

whence as before

$$s^2 = 0{\cdot}003\,092, \text{ with 30 d.f.}$$

This is not possible if there are other design constraints, such as a balancing over litters; an alternative is illustrated in §5.4. The contrast L'_1, deviation from parallelism, can be constructed as

$$\begin{aligned}
L'_1 &= (k-1)(S_1 - T_1) + (k-3)(S_2 - T_2) + \ldots - (k-1)(S_k - T_k) \quad (k \text{ even}), \\
L'_1 &= \tfrac{1}{2}(k-1)(S_1 - T_1) + \tfrac{1}{2}(k-3)(S_2 - T_2) \\
&\quad + \ldots - \tfrac{1}{2}(k-1)(S_k - T_k) \quad (k \text{ odd}).
\end{aligned} \tag{5.3.11}$$

Two validity tests are given by a t-test on the deviations from zero of L_p, L'_1, by means of their standard errors:

$$\text{SE of } L_p = sN^{1/2}, \tag{5.3.12}$$

and

$$\begin{aligned}
\text{SE of } L'_1 &= \{N(k^2-1)/3\}^{1/2} \quad (k \text{ even}), \\
\text{SE of } L'_1 &= \{N(k^2-1)/12\}^{1/2} \quad (k \text{ odd}).
\end{aligned} \tag{5.3.13}$$

This is exactly equivalent to a variance ratio test in the analysis of variance. Linearity tests could be made in similar form, using L_2, L'_2; if all tests are wanted, however, the analysis of variance is the best procedure.

Before evaluation of the fiducial limits of R, g should be calculated from equation (4.14.3):

$$\begin{aligned}
g &= \frac{Nt^2 s^2 (k^2 - 1)}{3L_1^2} \quad (k \text{ even}), \\
g &= \frac{Nt^2 s^2 (k^2 - 1)}{12L_1^2} \quad (k \text{ odd}).
\end{aligned} \tag{5.3.14}$$

Equation (4.14.2) then leads to:

$$R_L, R_U = \frac{Z_S}{Z_T} \text{antilog} \left[\frac{d(k^2-1)}{3L_1(1-g)} \left\{ L_p \pm ts \left[N \left\{ (1-g) + \frac{(k^2-1)L_p^2}{3L_1^2} \right\} \right]^{1/2} \right\} \right] \quad (k \text{ even}),$$

$$R_L, R_U = \frac{Z_S}{Z_T} \text{antilog} \left[\frac{d(k^2-1)}{6L_1(1-g)} \left\{ L_p \pm ts \left[N \left\{ (1-g) + \frac{(k^2-1)L_p^2}{12L_1^2} \right\} \right]^{1/2} \right\} \right] \quad (k \text{ odd}).$$
(5.3.15)

5.4 The (2, 2) assay

The most popular symmetric design is the (2, 2), of which an example was given in §4.15. It has grave limitations (Chapter 6), but here the concern is with the standard analysis. Table 5.4.1 contains coefficients analogous to those of Table 5.2.3 (cf. Table 4.16.2). As an alternative to analysis of variance (§4.16), the contrasts may be used directly. By equation (5.3.9),

$$R = \frac{Z_S}{Z_T} \text{antilog} \frac{dL_p}{L_1}. \tag{5.4.1}$$

For the validity tests,

$$\text{SE of } L_p = \text{SE of } L_1' = sN^{1/2}. \tag{5.4.2}$$

Moreover,

$$g = \frac{Nt^2s^2}{L_1^2}, \tag{5.4.3}$$

and from equations (5.3.15)

$$R_L, R_U = \frac{Z_S}{Z_T} \text{antilog} \left[\frac{d}{L_1(1-g)} \left\{ L_p \pm ts \left[N \left(1 - g + \frac{L_p^2}{L_1^2} \right) \right]^{1/2} \right\} \right], \tag{5.4.4}$$

which may be written

$$R_L, R_U = \frac{Z_S}{Z_T} \text{antilog} \left[\frac{d\{L_p L_1 \pm [Nt^2s^2(L_p^2 + L_1^2 - Nt^2s^2)]^{1/2}\}}{L_1^2 - Nt^2s^2} \right]. \tag{5.4.5}$$

TABLE 5.4.1 Coefficients of orthogonal contrasts for the (2, 2) design

Contrast	Dose				Divisor
	S_1	S_2	T_1	T_2	
L_p	−1	−1	1	1	$4n$
L_1	−1	1	−1	1	$4n$
L_1'	1	−1	−1	1	$4n$

De Beer and Sherwood (1945) reported the results of an assay of penicillin. The doses of S, not stated in the original publication, are here assumed to be 50 and 200 units per ml; the doses of T were a 1 in 4 dilution and the basic 'undiluted' preparation. Paper discs, damped with penicillin solution, were placed on

§ 5.4   SYMMETRIC DOSE-STRUCTURE   113

agar plates inoculated with *Bacillus subtilis*. The four plates used each had one circle of each of the four doses. The responses measured, diameters of the zones of growth, are shown in Table 5.4.2 in units of 0·25 mm so as to be expressed as integers.

In Table 5.4.2, contrasts have been formed for each plate separately as well as for dose totals. From the totals,

$$R = \frac{200}{1} \text{ antilog} \left(-\frac{155}{135} \log_{10} 4 \right)$$

$$= 200 \times 0·2036$$

$$= 40·7.$$

Thus the undiluted test preparation is estimated to contain 40·7 units per ml. The assay was in randomized blocks, each plate being a block; differences between plates must be eliminated before estimation of variance, and s^2 therefore cannot be calculated from within dose groups. Consideration of the structure of the analysis of variance shows that s^2 should have 9 d.f., and these can be obtained as comparisons between contrasts L_p, L_1, L_1' formed for each plate separately. From the separate values for L_p, the sum of squares is

$$(-42)^2 + (-41)^2 + (-38)^2 + (-34)^2 - \tfrac{1}{4}(-155)^2 = 38·75;$$

similarly from L_1 and L_1' are obtained 32·75 and 56·75 respectively, each with 3 d.f. Moreover, any contrast between entries in one of the columns L_p, L_1, L_1' in Table 5.4.2 is orthogonal with all contrasts between doses, between plates, and between entries in either of the other two columns (§4.16). The three sums of squares must therefore belong to the error component of the analysis of variance, and since they are mutually orthogonal, must account for all the 9 d.f. for that error. Each entry in the L_p, L_1, L_1' columns is formed by additions and subtractions among four responses, so that the mean square from each sum of squares would estimate $4\sigma^2$. Hence, on pooling the three,

$$4s^2 = \frac{38·75 + 32·75 + 56·75}{3 + 3 + 3}$$

and

$$s^2 = 3·562.$$

The reader should verify this by the usual computations, leading to the analysis of variance, Table 5.4.3, that is not strictly necessary here.

From equation (5.4.2),

$$L_p = -155 \pm 7·5,$$
$$L_1' = 7 \pm 7·5;$$

the squares of the ratio of L_p, L_1' to their SEs are the corresponding variance ratios in Table 5.4.3. The clear significance of L_p causes fear that the doses of T may have run beyond the region of linearity; indeed the upper dose of T gave smaller responses than the lower dose of S. Because L_1' shows no indication of

TABLE 5.4.2   Diameters of zones of inhibition in an assay of penicillin (units of 0·25 mm)

Plate	S (units per ml)		T (dilution)		L_p $= -S_1 - S_2 + T_1 + T_2$	L_1 $= -S_1 + S_2 - T_1 + T_2$	L'_1 $= S_1 - S_2 - T_1 + T_2$
	50	200	0·25	1·00			
I	92	108	68	90	−42	38	6
II	95	111	74	91	−41	33	1
III	93	108	72	91	−38	34	4
IV	90	107	75	88	−34	30	−4
Total	370	434	289	360	−155	135	7

§5.5   SYMMETRIC DOSE-STRUCTURE

TABLE 5.4.3   Analysis of variance for Table 5.4.2

Adjustment for mean		1870·56	
Nature of variation	d.f.	Sum of squares	Mean square
Preparations	1	1501·56	1501·56
Regression	1	1139·06	1139·06
Parallelism	1	3·07	3·07
Between doses	3	2643·69	
Between plates	3	24·69	
Error	9	32·06	3·562
Total	15	2700·44	

non-parallelism, and experience of linearity in penicillin assays based upon circles of growth inhibition has been good, acceptance of the assay as valid seems justifiable.

For 9 d.f., $t = 2·262$,

$$Nt^2 s^2 = 291·6,$$

and

$$g = \frac{291·6}{135^2}$$

$$= 0·0160.$$

Equation (5.4.5) then gives

$$R_L, R_U = 200 \text{ antilog} \left[\log 4 \times \left(\frac{-20\,925 \pm 3498}{17\,933} \right) \right]$$

$$= 200 \text{ antilog } \bar{1}·1801, 200 \text{ antilog } \bar{1}·4149$$

$$= 30·3, 52·0.$$

Thus the potency of T seems almost certain to lie between 30·3 units per ml and 52·0 units/ml. So small a g might have been neglected, and limits calculated from equation (5.4.4) with $g = 0$. The results, 31·2 and 53·2 units/ml, are not very different but are sufficiently so to make the little extra effort of equation (5.4.5) worth while.

5.5   The (3, 3) assay

Table 5.2.3 contains the coefficients for the five contrasts in their most useful form. By equation (5.3.9),

$$R = \frac{Z_S}{Z_T} \text{ antilog } \frac{4dL_p}{3L_1}. \tag{5.5.1}$$

Equations (5.3.12), (5.3.13) give

$$\text{SE of } L_p = sN^{1/2}, \tag{5.5.2}$$

$$\text{SE of } L'_1 = s(2N/3)^{1/2}, \tag{5.5.3}$$

and the two curvature contrasts may be tested for significance with the aid of

$$\text{SE of } L_2 = \text{SE of } L_2' = s(2N)^{1/2}. \tag{5.5.4}$$

From equation (5.3.14),

$$g = \frac{2Nt^2s^2}{3L_1^2}, \tag{5.5.5}$$

and equation (5.3.15) reduces to

$$R_L, R_U = \frac{Z_S}{Z_T} \text{antilog} \left[\frac{4d\{L_pL_1 \pm [Nt^2s^2(2L_p^2 + 3L_1^2 - 2Nt^2s^2)/3]^{1/2}\}}{3L_1^2 - 2Nt^2s^2} \right] \tag{5.5.6}$$

Applied to the vitamin B_{12} assay in §5.2, with $N = 36$ and $s^2 = 0.003\ 092$ (30 d.f.), these formulae give

$$L_p = 0.32 \pm 0.33$$
$$L_1' = -0.18 \pm 0.27$$
$$L_2 = 0.34 \pm 0.47$$
$$L_2' = -0.16 \pm 0.47.$$

The significance tests are the same as from Table 5.2.2. Also,

$$g = \frac{72 \times 2.042^2 \times 0.003\ 092}{3 \times 2.52^2}$$
$$= 0.0487$$

as before. Then

$$R = \frac{1.2}{6} \times \text{antilog} \frac{4(\log 1.5) \times 0.32}{3 \times 2.52}$$
$$= 0.214,$$

with limits at

$$R_L, R_U$$
$$= 0.2 \times \text{antilog} \left[\frac{4(\log 1.5)\{0.32 \times 2.52 \pm [0.1547(2 \times 0.32^2 + 3 \times 2.52^2 - 0.9283)]^{1/2}\}}{3 \times 2.52^2 - 0.9283} \right]$$
$$= 0.185, 0.250.$$

The results are exactly as in §5.2.

5.6 The (4, 4) assay

Similar formulae are easily constructed for higher values of k. Fisher and Yates (1963) have described the general method in connexion with their Table XXIII, from which the orthogonal coefficients may be taken. Table 5.6.1 relates to the next member of the series, the (4, 4) assay. In addition to the contrasts previously used, a component of cubic curvature, L_3, and another for the difference between cubics for the two preparations, L'_3, are now required. Significance of L_3 or L'_3 bears much the same interpretation as significance of L_2 or L'_2, though it conveys rather different information on the form of the non-linear response curve. The estimate of potency is

$$R = \frac{Z_S}{Z_T} \text{antilog} \frac{5dL_p}{L_1}, \tag{5.6.1}$$

and other formulae require substitution of $k = 4$ in equations (5.3.12)–(5.3.15).

TABLE 5.6.1 Coefficients of orthogonal contrasts for the $(4, 4)$ design

Contrast	S_1	S_2	S_3	S_4	T_1	T_2	T_3	T_4	Divisor
L_p	−1	−1	−1	−1	1	1	1	1	$8n$
L_1	−3	−1	1	3	−3	−1	1	3	$40n$
L'_1	3	1	−1	−3	−3	−1	1	3	$40n$
L_2	1	−1	−1	1	1	−1	−1	1	$8n$
L'_2	−1	1	1	−1	1	−1	−1	1	$8n$
L_3	−1	3	−3	1	−1	3	−3	1	$40n$
L'_3	1	−3	3	−1	−1	3	−3	1	$40n$

5.7 Aids to arithmetic

An investigator who conducts a large number of assays of the same simple design may fear the labour of statistical analysis, and will certainly wish for the most expeditious scheme of computing. Although statistical analysis seldom forms a major part of assay costs, it can be economically and operationally important in a series of routine assays. In 1938, Morrell, Chapman and Allmark could write: 'a calculating machine is an instrument not often found in a biological laboratory'. Today few pharmacologists or others concerned with bioassay lack access to desk calculators more sophisticated than, and very much cheaper than, any available in 1938. Large organizations will also have access to high-speed computers.

An important first step to reducing the arithmetical labour of routine assays is standardization. For a series of $(2, 2)$ or $(3, 3)$ assays all with the same number of subjects and the same dose-spacing, a standard form with spaces for the data and for each stage of calculation can save much time. All calculation can be presented as a sequence of simple arithmetical operations, which can be performed by an assistant with no knowledge of their meaning. Each completed analysis must be inspected and interpreted by someone qualified to understand it, lest unexpected peculiarities in a set of data escape detection.

The formulae in §§5.2–5.6 indicate how computations may be streamlined for symmetric parallel-line assays. Various authors have proposed similar formulae for specific types of assay. Naturally the $(2, 2)$ design has been most frequently discussed, and formulae essentially the same as those in §5.4 have been put forward by Bliss (1944b, c; 1945), Bliss and Marks (1939a, b), Finney (1944b), Gridgeman (1944a, b), Knudsen and Randall (1945), Schild (1942), Sherwood, Falco and de Beer (1944). Formulae for $(3, 3)$ and $(4, 4)$ assays given by

Sherwood (1947) have little novelty as they merely employ orthogonal coefficients in the same way as was suggested earlier by Bliss and Marks. Erroneous and extraordinary suggestions such as those of Osgood (1947) and Osgood and Graham (1947) can now be disregarded. This book is concerned with general principles and methods, not with schemes of computation specially devised for a particular assay design. A reader who masters the general methods will readily acquire skill in streamlined computations. Unfortunately almost every author has succumbed to the temptation to adopt an idiosyncratic notation, without thought of whether his use of symbols conflicts with the current usages for less specialized designs or by other authors.

A non-statistician unfamiliar with general theory is unwise to rely on his own judgement in deciding to use formulae such as have been mentioned above. The same is true, indeed, in the choice between alternative computer programs. He will find most attractive that procedure which appears to demand least arithmetical or other labour, but tempting simplifications may conceal limitations on applicability or serious loss of information. Writers on streamlined computations for bioassay sometimes omit all mention of the precision of estimates, or are content to assess fiducial limits with the aid of a formula based on equation (4.12.9), so neglecting g; validity tests also are usually neglected. A method that is safe for an antibiotic assay, because variances are low and validity well-established, may be completely untrustworthy for an assay of the same design that uses a much more variable dose–response relation. Moreover, the streamlined method can never take account of accidents that disturb symmetry. The vitamin B_{12} assay in §5.2 and the oestrone assay in §4.21 are almost the same in design, differing only in that the oestrone assay has more blocks and has some observations 'missing'; streamlined computing is suitable for the one, but not for the other.

For a computer to be used in bioassay calculations requires easy access to the computer and availability of programs. Before the computer age, and before even electromechanical calculators were normal equipment for bioassay laboratories, nomographs were popular with some assayists; they may still be good for rapid preliminary assessment. Nomographic estimation involves locating the values of L_p and L_1 on two scales (usually but not always linear), joining the points by a straightedge, and reading RZ_T/Z_S from the intersection of this join with a third scale. Each value of k, and each dose interval, d, needs its own nomograph, but a set for the most common arrangements could be formed. Additional scales can be constructed to aid evaluation of Var(M), and hence, if g is small, the fiducial limits for R. Nomographs for particular assays have been described by Anon (1946), Bliss (1946a), Harte (1948), Knudsen (1945a, 1950), Knudsen and Randall (1945), Knudsen, Smith, Vos and McClosky (1946), Koch (1947), and Sherwood (1951). Potency estimation from a nomograph suffers from arithmetical inaccuracy, although it can often be made as accurate as the data justify. More important criticisms are the neglect of validity tests and reliance on a variance formula. Developments by Healy (1949), Gridgeman (1951), and Leech and Grundy (1953) indicate how, with a little extra arithmetic, validity tests may be made and fiducial limits based upon Fieller's theorem may be found from a more elaborate nomograph. Healy illustrated his type of chart for a (3, 3) assay with dose ratio 1·5 ($k = 3, d = 0.176$); Gridgeman's may require more calculation for the first construction of the nomograph, but seems to be particularly convenient in use. Leech and Grundy extended Healy's ideas so as to apply them to randomized blocks.

Other writers have suggested simple pieces of machinery for special use in bioassay. De Beer (1941) proposed a type of protractor for direct reading of regression coefficients on a dose–response diagram. Goyan and Dufrenoy (1947), and Dufrenoy and Goyan (1947) described a device of movable scales which, when superposed on a dose–response diagram, enable the estimate of relative potency to be read. Lees (1949) described a rotating drum, to the surface of which is attached a specially prepared table, that replaces 'the conventional slide rule and semi-logarithmic paper techniques ... by the simple operation of turning two wheels on a machine'; his explanation is not very clear, but the machine appears to be closely related to a nomograph. All these devices suffer from the disadvantage that they deal only with estimation and take no account of validity or precision.

5.8 Range estimation for standard deviations

The calculation of sums of squares is quick and easy today, but temporary lack of access to computing aids can make welcome an alternative method of estimating σ, the SD per response. A common procedure is to estimate the SD from a range (Bliss 1946a; Knudsen, 1945a; Wood 1947b). The difference between the highest and the lowest of a set of values is clearly a measure of variability. If the distribution of individual deviations is normal, the best estimate that can be formed from this range is less efficient than the root mean square estimate, but the efficiency is high when the number of values involved is small. Table 5.8.1 shows the factor by which the range must be multiplied in order to estimate the SD; it relates only to a normal distribution, and to the small numbers of observations likely to be needed in bioassays.

TABLE 5.8.1 Estimation of a standard deviation from a range in a sample of normally distributed observations

No. of observations	Factor by which range must be multiplied
2	0·887
3	0·591
4	0·486
5	0·430
6	0·395
7	0·370
8	0·351
9	0·337
10	0·325

The ranges of the four values in the columns L_p, L_1, L_1' in Table 5.4.2 are 8, 8, 10 respectively, averaging 8·67. The true SD of each column is 2σ, since the values are compounded by additions and subtractions of four responses. Hence σ may be estimated by

$$s = \tfrac{1}{2} \times 0\cdot486 \times 8\cdot67$$
$$= 2\cdot11,$$

instead of 1·89 as in §5.4. An estimate so obtained should not be used in the formulae for tests of validity and fiducial limits, because these assume s^2 to be a mean square. Tests of significance can be based upon tables analogous to the t-table (Lord, 1947). To use these tables in deriving limits of error is less easily justified, though the values so obtained should be a reasonable indicator, provided that g is reasonably small. Those who use nomographic methods generally base assessments of precision on ranges.

The method described for a (2, 2) assay is easily extended to others. Its uncritical adoption is not recommended, because non-normality of distribution is likely to have more serious consequences for range estimation than for mean square estimation.

5.9 Distribution-free methods

In §2.5, distribution-free methods were mentioned. Several publications have described their use in bioassay, either as a simplification of arithmetic or as insurance against non-normality. Bennett (1969, 1970) proposed the simple expedient of ranking all the responses from largest to smallest and then analyzing the rank numbers essentially as described in Chapters 4, 5. Thus the fifth largest value of y is replaced by the number 5. General theory indicates that estimates and significance tests will behave approximately as for normally distributed responses. Van Strik (1961) earlier used essentially the same procedure in assays of progesteronal activity, in which proliferation in rabbit uteri was scored on a non-numerical graded scale.

The analysis of the simple integers that may be regarded as a particular metametric transformation of the original response is very speedy on a desk calculator, but of course does not represent any appreciable saving for a computer. Presumably some information is sacrificed by neglect of the numerical values of the responses, but this is known not to be very great even when true normality obtains; if the actual responses are markedly non-normal in distribution, the method may give some protection against grossly misleading conclusions. In practice, an analysis of ranks will often give estimates and other results not very different from those obtained by analysis of actual responses. Sen (1971, 1972) has described a more sophisticated type of distribution-free estimation of potency.

Distribution-free methods are not restricted to symmetric assay designs, but their relation to easy computation made their introduction here appropriate. The cautious will wish to make extensive use of them, and perhaps regular users of bioassay should occasionally check that analysis of typical data by methods which assume normality is giving much the same conclusions as distribution-free analysis. Nevertheless, for most practical purposes the use of standard methods and some appeal to the central limit theorem is as acceptable in bioassay as in other branches of statistics.

5.10 Control charts

If a standard pattern of assay has been adopted as a factory routine, a system of control charts should be instituted, in order to check that experimental conditions and processes are being maintained satisfactorily constant. This book is primarily concerned with bioassay in research problems, and no details of routine applications are presented. There are many books on industrial quality control; Bliss (1952a) and Bliss and Pabst (1955) have described methods particularly relevant to control charts for s^2, b, and s/b.

Knudsen and Randall (1945), discussing the practice of the United States Food and Drug Administration for cylinder plate antibiotic assays, advised that

> Separate control charts should be kept for 3-plate assays, 5-plate assays, etc., and as each assay is run, the values ... are plotted on the proper control chart. If they fall outside the control limits, some cause of trouble must be looked for, and, when it is found, the assay should be repeated. Contamination of the plates by bacteria from the air may be a cause of trouble, although this can usually be observed. A variation in thickness of the inoculated agar layer may result in excessive variation in zone size, and unapparent "leakers" where the leakage is symmetrical around one or more of the cups may also result in larger circular zones. Of course errors of dilution or weighing will not be pointed out by the control chart method.

Noel (1945) also made use of quality control. In a series of assays like that in §5.4, if the number of plates and the dose interval be constant, s^2, L_1, and L_1' ought all to remain in control, and each might be made the basis of a control chart. A paper by Jones (1945) is of interest. Knudsen (1945b) made a further suggestion about the use of control charts in biological laboratories, obviously of interest in connexion with bioassay: 'The weights of mature rats and the numbers of animals in each litter could be plotted on a control chart, to insure a continual supply of uniform animals and to spot any defects in diet and care that may be causing an untoward variation in the animals, or in the sizes of litters'. Cohen *et al.* (1959) presented a chart showing b excellently in control over a series of 39 assays of tetanus toxoids.

When the assay technique is satisfactorily in control, some gain may be expected from pooling estimates of s^2, and perhaps also of b, from several consecutive assays, so as to gain precision on these quantities (Gridgeman, 1944b). This may be particularly useful in assays for which sampling variation in s^2 or in L_1 is largely responsible for low precision in R. To give exact expression to errors of estimation after such pooling is not easy, but the obvious modifications in the degrees of freedom and in Var(b) are unlikely to be far wrong. If the control is completely satisfactory, a long series of results may be pooled in order to give what are effectively population values, σ^2 and β, instead of estimates s^2 and b.

5.11 Quadratic regression for parallel line assay

Suppose that the regression equation for S is quadratic rather than linear (§3.8):

$$Y_S = \alpha + \beta x + \gamma x^2. \tag{5.11.1}$$

Then if T responds similarly,

$$Y_T = \alpha + \beta(x + \log \rho) + \gamma(x + \log \rho)^2. \tag{5.11.2}$$

Elston (1965) noted that in some circumstances a simple estimation of potency is still possible. Cox (1972) described a slightly different approach to the same procedure, and Williams (1973) looked particularly at the combination of the two estimates that arise in the first stage of the analysis.

If a symmetric (k, k) design has been used for the assay, independent estimates of $\log \rho$ are obtainable from two ratios of contrasts, L_p/L_1 and L_1'/L_2. This is true for all $k > 2$, but of course for $k = 2$ no L_2 exists. The presentation that follows, in terms of $k = 4$, illustrates the general pattern. For $k = 4$, and with x scaled so as to take the values $-3, -1, 1, 3$, application of the coefficients in Table 5.6.1 to equations (5.11.1), (5.11.2) shows that

$$\begin{aligned} \mathrm{E}\,(L_p) &= 4n[\beta \log \rho + \gamma(\log \rho)^2], \\ \mathrm{E}\,(L_1) &= 40n(\beta + \gamma \log \rho), \\ \mathrm{E}\,(L_1') &= 40n\gamma \log \rho, \\ \mathrm{E}\,(L_2) &= 32n\gamma, \\ \mathrm{E}\,(L_2') &= \mathrm{E}\,(L_3) = \mathrm{E}\,(L_3') = 0. \end{aligned} \tag{5.11.3}$$

Because the 7 contrasts are orthogonal, it is evident that

$$M_1 = 10L_p/L_1 \tag{5.11.4}$$

and

$$M_2 = 4L_1'/5L_2 \tag{5.11.5}$$

are statistically independent estimates of $\log \rho$ obtainable simultaneously from the one assay, and that L_2', L_3, L_3' still provide information for validity tests. For other k, the multipliers of L_p/L_1 and L_1'/L_2 are different, but the essentials of the method remain. Indeed the theory can be extended to a wider class of assays in which the only requirements of symmetry are that S and T shall be tested at the same number of doses, with equal ratios between corresponding pairs of doses and equal numbers of subjects at corresponding doses.

The final estimate of $\log \rho$ requires combination of M_1, M_2. For this purpose, the general likelihood method of §14.3 is appropriate, because the same σ^2 is appropriate to M_1 and M_2. By standard formulae, the $(4, 4)$ assay has

§ 5.11   SYMMETRIC DOSE-STRUCTURE   123

$$\left.\begin{aligned}\mathrm{Var}(10L_p) &= 800n\sigma^2,\\ \mathrm{Var}(L_1) &= 40n\sigma^2,\\ \mathrm{Var}(4L_1') &= 640n\sigma^2,\\ \mathrm{Var}(5L_2) &= 200n\sigma^2.\end{aligned}\right\} \quad (5.11.6)$$

If both M_1 and M_2 are precisely estimated, which primarily means that the denominators are large relative to their standard errors, the weighted mean calculations of §14.1 should suffice. Note that if a quadratic is assumed where a linear regression is satisfactory, L_2 is likely to be close to zero and M_2 will be so imprecise that the potency estimate can be taken from M_1 alone, exactly as for a linear regression.

The symmetric (2, 2) assay presents special features. Because L_2 and therefore M_2 are not available, M_1 stands alone and potency is estimated exactly as for linear regression: equations (5.4.1), (5.4.5) still apply, as was noticed first by Gridgeman (1943) and Wood (1944a). No validity test is possible, because the remaining dose contrast, L_1', is now needed for estimating the extra parameter.

Constraints such as blocks for litter mates or for doses of antibiotics on one plate demand the obvious modifications to the analysis; they could disturb the theory if their effects were not simply additive. The simplicity of the analysis disappears completely if the regression is a polynomial of higher degree (Elston erroneously implied the contrary). The existence of two independent estimates from a symmetric assay and a quadratic regression is something of a mathematical accident, and is of limited practical use: knowledge that a non-linear regression is quadratic rather than of some other form is rare. If a regression is known to deviate only slightly from linearity, however, over a moderate range of doses a quadratic equation may be an adequate approximation that permits a simple valid analysis.

For an assay that lacks the symmetry necessary to make the method described here applicable, and for a regression that contains terms of higher degree or that is not polynomial, relative potency can still be estimated by a generalized least squares procedure. Equations analogous to (5.11.1), (5.11.2) are fitted to the data directly, with $\log \rho$ or ρ as one parameter to be estimated at the same time as all the others. Except for some special cases (cf. §16.5), this analysis is seldom wanted.

Elston's (1965) data (Table 5.11.1) provide a good example. They are analyzed here without any block constraints, in accordance with Elston's statement that the experiment was completely randomized. From Table 5.6.1 and the dose totals in Table 5.11.1,

$$L_p = -1\cdot 76,$$
$$L_1 = 19\cdot 90,$$
$$L_1' = 2\cdot 88,$$
$$L_2 = -2\cdot 90,$$

$L'_2 = 0.24,$

$L_3 = -1.70,$

$L'_3 = -1.44.$

Completion of the analysis of variance, Table 5.11.2, calls for no comment.

TABLE 5.11.1 Weights of mouse uteri (in $\log_{10}$ of mg) in an assay of gonadotrophins

Doses (in arbitrary units)							
S				T			
0.192	0.48	1.2	3.0	0.384	0.96	2.4	6.0
0.93	1.08	1.61	1.73	0.71	1.26	1.53	1.67
1.17	1.61	1.76	1.63	1.00	1.32	1.68	1.70
1.03	1.60	1.75	1.66	0.91	1.18	1.82	1.67
1.19	1.64	1.70	1.84	0.96	1.31	1.65	1.80
4.32	5.93	6.82	6.86	3.58	5.07	6.68	6.84

TABLE 5.11.2 Analysis of variance for Table 5.11.1

Adjustment for mean		66.4128	
Nature of variation	d.f.	Sum of squares	Mean square
L_p	1	0.0968	
L_1	1	2.4751	
L'_1	1	0.0518	
L_2	1	0.2628	
L'_2	1	0.0018	
L_3	1	0.0181	
L'_3	1	0.0130	
Between doses	7	2.9194	
Error	24	0.4182	0.01742
Total	31	3.3376	

If the log doses for S and T are both scaled to $-3, -1, 1, 3$ and M_1, M_2 are based on this scale, multiplication by 0.1990 (or $\frac{1}{2}\log_{10} 2.5$) is needed to put any M in terms of logarithms to base 10, and the antilog must be multiplied by 0.5 because doses of T were twice those of S. From equations (5.11.4) and (5.11.5),

$M_1 = -17.6/19.90 = -0.8844,$

$M_2 = -11.52/14.50 = -0.7945.$

Moreover, with $n = 4$, equations (5.11.6) give

$\text{Var}(10L_p) = 3200\sigma^2,$

$\text{Var}(L_1) = 160\sigma^2,$

$$\text{Var}(4L'_1) = 2560\sigma^2,$$

$$\text{Var}(5L_2) = 800\sigma^2.$$

Here σ^2 must be estimated by s^2 with 24 d.f. Fieller's theorem could be applied to either estimate separately in order to give its fiducial limits. However,

and
$$R_1 = 0.5 \text{ antilog} (-0.8844 \times 0.1990) = 0.333$$
$$R_2 = 0.5 \text{ antilog} (-0.7945 \times 0.1990) = 0.347$$

are obviously in very close agreement, and immediate use of the procedure of §14.3 is sensible.

The maximum of the likelihood function gives as a test statistic for agreement of M_1, M_2

$$F = 0.000\,347/0.017\,42.$$

This is so small relative to 4·26, the value for probability 0·95 with 1, 24 d.f., that no doubts arise about the hypothesis that M_1, M_2 are estimating the same parameter. Any significant difference here would put in question the validity of the quadratic regression for these data. At the minimum,

$$M = -0.8524.$$

Further use of the tabular $F = 4.26$, in accordance with §14.3, leads to the limits $-1.5441, -0.2459$. Hence

$$R = 0.5 \text{ antilog} (-0.8524 \times 0.1990) = 0.338,$$

with limits at

$$R_L = 0.246,$$

$$R_U = 0.447.$$

These are very similar to Cox's (1972) results, although he used weighted means and determined limits from an approximate variance. Limits calculated for R_1 alone are 0·228 and 0·475, and those for R_2 are 0·168 and 0·538. Evidently in this example the gain from using the information provided by the second estimate is small, but not negligible – the ratio of R_U to R_L is reduced from 2·08 to 1·82.

5.12 Computer programs

Anyone who has regular heavy demands for the analysis of bioassays will submit them to a computer. Even the occasional user of assays can code his results for a standard program with far less trouble than is involved in personally undertaking the detailed arithmetic. There is today little to be said for the short-cut and approximate methods mentioned in §5.7, for they represent negligible saving of time in electronic data processing. For assayists who do not have access to any large computer, programmable desk and pocket calculators undoubtedly can do much to help. Their cost is small relative to laboratory costs for assays, and to be without one is a false economy. They have the disadvantage of requiring special programming in languages of restricted capabilities and of not producing well-formatted printed output. They are a second-best, but not to be despised.

With a good computer system, an assay program can easily be written in any widely accepted language, though the process of writing, checking and documenting is time-consuming and not justifiable unless the program is intended for frequent use. In the interests of economy, programs should be written in high-level languages so as to be readily portable between computers, including computers from different manufacturers. The arithmetical part of the program will almost invariably make relatively small demands on the computer, even for the more complicated designs introduced in later chapters, and the amount of data in any one assay is never great; attention should be concentrated on convenience of input and informativeness of output.

Rather surprisingly, few general programs for bioassay have been reported. Of course, programs for analysis of variance of planned experiments can undertake the central calculations, but they are not ideally adapted for input of assay data and they do not complete their task by estimating relative potency or producing appropriate summaries for output. A computer is not being used to best advantage if preliminary organization of data and final summarizing of results must be regarded as separate jobs; these would use negligible amounts of computer time, but the need to perform them manually can be a serious obstacle to the adoption of good statistical processing for routine assays. Assayists will eventually demand either a range of special programs for particular types of assay or a comprehensive program package that might include almost the whole contents of this book. An alternative approach would be to program for bioassay within a general statistical system, GENSTAT perhaps being the obvious choice. Its disadvantage is that only the largest computers can mount such a system satisfactorily, and to invoke it for the simpler bioassay analyses may be expensive.

McArthur, Ulfelder and Finney (1966) described a program for a simple parallel line assay with no block constraints, similar to that described in §4.2. Although it has been used in a number of bioassay laboratories, it was very limited in its facilities. It has been superseded by a totally rewritten version, PARLIN (Finney, 1976a). This program, though not ideal, meets many needs. PARLIN will handle parallel line assays in randomized blocks or without block constraints; the numbers of preparations, doses and subjects per dose are effectively unlimited. In the absence of block restrictions, the numbers of subjects per dose need not be equal (cf. Table 4.2.1); for randomized blocks, the blocks need not be of the same size as long as each contains equal replication of all doses.

Tables 5.12.1–5.12.5 show how PARLIN handles the vitamin B_{12} assay of §5.2. Various headings are of course at the choice of the user. The first section of output (Table 5.12.1) reports back the data as a check on input; note that both doses and responses are entered as their absolute values before metametric transformation, so that these occasionally tedious preliminary calculations are taken from the user. Input for an assay requires only about 5 minutes of preparation, 20–30 punched cards or the equivalent from a keyboard terminal: apart

§ 5.12   SYMMETRIC DOSE-STRUCTURE

from the headings and the numerical values of doses and responses, the design must be specified as also must optional items of analysis and output.

```
                         TABLE   5.12.1
       +------------------------------------------------+
       :    DATA FOR      - BIOASSAY USING PARLIN 7  :
       +------------------------------------------------+
       SOURCE OF DATA:     EMERY, LEES AND TOOTILL, 'THE ANALYST', 76, P. 145
       NAME OF ASSAY:      VITAMIN B12, MICROBIOLOGICAL ASSAY
       DATE OF ASSAY:      1951 (TURBIDIMETRIC, LACTOBACILLUS LEICHMANNII)
       DATE OF ANALYSIS:   02/02/77, AT 21.23.59

       C O M M E N T:       A COMPLETELY RANDOMIZED PARALLEL LINE ASSAY

                      =*= FOR CURRENT DATA, THIS IS RUN NO.   1  =*=


              TABLE OF DATA (AS INITIALLY ENTERED)
              ------------------------------------

       PREPARATION    1:  CRYSTALLINE B12
                DOSE NO.   1  =      0.8000;    6 RESPONSES
             0.9300            0.8300            0.8400            0.5800            1.0600
             0.8600
                DOSE NO.   2  =      1.2000;    6 RESPONSES
             1.1200            1.1500            0.9800            0.7400            1.1200
             1.0500
                DOSE NO.   3  =      1.8000;    6 RESPONSES
             1.3800            1.2900            1.3000            1.2800            1.2800
             1.3200

       PREPARATION    2:  UNKNOWN
                DOSE NO.   1  =      4.0000;    6 RESPONSES
             0.8300            0.8700            0.9600            0.9200            0.8000
             1.0300
                DOSE NO.   2  =      6.0000;    6 RESPONSES
             1.1900            1.0800            0.9500            1.1300            1.0900
             1.0500
                DOSE NO.   3  =      9.0000;    6 RESPONSES
             1.3300            1.3300            1.3100            1.3400            1.2000
             1.3200
```

Table 5.12.2 records the metameters used for the statistical analysis. Although equation (5.2.1) is convenient for analysis on a desk calculator, the computer gains nothing from any special base of logarithms and may as well use common or natural logarithms. PARLIN can compute and use any response metameter among those defined by equation (15.6.2); this includes a choice of powers of the observed responses, but (for reasons explained in §15.6) here in the form

$$Y = 2(U^{1/2} - 1) \qquad (5.12.1)$$

instead of equation (3.9.6). One option available to the user of the program, to be adopted very rarely, is to nominate some responses for omission from the analysis. No such deletions are appropriate to the vitamin B_{12} data.

The most important results output appear as Table 5.12.3. The "activators" are a series of code numbers that guide the program through options on form of analysis and output; one point deserving mention here is that they permit a sequence of "runs" on the same data to be made with different metameters and different deletions. As a warning on gross errors, the smallest and largest responses among the data are listed. If the assay has no blocking, variance heterogeneity is examined, both by Bartlett's χ^2 and by output of mean squares for the lowest dose and for the highest dose of each preparation pooled over all prep-

```
                TABLE 5.12.2
    +------------------------------------------------+
    :  METAMETERS FOR - BIOASSAY USING PARLIN 7     :
    +------------------------------------------------+
                        =*= FOR CURRENT DATA, THIS IS RUN NO.  1 =*=

 :: METAMETER = ((RESPONSE-    0.0000)**(  0.5000)-1.0)/(  0.5000)

            TABLE OF METAMETERS FOR ANALYSIS
            --------------------------------

   PREPARATION    1:   CRYSTALLINE B12
       DOSE METAMETER NO.   1  =  -0.09691;   6 RESPONSES
         -0.0713         -0.1779         -0.1670         -0.4768         0.0591
         -0.1453
       DOSE METAMETER NO.   2  =   0.07918;   6 RESPONSES
          0.1166          0.1448         -0.0201         -0.2795         0.1166
          0.0494
       DOSE METAMETER NO.   3  =   0.25527;   6 RESPONSES
          0.3495          0.2716          0.2804          0.2627         0.2627
          0.2978

   PREPARATION    2:   UNKNOWN
       DOSE METAMETER NO.   1  =   0.60206;   6 RESPONSES
         -0.1779         -0.1345         -0.0404         -0.0817        -0.2111
          0.0298
       DOSE METAMETER NO.   2  =   0.77815;   6 RESPONSES
          0.1817          0.0785         -0.0506          0.1260         0.0881
          0.0494
       DOSE METAMETER NO.   3  =   0.95424;   6 RESPONSES
          0.3065          0.3065          0.2891          0.3152         0.1909
          0.2978

 NO RESPONSE METAMETERS HAVE BEEN DELETED
```

arations. Either of these pieces of information may draw attention to heteroscedasticity, especially of the kind that arises from a simple dependence of variance on expected response. Both here indicate very marked heterogeneity, the metameter adopted in the light of §3.9 apparently overcorrecting and producing an inverse relation between variance and expected response. This point did not emerge in §5.2; it casts doubts on the optimality of the analysis, but because of the symmetry of the assay in respect of dose intervals and numbers of subjects per dose, it is unlikely to have serious consequences. After summarized information on the error mean square, the regression coefficient and g, the program lists the potency of each test preparation relative to the standard, with limits at probabilities 0·95 and 0·99. The small numerical differences between this potency estimate and its 0·95 limits and those in §5.2 are solely due to the lesser numerical accuracy in $u^{1/2}$ and other steps of §5.2. Analysis of untransformed responses gives a potency estimate 0·2134, with limits at 0·1864, 0·2457 and 0·1775, 0·2593, differing only to a negligible extent.

One optional part of the output contains the details of analysis shown as Table 5.12.4. In routine assays, the analysis of variance will not be wanted every time; because of the difference between equations (3.9.6) and (5.12.1), mean squares in Table 5.12.4 are four times the corresponding entries in Table 5.2.2 (not exactly, since Table 5.2.1 contained response metameters only to 2 decimal places). Mean dose and response metameters for each preparation are available in case the user of the program has need to plot an exact dose–response diagram. The user can have a table summarizing information for each dose in the assay at

§ 5.12 SYMMETRIC DOSE-STRUCTURE

TABLE 5.12.3

```
+------------------------------------------------+
:  RESULTS FROM  -  BIOASSAY USING PARLIN 7      :
+------------------------------------------------+
                   =*= FOR CURRENT DATA, THIS IS RUN NO.   1 =*=

:: METAMETER = ((RESPONSE-     0.0000)**(  0.5000)-1.0)/(  0.5000)

ACTIVATORS USED WERE:
   KA1 = 3, KA2 = 2, KA3 = 0, KA4 = 3, KA5 = 3, KA6 =105, KA7 = 0

   MINIMUM RESPONSE METAMETER ANALYZED =    -0.4768
   MAXIMUM RESPONSE METAMETER ANALYZED =     0.3495
EXAMINATION OF VARIANCE HETEROGENEITY:
   BARTLETT KI-SQUARED =  16.8884, WITH       5 D.F.

   AVERAGE VARIANCE FOR LOWEST DOSES  =     0.019711, WITH   10 D.F.
   AVERAGE VARIANCE FOR HIGHEST DOSES =     0.001635, WITH   10 D.F.

REGRESSION COEFFICIENT:      B =       1.18919,  S.E.(B) =       0.12881
EFFECTIVE SD PER RESPONSE: S/B =       0.09344
INDEX OF REGRESSION SIGNIFICANCE: T**2*V(B)/B**2
    FOR PROBABILITY 0.95, G =   0.0489
    FOR PROBABILITY 0.99, G =   0.0887

P O T E N C Y    E S T I M A T E S    A N D    L I M I T S
-------------------------------------------------------------
PREPARATION      POTENCY       LIMITS(0.95)           LIMITS(0.99)

UNKNOWN
     2           0.2154      0.1859    0.2515      0.1762    0.2671
```

which the observed mean response metameter differs from the fitted regression line by more than a specified multiple of the standard error of the deviation, with the possibility of including every dose. If N_D subjects were tested at a dose with metameter x_D, for a preparation that had in all N_P subjects and a mean dose metameter x_P, the variance of the deviation is estimated to be

$$s^2 \left[\frac{1}{N_D} - \frac{1}{N_P} - \frac{(x_D - x_P)^2}{\Sigma S_{xx}} \right] \tag{5.12.2}$$

(Finney, 1946d). A similar tabulation for individual responses can be had. For a completely randomized design, the variance remains as in (5.12.2) with $N_D = 1$. For randomized blocks, the variance becomes

$$s^2 \left[1 - \frac{1}{N_P} - \frac{1}{N_B} + \frac{1}{N} - \frac{(x_D - x_P)^2}{\Sigma S_{xx}} \right], \tag{5.12.3}$$

where N_B is the number of responses (for all preparations) in the same block and N is the total number of responses in the assay. The table for dose means can help in assessing the evidence for curvature of the dose–response relation, or for anomalous behaviour at a particular dose; that for individual responses may detect single anomalous values.

The final option is for output of a crude dose–response diagram. By appropriate horizontal shifts, the separate regression lines are superposed so that the

TABLE 5.12.4

```
+-------------------------------------------------+
:   ANALYSIS FOR  - BIOASSAY USING PARLIN 7       :
+-------------------------------------------------+
```

=*= FOR CURRENT DATA, THIS IS RUN NO. 1 =*=

:: METAMETER = ((RESPONSE-   0.0000)**(0.5000)-1.0)/(0.5000)

ANALYSIS OF VARIANCE

SOURCE OF VARIATION	D.F.	SUM OF SQUARES	MEAN SQUARE
PREPARATIONS	1	0.013222	0.013222
REGRESSION	1	1.052416	1.052416
PARALLELISM	1	0.006078	0.006078
ALL QUADRATICS	2	0.007249	0.003624
DOSES	5	1.078965	
ERROR	30	0.370403	0.01234677
T O T A L	35	1.449368	

SUMMARY OF INTERMEDIATE CALCULATIONS

S**2 =   0.012347, WITH   30 D.F.

B =   1.189187, WITH VARIANCE FACTOR   1.343734

PREPARATION	XBAR(1)-XBAR(I)	YBAR(I)-YBAR(1)	VARIANCE FACTOR
2	-0.698970	0.038329	0.111111

PREPARATION MEANS, FOR USE IN PLOTTING

	PREPARATION STANDARD LISTED FIRST	MEAN OF X	MEAN OF Y	YBAR-B*XBAR
1	CRYSTALLINE B12	0.0792	0.0485	-0.0456
1	UNKNOWN	0.7782	0.0868	-0.8385

INDIVIDUAL RESPONSES AND DEVIATIONS

(ONLY FOR DEVIATIONS EXCEEDING 2.0 TIMES STANDARD ERROR)

DOSE	RESPONSE NO.	X	Y	EXPTN	DEVN	S.E.
STANDARD:	CRYSTALLINE B12					
1	4	-0.097	-0.477	-0.161	-0.316	0.106
	5		0.059	-0.161	0.220	0.106
2	4	0.079	-0.280	0.049	-0.328	0.108

RESPONSE MEANS AND VARIANCES

(ONLY FOR DEVIATIONS EXCEEDING 0.5 TIMES STANDARD ERROR)

DOSE	RESPONSES	X	YBAR	EXPTN	VARIANCE	DEVN	S.E.
STANDARD:	CRYSTALLINE B12						
2	6	0.079	0.021	0.049	0.025282	-0.027	0.037
3	6	0.255	0.287	0.258	0.001096	0.030	0.029
PREPARATION	2: UNKNOWN						
1	6	0.602	-0.103	-0.123	0.008061	0.020	0.029

general pattern of deviations can be better displayed. This also can give warning of curvature. Within wide limits, the size of frame for the diagram can be chosen. Of course accuracy of representation is restricted by the size of the unit type space on the printing device.

§ 5.12   SYMMETRIC DOSE-STRUCTURE   131

TABLE 5.12.5

```
+----------------------------------------------+
: DIAGRAM FOR   - BIOASSAY USING PARLIN 7      :
+----------------------------------------------+
                =*= FOR CURRENT DATA, THIS IS RUN NO.  1 =*=
:: METAMETER = ((RESPONSE-   0.0000)**(  0.5000)-1.0)/(  0.5000)
```

[Plot: RESPONSES (vertical axis) vs STAGGERED DOSES (horizontal axis), showing an approximately linear increasing series of points from lower-left to upper-right, with markers "1" at lower-left, middle, and upper-right.]

```
1 UNIT HORIZONTALLY =   0.0045 UNITS OF DOSE METAMETER
1 UNIT VERTICALLY   =   0.0084 UNITS OF RESPONSE METAMETER
```

PARLIN was first written from a keyboard terminal to EMAS, the Edinburgh multi-access computer system. It is therefore readily adapted for on-line use, with output returned directly or off-line, though the exact manner of achieving this will depend upon the multi-access system on which it is to run. During the preparation of this chapter, versions of Tables 5.12.1–5.12.5 were output at the terminal in about 30 minutes from when the author began to prepare the input file; in routine use, the time could be reduced. The tables were complete in every way, though subsequently the input was modified slightly in order to improve them for presentation here. With batch processing, the processor time will usually be only a few seconds, and the delay to the user is determined entirely by the turn-round characteristics of the computer system. Difficulty can arise because of delays in access to a computer. For example, a research investigator who makes a standard assay daily and needs the results from each day to determine his next day's work will suffer if the computer queue involves an 18-hour delay. He may be wise to invest in a small desk-top programmable calculator; he may even accustom himself to a spell of concentrated arithmetic on a daily commuting journey! Systematic computation is now so cheap, however, that all who frequently use bioassays (whether as part of the medical care of patients, as a step in the control of pharmaceutical and similar products, or as an adjunct to pharmacological research) should seek to organize their work and their access to a computer so as to analyze data by good programs with low costs and little trouble.

PARLIN can certainly be improved and extended. The limitation to a constant variance could be removed; the most natural generalization is to a variance functionally dependent upon the expected response, which increases the computational complexity because some form of iteration is required (Chapter 16). There is no difficulty in extending the program to other assay designs, such as those discussed in Chapters 9–11. Indeed, as already stated, the present version is applicable to 'multiple assays' in which several test preparations are simultaneously compared with one standard. At the time of writing, no clear demand has appeared for a program to handle particular types of incomplete blocks or the occasional missing observation that destroys orthogonality; an entirely general program would require more cumbersome input and would burden the performance of simple standard analyses with the overheads of seldom needed facilities.

Another desirable development is improved monitoring of the data in order to draw attention to anomalous responses (possibly consequences of error in data preparation, copying mistakes in recording, or unsuspected irregularities of experimentation), and to systematic or erratic differences between variance estimates at different doses. The relevant parts of Tables 5.12.3–5.12.5 are but a crude attempt to make the computer assume some of the duties of organized scanning of data that have properly been regarded as the responsibility of statisticians. Ordinarily, information of this kind should appear in the output only if warning indicators surpass some agreed thresholds.

The present version of PARLIN is written in FORTRAN IV. Its 1400 instructions occupy about 102K bytes of core store, and the program makes no use of any other storage device.

6
Principles of planning an assay

6.1 The purpose of an assay

Despite the emphasis that must be placed on techniques for calculating tests of significance, standard errors, and fiducial limits, nothing must be allowed to obscure the primary objective of every biological assay:

(i) To estimate the potency of the test preparation. Usually the potency will be assessed relative to a standard, but an example of estimation in absolute terms is discussed in § 20.5. The estimate itself is of little use unless the investigator is able:

(ii) To ensure adequate precision for the potency estimate. In other words, the deviation of the estimate from the true value must be almost certainly too small to affect any action based upon the assay. Vague assurance is seldom sufficient to give an investigator confidence in the precision of his assay; he must plan so as

(iii) To obtain a numerical assessment of the precision of the potency estimate.

A valid estimate needs to be more than merely the result of applying a formula to a set of observations; it must also be correctly interpretable as a property of the test preparation. Unless the investigator's belief in the validity of his assay *a priori* is so strong as to make the precaution unnecessary, he will wish

(iv) To include tests of the fundamental validity of the assay and of the validity of the statistical analysis used in calculating the potency estimate.

The implications are discussed more thoroughly in Chapter 15.

6.2 Validity

An assay is valid only if it leads to a consistent estimate of the parameter ρ. 'Consistency' here has the strict statistical sense of convergence in probability (§ 3.12; Cramér, 1946; Kendall and Stuart, 1973): if the size of the assay were increased by adding more and more subjects at each dose, the estimate, R, would, in the language of mathematical statistics, 'almost certainly' or 'with probability approaching unity' approach the true value, ρ, as a limit.

In an analytic dilution assay, T is identical with S except for a different amount of inert diluent. Any reasonable method of estimating equally effective doses will lead to a valid estimate of ρ, and the choice may be made entirely from considerations of precision and economy. If the conditions for an analytic assay are in doubt – perhaps because T contains impurities or other constituents that may affect the response – the assay should be so designed as to permit detection of disturbances. No statistical analysis can uncover the action of constituents of T that produce responses related to dose in the same way as for the substance to be estimated. Wood (1946a) wrote (of slope ratio assays): 'If, for example, in a riboflavine assay the test preparation should happen to contain not only riboflavine, but also some other growth stimulating factor, and if this factor stimulated growth proportionally to the dosage at all dosage levels, no statistical test and no method of calculating the result could possibly detect anything suspicious in the result obtained. The combined riboflavine and other factor would be estimated as riboflavine.' Validity tests, however, can give some chance of detecting disturbances in the character of the response; for parallel line assays, the test of parallelism is particularly important. In a comparative dilution assay, for which the effective constituents of the two preparations are not chemically identical, a test of deviations from the hypothesis of similarity is an essential check on the validity of all conclusions.

Most methods of statistical analysis of assays depend upon representation of the dose–response relation (in metametric form) by a simple algebraic equation, usually a straight line. Though an assay be fundamentally valid, the statistical analysis will be invalid if the data show this representation to be wrong. For many of the commoner assay techniques, experience has taught the form of equation that is a reasonably adequate representation of the truth; even then, linearity or similar validity tests are valuable safeguards against the misuse of data from an assay with an unsuspected abnormality. If sufficient experience of an assay technique is lacking, validity tests must be incorporated into the design. Small deviations from an ideal relation may have no great effect, but serious discrepancies are a warning that a better specification of the model must be sought, and statistical analysis based upon that, before valid inferences can be drawn.

6.3 The economics of design

By the design of an assay is meant the set of doses to be tested, the rules for allocating subjects to doses, the arrangement of all other experimental constraints such as litter-mate control, the order in which doses are given or measurements made, the subdivision of labour between collaborating observers, and other matters relating to logical structure rather than to technical details of experimentation. Choice of a design involves compromise between the conflicting interests of the precision of estimation, the quality of validity tests, and the convenience, simplicity, and inexpensiveness of the experimental and statistical procedures; these last may be said to relate to the *cost* of the assay, whether cost be measured in monetary units or on any other scale that allows the total expenditures of time, labour, and materials to be compounded into a single figure. The investigator will wish either to obtain the best possible results, in respect of precision and power of validity tests, for a specified cost, or to minimize the cost of obtaining specified precision and power. The two requirements, different aspects of the same conditions for optimality, are never fulfilled perfectly. Nevertheless, useful general indications can be based upon even a slight previous knowledge of the relevant quantities.

Full analysis of the economics for all the complicated types of design that can be contrived is impracticable, but consideration of the principles governing choice of the number and spacing of doses for the simplest types of assay gives useful guidance in other situations. For a parallel line assay, the demands that, at a given cost, both the most reliable estimate of ρ and an assessment of its precision shall be obtained are compatible (for a slope ratio assay this is not true: § 8.1). On the other hand, few assays can provide adequate validity tests unless the design be of less than maximal efficiency in respect of estimating ρ. A dilution assay will rarely be planned without prior belief in the condition of similarity. Under this condition and with a prodigal expenditure of time, labour, and materials, any desired reliability of estimation coupled with any desired power of validity tests can be obtained. In practice, limited resources require that the investigator choose dose levels and numbers of subjects at each dose in relation to any pre-existing knowledge of relative potency and of the appropriate metametric transformations, and to his degree of certainty in respect of monotony and similarity. If previous experience makes him almost certain of his choice of metameters, of the range of doses over which a linear regression obtains, and of similarity, he need not reduce his precision by insistence on powerful tests of validity. If he lacks this certainty, some sacrifice of precision is necessary, but he must not lose sight of his primary object.

6.4 Pilot investigations

In many types of statistical inquiry, the optimal design cannot be determined unless the answer is already known. The conditions for maximum precision of a bioassay themselves involve ρ. An assay is seldom undertaken without some knowledge of the potency of T, and even a very little information can greatly help the planning. If absolutely nothing is known about ρ, as in the first assay of material from a new source, the investigator should begin with a small *pilot assay*, in which a very few responses suffice to indicate the order of magnitude of ρ. He should choose doses so as to explore the widest conceivable range for the potency: even if ρ is located only within a ten-fold range on either side, plans for the main assay will be greatly assisted.

In general, data from the pilot assay will not be combined with those from the main assay, and so will not contribute directly to the precision of the potency estimate. What then is the best division of a fixed total number of subjects, N, between the pilot and the main assays? No simple answer can be given, though something between $N/10$ and $N/4$ is suggested as a working rule. If the experimental technique allows each response to be measured before treatment of the next subject begins, sequential methods of statistical analysis may facilitate combining the two stages, so that the early tests in the main assay are used as guides to progressive modifications in design. Staircase methods for quantal responses (§ 19.6) are of this kind, and analogous procedures for quantitative responses might be developed.

6.5 Symmetry

Other things being equal, a symmetric design ought always to be chosen. For parallel line assays, the basic pattern is that of the symmetric (k, k) point, k doses of each preparation equally spaced on the log scale and with equal numbers of responses measured at each dose. On this may be imposed complications appropriate to special needs. For example, an expectation of large differences between litters or other natural groupings of the subjects can be accommodated by block restrictions; the simplest form is that of the randomized block design, in which group differences are balanced over doses and so do not affect the precision of the potency estimate. If tests must be made on two or more days, or by several observers, again balance may be secured by the use of randomized blocks. Limitations of resources may be an obstacle to complete symmetry, and compel the use of, say, a balanced incomplete block design (§§ 9.5–9.7). When symmetry is unachievable with available resources there is no objection to an unsymmetric design; the oestrone assay in § 4.21, for example, has a legitimate design, but the computations are more laborious than if the gaps had been filled. On the other hand, unnecessary departures from symmetry usually bring some disadvantages; a small gain in precision or in one validity test may compensate for a loss on something else, but much the same flexibility comes by choice of k. The vitamin D_3 assay in § 4.2 was poor in design; because N_S and N_T were nearly equal and, in spite of badly chosen doses, the mean difference in log dose was almost equal to M, its flaws were not disastrous.

Symmetric designs are usually the easiest to use and are certainly the easiest for statistical analysis. Of course, accidents during an assay may convert a symmetric design into an unsymmetric.

6.6 The cost of statistical analysis

The labour and cost of statistical analysis is generally only a small part of the total expenditure on an assay; the marginal savings in cost of analysis that might be effected by modifications in design are seldom important. This will always be true of an assay carried out as a research project, in which the cost of securing and measuring each response will be high: every effort should be made to design the assay so that each response is as useful as possible, and considerations of computational labour should not prevent a fully efficient analysis. Even in routine assays of a preparation manufactured in large batches, in which the whole assay technique may be simplified and 'streamlined', suggestions for omitting some parts of the statistical analysis should be viewed with suspicion lest they be false economies. The amount of arithmetic required in such a routine assay can be made quite small by careful arrangement of the calculations (§§ 5.4–5.8). Such devices as estimating standard deviations from ranges have been popular, but today the low cost of computation casts doubt on the merits of adopting an inefficient and non-standard scheme of calculation in order to reduce the arithmetic.

A biochemist, pharmacologist, or microbiologist whose own statistical expertise is small will perhaps object to some of the designs in later chapters:

however theoretically efficient they may be, they are valueless to him, because when he had obtained the data he would have no idea how to analyze them. This difficulty illustrates the need for close collaboration between the experimental scientist and the statistician. To reject a good design for such a reason would be short-sighted; the right policy is surely either to learn how to analyze the data or to obtain assistance from a professional statistician. Statistical science is one of the precision instruments available to the experimenter, who, if he is to make proper use of the knowledge at his disposal, must either learn to handle it himself or find someone else to do so for him. Many investigators will put themselves to great trouble in acquiring skill with a difficult biological or chemical technique, yet deny themselves the benefits of statistical techniques that they consider beyond their understanding. They are the losers, even though the fault may lie in part with statisticians who fail to make their methods sufficiently clear to the non-mathematician. However restricted his mathematical experience, any scientist should be able to appreciate the principles of experimental design and to learn the basic techniques of statistical analysis; on more abstruse matters, he can then be content to accept help from a statistician.

A more serious problem arises when the experimental design that appears theoretically ideal for a certain purpose is so complex in its structure that its execution in the laboratory would be technically difficult or attended by the risk of frequent mistakes. This may happen if a design requires a complicated sequence of doses to be applied to each subject, with different sequences for different subjects. Mistakes in dosing could destroy the intended symmetry; even if the experimental programme were carried out faultlessly, the precision might be much less than was expected because of the difficulties of running the experiment. In framing his advice, the statistician needs to remember that a simple design can give better results than one for which complexity defeats its own ends.

6.7 Randomization

In all subsequent discussion, every choice in the selection of a particular configuration for a design, or in the allocation of subjects to a place in the design, will be assumed to be made at random unless there is explicit statement to the contrary. In the assay of § 5.2, the 36 tubes would be divided into six sets of six entirely at random. In a randomized block design that uses litters as blocks, one member of each litter is selected at random for the first dose, a second member at random for the second dose, and so on. If the design is based upon a Latin square, that square will be selected at random from all squares of the same size, or at least from all of a particular transformation set (Fisher and Yates, 1963). Each randomization should be made independently of others in the same assay; each of a series of assays of the same design should have its own randomization. By randomization is always meant a strict process of selection by lot, using a table of random numbers or equivalent machinery. Haphazard selection according to the whim of the investigator must never be regarded as randomization. The order in which the subjects receive their doses should also be random. If all subjects at one dose are treated in succession, a correlation between their responses may be introduced, and may pass undetected even though it affects the true experimental errors. To give doses in random order, however, may introduce a risk of gross blunders, and some discretion must be exercised in the enforcement of this requirement (cf. § 11.3).

The reasons for randomization are not peculiar to biological assay, but are fundamental to the probabilistic interpretation of statistical data (Fisher, 1966). Any departure from randomization may seriously affect the validity of experimental comparisons. Emmens (1948, § 6.1) quotes an instance of selecting mice from a cage; an assistant omitted strict randomization and, as a result, the weights of the mice and the order in which they were taken from the cage were significantly correlated. A selection of insects for tests of insecticides which allows the more active individuals in a culture a greater chance of appearance in the first batches may produce heterogeneity of sex-ratio, or of age distribution, between batches (§ 4.20). Deliberate omission of a randomization is equivalent to an assertion that measurements will be independent of the systematic element introduced, or at least that any correlation is negligible in comparison with the experimental errors. Some assayists place replicate tubes for microbiological assays adjacently in the incubator, on the assumption that position does not affect bacterial growth. Better procedures are to number the tubes for identification and then to randomize, or to immerse tubes in a water-bath with forced circulation providing more accurate thermostatic control than is possible when they are surrounded by air. Wood (1950, personal communication) pointed out that to move the tubes round the incubator in a revolving holder enables local environmental differences to be balanced even in an incubator of poor thermostatic quality. Responsibility for a decision to omit a randomization rests with the investigator: the data rarely permit the statistician to check the assumption that no bias is introduced, though, without strong supporting evidence from previous experiments, his experience will make him sceptical. Romani, Robertson and Diczfalusy (1976) illustrate how heterogeneity between estimates of the same potency in replicate assays was a feature of their assays of luteinizing hormone, until firm insistence on randomization at various stages caused its disappearance.

The agar-plate technique for assaying antibiotics can illustrate the consequences of a failure to randomize. In an assay such as that discussed in § 5.4, the circles are placed symmetrically on the plate at angles of $0°, 90°, 180°, 270°$ from a fixed radius. If the four dose levels were assigned to the four positions on a plate entirely randomly and independently for each plate, the contrasts L_p, L_1, L_1' would necessarily have equal variances. An investigator tempted by the convenience of standardization, however, might always take the order round the plate as S_1, T_1, S_2, T_2. The fact that S_1, S_2 and T_1, T_2 were always further apart than other pairs of circles would make the mean squares derived from L_p, L_1, L_1' have expectations $\sigma^2(1 + \phi_2 - 2\phi_1)$, $\sigma^2(1 - \phi_2)$, $\sigma^2(1 - \phi_2)$ respectively, where ϕ_1 and ϕ_2 are the correlation coefficients between responses for adjacent circles and between responses for non-adjacent circles. Since ϕ_1 would normally exceed ϕ_2, the variance of L_p would be less than that of L_1 or L_1'. The variances would have to be estimated separately, using contrasts from each plate as in Table 5.4.2; validity tests and fiducial limits would have to make allowance for this complication, the latter by use of the first analogue of Fieller's theorem. Because of the reduced numbers of d.f. for estimates of variance, even with variances as small as are usually found for penicillin responses, an assay sufficiently precise for practical purposes would scarcely be obtained with less than eight plates.

In a series of penicillin assays, of design similar to that used by de Beer and Sherwood (§ 5.4), Knudsen and Randall (1945) found evidence of heterogeneity of variability in the contrasts L_p, L_1. The order of doses was systematic,[*] S_1, S_2, T_2, T_1 on each plate, for which the variances of L_p, L_1, L'_1 must be $\sigma^2(1-\phi_2)$, $\sigma^2(1-\phi_2)$, $\sigma^2(1+\phi_2-2\phi_1)$ respectively. Knudsen and Randall used range estimates of standard deviations (§ 5.8), and showed clearly that the range for L_1 exceeded that for L_p, but lack of randomization cannot be responsible for this difference. Non-normality of the distribution of responses may have caused inequality of the mean ranges even though the mean variances must be equal – not a very plausible explanation! Examination of L'_1, a contrast not discussed by the authors, might give some clue to what happened (see also § 14.4). Enough has been said to emphasize the dangers of assuming that randomization is an unimportant refinement or an unnecessary complication.

6.8 Parallel line assays

The remainder of this chapter is concerned only with parallel line assays. The optimal design for a projected assay depends upon existing knowledge about S and T, and any rules formulated must permit modification to meet special circumstances. Nevertheless, a common situation, leading to a useful guide, is that

(a) past experience (or preliminary study of responses to S) has shown a linear regression on log dose to be a satisfactory approximation over a moderately wide range of doses;

(b) experience or study has indicated roughly the magnitudes of the regression coefficient and the standard deviation of responses about the regression (though any change in the source of the subjects or the conditions of experimentation may affect these quantities);

(c) from a pilot assay, or any other source of information, a rough estimate, R_0, of ρ is available.

Sheps and Hendrie (1958) prepared a table to assist the planning of an assay that conforms to prerequisites of precision, but the advice below is somewhat more general.

The assayist usually wishes to know how best to deploy a total of N subjects. A major consideration is the error of estimation: to be useful, an assay must give a high probability that R is close to the true value ρ. From (4.14.2), the quarter-square of the fiducial interval for M is

---

[*] L.F. Knudsen (personal communication).

$$I = \frac{t^2 s^2}{b^2 (1-g)^2} \left[(1-g) \left(\frac{1}{N_S} + \frac{1}{N_T} \right) + \frac{(M - \bar{x}_S + \bar{x}_T)^2}{\Sigma S_{xx}} \right]. \tag{6.8.1}$$

The aim is to keep I small, with the restriction that $N_S + N_T = N$ (Bliss, 1950). An ideal design will therefore be so chosen that

- (i) t is small;
- (ii) s is small;
- (iii) b is large;
- (iv) g is small;
- (v) $\left(\frac{1}{N_S} + \frac{1}{N_T} \right)$ is small;
- (vi) $(M - \bar{x}_S + \bar{x}_T)$ is small;
- (vii) ΣS_{xx} is large.

This list is formidable, but not all items are independent.

(i) If the probability level for the fiducial limits is supposed fixed by external considerations, the value of t is determined by the number of degrees of freedom for s^2. A minimum of 10 d.f. for t is desirable, but increase beyond that is seldom important by comparison with the other factors. If the N subjects are arranged in a (k, k) design without further constraints, t has $(N - 2k)$ d.f.; if the subjects are divided into r blocks, this is reduced to $(N - 2k - r + 1)$. Thus an assay with N less than 20 is rarely satisfactory.

(ii) The variance, estimated by s^2, depends upon homogeneity of subjects, good choice of design, and care in the conduct of the assay. Use of a highly inbred line of animals can be undesirable, because of reduced vigour, and possibly increased variability in some physiological responses, and F_1 hybrids of inbred lines may give lower values of s^2 (McLaren and Michie, 1954; Biggers and Claringbold, 1954). Control of environment and nutrition, before and during an assay, and accuracy in experimental technique are important. Blocks of litter-mates (§ 6.7, Chapter 9) and other constraints may eliminate irrelevant variability.

(iii) The regression coefficient, b, is the increase in response per unit increase in dose, in terms of the metameters; it measures the responsiveness of the subjects to changes in dose. Any breeding of subjects for assays should aim at high values of b as well as low values of s.

(iv) Since

$$g = \frac{t^2 s^2}{b^2 \Sigma S_{xx}},$$

all steps based upon headings (i), (ii), (iii) and (vii) also benefit g. Whereas the width of the fiducial interval is directly proportional to s, t, and $1/b$, for g all that is necessary is a reasonably small value, say 0·05, and further reduction below this gives a negligible return. In a well-planned assay, the second term in (6.8.1) is small by comparison with the first, and I is therefore inversely proportional to $(1-g)$; for small g, the width of the fiducial interval is approximately proportional to $(1+\tfrac{1}{2}g)$. If the second term in (6.8.1) is not small, the factor of proportionality increases but cannot exceed $1/(1-g)$.

(v) For a fixed N, $\left(\dfrac{1}{N_S}+\dfrac{1}{N_T}\right)$ has its minimum at $4/N$ when $N_S = N_T = \tfrac{1}{2}N$; subjects should therefore be divided equally between S and T unless there are strong reasons to the contrary. A small deviation from equality will cause a trivial loss, but, except in the simplest designs, will not easily fit in with the structure and will increase the labour of statistical analysis. The design of the oestrone assay in § 4.21, a good attempt to come near to symmetry with awkward material, is obviously better than if all incomplete litters had been rejected (in which case, only eight subjects could have been used), and better than if all the missing entries had occurred for the same dose; the statistical analysis (§ 4.23) is much more complicated than that in §§ 4.15–4.18. Sometimes, rejection of a few subjects superfluous to a good design may be preferable to adoption of an inferior design that uses all available subjects. Of course, these remarks relate to self-contained assays of a single test preparation. If data from previous tests of S can be used in the current assay (§ 5.10), or if several test preparations are to be assayed simultaneously (Chapter 11), the situation is different.

(vi) The aim of making $(M-\bar{x}_S+\bar{x}_T)$ small in absolute magnitude will best be achieved by first choosing doses for S and then dividing each by R_0 to give corresponding doses of T such that

$$x_T = x_S - \log R_0. \tag{6.8.2}$$

If equal numbers of subjects are assigned to corresponding doses,

$$M - \bar{x}_S + \bar{x}_T = M - \log R_0, \tag{6.8.3}$$

an expression which will not be far from zero unless R_0 is a bad estimate.

(vii) Since the largest contributions to S_{xx} come from doses far from $\bar{x}$, ΣS_{xx} will be maximized by testing all subjects at the extremes of the region of linearity. Even if this could be done satisfactorily for S, an equal range of x for T (about which less is known) would involve serious risk that doses might fall outside the region of linearity. To use unequal ranges of x for the two preparations is generally inconvenient. Instead, the dose range for S should be made a little less than the maximum, in such a way that equation (6.8.2) gives doses for T that are unlikely to be too extreme for the linear regression.

The maximum value for ΣS_{xx} will then be given by a $(2, 2)$ design with $N/4$ subjects at the highest and lowest doses of each preparation. The validity tests can be improved only at the expense of ΣS_{xx} and g; use of a symmetric $(3, 3)$ design, for example, will reduce ΣS_{xx} to $\tfrac{2}{3}$ of its maximum. If the choice of doses succeeds in making $(M - \bar{x}_S + \bar{x}_T)$ small, the second term in (6.8.1) can suffer a decrease in ΣS_{xx} without affecting I seriously, and control of g may often be effected in other ways. In some types of assay, however, the responses are so variable that any decrease in ΣS_{xx} has unfortunate consequences for g. This should not in itself be regarded as justification for the use of a $(2, 2)$ design rather than a $(3, 3)$: the large g, characteristic of the technique, is better countered by increasing N. Almost without exception, a potency estimate of low precision with a reasonable assurance of validity is preferable to one of apparently higher precision rendered entirely irrelevant by invalidity. Any contention that, in a particular assay, the precision is so low that careful attention to validity is unnecessary would be better replaced by an admission that the assay is worthless because of the inherent variability.

6.9 Symmetric (k, k) designs

If an assay is to have validity tests, some sacrifice of the optimality conditions in § 6.8 is unavoidable. This can be examined quantitatively for symmetric designs. A symmetric (k, k) design with equal numbers of subjects at doses equally spaced on the log scale has $N/2k$ subjects at each dose. If the highest and lowest doses of either preparation differ by X on the log scale, the formulae of § 5.3 have

$$\left. \begin{array}{l} d = \dfrac{X}{2(k-1)} \quad (k \text{ even}) \\ \\ d = \dfrac{X}{(k-1)} \quad (k \text{ odd}) \end{array} \right\} \tag{6.9.1}$$

The values of $(x - \bar{x})$ are those of § 5.3 multiplied by d, and therefore, for all k,

$$\Sigma S_{xx} = \frac{1}{12} NX^2(k+1)/(k-1). \tag{6.9.2}$$

Write

$$A = \frac{12t^2 s^2}{Nb^2 X^2}, \tag{6.9.3}$$

and

$$h = (M - \bar{x}_S + \bar{x}_T)/X. \tag{6.9.4}$$

Then simple algebra shows that

$$g = A(k-1)/(k+1) \tag{6.9.5}$$

and, from (6.8.1),

$$I = \frac{AX^2(k+1)[k+1-(k-1)(A-3h^2)]}{3[k+1-A(k-1)]^2}. \tag{6.9.6}$$

Note that A and h have been defined so as not to involve k; h is the deviation of M from the approximation used in designing the assay, expressed as a fraction of the dose range X. When A is small enough for terms in A^2 or higher powers to be

§ 6.9   PRINCIPLES OF PLANNING AN ASSAY

TABLE 6.9.1   Percentage precision of the symmetric (k, k) design relative to the optimal $(2, 2)$; (g is assumed small)

Values of k	Values of $|h|$						
	0·0	0·1	0·2	0·3	0·5	1·0	1·5
2	100	99	96	92	80	50	31
3	100	99	94	88	73	40	23
4	100	98	93	86	69	36	20
5	100	98	93	85	67	33	18
10	100	98	91	82	62	29	15
Limit	100	97	89	79	57	25	13

TABLE 6.9.2   Percentage reliability of the symmetric (k, k) design relative to the optimal $(2, 2)$; $(A = 0.25)$

Values of k	Values of $|h|$						
	0·0	0·1	0·2	0·3	0·5	1·0	1·5
2	100	99	96	91	79	48	29
3	95	94	89	83	67	35	20
4	93	91	85	78	61	30	16
5	91	89	83	75	57	27	14
10	87	84	77	68	49	21	11
Limit	82	79	71	60	41	16	8

TABLE 6.9.3   Percentage reliability of the symmetric (k, k) design relative to the optimal $(2, 2)$; $(A = 0.5)$

Values of k	Values of $|h|$						
	0·0	0·1	0·2	0·3	0·5	1·0	1·5
2	100	99	95	90	77	45	27
3	90	88	83	76	60	30	16
4	84	82	76	68	51	24	12
5	80	78	71	63	46	20	10
10	71	68	61	52	35	14	7
Limit	60	57	48	39	24	9	4

TABLE 6.9.4   Percentage reliability of the symmetric (k, k) design relative to the optimal $(2, 2)$; $(A = 1.0)$

Values of k	Values of $|h|$						
	0·0	0·1	0·2	0·3	0·5	1·0	1·5
2	100	99	94	88	73	40	23
3	75	73	67	59	43	19	10
4	60	57	51	43	28	11	5
5	50	47	40	32	20	7	3
10	27	24	18	12	6	2	1
Limit	0	0	0	0	0	0	0

neglected, this is equivalent to the approximation

$$\text{Var}(M) = \frac{4s^2}{Nb^2}\left[1 + 3h^2\left(\frac{k-1}{k+1}\right)\right]. \tag{6.9.7}$$

For any fixed A and h, I increases as k increases; therefore, $k = 2$ minimizes I (since $k = 1$ is not allowable). The quality of any other assay relative to the optimal with $k = 2$, $h = 0$ may be represented by the *relative reliability*, the inverse ratio of values of I. Provided that both designs are large enough for t to be practically independent of its degrees of freedom, this is

$$\text{Rel} = \frac{3[(k+1) - A(k-1)]^2}{(3-A)(k+1)[(k+1) - (A - 3h^2)(k-1)]}. \tag{6.9.8}$$

This definition (Finney, 1947a) is not entirely satisfying, as the value depends (through t) on the probability level chosen for the fiducial limits, but by consideration of various values of A some idea can be gained of how great an effect a change in k has on the estimation of ρ. When A is sufficiently small for $\text{Var}(M)$ to be used, (6.9.8) simplifies to the ratio of asymptotic variances:

$$\text{Rel} = \frac{k+1}{k+1+3h^2(k-1)}, \tag{6.9.9}$$

a relative precision that is independent of any particular t. Table 6.9.1 shows values for the relative precision in (6.9.9), calculated for various k, h; Tables 6.9.2, 6.9.3 and 6.9.4 show the relative reliability for three typical values of A.

If h is small in absolute magnitude, the ill effects of using a large k are slight. An assay in which a bad choice of doses, attributable to poor information on R_0, causes $|h|$ to exceed 0·3 is seldom satisfactory; its analysis of variance is likely to show a significant difference between preparations, a recognized danger signal for invalidity (§ 4.7). The (3, 3) assay of vitamin B_{12} (§ 5.2) had the excellent values $h = 0\cdot085$ and $A = 0\cdot1$; nothing appreciable would have been gained by arranging the 3 subjects in a (2, 2) assay, and little would have been lost if they had been distributed between a larger number of doses. In the penicillin assay (§ 5.4), even though g was small, $h = -1\cdot1$, and equation (6.9.9) shows that the precision has been reduced to 45 percent of the maximum by the poor choice of doses: had a (3, 3) design been used, the precision would have fallen to 36 percent.

If $A < 0\cdot25$, the tables show the value of h as the most serious cause of reduced reliability; if $|h|$ is also small, the number of dose levels is not very important up to $k = 5$. For $A = 0\cdot5$, however, the fall in reliability as k increases is much more marked than for $A = 0$. If A is large, any increase in k reduces the reliability seriously even when $|h|$ is small. Although an assay of reasonable size, using subjects and responses of the low variability desirable, should not have A greater than 0·2, much larger values are sometimes encountered.

6.10 Choice of k

If the statistical validity of the analysis of an assay were in no doubt, the (2, 2) design would always be chosen. Only when experience of an assay technique has

§ 6.10   PRINCIPLES OF PLANNING AN ASSAY   145

been great, however, can statistical validity over the range of doses tested be practically certain. In general, the contention that at least 3 doses of each preparation should be used must still be upheld: the increase in fiducial range is a small price to pay for the gain of linearity tests, even though $|h|$ and A may both be large. On the other hand, assuming preliminary investigations into the character of the response curve and the choice of metameters (§ 3.5) to have been conducted, there is little to be said for using more than 3 levels. The loss in reliability as k increases makes clear the harm that results from confusion of purpose. Too often, investigators fail to distinguish between the need to study the response curve as a preliminary to the establishment of an assay procedure and the need for efficient allocation of subjects in an assay. For complex assay designs, $k = 4$ is sometimes useful; because $4 = 2^2$, more satisfactory factorial and confounding schemes can be based upon 4 than upon 3 (§ 9.9), but a larger k is rarely required (§ 4.20). The practice of using a few subjects at each of many dose levels, undoubtedly essential for a proper study of a response curve, is not appropriate to assay. The argument in §§ 6.8 and 6.9 was based upon the simplest designs, in which the subjects tested at each dose are randomly selected from all available. The essential features of the conclusions are applicable to designs of more complex character, in which block constraints are introduced. Even in the incomplete block designs described in Chapter 9, the same considerations govern the choice of k.

These points are well illustrated by consideration of modifications of the uterine weight assay for oestrone. The assay data in §§ 4.21–4.23 led to

$$s^2 = 551 \cdot 1$$
and
$$bX = 16 \cdot 00 \times 2$$
$$= 32 \cdot 00.$$

Suppose that a $(2, 2)$ assay were to be conducted with 24 rats in 6 litters of 4, the dose range being the same as in Chapter 4. The value of A might be expected to be about

$$A = \frac{12 \times (2 \cdot 131)^2 \times 551 \cdot 1}{24 \times (32 \cdot 00)^2}$$
$$= 1 \cdot 2.$$

Equation (6.9.8) shows that a $(3, 3)$ assay with the same total number of subjects would have a relative reliability of 67 percent if $|h|$ were negligible and only 34 percent if $|h|$ were as much as 0·5. This assumes that only intra-litter variance is involved, as would be true if 4 litters of 6 rats were available; if the subjects to be used were still 6 litters of 4, an incomplete block design (Chapter 9) would be necessary but might cause a further loss in reliability (§ 9.9). Thus at best the $(3, 3)$ design will have a 22 percent wider fiducial range for M, and in unfavourable circumstances the loss may be greater. This price must be paid if adequate validity tests are to be incorporated into the assay. A $(4, 4)$ design would increase the fiducial range by 46 percent at least, but such extravagance is rarely needed.

6.11 Comparison of assay techniques

Earlier sections have been concerned with the purely statistical aspects of assay design – the effects of the structure of the experiment on its efficiency as a method of assay. Implicit was the assumption that all comparisons referred to alternative designs for the same subjects, using the same manipulative techniques and making the same measurements. The only permissible variations are in the allocation of subjects to doses. Most of the factors listed in § 6.8 depend solely on structure, and not on the choice of a response for measurement; s, b, and g, being functions of the particular responses measured, may be altered by a change in the stock of subjects or in the property of the subjects that is chosen for measurement.

In a good assay, for which g and $|h|$ are both small, not only is a variance formula for M permissible but it may be written approximately

$$\text{Var}(M) = \frac{s^2}{b^2}\left(\frac{1}{N_S} + \frac{1}{N_T}\right). \tag{6.11.1}$$

Hence s/b, which is measured on the same scale as x, may be regarded as an effective standard deviation per response in respect of the measure of log potency; equation (6.11.1) has the form of a variance of the difference between means of N_S and N_T observations. Alternative experimental techniques for the estimation of the potency of a particular stimulus may therefore be compared in terms of the magnitudes of s/b. The physical dimensions of this quantity are those of a log potency, and before values from different sources are compared they must be converted to the same base of logarithms (10 or other) for their dose metameters. Apart from complications caused by g and h, the number of subjects required in order to secure a specified precision of estimation is proportional to the square of s/b. Choice of technique cannot be based entirely on this quantity, as unless the range of linearity of response, X, is reasonably large, the variance formula (6.11.1) cannot be used, but at least it is a major factor to be considered. Bliss and Cattell (1943) summarized values of s/b (base 10) ranging from 0·03 to 0·5 in a series of 45 parallel line assays for various drugs; Jones (1945) reported two long series of values obtained in routine tests of vitamin D using line test and X-ray methods. Somers (1950) discussed alternative techniques for the assay of thyroid preparations in terms of values of s/b.

The costs of alternative techniques must be considered in conjunction with their precisions. An assay using cats may have a lower value of s/b than a similar technique using rats, but against this must be balanced the higher cost of a cat. Again, for any one species of subject, the stimulus may have a general effect on many measurable characteristics as well as a specific effect on a particular organ. A response specific to the stimulus may carry less risk of invalidity because its magnitude is less likely to be modified by other constituents of the test preparation. It is therefore likely to show greater responsiveness to change in dose, and so to give a smaller value of s/b, but a general body measurement can be made without destruction of the subject. The experimenter must choose the species and characteristics that are 'best' in the widest sense, balancing the more

§ 6.11 PRINCIPLES OF PLANNING AN ASSAY

costly animal or the more laborious measurement against its possibly higher inherent precision. In some circumstances, human volunteers can be used as subjects, without danger and with the advantage of coming closer to the conditions of clinical use of a drug, which may be important if there is any doubt about the analytical dilution property; Mongar (1959), Myerscough and Schild (1958), and Schild (1959) have given examples. Every new measurement or assay technique should have its own study of the dose–response relation, although often the metametric transformation can have the simplicity of equation (3.13.27).

Sometimes several different measures of response to each dose can be measured on each subject, so as to specify a multivariate response $y_1, y_2, y_3, \ldots$. The possibility of improving assay precision by a combination of these involves the construction of a *discriminant function* (Chapter 13).

7
Slope ratio assays

7.1 The power dose metameter

The assays in Chapters 4 to 6 used a logarithmic dose metameter, corresponding to $\lambda = 0$ in §3.7. The alternative form of regression with $\lambda \neq 0$ must now be considered. As in §3.7, this requires the dose metameter

$$x = z^\lambda, \qquad (7.1.1)$$

in terms of which the regression lines for the two preparations are

$$\left.\begin{array}{l} Y_S = \alpha + \beta x, \\ Y_T = \alpha + \beta \rho^\lambda x. \end{array}\right\} \qquad (7.1.2)$$

These equations represent two lines, of slopes β and $\beta \rho^\lambda$, intersecting at $x = 0$: obviously the expected response to zero dose must be the same for both preparations. If b_S, b_T are estimates of the regression coefficients,

$$R = \left(\frac{b_T}{b_S}\right)^{1/\lambda} \qquad (7.1.3)$$

is an estimate of the relative potency, whence comes the name *slope ratio assay*. Throughout this book, $\lambda = 1$ will be assumed, because in all current applications, it appears to be an adequate approximation to the truth. The restriction is not very serious. Results appropriate to any other λ can be derived by applying the same methods to give estimates of R^λ and its fiducial limits, and then raising each to the power $1/\lambda$. If λ itself had to be estimated for each assay, instead of being regarded as a part of the definition of the dose metameter with a value determined from preliminary investigations, the analysis would be more complicated. The method of calculation for that problem will not be described here; rarely would data from one assay suffice to estimate λ with reasonable precision.

Though their discussion did not explicitly recognize the nature of the analysis, Birch and Harris (1934) appear to have been the first to publish a slope ratio assay.[*] They found the duration of cure of bradycardia in vitamin B_1 deficient rats to be directly proportional to dose of vitamin. They therefore estimated the potency of a test preparation by adjusting dose scales until its response curve coincided with that for the standard, essentially the procedure now used but performed graphically rather than arithmetically.

Equations (7.1.2) were first systematically discussed for vitamins (Burn *et al.*,

---

[*] Mr L.E. Hudson has told me that, as early as 1908, W.S. Gosset (better known as 'Student') was using analogous methods with $\lambda = 2$ for estimating the effect of hops on the life of beer.

§ 7.5 SLOPE RATIO ASSAYS

three mean squares might have been formed by pooling in sets of 3 d.f. from low responses, high responses, and the remainder.

The sum of squares for deviations from the linear regression equation gives a composite validity test relating to linearity of the regressions and intersection of the two lines in a point estimated by the blanks. Because this sum of squares does not subdivide into components as readily as does the corresponding sum of squares in a parallel line assay, users of an assay such as that under discussion often rest content with the composite test. In the present instance, the ratio of the mean square to that for error is 2·4, as compared with 3·4 for probability 0·05 (Appendix Table II). Though not significant, the ratio is sufficiently large to arouse the suspicion that at least one large component may be concealed in the sum of squares for deviations, and a full analysis should be undertaken.

As for parallel line assays, linearity is a requirement for statistical validity; intersection of the true regressions at $x = 0$ is a requirement for fundamental validity, deriving from the hypothesis of similarity and analogous to parallelism in parallel line assays. From the sum of squares for deviations from regression, two components, to be briefly described as 'Blanks' and 'Intersection', each with 1 d.f., may be separated. First suggested by Bliss (1947b), these were discussed more generally by Finney (1951a). One represents the deviation of the mean response to zero dose from a new version of equation (7.2.1) fitted to the data from the non-zero doses only; it provides a test of whether the equation remains valid down to zero dose. The second component depends upon the difference in the expected responses for zero dose calculated from lines fitted to the two preparations separately, and thus it tests fundamental validity (Wood, 1945). The remaining degrees of freedom comprise the curvature components for the two preparations separately, excluding zero dose.

For an unsymmetric design, computation of the squares for blanks and intersection is tedious. The easiest general procedure is to introduce a new variate, x_0, defined to have the value unity for the blanks, zero elsewhere. Calculate then a multiple linear regression on x_S, x_T, x_0: equations (7.4.3) extend to

$$\left.\begin{aligned}0{\cdot}1500 b_S - 0{\cdot}7500 b_T - 0{\cdot}1667 b_0 &= 2{\cdot}3400, \\ -0{\cdot}7500 b_S + 10{\cdot}0000 b_T - 1{\cdot}0000 b_0 &= 7{\cdot}6500, \\ -0{\cdot}1667 b_S - 1{\cdot}0000 b_T + 1{\cdot}7778 b_0 &= -8{\cdot}6667.\end{aligned}\right\} \quad (7.5.1)$$

The values of b_S, b_T are no longer those of equations (7.4.4), but the same notation is retained. The solutions of equations (7.5.1) are

$$\left.\begin{aligned}b_S &= 30{\cdot}1253, \\ b_T &= 2{\cdot}9874, \\ b_0 &= -0{\cdot}3704;\end{aligned}\right\} \quad (7.5.2)$$

note the necessity for a large number of decimal places in order to give sufficient accuracy in the next calculation. The sum of squares for the new regression is

$2\cdot3400b_S + 7\cdot6500b_T - 8\cdot6667b_0 = 96\cdot557.$

The difference between this and the sum of squares with 2 d.f. in equation (7.4.5) is the required component for blanks. That for intersection is obtained indirectly by finding the residual sum of squares after fitting two separate linear regressions to the non-zero doses. By calculation from the dose totals for the five doses of S and three of T, the sums of squares between doses are $46\cdot296$ and $8\cdot143$, with 4 d.f. and 2 d.f. respectively. Linear regressions account for

and
$$(-2 \times 6\cdot7 - 9\cdot7 + 15\cdot7 + 2 \times 18\cdot9)^2/20 = 46\cdot208$$
$$(-9\cdot7 + 15\cdot4)^2/4 = 8\cdot122$$

respectively. The residual for all types of curvature is therefore

$$(46\cdot296 - 46\cdot208) + (8\cdot143 - 8\cdot122) = 0\cdot109$$

with 4 d.f., which could be further split into quadratic, cubic, and other components. The square for intersection is now obtained by subtraction ($0\cdot278 - 0\cdot144 - 0\cdot109$) and Table 7.5.1 is completed.

TABLE 7.5.1 Complete analysis of variance for Table 7.3.1

Adjustment for mean		602·045	
Nature of variation	d.f.	Sum of squares	Mean square
Regression	2	96·412	
Blanks	1	0·145	0·145
Intersection	1	0·025	0·025
Curvature	4	0·108	0·027
Between doses	8	96·690	
Error	9	0·175	0·0194
Total	17	96·865	

The mean square for blanks is significantly greater than error – clear evidence of invalidity. Fortunately, this is only statistical invalidity, the mean response for the blanks being appreciably lower than is predicted by fitting equation (7.2.1) to the other data. Because a slight curvature at very low doses is not uncommon, Wood (1946a) suggested that, for riboflavin assays using *Lactobacillus helveticus* as test subject, a small amount of the standard preparation (say 0·03 µg per tube) might be added to the basal medium, and only quantities in excess of this regarded as experimental doses. The conventional zero dose should then be brought on to the linear portions of both response curves. For a true analytic dilution assay, this procedure seems unexceptionable and adaptable to assays of other materials. For a comparative assay, the risk of chemical or biological complications through mixing the standard and test preparations must be considered.

The practice of placing replicate tubes of a dose adjacent to one another in the incubator has been criticized in §6.7 (cf. §§4.19, 4.20): it can be one reason for

§ 7.6 SLOPE RATIO ASSAYS 155

variance heterogeneity or for the mean square for deviation from regression being larger than the error mean square. Such a flaw casts doubt on any assessment of fiducial limits of the relative potency. Similar considerations apply to titrating the acidity after incubation. To match duplicate tubes with one another for colour, instead of independently with a standard, may produce a correlation of response measurements and underestimation of the error variance. Experimenters are tempted to believe that consecutive treatments of, or consecutive measurements on, duplicate tubes give independent observations, and to ignore the possible correlation of subjective errors; the risk of biases that may pass unsuspected is not negligible. In the present assay, curvature at very low doses seems a more plausible explanation of the evidence for invalidity, since only the square for blanks is much greater than the error mean square.

7.6 Potency estimation

In view of the anomalous behaviour of the responses at zero dose (Table 7.5.1), estimation of potency from the non-zero dose levels seems desirable. Omission of the blanks makes a trivial difference to R itself, but it affects the precision; use of the blanks would give an apparently more precise but possibly misleading estimate. The natural method of computation is to proceed exactly as in §7.4 but to use only the data from the 16 tubes. As is obvious from theoretical considerations, the values of b_S, b_T are those in equations (7.5.2). Therefore

$$R = 2 \cdot 987/30 \cdot 125 \qquad (7.6.1)$$
$$= 0 \cdot 0992;$$

adjustment for the dilution then gives 496 µg per g as the estimated potency of the meat extract.

Evaluation of variances requires the inverse matrix of the coefficients in equations (7.5.1); as in §4.23, this is found to be

$$\mathbf{V} = \begin{pmatrix} 16 \cdot 9663 & 1 \cdot 51685 & 2 \cdot 44382 \\ 1 \cdot 51685 & 0 \cdot 241573 & 0 \cdot 278090 \\ 2 \cdot 44382 & 0 \cdot 278090 & 0 \cdot 948034 \end{pmatrix}. \qquad (7.6.2)$$

In routine computation, $\mathbf{V}$ would be found first, and the regression coefficients obtained from it, as

$$b_S = 2 \cdot 3400 v_{11} + 7 \cdot 6500 v_{12} - 8 \cdot 6667 v_{13}, \text{ etc.} \qquad (7.6.3)$$

Multiplication of s^2 by v_{11}, v_{22}, and v_{12} in turn gives the variances of b_S, b_T and their covariance. For s^2, there appears to be no objection to using the error mean square in Table 7.5.1, in spite of its inclusion of 1 d.f. from the blanks:

$$s^2 = 0 \cdot 0194. \qquad (7.6.4)$$

Fieller's theorem (§4.12) then gives the fiducial limits of R as

$$R_L, R_U = \left[R - \frac{gv_{12}}{v_{11}} \pm \frac{ts}{b_S}\left\{v_{22} - 2Rv_{12} + R^2 v_{11} - g\left(v_{22} - \frac{v_{12}^2}{v_{11}}\right)\right\}^{1/2}\right] \bigg/ (1-g), \qquad (7.6.5)$$

where

$$g = \frac{t^2 s^2 v_{11}}{b_S^2}. \qquad (7.6.6)$$

In a good microbiological assay, the variation between replicate tubes should be relatively much less than that between animals in a macrobiological assay. If g is negligible, fiducial limits may be based upon the variance formula

$$\text{Var}(R) = \frac{s^2}{b_S^2}[v_{22} - 2Rv_{12} + R^2 v_{11}]. \qquad (7.6.7)$$

In the example under discussion,

$$g = \frac{(2 \cdot 262)^2 \times 0 \cdot 0194 \times 16 \cdot 966}{(30 \cdot 125)^2}$$

$$= 0 \cdot 0019,$$

which is sufficiently small to be neglected. From (7.6.5), the fiducial limits are

$$R_L, R_U = \left[0 \cdot 099\,15 - 0 \cdot 000\,17 \pm \frac{2 \cdot 262}{30 \cdot 125}\{0 \cdot 0194 \times (0 \cdot 241\,57 - 0 \cdot 300\,79 + 0 \cdot 166\,79 - 0 \cdot 000\,20)\}^{1/2}\right] \bigg/ 0 \cdot 9981$$

$$= 0 \cdot 0957, 0 \cdot 1026.$$

Since the meat extract was diluted 5000-fold for use as a test preparation, it is estimated to contain 496 µg per g, with fiducial limits at 478 µg and 513 µg per g. The same result may be obtained by using equation (7.6.7) to give the SE of 7·6 µg per g to the potency estimate. Had the indications of invalidity been ignored and the data from the blanks used, R would have been obtained from equations (7.4.4); the estimate of potency would then have been 498 ± 7·3 µg per g, with limits at 482 µg and 514 µg per g. In spite of the significance of Blanks in Table 7.5.1, the difference in conclusion is clearly unimportant here.

7.7 General formulae

The general variance and covariance matrix may be written

$$\mathbf{V} = \begin{pmatrix} \dfrac{S_{x_T x_T}}{\Delta} & -\dfrac{S_{x_S x_T}}{\Delta} \\ -\dfrac{S_{x_S x_T}}{\Delta} & \dfrac{S_{x_S x_S}}{\Delta} \end{pmatrix}, \qquad (7.7.1)$$

where

$$\Delta = S_{x_S x_S} S_{x_T x_T} - (S_{x_S x_T})^2. \qquad (7.7.2)$$

The regression coefficients, solutions of equations (7.2.3), are

$$b_S = v_{11}S_{x_Sy} + v_{12}S_{x_Ty},$$
$$b_T = v_{12}S_{x_Sy} + v_{22}S_{x_Ty}. \quad (7.7.3)$$

Equation (7.2.4) gives R. If g, equation (7.6.6), is small, the variance formula

$$\text{Var}(R) = \frac{s^2}{b_S^2 \Delta}(S_{x_S x_S} + 2RS_{x_S x_T} + R^2 S_{x_T x_T}) \quad (7.7.4)$$

may be used; if g is not small, equation (7.6.5) is needed. An analysis of variance like that in Table 7.4.1 can be constructed, with the aid of formula (7.4.5) for the sum of squares that is accounted for by the regression. If the replicates are not classified into blocks, the error sum of squares can be calculated directly from the pooled variation within doses. The usual procedure of analysis of variance for the elimination of block effects would be necessary, for example, if the tubes in the assay just discussed had been arranged in the incubator in two randomized blocks of nine.

These formulae apply to any spacing of doses, to any numbers of tubes at each dose, and whether or not tests are made at zero dose. If the sum of squares for deviations from regression is to be subdivided as in Table 7.5.1, the formulae may be applied twice, once including and once excluding the blanks, so as to give the 1 d.f. for blanks as the difference between two residual sums of squares, but the procedure in §7.5 is preferable.

Important though the general formulae are, symmetric designs should be adopted whenever possible, because of their efficiency and their relative simplicity of execution and analysis.

7.8 The symmetric $(1, k, k)$-point design

A slope ratio design analogous to the (k, k) for parallel lines is the $(1, k, k)$; in its symmetric form, this has equal numbers of subjects at zero dose and at k equally spaced doses of each preparation. Without loss of generality, the scales may be so chosen that the highest dose of each preparation is unity; if on the original scales these doses are X_S, X_T, the relative potency calculated on the conventional scales must finally be multiplied by X_S/X_T. The total number of subjects,

$$N = n(2k + 1), \quad (7.8.1)$$

includes n at zero dose and n at doses $\frac{1}{k}, \frac{2}{k}, \ldots, \frac{k-1}{k}, 1$ of each preparation.

The algebra may be developed as in §5.3. If $C, S_1, S_2, \ldots, S_k, T_1, \ldots, T_k$ represent dose totals for the blanks and for the two preparations,

$$S_{x_S y} = \frac{1}{k}(S_1 + 2S_2 + \ldots + kS_k) - (k+1)Sy/(4k+2),$$
$$S_{x_T y} = \frac{1}{k}(T_1 + 2T_2 + \ldots + kT_k) - (k+1)Sy/(4k+2). \quad (7.8.2)$$

Equation (7.7.1) then becomes

$$\mathbf{V} = \frac{3k}{N(k+1)(k^2+k+1)} \begin{pmatrix} 5k^2+5k+2 & 3k(k+1) \\ 3k(k+1) & 5k^2+5k+2 \end{pmatrix}, \quad (7.8.3)$$

the factor outside the matrix being understood as multiplying each of the elements. Equations (7.7.3) give the regression coefficients. The components of the analysis of variance for blanks and intersection are given by two orthogonal contrasts, L_B and L_I, defined by

$$L_B = k(k-1)C - (2k-2)(S_1+T_1) - (2k-5)(S_2+T_2)$$
$$- (2k-8)(S_3+T_3) + \ldots + (k-1)(S_k+T_k) \quad (7.8.4)$$

for which the divisor is

$$Nk(k-1)(k^2+k+1)/(2k+1), \quad (7.8.5)$$

and

$$L_I = (2k-2)(S_1-T_1) + (2k-5)(S_2-T_2) + (2k-8)(S_3-T_3) + \ldots$$
$$- (k-1)(S_k-T_k) \quad (7.8.6)$$

for which the divisor is

$$Nk(k-1). \quad (7.8.7)$$

The sum of squares for the remaining $(2k-4)$ d.f. between doses is then found by subtraction of

$$b_S S_{x_S y} + b_T S_{x_T y}, \quad (2 \text{ d.f.}), \quad (7.4.5)$$

$$\frac{(2k+1)L_B^2}{Nk(k-1)(k^2+k+1)} \quad (1 \text{ d.f.}), \quad (7.8.8)$$

and

$$\frac{L_I^2}{Nk(k-1)} \quad (1 \text{ d.f.}) \quad (7.8.9)$$

from the complete sum of squares between doses with $2k$ d.f.; it may be further partitioned into quadratic, cubic and higher-order components for each preparation, by applying to the S totals and the T totals separately the same orthogonal coefficients as were discussed in Chapter 5.

In the absence of evidence of invalidity, the estimate and its fiducial limits are assessed by equations (7.2.4) and (7.6.5). The variances and covariance of the regression coefficients are obtained from $\mathbf{V}$ as

$$\left.\begin{aligned} \text{Var}(b_S) &= s^2 v_{11}, \\ \text{Var}(b_T) &= s^2 v_{22}, \\ \text{Cov}(b_S, b_T) &= s^2 v_{12}, \end{aligned}\right\} \quad (7.8.10)$$

whence

$$g = \frac{3t^2 s^2 k(5k^2+5k+2)}{N b_S^2 (k+1)(k^2+k+1)}. \quad (7.8.11)$$

§ 7.10   SLOPE RATIO ASSAYS

The formula for Var(R) and the fiducial limits can be written in terms of k, but they do not take any particularly simple form and the general expressions are more easily remembered.

If the assay is statistically invalid because of the significance of L_B, the data for the blanks may be rejected; the remainder may be treated as a $(0, k, k)$ design, with a new analysis as described in § 7.12, remembering that the total number of subjects is now only $2nk$ or $2Nk/(2k+1)$.

7.9 The (1, 1, 1) assay

The simplest special case, the $(1, 1, 1)$ design, is even less satisfactory than is the $(2, 2)$ parallel line assay. Although it provides no validity tests of any kind, it is of interest as a standard of comparison, since, if the experimenter were certain of validity *a priori*, it would lead to the most reliable estimate of potency. In any $(1, 1, 1)$ assay, the regression lines are obtained by joining the points representing the mean S and T responses to the point for the blanks, so giving a perfect fit of the data to the regression equations; no degrees of freedom remain for tests of statistical or fundamental validity.

In the symmetric design, $N/3$ subjects are tested at zero dose and at unit dose of S, T. If the totals of the responses at the three doses are C, S, T, the general equations of § 7.8 agree that

$$b_S = 3(S_1 - C)/N,$$
$$b_T = 3(T_1 - C)/N,$$

and

$$R = (T_1 - C)/(S_1 - C). \qquad (7.9.2)$$

(7.9.1)

Moreover,

$$g = \frac{6t^2 s^2}{Nb_S^2}$$
$$= \frac{2Nt^2 s^2}{3(S_1 - C)^2}. \qquad (7.9.3)$$

The fiducial limits might be expressed directly in terms of the totals C, S_1, T_1, but they are more conveniently written

$$R_L, R_U = \left[R - \tfrac{1}{2}g \pm \frac{t}{b_S} \left\{ \frac{3s^2}{2N}(4 - 4R + 4R^2 - 3g) \right\}^{1/2} \right] \bigg/ (1-g). \qquad (7.9.4)$$

When g is small, a satisfactory approximation is

$$\text{Var}(R) = \frac{6s^2(1 - R + R^2)}{Nb_S^2}. \qquad (7.9.5)$$

7.10 The (1, 2, 2) assay

The $(1, 2, 2)$ design is usually preferable to the $(1, 1, 1)$, though its advantages are achieved at the cost of a reduction in relative reliability (Chapter 8). The symmetric form is perhaps the most useful of all slope ratio designs. To each of the five doses (zero and $\frac{1}{2}$, 1 unit of each preparation) $N/5$ subjects are assigned. Equation (7.8.3) becomes

$$\mathbf{V} = \frac{4}{7N} \begin{pmatrix} 16 & 9 \\ 9 & 16 \end{pmatrix}. \qquad (7.10.1)$$

The regression coefficients can be calculated from equations (7.7.3) and the corresponding sum of squares from (7.4.5).

Alternatively, the calculations may be made directly from contrasts between responses. This is analogous to the procedure in Chapter 5, but the contrasts for b_S and b_T are not mutually orthogonal (§ 4.16); consequently, the two regression coefficients cannot be made to give independent squares for the analysis of variance, and use of formula (7.4.5) is unavoidable. The remaining dose contrasts can be subdivided into components orthogonal with one another and with b_S and b_T as in § 7.8. In Table 7.10.1, the first two divisors are to be used only for forming b_S and b_T, and the others only for forming the appropriate squares. The reader should verify the non-orthogonality of b_S and b_T, and the orthogonality of every other pair. The contrasts L_B and L_I, together with the regression, account for the whole of the sum of squares between doses, as will be numerically verified in Tables 7.10.2 and 7.10.3.

TABLE 7.10.1   Coefficients of regression and orthogonal contrasts for the (1, 2, 2) design

Contrast	C	S_1	S_2	T_1	T_2	Divisor
b_S	-15	1	17	-6	3	$35n/2$
b_T	-15	-6	3	1	17	$35n/2$
L_B	2	-2	1	-2	1	$14n$
L_I	0	2	-1	-2	1	$10n$

The analysis of variance analogous to Table 7.5.1 may now be completed. If examination of L_B, L_I and any other relevant tests does not indicate invalidity, the fiducial limits to R are

$$R_L, R_U = \left[R - \frac{9g}{16} \pm \frac{t}{b_S} \left\{ \frac{8s^2}{7N} \left(8 - 9R + 8R^2 - \frac{175g}{32} \right) \right\}^{1/2} \right] \Big/ (1-g), \qquad (7.10.2)$$

where

$$g = \frac{64t^2 s^2}{7N b_S^2}. \qquad (7.10.3)$$

When g is small enough to be ignored,

$$\mathrm{Var}(R) = \frac{8s^2 (8 - 9R + 8R^2)}{7N b_S^2}. \qquad (7.10.4)$$

Formulae for unsymmetric (1, 2, 2) designs (Wood and Finney, 1946) are rarely needed and will not be reproduced here.

§ 7.10   SLOPE RATIO ASSAYS   161

The computational simplicity of the symmetric (1, 2, 2) design may be illustrated by an assay of a sample of malt for its riboflavin content, using *Lactobacillus helveticus* as the test organism and titrating for acidity with sodium hydroxide. The data in Table 7.10.2 were reported by Wood (1946a), who used 20 tubes, 4 each for blanks, 0·1, 0·2 μg of standard riboflavin and 0·025, 0·05 g malt per tube. Responses were measured to the nearest 0·05 ml: for arithmetical convenience, this may be taken as the unit of response, so that all values of y are integers.

TABLE 7.10.2 Responses in an assay of riboflavin in malt (measured in units of 0·05 ml N/10 NaOH)
Standard preparation: 1 unit = 0·2 μg riboflavin
Test preparation: 1 unit = 0·05 g malt

Blanks	Standard		Test	
$x_S = 0$ $x_T = 0$	$x_S = \frac{1}{2}$ $x_T = 0$	$x_S = 1$ $x_T = 0$	$x_S = 0$ $x_T = \frac{1}{2}$	$x_S = 0$ $x_T = 1$
38	97	167	80	121
45	100	164	88	124
40	105	159	90	122
44	98	156	82	122
167	400	646	340	489

Equations (7.2.3) become

$$3 \cdot 2 b_S - 1 \cdot 8 b_T = 233 \cdot 4,$$
$$-1 \cdot 8 b_S + 3 \cdot 2 b_T = 46 \cdot 4.$$

Alternatively, for the dose totals in Table 7.10.2, the contrasts defined in Table 7.10.1 give

$$b_S = 8304/70 = 118 \cdot 629,$$
$$b_T = 5686/70 = 81 \cdot 229.$$

The sum of squares for regression is therefore

$$118 \cdot 629 \times 233 \cdot 4 + 81 \cdot 229 \times 46 \cdot 4 = 31\,456 \cdot 9.$$

Again using Table 7.10.1,

$$L_B = -11,$$
$$L_I = -37,$$

which make contributions $11^2/56$, $37^2/40$ to the analysis of variance in Table 7.10.3. Direct computation of the sum of squares between doses (4 d.f.) checks the total of these three items. The error sum of squares may be obtained by subtraction, or by pooling contributions, each with 3 d.f., from the columns of Table 7.10.2. The mean squares from these five contributions, 10·9, 12·7, 24·3,

22·7 and 1·6, show no association with the magnitude of the mean response; randomization had been strictly conducted, so that the small component from the higher dose of malt has no obvious explanation, and Bartlett's test discloses no heterogeneity. Table 7.10.3 gives no indication of invalidity in respect of blanks or intersection.

TABLE 7.10.3 Analysis of variance for Table 7.10.2

Adjustment for mean		208 488·2	
Nature of variation	d.f.	Sum of squares	Mean square
Regression	2	31 456·9	
Blanks	1	2·2	2·2
Intersection	1	34·2	34·2
Between doses	4	31 493·3	
Error	15	216·5	14·43
Total	19	31 709·8	

In the units for analysis, the estimate of potency is, by equation (7.2.4),

$$R = 81·229/118·629$$
$$= 0·6847.$$

Equation (7.10.3) gives

$$g = \frac{16 \times (2·131)^2 \times 14·43}{35 \times (118·63)^2} = 0·0021,$$

so small that the standard error of R, 0·0181, could safely be used. Equation (7.10.2) gives

$$R_L, R_U = 0·6464, 0·7235$$

as the fiducial limits. In order to express the results as μg riboflavin per g malt, they must be multiplied by the ratio of units, 0·2/0·05. The conclusion is that the malt contains 2·74 μg riboflavin per g with fiducial limits at 2·59 and 2·89 μg per g.

7.11 Other $(1, k, k)$ assays

Designs with higher values of k are seldom chosen. As shown in Chapter 8, they are appreciably less efficient, and their additional validity tests are not often needed. Uncertainty about the upper limit of the range of linearity occasionally makes desirable an assay with one or two high dose levels that can be rejected from the analysis if they are clearly beyond the linear region.

For the $(1, 3, 3)$ design,

$$V = \frac{9}{26N} \begin{pmatrix} 31 & 18 \\ 18 & 31 \end{pmatrix}. \qquad (7.11.1)$$

§ 7.11 SLOPE RATIO ASSAYS 163

Table 7.11.1 shows the non-orthogonal contrasts for the regression coefficients. The remaining 4 d.f. between doses may be divided orthogonally into blanks, intersection, and a quadratic component for each preparation. The contrasts are shown in Table 7.11.1; since the regression curves are not parallel, no meaning attaches to average measures of curvature, and the use of separate components for the two preparations seems preferable. The divisors are used to give the magnitudes of b_S and b_T, and to give the squares for the other contrasts; the sum of squares for the regression must as usual be calculated from formula (7.4.5).

TABLE 7.11.1 Coefficients of regression and orthogonal contrasts for the (1, 3, 3) design

Contrast	C	S_1	S_2	S_3	T_1	T_2	T_3	Divisor
b_S	−42	−11	20	51	−24	−6	12	$182n/3$
b_T	−42	−24	−6	12	−11	20	51	$182n/3$
L_B	6	−4	−1	2	−4	−1	2	$78n$
L_I	0	4	1	−2	−4	−1	2	$42n$
L_{2S}	0	1	−2	1	0	0	0	$6n$
L_{2T}	0	0	0	0	1	−2	1	$6n$

The (1, 4, 4) design is easily incorporated into a programme that ordinarily uses (1, 2, 2), as it requires only that additional tests be made at doses one-quarter and three-quarters of the highest for each preparation. For this design,

$$\mathbf{V} = \frac{24}{35N} \begin{pmatrix} 17 & 10 \\ 10 & 17 \end{pmatrix}. \qquad (7.11.2)$$

The regression and other contrasts are in Table 7.11.2; a factor 3 has been removed from L_B and L_I. The additional degrees of freedom are associated with cubic components for the two preparations, L_{3S} and L_{3T}. The relation of the quadratic and cubic components to L_2 and L_3 in Table 5.6.1 should be clear. For most purposes, a composite test of residual curvature with 4 d.f. will suffice; the sum of squares is obtained by subtracting the regression, blanks, and intersection components from the total between doses.

TABLE 7.11.2 Coefficients of regression and orthogonal contrasts for the (1, 4, 4) design

Contrast	C	S_1	S_2	S_3	S_4	T_1	T_2	T_3	T_4	Divisor
b_S	−30	−13	4	21	38	−20	−10	0	10	$105n/2$
b_T	−30	−20	−10	0	10	−13	4	21	38	$105n/2$
L_B	4	−2	−1	0	1	−2	−1	0	1	$28n$
L_I	0	2	1	0	−1	−2	−1	0	1	$12n$
L_{2S}	0	1	−1	−1	1	0	0	0	0	$4n$
L_{2T}	0	0	0	0	0	1	−1	−1	1	$4n$
L_{3S}	0	−1	3	−3	1	0	0	0	0	$20n$
L_{3T}	0	0	0	0	0	−1	3	−3	1	$20n$

7.12 The symmetric $(0, k, k)$-point design

A second type of symmetric slope ratio assay has no blanks, but instead has k doses equally spaced between c and 1 for each preparation, c being taken as small yet sufficiently large to avoid curvature. Thus the doses are

$$c, c + \frac{1-c}{k-1}, c + \frac{2(1-c)}{k-1}, \ldots, c + \frac{(k-2)(1-c)}{k-1}, 1,$$

with $N/2k$ subjects at each. This design would not be chosen if the linear regression of response on dose were believed to hold down to zero dose, for it would then fail to use the whole range of linearity and so would give results less precise than the best obtainable. For an assay based upon a response that is known to depart from a linear regression at low doses, a $(0, k, k)$ design can be a good choice; the value of c might perhaps be about 0·1.

General formulae involve both c and k, and offer no particular advantages over the complete regression calculations. One special case is that of a $(1, k, k)$ converted into a $(0, k, k)$ by the necessity of rejecting the tests on the blanks, because of significant deviation from the linear regression equation (cf. § 7.8). For this design, $c = 1/k$, and N is now the original N multiplied by $2k/(2k+1)$. Hence

$$\mathbf{V} = \frac{6k^2}{N(k^2-1)(2k+1)} \begin{pmatrix} 5k+1 & 3(k+1) \\ 3(k+1) & 5k+1 \end{pmatrix}. \quad (7.12.1)$$

The blanks contrast no longer occurs; that for intersection is still L_I as defined by (7.8.6), with divisor $\frac{1}{2}N(k-1)(2k+1)$ in terms of the new N.

TABLE 7.12.1 Coefficients of regression and orthogonal contrasts for the (0, 2, 2) design with $c = \frac{1}{2}$

Contrast	S_1	S_2	T_1	T_2	Divisor
b_S	−4	7	−6	3	$5n$
b_T	−6	3	−4	7	$5n$
L_I	2	−1	−2	1	$10n$

TABLE 7.12.2 Coefficients of regression and orthogonal contrasts for the (0, 3, 3) design with $c = \frac{1}{3}$

Contrast	S_1	S_2	S_3	T_1	T_2	T_3	Divisor
b_S	−3	1	5	−4	−1	2	$14n/3$
b_T	−4	−1	2	−3	1	5	$14n/3$
L_I	4	1	−2	−4	−1	2	$42n$
L_{2S}	1	−2	1	0	0	0	$6n$
L_{2T}	0	0	0	1	−2	1	$6n$

For ease of reference, the contrasts required for $k = 2, 3, 4$ in designs resulting from rejection of the blanks in a $(1, k, k)$ assay are summarized in Tables 7.12.1–7.12.3; these are arranged to correspond to Tables 7.10.1, 7.11.1, and 7.11.2. The variance matrices are

TABLE 7.12.3 Coefficients of regression and orthogonal contrasts for the (0, 4, 4) design with $c = \frac{1}{4}$

Contrast	S_1	S_2	S_3	S_4	T_1	T_2	T_3	T_4	Divisor
b_S	−8	−1	6	13	−10	−5	0	5	$15n$
b_T	−10	−5	0	5	−8	−1	6	13	$15n$
L_I	2	1	0	−1	−2	−1	0	1	$12n$
L_{2S}	1	−1	−1	1	0	0	0	0	$4n$
L_{2T}	0	0	0	0	1	−1	−1	1	$4n$
L_{3S}	−1	3	−3	1	0	0	0	0	$20n$
L_{3T}	0	0	0	0	−1	3	−3	1	$20n$

$$\mathbf{V} = \frac{8}{5N}\begin{pmatrix} 11 & 9 \\ 9 & 11 \end{pmatrix}, \qquad (7.12.2)$$

$$\mathbf{V} = \frac{27}{7N}\begin{pmatrix} 4 & 3 \\ 3 & 4 \end{pmatrix}, \qquad (7.12.3)$$

and

$$\mathbf{V} = \frac{32}{15N}\begin{pmatrix} 7 & 5 \\ 5 & 7 \end{pmatrix} \qquad (7.12.4)$$

respectively. In the formulae, N always represents the total number of subjects in the assay as analyzed, and n the number of subjects per dose, so that equation (7.8.1) must be replaced by

$$N = 2nk. \qquad (7.12.5)$$

7.13 Routine assays

With the obvious modifications, much of §§ 5.7–5.10 applies to slope ratio assays. If a particular design is used frequently as a routine, the calculations should be standardized and reduced to a minimal labour consistent with extracting adequate information. For desk calculation of any of the symmetric designs, the formulae in earlier sections are easily applied systematically; a computer program that both draws attention to any evidence of invalidity and completes the potency estimation is simple to write. Nomographs could be developed. Approximations such as range estimation of standard deviations (Wood, 1947b) should have the same status as for parallel line assays.

For routine assays, control charts should prove a valuable guard against unsuspected changes in experimental conditions. Control charts might be set up for s^2 and b_S, and also for L_B and L_I or for the ratios of these two quantities to their standard errors. In order to give better values for s^2 and b_S, some pooling of estimates from previous assays might be permitted in a series of assays showing satisfactory control. The suggestions in § 5.10 are readily adapted.

7.14 Other slope ratio problems

Whereas the study of parallel regression lines preceded their special use in biological assays, development of statistical methods for concurrent pencils of regression lines seems to have begun with slope ratio assays. Neither the designs nor the arrangement of the analysis of variance suited to assays are necessarily the best in other circumstances. Claringbold (1959) pointed out that, when chief interest lies in tests of significance of slope differences, sets of mutually orthogonal contrasts may be preferable and also that the blanks may give little relevant information. His valuable paper should be seen by any who have problems allied to the slope ratio situation but not strictly of an assay type. Other special methods of analysis appropriate to pencils of lines have been well presented by Williams (1959).

8
Efficiency in slope ratio assays

8.1 General principles

The principles of assay design described in §§ 6.1–6.7 are as relevant and as important in slope ratio as in parallel line assays. Their application leads to different advice, because the formulae expressing precision and reliability are different. The requirements of good design for slope ratio assays are here discussed under the assumption that conditions (a), (b), and (c) of § 6.8 again apply, except that the regression is now known to be linear on the absolute measure of dose. The problem for the assayist is still that of making the best use of a total of N subjects.

Suppose that the highest doses of S, T used in an assay are X_S, X_T. As in § 7.7, define v_{11}, v_{12}, v_{22} to be the elements of the variance matrix after rescaling all doses so that the highest dose of each preparation is unity; that is to say, these quantities relate to an assay in which all doses of S have been divided by X_S, all doses of T by X_T. Then, for the assay as actually performed,

$$\mathbf{V} = \begin{pmatrix} \dfrac{v_{11}}{X_S^2} & \dfrac{v_{12}}{X_S X_T} \\ \dfrac{v_{12}}{X_S X_T} & \dfrac{v_{22}}{X_T^2} \end{pmatrix}. \qquad (8.1.1)$$

This definition enables the effect of the range of doses to be kept distinct from that of changing the distribution of doses over their range. From equation (7.6.5) the quarter-square of the fiducial interval for R is

$$I = \frac{t^2 s^2}{b_S^2(1-g)^2}\left[\frac{v_{22}}{X_T^2} - \frac{2R v_{12}}{X_S X_T} + \frac{R^2 v_{11}}{X_S^2} - \frac{g}{X_T^2}\left(v_{22} - \frac{v_{12}^2}{v_{11}}\right)\right], \qquad (8.1.2)$$

where

$$g = \frac{t^2 s^2 v_{11}}{b_S^2 X_S^2}. \qquad (8.1.3)$$

Equation (8.1.2) may be written

$$I = \frac{R^2 t^2 s^2}{b_S^2 X_S^2 h^2 (1-g)^2}\left[v_{22} - 2h v_{12} + h^2 v_{11} - g\left(v_{22} - \frac{v_{12}^2}{v_{11}}\right)\right]; \qquad (8.1.4)$$

the quantity

$$h = R X_T / X_S, \qquad (8.1.5)$$

the ratio of estimated relative potency to relative magnitudes of the highest doses, plays a part similar to that of h in § 6.9. Since R, a property of the preparations, is not at the investigator's choice, his influence on the reliability of the estimate will be restricted to attempts to ensure that

(i) t is small;
(ii) s is small;
(iii) $b_S X_S$ is large;
(iv) g is small;
(v) h is large;
(vi) $\left[v_{22} - 2h v_{12} + h^2 v_{11} - g \left(v_{22} - \dfrac{v_{12}^2}{v_{11}} \right) \right]$ is small.

Several of these are similar to the requirements listed in § 6.8, and need not be discussed again in detail.

(i) The need of having enough degrees of freedom for t again indicates that N should exceed 20, unless, as may happen in microbiological assays, s is exceedingly small.

(ii) Genetic and environmental homogeneity of subjects are important to the control of the size of s; in some types of assay, a randomization restricted by block constraints will eliminate irrelevant variation.

(iii) In parallel line assays, a high value of b is desirable: in slope ratio assays, an increase in b_S is of no use if it is accompanied by a compensating reduction in the upper limit of doses for which the regression is linear. For a slope ratio assay, X_S should be chosen to have the largest possible value consistent with existing evidence of linearity of regression, and efforts to improve the subjects or the conditions of experiment should be directed at increasing the total increment in response, $b_S X_S$, between zero dose and X_S. Whether or not curvature prevents the use of very small doses does not affect this stage of the argument.

(iv) Equation (8.1.3) shows g to be reduced by measures taken under headings (i), (ii), (iii). A small value of v_{11}, ensured by a wide spread of doses between 0 and X_S, both reduces g and is desirable under heading (vi). The width of the fiducial interval will be approximately proportional to $1/(1-g)$, so that any reduction in g below 0·05 can give little benefit.

(v) Since R is outside control and X_S has been chosen to be as large as possible, h can be increased only by making X_T large. The greatest value consistent with the need for remaining on the linear portion of the regression for T is X_S/ρ. In practice, X_T should be taken a little smaller than X_S/R_0, where R_0 again is a preliminary rough estimate of ρ, in order to guard against the risk of non-linearity caused by an accidentally low value of R_0. Ideally, h will be near to, but rather less than, unity.

(vi) This expression will be minimized by using extremes of dose, so concentrating the subjects at X_S, X_T and the lowest doses of the two preparations that do not go outside the region of linearity at that end of the scale. If linearity extends to zero dose, tests at that level are desirable because subjects then do double duty by occurring at the lower extreme of dose for each preparation. The ideal distribution of subjects between these doses will not be discussed here; allocation of some subjects to intermediate doses conflicts with the ideals of reliability, but is essential if any validity tests are to be made, and should be regarded as a prerequisite of a good assay unless validity is certain *a priori* (§ 8.4).

8.2 Symmetric $(1, k, k)$ designs

Considerations of symmetry suggest that a $(1, k, k)$ design should be adopted for any assay in which tests on blanks can be included. Departure from symmetry is discussed in § 8.3. For the symmetric design having X_S, X_T as the two highest doses, equation (7.8.3) may be modified to

$$\mathbf{V} = \begin{pmatrix} \dfrac{K_1}{NX_S^2} & \dfrac{K_2}{NX_S X_T} \\ \dfrac{K_2}{NX_S X_T} & \dfrac{K_1}{NX_T^2} \end{pmatrix}, \quad (8.2.1)$$

in which

$$K_1 = \frac{3k(5k^2 + 5k + 2)}{(k+1)(k^2 + k + 1)}, \quad (8.2.2)$$

$$K_2 = \frac{9k^2}{k^2 + k + 1}. \quad (8.2.3)$$

Now

$$g = AK_1/15, \quad (8.2.4)$$

where

$$A = \frac{15 t^2 s^2}{N b_S^2 X_S^2}, \quad (8.2.5)$$

and equation (8.1.4) can be written

$$I = \frac{AR^2 [15 K_1 (1 + h^2) - 30 K_2 h - A(K_1^2 - K_2^2)]}{h^2 (15 - AK_1)^2}. \quad (8.2.6)$$

Note that k enters this expression only in K_1 and K_2. When A is small enough for terms in A^2 or higher powers to be neglected, (8.2.6) is equivalent to the approximation

$$\text{Var}(R) = \frac{s^2}{N b_S^2 X_T^2} [K_1 (1 + h^2) - 2K_2 h], \quad (8.2.7)$$

the form here taken by equation (7.6.7).

Whatever the values of A and h, I increases as k increases and is therefore minimized by $k = 1$. The quality of any other assay relative to the optimal with

TABLE 8.2.1 Percentage precision of the symmetric $(1, k, k)$ design, relative to the optimal symmetric $(1, 1, 1)$, (g is assumed small)

Values of k	Values of h						
	1·2	1·0	0·9	0·8	0·7	0·5	0·3
1	116	100	89	76	62	33	11
2	87	75	67	57	46	24	8
3	77	67	59	50	40	21	7
4	72	63	56	47	38	19	6
5	69	60	53	45	36	19	6
10	63	55	49	41	33	17	5
Limit	58	50	44	38	30	15	5

TABLE 8.2.2 Percentage reliability of the symmetric $(1, k, k)$ design relative to the optimal symmetric $(1, 1, 1)$, $(A = 0·25)$

Values of k	Values of h						
	1·2	1·0	0·9	0·8	0·7	0·5	0·3
1	114	100	90	77	63	34	12
2	79	70	63	54	44	23	7
3	67	60	54	47	38	20	6
4	61	55	50	43	35	18	6
5	58	52	47	40	33	17	5
10	51	46	42	36	29	15	5
Limit	44	40	37	32	26	13	4

TABLE 8.2.3 Percentage reliability of the symmetric $(1, k, k)$ design relative to the optimal symmetric $(1, 1, 1)$, $(A = 0·5)$

Values of k	Values of h						
	1·2	1·0	0·9	0·8	0·7	0·5	0·3
1	112	100	91	79	65	35	12
2	69	63	58	51	42	22	7
3	55	51	47	41	34	18	5
4	48	45	42	37	30	16	5
5	43	41	38	34	28	14	4
10	36	34	32	29	24	12	4
Limit	28	28	26	24	20	10	3

TABLE 8.2.4 Percentage reliability of the symmetric $(1, k, k)$ design relative to the optimal symmetric $(1, 1, 1)$, $(A = 1·0)$

Values of k	Values of h						
	1·2	1·0	0·9	0·8	0·7	0·5	0·3
1	107	100	93	83	70	39	13
2	42	42	41	39	33	18	5
3	22	24	25	24	21	11	3
4	14	16	17	17	15	8	2
5	9	11	12	13	12	6	1
10	3	4	4	5	5	3	0
Limit	0	0	0	0	0	0	0

$k = 1$, $h = 1 \cdot 0$ may again be represented by the relative reliability (§ 6.9), the inverse ratio of values of I. Provided that the designs compared are large enough for t to be practically independent of its degrees of freedom, this is

$$\text{Rel} = \frac{h^2(10 - 3A)(15 - AK_1)^2}{(5 - 2A)^2[15K_1(1 + h^2) - 30K_2 h - A(K_1^2 - K_2^2)]}. \qquad (8.2.8)$$

When A is sufficiently small for $\text{Var}(R)$ to be used, (8.2.8) simplifies to

$$\text{Rel} = 6h^2/(K_1 - 2K_2 h + K_1 h^2). \qquad (8.2.9)$$

Table 8.2.1 shows the relative precision in (8.2.9) for various k, h; Tables 8.2.2, 8.2.3, 8.2.4 show the relative reliability for three typical values of A, all at probability 0·95. If $A = 1 \cdot 0$, the factor $(15 - AK_1)$ will tend to zero as k becomes large, and for larger A the reliability will fall to zero even for quite small k. In microbiological assays, such values of A ought never to be encountered; the riboflavin assay in § 7.10, for example, had A only about 0·003.

Even when A is small, the loss in reliability as the number of dose levels increases is more serious than for parallel line assays. Low values of h also seriously reduce reliability. These trends are accentuated at large values of A.

8.3 The asymmetric (1, 1, 1) design

If equation (7.7.4) is written for the special case of the (1, 1, 1) design with n_0, n_S, n_T subjects at the three doses $(n_0 + n_S + n_T = N)$, $\text{Var}(R)$ is easily seen to be minimized by

$$\frac{n_0}{1 - h} = \frac{n_S}{h} = n_T = \tfrac{1}{2}N. \qquad (8.3.1)$$

Under the optimal condition of $h = 1 \cdot 0$, this leads to the surprising conclusion that $\tfrac{1}{2}N$ subjects should be assigned to each preparation and none to the blanks! Obviously such an arrangement could not be satisfactory, since it cannot give any information on b_S and b_T. The explanation lies in the neglect of g; as n_0 is reduced to zero, g increases without limit.

Proper examination of the efficiency requires use of the full formula for fiducial limits. Of chief interest is the semi-symmetric design having equal numbers of subjects for S and T:

$$\left. \begin{array}{l} n_0 = Np, \\ n_S = n_T = \tfrac{1}{2}N(1-p), \end{array} \right\} \qquad (8.3.2)$$

where for maximum reliability the fraction p will presumably be less than $\tfrac{1}{3}$. Equation (8.2.6) still applies, but with

$$K_1 = \frac{1+p}{p(1-p)}, \qquad (8.3.3)$$

$$K_2 = \frac{1}{p}. \qquad (8.3.4)$$

The optimal fully symmetric design, which has $h = 1 \cdot 0$ and $p = \frac{1}{3}$, was taken as the standard in § 8.2 and may be so used again here. With the changed definitions of K_1 and K_2, the relative reliability of the semi-symmetric design is still given by equation (8.2.8). This expression, evaluated for $p = 0 \cdot 5, \frac{1}{3}, 0 \cdot 2, 0 \cdot 1, 0 \cdot 05$, is tabulated in Tables 8.3.1–8.3.4.

TABLE 8.3.1 Percentage precision of the semi-symmetric (1, 1, 1) design relative to the optimal symmetric (1, 1, 1), (g is assumed small)

Values of p	Values of h						
	1·2	1·0	0·9	0·8	0·7	0·5	0·3
0·5	88	75	67	58	48	27	10
$\frac{1}{3}$	116	100	89	76	62	33	11
0·2	137	120	106	89	70	34	10
0·1	148	135	118	95	70	28	7
0·05	146	143	121	90	60	20	4

TABLE 8.3.2 Percentage reliability of the semi-symmetric (1, 1, 1) design relative to the optimal symmetric (1, 1, 1), ($A = 0 \cdot 25$)

Values of p	Values of h						
	1·2	1·0	0·9	0·8	0·7	0·5	0·3
0·5	86	74	67	58	49	28	10
$\frac{1}{3}$	114	100	90	77	63	34	12
0·2	131	117	105	89	70	34	10
0·1	125	120	107	86	63	24	6
0·05	88	100	87	63	39	11	2

TABLE 8.3.3 Percentage reliability of the semi-symmetric (1, 1, 1) design relative to the optimal symmetric (1, 1, 1), ($A = 0 \cdot 5$)

Values of p	Values of h						
	1·2	1·0	0·9	0·8	0·7	0·5	0·3
0·5	84	74	67	59	49	29	11
$\frac{1}{3}$	112	100	91	79	65	35	12
0·2	123	113	103	88	70	34	10
0·1	96	100	92	75	53	19	4
0·05	27	44	42	27	14	3	1

TABLE 8.3.4 Percentage reliability of the semi-symmetric (1, 1, 1) design relative to the optimal symmetric (1, 1, 1), ($A = 1 \cdot 0$)

Values of p	Values of h						
	1·2	1·0	0·9	0·8	0·7	0·5	0·3
0·5	78	72	66	60	51	31	12
$\frac{1}{3}$	107	100	93	83	70	39	13
0·2	100	100	95	84	68	32	8
0·1	23	35	39	34	21	5	1
0·05	0	0	0	0	0	0	0

Table 8.3.1 ignores g, and the steady increase in the column for $h = 1 \cdot 0$ as p decreases must not be misunderstood; the increase in K_1 as p decreases can combine with even a very small value of A to give an appreciable g. Nevertheless, if h is near to unity and A is not unduly large, reliability is increased by taking p about 0·2 or 0·1; thus, for $A = 0 \cdot 2$, as much as 23 percent may be gained by allotting only one-tenth of the subjects to zero dose instead of one-third. This gain will be reduced if misjudgment of doses leads to the actual value of h being much smaller than that used in planning the assay. If also A is much larger than was expected, the effect of having assigned very few subjects to the blanks may be disastrous, as comparisons between $p = \frac{1}{3}$ and $p = 0 \cdot 1$ in Table 8.3.4 show. The simplicity of the fully symmetric design, and the fact that its reliability is seldom much below the maximum, therefore make it usually preferable to the semi-symmetric. Perhaps the most valuable alternative is that with $p = 0 \cdot 2$, $n_0 = N/5$, $n_S = n_T = 2N/5$. This is convenient because so similar to the symmetric $(1, 2, 2)$ design, the only change being to place on each double dose the subjects that are normally given a single dose; unless A is very large, even a small value of h will not make it less reliable than the fully symmetric design.

8.4 The choice of k

As a guide to the choice of k, little need be added to § 6.10, for in this respect slope ratio assays are very similar to parallel line assays. If there were no doubts of statistical validity, a $(1, 1, 1)$ design would always be chosen, ideally with about twice as many subjects for each non-zero dose as for the blanks. Unfortunately such certainty of validity is rare, and the conflict between precision and validity tests must be resolved by choosing $k > 1$. In general assay practice, a value greater than 2 is seldom desirable; as suggested in § 7.11, when a larger number of doses is wanted, the $(1, 4, 4)$ design is about the best.

8.5 Symmetric $(0, k, k)$ designs

If knowledge of non-linearity or any other circumstance discourages the inclusion of tests on blanks, the natural design to adopt is a symmetric $(0, k, k)$ (§ 7.12) with its lowest doses cX_S, cX_T, as small as the investigator dare risk. The reliability depends upon c as well as upon k. As an illustration, the case of $c = 0 \cdot 1$ has been studied, for which the exclusion of low doses removes one-tenth of the dose range from experimental use. For purposes of comparison, the symmetric $(1, 1, 1)$ design with $h = 1 \cdot 0$ is again taken as the standard of reliability, even though, when a common-zero design is not permissible, the best symmetric scheme is a $(0, 2, 2)$; the advantage of this procedure is that the relative reliability also measures the loss due to the shortening of the available dose range. Again equation (8.2.8) applies, but with

$$K_1 = \frac{200(k-1)(175k-67)}{27(k+1)(74k-47)} \tag{8.5.1}$$

and

$$K_2 = \frac{24\,200(k-1)^2}{27(k+1)(74k-47)}. \tag{8.5.2}$$

Tables 8.5.1–8.5.4 show numerical values comparable with those in Tables 8.2.1–8.2.4.

TABLE 8.5.1 Percentage precision of the symmetric $(0, k, k)$ design relative to the optimal symmetric $(1, 1, 1)$ ($c = 0.1$, g is assumed small)

Values of k	Values of h						
	1·2	1·0	0·9	0·8	0·7	0·5	0·3
2	88	76	68	58	48	26	9
3	76	66	58	50	40	21	7
4	72	62	55	47	38	20	6
5	70	61	54	46	37	19	6
10	66	58	51	43	34	17	5
Limit	63	56	49	41	32	15	5

TABLE 8.5.2 Percentage reliability of the symmetric $(0, k, k)$ design relative to the optimal symmetric $(1, 1, 1)$ ($c \doteq 0.1$, $A = 0.25$)

Values of k	Values of h						
	1·2	1·0	0·9	0·8	0·7	0·5	0·3
2	85	74	66	58	48	27	9
3	68	60	54	47	38	21	7
4	61	55	50	43	35	18	6
5	58	52	47	41	33	17	5
10	51	47	43	37	29	14	4
Limit	45	42	38	33	26	12	3

TABLE 8.5.3 Percentage reliability of the symmetric $(0, k, k)$ design relative to the optimal symmetric $(1, 1, 1)$ ($c = 0.1$, $A = 0.5$)

Values of k	Values of h						
	1·2	1·0	0·9	0·8	0·7	0·5	0·3
2	80	71	65	57	48	27	10
3	58	53	49	43	36	19	6
4	49	46	42	37	31	16	5
5	44	41	39	34	28	14	4
10	34	33	32	28	23	11	3
Limit	24	25	24	22	18	8	2

TABLE 8.5.4 Percentage reliability of the symmetric $(0, k, k)$ design relative to the optimal symmetric $(1, 1, 1)$ ($c = 0.1$, $A = 1.0$)

Values of k	Values of h						
	1·2	1·0	0·9	0·8	0·7	0·5	0·3
2	68	64	60	54	47	27	10
3	31	32	31	29	26	15	4
4	16	18	19	19	17	9	2
5	9	11	12	12	12	6	1
10	0	0	0	1	1	0	0
Limit	0	0	0	0	0	0	0

A symmetric $(0, k, k)$ design with $c = 0 \cdot 1$ and $h = 1 \cdot 0$ has at best 76 percent reliability relative to the symmetric $(1, 1, 1)$, and the loss in reliability increases rapidly as k or A is increased. Comparison with Tables 8.2.1–8.2.4 reveals the surprising fact that, for low values of A, the reliabilities for corresponding k are almost equal; indeed the $(0, k, k)$ may have a slight advantage, presumably because it allots more subjects to high doses. If c were larger, this would not be true. If A is as large as $1 \cdot 0$, the $(0, k, k)$ design is appreciably more reliable than the $(1, k, k)$ for small k; for large k, the situation is reversed, but both reliabilities are then so low that neither assay is of practical use.

In the light of the last paragraph, what has been said in (vi) of § 8.1 about blanks may be explained further. The merit of using blanks appears to be not that they increase reliability but that, when there is reasonable hope of linearity, they represent a cheap method of obtaining validity tests without sacrifice of reliability. The $(0, 2, 2)$ has the highest reliability of all $(0, k, k)$ designs, but it permits no test of statistical validity (the 'intersection' contrast gives a test of fundamental validity). As Tables 8.5.1–8.5.4 show, its replacement by a $(0, 3, 3)$ in order to obtain linearity tests involves a serious loss of reliability, especially if A is large. Much the same end can be achieved at a lesser cost in reliability by using a $(1, 2, 2)$ design, for test of the 'blanks' contrast is in reality a test of linearity of regression. From this point of view, Tables 8.5.1–8.5.4 are most interestingly used to compare a $(0, k + 1, k + 1)$ design with the corresponding $(1, k, k)$ in Tables 8.2.1–8.2.4.

8.6 Comparison of assay techniques

The argument of §§ 8.1–8.5 has concentrated on the efficiencies of different experimental designs that use the same subjects, techniques, and measurements. Just as for parallel line assays (§ 6.11), questions may arise as to which of several assay techniques for estimating the same potency is preferable. Equation (8.1.4) shows that in a good assay, with g small and h near to unity, the standard error of R is proportional to $s/b_S X_S$. This is a measure of the intrinsic quality of the technique for the purposes of assay, whereas its multiplier depends only upon the number of subjects used and their distribution between doses. It plays a part analogous to that of s/b in parallel line assays, though it cannot be given a simple interpretation. Its physical dimensions are those of a pure ratio, so that not only does it provide a scale of comparison for alternative assays of the same preparation but it also enables the efficiencies of slope ratio assays for entirely different purposes to be compared. The value of X_S used is assumed to be the largest possible; data that do not extend over the whole region of linearity can therefore give only an upper limit for $s/b_S X_S$. Apart from complications introduced by g and h, with a particular set of doses and proportional allocation between doses, the number of subjects needed in order to secure a specified precision is inversely proportional to the square of $s/b_S X_S$. Although this intrinsic error enables the economics of alternative slope ratio techniques to be compared, no simple comparison between a parallel line assay and a slope ratio assay for estimation of the potency of the same preparation can be made.

The remarks on the choice of a response, in the second half of § 6.11, apply equally to slope ratio assays.

9
Incomplete block designs

9.1 Design in bioassay

The theory of experimental design, developed largely for agriculture, finds application in many other branches of research and technology. The literature cannot be discussed thoroughly here, even in its uses in bioassay. Early chapters of Fisher's classic work *The Design of Experiments* (1966) make an excellent introduction. The statistician should be familiar with general texts, such as Cochran and Cox (1957), Finney (1959a), John (1971), Kempthorne (1952), Yates (1937), and John and Quenouille (1977); Fisher and Yates (1963) contains much useful information, and Cox (1958) is especially good on more practical aspects. The biologist need not have the same expert knowledge, but he should understand the basic principles and know the characteristics of the more important designs. Whatever the field of application, the statistician should regard it as his duty to tailor a design to the special requirements of an investigation rather than to rely upon published lists of standard designs; nevertheless, the catalogues now available are so extensive that often little or no modification is needed to fit a design to a particular purpose.

Considerations of symmetry, number of dose levels, choice of doses, magnitude of regression coefficients, replication, and so on, though relevant to all assays based on quantitative responses (Chapters 6 and 8), are not in themselves sufficient to determine the ideal design for an assay. The subjects may be limited in total number and in their classification into litters or other appropriate blocks; the number of tests performable on one day or by one operator may be restricted; the total amount of test preparation available may be small; these and other constraints will often influence recommendation of the design likely to give the best results. When constraints are not severe, the basic designs of complete randomization and randomized blocks are usually optimal. If two blocking systems compete for attention, a Latin square may be the answer (cf. §§ 2.8, 11.3). For example, Long, Miles and Perry (1954) used 8 guinea pigs and 4 sites of intradermal injection on each flank in a (2, 2) assay of tuberculin. They modified an 8 × 8 Latin square for sites and guinea pigs by allowing each of the doses to correspond to two of its letters, so balancing the design for two sources of variation. Fisher (1949) described another tuberculin assay in which Latin square balancing over sites was used for 120 cows. He adopted an ingenious but complicated method of analysis to overcome difficulties arising because variance was obviously correlated with response; an analysis using a logarithmic or square-root response metameter and the standard methods of this book would have given much the same conclusions.

Constraints on a design occasionally produce a situation in which blocks

TABLE 9.3.2 Sums and differences of responses for litter-mates in Table 9.3.1

	S_1+T_1	S_2+T_2	S_3+T_3	T_1-S_1	T_2-S_2	T_3-S_3
	35	78	88	-5	-8	-14
	30	67	79	-14	-7	-11
	36	75	69	0	-1	-23
	40	77	92	-12	-3	-4
	45	78	94	-15	-16	-16
	64	62	81	6	8	11
	81	89	90	-13	7	-8
	36	90	102	4	4	0
	35	53	79	3	-17	-7
	53	60	55	-7	0	1
Totals	455	729	829	-53	-33	-71

TABLE 9.3.3 Analysis of variance for Table 9.3.1

Adjustment for mean		67 536	
Nature of variation	d.f.	Sum of squares	Mean square
Regression	1	3 497	3497
Quadratic	1	252	252
Error (1)	27	2 663	98·63
Between litters	29	6 412	
Preparations	1	411	411
Parallelism	1	8	8
Difference of quadratics	1	28	28
Error (2)	27	1 062	39·33
Total	59	7 921	

described, and the two error sums of squares are found as residuals. The easiest procedure is to tabulate sums and differences of pairs of responses for each litter, as in Table 9.3.2. The sum of squares of deviations of the thirty litter totals, divided by 2, is the total sum of squares for the inter-litter section of the analysis. The sum of squares of the thirty differences within litters (with no adjustment for a mean), divided by 2, is the total sum of squares within litters. These two make up the complete sum of squares of deviations with 59 d.f. By application of the coefficients in Table 5.2.3 to the dose totals in Table 9.3.1, or directly from the litter totals and differences in Table 9.3.2:

$$
\begin{aligned}
&&&&&\text{Divisor}\\
L_p &= -53-33-71 &&= -157 & 60\\
L_1 &= -455+829 &&= 374 & 40\\
L_1' &= 53-71 &&= -18 & 40\\
L_2 &= 455-2\times 729+829 &&= -174 & 120\\
L_2' &= -53+2\times 33-71 &&= -58 & 120
\end{aligned}
$$

The squares for the dose contrasts are formed exactly as in § 5.2, and the analysis of variance, Table 9.3.3, may then be completed.

§ 9.3 INCOMPLETE BLOCK DESIGNS

Each validity test requires comparison of the mean square with the error mean square in the same section of the analysis. The mean square for preparations is significantly greater than the error within litters; the parallelism mean square is small, and that for quadratic curvature is not sufficiently in excess of the error between litters to be alarming. Taken together, the tests do not suggest invalidity, and, provided that the investigator began in the belief that he was performing an analytic dilution assay (§ 15.5), estimation may proceed.

The value of R might be calculated from $\bar{y}_S$, $\bar{y}_T$, b, and M in the standard manner, but calculation may be simplified by use of equation (5.5.1), the applicability of which is not disturbed by the confounding since the definitions of L_p, L_1 are unaltered. Hence

$$R = \frac{Z_S}{Z_T} \text{antilog}\left\{\frac{4L_p}{3L_1}\log\left(\frac{5}{3}\right)\right\}$$

$$= \frac{2\cdot 5}{1\cdot 25} \text{antilog}\left\{\frac{-4 \times 157 \times 0\cdot 2218}{3 \times 374}\right\}$$

$$= 2 \text{ antilog } \bar{1}\cdot 8759$$

$$= 1\cdot 503.$$

The fiducial limits cannot be computed directly from equation (5.5.6), because the variances of L_p and L_1 are based upon different mean squares. In fact, if s_1^2, s_2^2 represent the error mean squares between and within litters, then

$$\text{Var}(L_p) = 60 s_2^2,$$

$$\text{Var}(L_1) = 40 s_1^2.$$

Limits to L_p/L_1 must therefore be obtained from the first analogue of Fieller's theorem (§ 4.13). Since s_1^2 and s_2^2 are both based on 27 d.f., the value of the d-statistic is almost independent of the angle θ, and as a first approximation may be taken as 2·05 (Appendix Table III; Fisher and Yates, 1963, Table V_1). Hence

$$g = \frac{d^2 \text{Var}(L_1)}{L_1^2} \quad \text{by equation (4.13.13)}$$

$$= \frac{(2\cdot 05)^2 \times 40 \times 98\cdot 63}{(374)^2}$$

$$= 0\cdot 1185.$$

Limits to L_p/L_1, whose numerical value is $-0\cdot 4198$, are then given by equation (4.13.12);

$$\left[-0\cdot 4198 \pm \frac{2\cdot 05}{374}\{0\cdot 8815 \times 60 \times 39\cdot 33 + (0\cdot 4198)^2 \times 40 \times 98\cdot 63\}^{1/2}\right]\bigg/0\cdot 8815$$

$$= -0\cdot 8039, -0\cdot 1486.$$

These are in reality only approximations, to be used in equation (4.13.10) in order to give angles θ_L, θ_U for the two limits:

$$\tan \theta_L = \frac{1}{0\cdot 8039} \left(\frac{60 s_2^2}{40 s_1^2}\right)^{1/2} = 0\cdot 96,$$

$$\tan \theta_U = \frac{1}{0\cdot 1486} \left(\frac{60 s_2^2}{40 s_1^2}\right)^{1/2} = 5\cdot 20,$$

whence $\theta_L = 44°$, $\theta_U = 79°$. Interpolation in Fisher and Yates's Table V_1 gives d for these angles as $2\cdot 046$ and $2\cdot 051$, values so close to $2\cdot 05$ as to make recalculation unnecessary; the result of recalculation of the two limits, each with its own values of d and g, is to give $-0\cdot 8029$, $-0\cdot 1485$. The limits may then be inserted in place of L_p/L_1 in the formula for R to give

$$R_L = 2 \text{ antilog } \bar{1}\cdot 7625 = 1\cdot 158,$$

$$R_U = 2 \text{ antilog } \bar{1}\cdot 9561 = 1\cdot 808.$$

The test preparation is therefore estimated to contain $1\cdot 50$ units of vitamin A per mg, with limits at $1\cdot 16$ and $1\cdot 81$ units/mg. The estimate of $2\cdot 0$ units/mg used in determining the doses of the test preparation appears to have been substantially wrong, as was demonstrated by the significance of the component for preparations in Table 9.3.3.

9.4 Criticism of the design for the assay of vitamin A

In Gridgeman's vitamin A assay, estimating the regression coefficient from differences between litters has lowered the reliability of the estimated potency relative to what could have been achieved had litters of six been available. Inspection of § 9.3 shows that even had s_1^2 been as small as s_2^2, the reduction of g from $0\cdot 118$ to about $0\cdot 047$ would narrow the fiducial range only by about 5 percent. Consequently in this instance the possible gain from an alternative design is not great. Nevertheless, a design having one of L_p and L_1 within and the other between litters, is undesirable. Usually s_1^2 will exceed s_2^2; in an assay with relatively higher variability than that in § 9.3, the value of g might be critical for reliability. The laborious calculations for assessment of fiducial limits are an additional objection to designs of the type discussed, though not a very serious one because dependence of the d-deviate upon the angle θ is slight when s_1^2, s_2^2 have equal degrees of freedom.

Provided that proper randomizations have been made, a value of L_1 measured as an inter-litter contrast estimates exactly the same quantity as if it were intra-litter. In the vitamin A assay, $L_1/40$ is an estimate of the increase in response per unit increase in dose (the unit being an increase by a factor of 5/3), for animals chosen at random from the population. The fact that L_p is an intra-litter contrast does not invalidate use of L_p/L_1 in estimating relative potency. Any departure from proper randomization of litters, destroys this validity: the litters assigned to doses S_1, T_1 must be 10 selected at random from the 30 available. Deliberate non-random allocation of litters should not be condoned. Yet departure from randomness does sometimes occur, a further reason for investigating designs in which L_p and L_1 are both intra-litter contrasts.

9.5 A balanced incomplete block assay of vitamin A

The objections to the design used in § 9.3 can be overcome by *balanced incomplete blocks*. Such designs may have any convenient number of subjects per block, with the condition that the number of blocks in which a named pair of doses both occur is the same for every possible pair. For a fixed number of subjects per block and a fixed number of doses to be tested, balance introduces restrictions on the number of replications of each dose (Fisher and Yates, 1963, Introduction; Cochran and Cox, 1957; Kempthorne 1952). Some balanced incomplete block designs suitable for biological assays are listed in § 9.7. For any such, the mean difference in response between two doses is measurable by a contrast within blocks, with precision independent of the particular doses; consequently L_p, L_1 can be formed orthogonal with block differences, as also can the other contrasts L'_1, L_2, L'_2. In addition, if the allocation to blocks has been randomized, independent contrasts measuring the same features of the dose–response relations can be formed between blocks. Thus a second estimate of relative potency is obtainable, though it is likely to be of lower precision than that calculated within blocks. The two estimates may be combined in order to summarize the whole information from the assay.

An artificial example will be discussed (Table 9.5.1). This purports to be the result of a second vitamin A assay, but with all possible types of block instead of only three. The same six doses are used, and each of the 15 possible pairs of doses occurs in two of the 30 blocks. For illustrative purposes, numerical values of the responses have been constructed so as to have the same totals for the 10 animals at each dose and about the same variances between and within blocks as in § 9.3.

Although the analysis of variance can be calculated by the general method for balanced incomplete blocks, its nature may be made clearer to those unfamiliar with complex experimental designs by developing it from first principles. The subdivision of the total sum of squares of deviations into 29 d.f. between litters and 30 d.f. within litters proceeds exactly as in § 9.3, using sums and differences for the two animals in each litter. Complications arise when the dose contrasts are considered, for all of those in Table 5.2.3 are partially confounded. Consequently, both sections of the analysis contribute information on each contrast. The components of the sum of squares might be found by a multiple regression procedure as in § 4.23. Five dummy variates take the values of the coefficients of the five contrasts in Table 5.2.3; the multiple regression calculations for the litter-mate sums then give the inter-litter components, and those for the differences the intra-litter components. The symmetry of the design makes evaluation of the regression coefficients very simple, since each set of five corresponds to a set of mutually orthogonal contrasts.

Two modified sets of orthogonal contrasts can simplify the calculations. Table 9.5.2 contains the coefficients for inter-block dose contrasts. The first

TABLE 9.5.1 Artificial data for weight increases (g) of male rats during three weeks of dosage of vitamin A
(Blocks are denoted by numbers I to XXX)

Block no.	S_1 0·9 units	S_2 1·5 units	S_3 2·5 units	T_1 0·45 mg	T_2 0·75 mg	T_3 1·25 mg
I	20	33	–	–	–	–
II	18	36	–	–	–	–
III	16	–	44	–	–	–
IV	22	–	33	–	–	–
V	29	–	–	35	–	–
VI	26	–	–	14	–	–
VII	47	–	–	–	48	–
VIII	30	–	–	–	30	–
IX	16	–	–	–	–	38
X	30	–	–	–	–	41
XI	–	40	47	–	–	–
XII	–	40	59	–	–	–
XIII	–	35	–	4	–	–
XIV	–	47	–	16	–	–
XV	–	27	–	–	35	–
XVI	–	43	–	–	35	–
XVII	–	43	–	–	–	50
XVIII	–	37	–	–	–	33
XIX	–	–	44	26	–	–
XX	–	–	48	28	–	–
XXI	–	–	35	–	43	–
XXII	–	–	43	–	33	–
XXIII	–	–	46	–	–	23
XXIV	–	–	51	–	–	51
XXV	–	–	–	20	37	–
XXVI	–	–	–	12	30	–
XXVII	–	–	–	21	–	33
XXVIII	–	–	–	25	–	40
XXIX	–	–	–	–	39	43
XXX	–	–	–	–	18	27

column shows the 15 types of total that can be formed from litters, and the totals of four responses from the two litters of each type are entered in the second column. The coefficients for $L_p, L_1, L_1', L_2, L_2'$ may be obtained by adding the coefficients in Table 5.2.3 for the two doses in each litter; thus, in the L_2' column, successive entries are obtained as $(-1+2), (-1-1), (-1+1)$, etc. A factor of 2 could be removed from the L_p column, but the explanation is clearer if this is retained. The divisors are obtained, as in § 4.16, by summing the squares of coefficients and noting that each entry in the second column is a total of four responses:

$$[(-2)^2 + (-2)^2 + (-2)^2 + 2^2 + 2^2 + 2^2] \times 4 = 96.$$

Every two columns are orthogonal: therefore, the square of each contrast may be divided by the corresponding divisor, entered in the analysis of variance (Table 9.5.4), and subtracted from the sum of squares between litters, in order to leave an error sum of squares for inter-litter variation. The six components of the inter-litter section of the analysis are exactly as given by the regression technique described in the preceding paragraph.

TABLE 9.5.2 Orthogonal coefficients for contrasts between litters for Table 9.4.1

Doses	Totals	L_p	L_1	L_1'	L_2	L_2'
$S_1 + S_2$	107	-2	-1	1	-1	1
$S_1 + S_3$	115	-2	–	–	2	-2
$S_1 + T_1$	104	–	-2	–	2	–
$S_1 + T_2$	155	–	-1	1	-1	-3
$S_1 + T_3$	125	–	–	2	2	–
$S_2 + S_3$	186	-2	1	-1	-1	1
$S_2 + T_1$	102	–	-1	-1	-1	3
$S_2 + T_2$	140	–	–	–	-4	–
$S_2 + T_3$	163	–	1	1	-1	3
$S_3 + T_1$	146	–	–	-2	2	–
$S_3 + T_2$	154	–	1	-1	-1	-3
$S_3 + T_3$	171	–	2	–	2	–
$T_1 + T_2$	99	2	-1	-1	-1	-1
$T_1 + T_3$	119	2	–	–	2	2
$T_2 + T_3$	127	2	1	1	-1	-1
Divisor		96	64	64	192	192
Sum		-126	301	-31	-93	-57

TABLE 9.5.3 Orthogonal coefficients for contrasts within litters for Table 9.4.1

Doses	Totals	L_p	L_1	L_1'	L_2	L_2'
$S_2 - S_1$	31	–	1	-1	-3	3
$S_3 - S_1$	39	–	2	-2	–	–
$T_1 - S_1$	-6	2	–	-2	–	2
$T_2 - S_1$	1	2	1	-1	-3	-1
$T_3 - S_1$	33	2	2	–	–	2
$S_3 - S_2$	26	–	1	-1	3	-3
$T_1 - S_2$	-62	2	-1	-1	3	-1
$T_2 - S_2$	0	2	–	–	–	-4
$T_3 - S_2$	3	2	1	1	3	-1
$T_1 - S_3$	-38	2	-2	–	–	2
$T_2 - S_3$	-2	2	-1	1	-3	-1
$T_3 - S_3$	-23	2	–	2	–	2
$T_2 - T_1$	35	–	1	1	-3	-3
$T_3 - T_1$	27	–	2	2	–	–
$T_3 - T_2$	13	–	1	1	3	3
Divisor		144	96	96	288	288
Sum		-188	447	-5	-255	-59

The intra-litter section of the analysis of variance may be similarly calculated. Table 9.5.3 shows the 15 types of difference between the two doses assigned to a litter and the totals of these differences for the two litters of each type. The coefficients for the contrasts are the differences of the coefficients in Table 5.2.3 corresponding to the pair of doses in a litter. In Table 9.5.3, the coefficients need not add to zero in each column, because each row of the table itself refers to a difference between two pairs of responses and the condition for a contrast is maintained. Again the contrasts are mutually orthogonal. The

TABLE 9.5.4 Analysis of variance for Table 9.4.1

Adjustment for mean		67 536	
Nature of variation	d.f.	Sum of squares	Mean square
Preparations	1	165	165
Regression	1	1 416	1416
Parallelism	1	15	15
Quadratic	1	45	45
Difference of quadratics	1	17	17
Error (1)	24	2 590	107·92
Between litters	29	4 248	
Preparations	1	245	245
Regression	1	2 081	2081
Parallelism	1	0	0
Quadratic	1	226	226
Difference of quadratics	1	12	12
Error (2)	25	1 039	41·56
Total	59	7 851	

divisors are found as for Table 9.5.2, and the analysis of variance may then be completed. Note that the sum of values for any pair of corresponding contrasts in Tables 9.5.2 and 9.5.3 is twice the corresponding value in the analysis of § 9.3 ($-126 - 188 = 2 \times 157$): dose totals are identical for the two sets of data, and in the present analysis each response is used twice.

If the error variance between litters, s_1^2, were much larger than that within litters, s_2^2, the first part of the analysis would be uninformative and all conclusions would be based on the second. Here s_1^2 is only 2·6 times s_2^2, but conclusions will be first obtained from the intra-litter analysis alone. For genuine data, validity tests would be made in the usual manner. In Table 9.5.4, both the preparations and the quadratic mean square are significantly greater than s_2^2, but this is solely because the artificial example is insufficiently realistic; the very small mean square for parallelism is more exactly 0·26, a value not significantly less than s_2^2. The coefficients in Table 9.5.3 show L_p to comprise the equivalent of 36 responses to T minus 36 responses to S. Therefore, in terms of intra-litter estimation,

$$\bar{y}_T - \bar{y}_S = L_p/36 = -5 \cdot 2222, \tag{9.5.1}$$

with a variance

$$\text{Var}(y_T - y_S) = 144 s_2^2/36^2 = s_2^2/9. \tag{9.5.2}$$

Similarly, L_1 is in essence a difference between 24 responses to high doses and 24 to low doses; because the responses to middle doses that occur in its definition (introduced in order to eliminate litter differences) have S_2 and T_2 as often with negative as with positive signs, their total expectation is zero. The ratio of successive doses was $\frac{5}{3}$, and therefore (§ 5.3) the logarithm to base $\frac{5}{3}$ is the convenient dose metameter. On this scale, the difference between high and low doses is 2; consequently the expected value of L_1 is 2×24 times the true regression

coefficient. Hence the regression coefficient is estimated as
$$b = L_1/48 = 9\cdot3125, \tag{9.5.3}$$
with variance
$$\text{Var}(b) = 96s_2^2/48^2 = s_2^2/24. \tag{9.5.4}$$
An estimate of relative potency may be formed from equation (4.11.5)
$$M = \bar{x}_S - \bar{x}_T - \frac{5\cdot2222}{9\cdot3125}$$
$$= \bar{x}_S - \bar{x}_T - 0\cdot5608.$$
Moreover, since $t = 2\cdot060$ for 25 d.f.,
$$g = \frac{(2\cdot060)^2 \times 41\cdot56}{(9\cdot3125)^2 \times 24} = 0\cdot0847,$$
and application of Fieller's theorem to the ratio $(\bar{y}_T - \bar{y}_S)/b$ gives
$$M_L, M_U = \bar{x}_S - \bar{x}_T - \left[0\cdot5608 \pm \frac{2\cdot060 s_2}{9\cdot3125}\left\{\frac{0\cdot9153}{9} + \frac{(0\cdot5608)^2}{24}\right\}^{1/2}\right]/0\cdot9153$$
$$= \bar{x}_S - \bar{x}_T - 1\cdot1406, \quad \bar{x}_S - \bar{x}_T - 0\cdot0848.$$

Multiplication by $\log \frac{5}{3}$, or $0\cdot22185$, converts these quantities to logarithms to base 10. As in § 9.3,
$$\left. \begin{array}{l} R = 2 \text{ antilog } \bar{1}\cdot8756 = 1\cdot502, \\ R_L = 2 \text{ antilog } \bar{1}\cdot7470 = 1\cdot117, \\ R_U = 2 \text{ antilog } \bar{1}\cdot9812 = 1\cdot915. \end{array} \right\} \tag{9.5.5}$$

Contrasts between litters provide entirely independent information on differences between doses, and hence on relative potency. The first part of Table 9.5.4 shows no indications of invalidity, and calculations similar to those just described give a second potency estimate. From Table 9.5.2, L_p is seen to be a difference between two sets of 24 responses, whence
$$\bar{y}_T - \bar{y}_S = \frac{L_p}{24} = -5\cdot2500, \tag{9.5.6}$$
with
$$\text{Var}(\bar{y}_T - \bar{y}_S) = 96s_1^2/24^2 = s_1^2/6. \tag{9.5.7}$$
A similar argument shows that
$$b = \frac{L_1}{32} = 9\cdot4062, \tag{9.5.8}$$
with
$$\text{Var}(b) = s_1^2/16. \tag{9.5.9}$$
Hence

$$M = \bar{x}_S - \bar{x}_T - 0.5581.$$

For 24 d.f., $t = 2.064$ leads to

$$g = 0.3248,$$

and Fieller's theorem applied to $(\bar{y}_T - \bar{y}_S)/b$ gives

$$M_L, M_U = \bar{x}_S - \bar{x}_T - 2.0532, \quad \bar{x}_S - \bar{x}_T + 0.4000.$$

The estimate and its fiducial limits are

$$\left. \begin{aligned} R &= 2 \text{ antilog } \bar{1}\cdot 8762 = 1\cdot 504, \\ R_L &= 2 \text{ antilog } \bar{1}\cdot 5445 = 0\cdot 701, \\ R_U &= 2 \text{ antilog } 0\cdot 0887 = 2\cdot 454. \end{aligned} \right\} \quad (9.5.10)$$

The very close agreement between the two values of R is attributable to the artificiality of the data; the two are statistically as truly independent of one another as if they had been derived from separate experiments. As might be expected, the second estimate is much less reliable than the first, yet it certainly conveys some information and a combined estimate should be better than the first alone. A method somewhat different from those of Chapter 14 seems appropriate.

Had the two error variances in Table 9.5.4 been equal, the variances of corresponding quantities estimated from intra- and inter-block contrasts would have been in the ratio 2:3, as is evident from equations (9.5.2), (9.5.4), (9.5.7) and (9.5.9). This may be expressed by the statement that 60 percent of the information on contrasts between doses is unconfounded and 40 percent is confounded between blocks. The variance of any intra-block contrast is equal to the variance formula in an unconfounded experiment with the same replication divided by 0.6; division by 0.4 gives the corresponding inter-block variance. For example, in a randomized block (3, 3) assay with 10 replicates, the variance of a mean response would be $s^2/10$; hence

$$\begin{aligned} \text{Var}(\bar{y}_T - \bar{y}_S) &= \text{Var}[\tfrac{1}{3}(\bar{y}_{T1} + \bar{y}_{T2} + \bar{y}_{T3}) - \tfrac{1}{3}(\bar{y}_{S1} + \bar{y}_{S2} + \bar{y}_{S3})] \\ &= \frac{6}{9} \times \frac{s^2}{10} \\ &= s^2/15. \end{aligned} \quad (9.5.11)$$

Then for the intra-block estimate the variance is

$$(s_2^2/15) \div 0\cdot 6 = s_2^2/9,$$

and for the inter-block,

$$(s_1^2/15) \div 0\cdot 4 = s_1^2/6,$$

so confirming (9.5.2) and (9.5.7).

A first step towards combined estimation should be the application of composite validity tests. From Table 9.5.2, the inter-litter L_2 comprises the equivalent of 48 responses to S_1, S_3, T_1, T_3 taken positively, 16 more of these taken negatively, and 32 responses to S_2, T_2 taken negatively. In units of a mean difference between rats, the measure of curvature is $L_2/32$ with variance

$$192 s_1^2/32^2 = 3 s_1^2/16.$$

Correspondingly from Table 9.5.3 for intra-litter contrasts the measure of curvature is $L_2/48$ with variance

$$288 s_2^2/48^2 = s_2^2/8.$$

Again the rule of the previous paragraph could be used. The Behrens distribution (§ 4.13) then provides the test procedure. Equations (4.13.5) and (4.13.6) give the weighted measure of curvature

$$\frac{-\dfrac{93}{32} \times \dfrac{16}{3 s_1^2} - \dfrac{255}{48} \times \dfrac{8}{s_2^2}}{\dfrac{16}{3 s_1^2} + \dfrac{8}{s_2^2}} = -\frac{0 \cdot 1436 + 1 \cdot 0226}{0 \cdot 0494 + 0 \cdot 1925}$$

$$= -4 \cdot 821, \tag{9.5.12}$$

with variance

$$\frac{1}{0 \cdot 0494 + 0 \cdot 1925} = 4 \cdot 134 \tag{9.5.13}$$

$$= (2 \cdot 033)^2.$$

From (4.13.8),

$$\tan \theta = \left(\frac{2 s_2^2}{3 s_1^2} \right)^{1/2} \tag{9.5.14}$$

$$= 0 \cdot 51,$$

whence

$$\theta = 27°.$$

By interpolation in Fisher and Yates's Table V_1, the 0·05 level of d for 24, 25 d.f. and 27° is 2·056; evidently the mean weighted curvature, equation (9.5.12), is more than 2·056 times its standard error, and the assay shows significant deviation from linearity. This and other similar validity tests are of no real interest for artificial data.

Ignoring now any indications of invalidity, only two sets of data are to be combined and these conform to conditions of symmetry set out in § 4.13. If $(\bar{y}_T - \bar{y}_S)$ and b for the two sections of the analysis, as defined by equations (9.5.1), (9.5.3), (9.5.6) and (9.5.8), play the roles of a_2, a_1, b_2, b_1 in the second analogue of Fieller's theorem, then, in the notation of § 4.13,

$v_{11} = \tfrac{1}{6},$

$v_{12} = 0,$

$v_{22} = \tfrac{1}{16},$

$k = \tfrac{2}{3},$

and, by equation (4.13.22),

$$s^2 = \left(\frac{1}{s_1^2} + \frac{3}{2s_2^2}\right)^{-1}$$

$$= 1/(0\cdot 009\ 27 + 0\cdot 036\ 09)$$

$$= 22\cdot 05.$$

The mean values of the response difference and the regression coefficient are

$$\bar{y}_T - \bar{y}_S = \frac{-5\cdot 2500 \times 0\cdot 009\ 27 - 5\cdot 2222 \times 0\cdot 036\ 09}{0\cdot 009\ 27 + 0\cdot 036\ 09}$$

$$= -5\cdot 2279,$$

$$b = \frac{9\cdot 4062 \times 0\cdot 009\ 27 + 9\cdot 3125 \times 0\cdot 036\ 09}{0\cdot 009\ 27 + 0\cdot 036\ 09}$$

$$= 9\cdot 3316.$$

Hence

$$M = -\frac{5\cdot 2279}{9\cdot 3316}$$

$$= -0\cdot 5602.$$

Moreover

$$\mathrm{Var}(\bar{y}_T - \bar{y}_S) = \frac{s^2}{6},$$

$$\mathrm{Var}(b) = \frac{s^2}{16},$$

and the orthogonality of the design ensures that the covariance is zero. By equation (4.13.21),

$$\tan \theta = \left(\frac{2s_2^2}{3s_1^2}\right)^{1/2},$$

a repetition of equation (9.5.14) which therefore again gives 2·056 as the 0·05 d-deviate. Equation (4.13.24) becomes

$$g = \frac{(2 \cdot 056)^2 \times 22 \cdot 05}{(9 \cdot 3316)^2 \times 16}$$
$$= 0 \cdot 0669,$$

and, by substitution in equation (4.13.23),

$$M_L, M_U = \bar{x}_S - \bar{x}_T - \left[0 \cdot 5602 \pm \frac{2 \cdot 056 s}{9 \cdot 3316} \left\{ \frac{0 \cdot 9331}{6} + \frac{(0 \cdot 5602)^2}{16} \right\}^{1/2} \right] \bigg/ 0 \cdot 9331$$
$$= \bar{x}_S - \bar{x}_T - 1 \cdot 0644, \quad \bar{x}_S - \bar{x}_T - 0 \cdot 1363.$$

After the usual conversion of the base of logarithms,

$$\left. \begin{array}{l} R = 2 \text{ antilog } \bar{1} \cdot 8757 = 1 \cdot 502, \\ R_L = 2 \text{ antilog } \bar{1} \cdot 7639 = 1 \cdot 161, \\ R_U = 2 \text{ antilog } \bar{1} \cdot 9698 = 1 \cdot 866. \end{array} \right\} \tag{9.5.15}$$

Thus Table 9.5.4 leads to an estimate of potency of 1·50 units of vitamin A per mg, with fiducial limits at 1·16 and 1·87 units per mg. The fiducial interval is 12 percent narrower than that from intra-litter analysis alone, and the additional calculations are not heavy. However, Fisher (1961b) indicated the flaw in applying Behrens's theory to weighted means and showed that the effective precision may be lower than at first appears, especially when degrees of freedom are few. The 8 percent increase in width of interval relative to that in § 9.3 results from insistence that all dose contrasts be equally confounded. Had inter-litter variance been relatively greater, the loss from partial confounding of $(\bar{y}_T - \bar{y}_S)$ might have more than balanced the greater precision of b; in this instance, the only gains from the balanced incomplete block design are in the power of the curvature test and the fact that numerator and denominator of M are contrasts of the same type.

9.6 Statistical analysis of balanced incomplete block designs

If inter-block and intra-block information are to be legitimately used in the analysis of an assay, random allocation of the sets of doses to blocks is essential. If the only randomization has been that within blocks, interpretation and estimation must be restricted to the intra-block analysis. When a fully randomized incomplete block design is used in an ordinary experiment for the comparison of treatments, the results are usually summarized in terms of weighted means of inter- and intra-block estimates of treatment differences. The method of statistical analysis in § 9.5 might therefore have been expected to be identical with that first described by Yates (Cochran and Cox, 1957, Chapter 11; Fisher and Yates, 1963), except for the final stages relating to calculation of M and its limits. Although the potency estimate was calculated according to the principles of Chapter 14, the symmetry of the design ensures that the result is the same as if equation (5.5.1) had been applied to weighted mean responses for the six doses.

One small difference from Yates's method appears in the analysis of variance.

Yates ingeniously obtained additional information on the inter-block variance from the sum of squares between treatments (i.e. between doses) in the inter-block section of the analysis, a valuable device in experiments for which the inter-block error has few d.f. The disadvantages are that the fiducial limit formulae do not apply and that the structure of the analysis is less easily understood. In most biological assays with balanced incomplete designs, as in § 9.5, s_1^2 will have enough d.f. for this refinement to be ignored. If preferred, Yates's method may be followed exactly until adjusted mean responses to each dose have been formed; an estimate of potency is then calculated as if these were responses in an assay with one subject per dose. Provided that the number of degrees of freedom involved is reasonably large, a deviate taken from Fisher and Yates's Table V_1 as roughly representative of the appropriate part of the table might perhaps be used in equation (4.13.23) without great harm. For Table 9.5.1, for example, the Yates analysis with $d = 2.06$ gives $R = 1.503$ with limits at 1·166, 1·860, which scarcely differ from the results of the other analysis.

Bliss (1947a) published details of the statistical analysis, by Yates's method, of two balanced incomplete block designs, both of which are also cross-over designs (Chapter 10). The rather simpler computations needed for the procedure of § 9.5 are described in §§ 10.6 and 12.5.

9.7 Catalogue of balanced incomplete block designs

No general rules exist for constructing all balanced incomplete block designs, but published lists (Cochran and Cox, 1957, Tables 9.5, 11.3; Fisher and Yates, 1963, Tables XVII–XIX) include all that are likely to be required in biological assay. Indeed, if the recommendations in Chapters 6 and 8 on numbers of doses are followed, only the simplest designs will be needed, except perhaps for multiple assays (Chapter 11). The most useful designs for parallel line and slope ratio assays are here classified according to the numbers of dose levels in a symmetric assay; they may be used for unsymmetric sets of doses if required. If greater replication than the stated minimal number of blocks is desired, the basic block pattern should be repeated.

(i) *(2, 2) designs* The design for blocks of 2 uses all possible blocks of this size, the property that defines an *unreduced* design. Six blocks are $S_1, S_2; S_1, T_1;$ $S_1, T_2; S_2, T_1; S_2, T_2; T_1, T_2$; these are repeated on further sets of six blocks, if more are available. The design is very similar to that discussed in § 9.5, with the simplification that it has only four doses instead of six. For blocks of 3, a set of four blocks is obtained by omitting each dose in turn; again the arrangement is repeated if possible.

(ii) *(3, 3) designs* The only design for blocks of 2 consists of a set of 15 blocks making up all possible pairs (§ 9.5). For blocks of 3, a design in 10 blocks is possible, this being half the total number of different blocks of 3 that can be formed (Table 9.7.1); each dose is used for five subjects, and each pair of doses occurs in two blocks. If 20 blocks were available, all possible combinations of three doses would be used, but the design in Table 9.7.1 is equally amenable to

§ 9.7 INCOMPLETE BLOCK DESIGNS

TABLE 9.7.1 Balanced incomplete block (3, 3) design in 10 blocks of three

Block no.	S_1	S_2	S_3	T_1	T_2	T_3
I	X	X		X		
II	X	X				X
III	X		X		X	
IV	X		X			X
V	X			X	X	
VI		X	X	X		
VII		X	X		X	
VIII		X			X	X
IX			X	X		X
X				X	X	X

TABLE 9.7.2 Balanced incomplete block (4, 4) design in 14 blocks of four

Block no.	S_1	S_2	S_3	S_4	T_1	T_2	T_3	T_4
I	X	X	X	X				
II	X	X			X	X		
III	X	X					X	X
IV	X		X		X	X		
V	X		X			X		X
VI	X			X	X			X
VII	X			X		X	X	
VIII		X	X		X			X
IX		X	X			X	X	
X		X		X	X		X	
XI		X		X		X		X
XII			X	X	X	X		
XIII			X	X			X	X
XIV					X	X	X	X

TABLE 9.7.3 Balanced incomplete block (1, 3, 3) design in 7 blocks of three

Block no.	C	S_1	S_2	S_3	T_1	T_2	T_3
I	X	X	X				
II	X			X	X		
III	X					X	X
IV		X		X		X	
V		X			X		X
VI			X	X			X
VII			X		X	X	

analysis by the standard procedure. The only design for blocks of 4 is unreduced; formed by omission of every pair of doses in turn, it requires 15 blocks. For blocks of 5, a set of six blocks is given by omitting each dose in turn.

(iii) *(4, 4) designs* For blocks of 2, 3, 5, 6 or 7, the only possible designs are

TABLE 9.7.4   Balanced incomplete block (1, 4, 4) design in 12 blocks of three

Block no.	C	S_1	S_2	S_3	S_4	T_1	T_2	T_3	T_4
I	×	×					×		
II	×		×					×	
III	×			×		×			
IV	×				×				×
V		×	×	×					
VI		×			×	×			
VII		×						×	×
VIII			×		×		×		
IX			×			×			×
X				×	×			×	
XI			×				×		×
XII						×	×	×	

TABLE 9.7.5   Balanced incomplete block (1, 4, 4) design in 18 blocks of four

Block no.	C	S_1	S_2	S_3	S_4	T_1	T_2	T_3	T_4
I	×	×		×				×	
II	×	×			×	×			
III	×	×					×		×
IV	×		×	×	×				
V	×		×			×	×		
VI	×		×					×	×
VII	×			×		×			
VIII	×				×		×	×	
IX		×	×	×			×		
X		×	×		×			×	
XI		×	×			×			×
XII	×			×	×				
XIII						×	×	×	
XIV			×	×		×		×	
XV			×		×		×		×
XVI				×	×	×	×		
XVII				×			×	×	×
XVIII					×	×		×	×

unreduced; hence the number of blocks must be a multiple of 28, 56, 56, 28 or 8 respectively. For blocks of 4, a design in 14 blocks, with only seven subjects at each dose, is possible (Table 9.7.2).

Designs for larger numbers of doses might be needed for multiple assays, but will not be listed here. Slope ratio assays seem less commonly to require blocking, but some $(1, k, k)$ designs are described here.

(iv) $(1, 1, 1)$ *designs*   Omission of each dose in turn gives a set of three blocks of 2, which may be replicated as often as desired in order to give an experiment of reasonable precision.

(v) $(1, 2, 2)$ *designs*   Designs in blocks of 2, 3, or 4 are unreduced; the number of blocks must therefore be a multiple of 10, 10, or 5 respectively.

§ 9.8   INCOMPLETE BLOCK DESIGNS

(vi) $(1,3,3)$ *designs* As stated in § 7.11, those designs are not likely to be wanted often. For blocks of 2, 5, or 6 only the unreduced type is possible. For blocks of 3, the design in Table 9.7.3 is balanced in seven blocks, instead of the 35 that would be required if all possible sets of 3 were used. The complement of this, consisting of doses omitted from each block, is a design for blocks of 4.

(vii) $(1,4,4)$ *designs* For blocks of 2, 7, or 8, only unreduced designs are possible. For blocks of 3, the 12 blocks in Table 9.7.4 form a balanced design with four subjects on each dose; the complement, the omitted doses, is a design in blocks of 6. For blocks of 4, Table 9.7.5 shows a balanced design in 18 blocks with eight subjects on each dose; the complement is a design in blocks of 5.

What is to be done if the number of blocks available is less than the minimum number for a balanced incomplete block design with the required number of doses, or is not an exact multiple of this minimum? The design constraints are inflexible; lesser symmetry brings contrasts of unequal precision and more complex calculations, though a computer will still remove most of the latter difficulty. Partially balanced designs (Cochran and Cox, 1957) are useful; abandonment of balance, and adoption of some simpler confounding scheme on the lines of § 9.8 will often be preferred.

A type of design that might occasionally be needed uses 'super-complete' blocks. If for a particular assay a $(2, 2)$ scheme of doses were considered desirable in the light of the arguments of Chapter 6, the fact that the most convenient blocks contain six subjects should not be regarded as a reason for adopting instead a $(3, 3)$ design. The better procedure is to make each block consist of one subject per dose plus two extra subjects assigned to doses in accordance with a balanced incomplete block scheme for blocks of two. The statistical analysis would present no special difficulties (cf. § 14.4).

9.8   Unbalanced confounding in the assay of vitamin A

The objections to the design in § 9.3 can be overcome differently, by seeking an alternative scheme of confounding that will give L_p and L_1 entirely as intra-block contrasts. Any such design must relegate some validity tests to dependence upon inter-block contrasts. As the example that follows illustrates, the computations are much simpler than those of §§ 9.3 and 9.5, because Fieller's theorem can be used in its original form. Considerations of simplicity of statistical analysis naturally appeal to the assayist, but (§ 6.6) should not persuade him to adopt a design other than the best for his purpose. If he can afford some sacrifice on validity tests, the designs described in this and the next section are excellent: if his knowledge of his materials and experimental conditions are insufficient, he must be content with a more laborious balanced incomplete block design.

For a $(3, 3)$ design in blocks of two, a very simple modification makes the required change in confounding. Suppose that three types of block are used again, but that, instead of the block constitutions in § 9.3, they are made up as $S_1, T_3; S_2, T_2; S_3, T_1$: the contrasts L_1 and L'_1 now change places in the analysis, the former becoming intra-block and the latter inter-block. In other respects the

TABLE 9.8.1 Second set of artificial data for weight increases (g) of male rats during three weeks of dosage with vitamin A
(Each pair of columns contains the responses for ten litters)

S_1 0·9 units	T_3 1·25 mg	S_2 1·5 units	T_2 0·75 mg	S_3 2·5 units	T_1 0·45 mg
20	33	43	35	51	19
22	26	37	30	45	17
18	35	38	37	46	5
26	32	40	37	48	27
30	32	47	31	55	21
29	53	27	35	35	28
47	52	41	48	49	23
16	38	43	47	51	33
16	37	35	18	43	18
30	41	30	30	27	10
254	379	381	348	450	201

TABLE 9.8.2 Analysis of variance for Table 9.8.1

Adjustment for mean		67 536	
Nature of variation	d.f.	Sum of squares	Mean square
Parallelism	1	8	8
Quadratic	1	252	252
Error (1)	27	2 680	99·26
Between litters	29	2 940	
Preparations	1	411	411
Regression	1	3 497	3497
Difference of quadratics	1	28	28
Error (2)	27	1 067	39·52
Total	59	7 943	

analysis is unaltered, so that the transfer of the regression contrast to the intrablock analysis is accomplished at the price of reduced power in the test of parallelism. An artificial example will emphasize the similarity to the analysis in § 9.3. The data in Table 9.8.1 purport to be from an assay of this design used in exactly the same circumstances as was the assay of vitamin A in Table 9.3.1; numerical values have been constructed so as to have the same dose totals and about the same variances between and within blocks.

As for Table 9.3.1, the device of forming sums and differences of pairs aids the subdivision of the total sum of squares into the components between and within litters. Since the dose totals are the same as before, the values of L_p, L_1, L'_1, L_2 and L'_2 are unaltered, but, as already stated, L_p, L_1 and L'_2 are now orthogonal with litter contrasts. The analysis of variance is shown in Table 9.8.2.

This artificial example needs no discussion of validity tests. The potency estimate, calculated exactly as in § 9.3, is again

§ 9.9　　　INCOMPLETE BLOCK DESIGNS　　　197

$$R = 1\cdot 503.$$

Since L_p and L_1 are both intra-block contrasts, only the one variance,

$$s_2^2 = 39\cdot 52$$

enters into the fiducial limit calculations. In the simple form of Fieller's theorem,

$$g = \frac{(2\cdot 052)^2 \times 40 \times 39\cdot 52}{374^2}$$

$$= 0\cdot 0476,$$

a substantial improvement on the value of $0\cdot 1185$ in § 9.3. Fiducial limits to L_p/L_1 may then be calculated, and in the usual manner converted into limits for R. The results are

$$R_L = 1\cdot 216,$$

$$R_U = 1\cdot 805.$$

Equations (5.5.1) and (5.5.6) may legitimately be applied to the calculation of R, R_L and R_U when, as here, L_p and L_1 are completely unconfounded.

The fiducial range, about 10 percent narrower than that in § 9.3, benefits from the reduction in g and from the reduction in the second term within the square root in the fiducial limit formula. The change is not dramatic, because the two error mean squares in the analysis of variance are of the same order of magnitude; the gain perhaps scarcely outweighs the loss of power on the test of parallelism. On other occasions, the change of design might make all the difference between a hopelessly imprecise estimate of potency and one of practical value.

9.9 Catalogue of confounded designs

The design in § 9.8 is one of a series for (k, k) parallel line assays, all of which leave L_p, L_1 unconfounded with blocks. If the investigator can afford some loss of power in validity tests, they are preferable to balanced incomplete blocks. The designs that follow are appropriate for assays with symmetric sets of regularly spaced doses.

(i) *(2, 2) designs*　　The balanced design for blocks of 2 requires sets of six blocks. A design using only two of these block types, with pairs of blocks assigned to S_1, T_2, and S_2, T_1, gives L_p, L_1 as intra-block contrasts and confounds L_1' as an inter-block contrast. The squares are computed in the usual manner, as illustrated in § 4.16, but are allocated to the appropriate section of the analysis of variance table as in § 9.8.

This design has been used to assay preparations of plant viruses. Price (1946) described an assay using two leaves from each plant with the halves of each leaf regarded as separate 'subjects'; the two halves of one leaf received doses S_1, T_1, and the two halves of the other leaf from the same plant received S_2, T_2, so confounding the regression contrast with differences between leaves. He noted the improvement of assigning a randomly chosen leaf to S_1, T_2 and the other on the same plant to S_2, T_1. Use of two leaves from each plant enables the confounded parallelism contrast to have the precision of an intra-plant comparison. The leaves and the two sides of each leaf would be numbered and labelled arbitrarily. Table 9.9.1 shows part of a suitably randomized design, for which

TABLE 9.9.1 Randomized (2, 2) design for the assay of a plant virus (L_p and L_1 unconfounded)

Plant no.	Leaf 1		Leaf 2	
	Left half	Right half	Left half	Right half
I	S_2	T_1	T_2	S_1
II	T_2	S_1	T_1	S_2
III	S_1	T_2	T_1	S_2
IV	T_2	S_1	S_2	T_1
V	T_1	S_2	T_2	S_1
⋮	⋮	⋮	⋮	⋮

TABLE 9.9.2 Partition of degrees of freedom in the analysis of variance for Table 9.9.1 with K plants

Nature of variation	d.f.
Between plants	$K - 1$
Parallelism	1
Error (1)	$K - 1$
Between leaves	$2K - 1$
Preparations	1
Regression	1
Error (2)	$2K - 2$
Total	$4K - 1$

the analysis of variance would have its degrees of freedom partitioned as in Table 9.9.2. If leaf 2 is always taken higher on the plant than leaf 1, if the 'left half' of a leaf is always defined as the half nearer to the light, or if in some similar way these names represent identifiable positions instead of arbitrary distinctions, the number of plants should be a multiple of 2 or of 4. The design can then be balanced for positional effects; components for these and for their interactions with dose effects must be isolated in the analysis of variance. Price (1945) and Price and Spencer (1943) have presented results from such assays.

(ii) *(3, 3) designs* The design for blocks of 2 (cf. §9.8) has block types S_1, T_3; S_2, T_2; S_3, T_1. The complementary design in blocks of 4, given by omitting each of these pairs of doses in turn, has the three block types shown in Table 9.9.3. This valuable and easily analyzed design would be ideal, for example, in a cylinder-plate assay for antibiotics in which not more than four tests could be

TABLE 9.9.3 $(3, 3)$ design in $3K$ blocks of four (L_p and L_1 unconfounded)

Block type	S_1	S_2	S_3	T_1	T_2	T_3
I	X	X			X	X
II	X		X	X		X
III		X	X	X	X	

accommodated on one plate. Evidently L_p, L_1, L_2' are completely free of confounding, and are therefore estimated in the usual manner from dose totals. The parallelism and quadratic contrasts are partially confounded: the intra-block estimates must be based on

$$L_1' = L_1'(\text{crude}) + \tfrac{1}{2}(B_{III} - B_I), \tag{9.9.1}$$

$$L_2 = L_2(\text{crude}) + \tfrac{1}{2}(B_I - 2B_{II} + B_{III}), \tag{9.9.2}$$

where the crude values are formed with the coefficients in Table 5.2.3 and B_I, B_{II}, B_{III} represent totals of all blocks of the three types. If the assay has K blocks of each type, the squares for dose contrasts in the analysis of variance are:

Preparations: $L_p^2/12K,$
Regression: $L_1^2/8K,$
Parallelism: $L_1'^2/6K,$
Quadratic: $L_2^2/18K,$
Difference of quadratics: $L_2'^2/24K.$

The principle is to add to the crude L_1', L_2 multiples of block totals determined so as to make each contrast orthogonal with all block contrasts. The standard rule for the divisor of the square of a contrast is applied after expression of the contrast in terms of the $12K$ individual responses.

From the inter-block section of the analysis, independent estimates of the partially confounded contrasts can be obtained as

$$L_1' = -\tfrac{1}{2}(B_{III} - B_I), \tag{9.9.3}$$

$$L_2 = -\tfrac{1}{2}(B_I - 2B_{II} + B_{III}). \tag{9.9.4}$$

The squares, $L_1'^2/2K$ and $L_2^2/6K$, correspond to 2 of the $(3K - 1)$ d.f. between blocks. However, even if the inter-block variance were no greater than the intra-block, only one-quarter of the information on L_1', L_2, would reside in the inter-block contrasts; if the inter-block variance is much larger, this part of the information is trivial. For most purposes, the intra-block analysis suffices; except for the modified definitions of L_1', L_2, it is as simple as if complete blocks of 6 were used.

If the block size is odd, partial confounding of L_p is unavoidable. Nevertheless, more precise estimation of L_p, L_1 than in the balanced incomplete block design is possible, at the usual price of less powerful validity tests. A design in blocks of 3 should have the two block types in Table 9.9.4; L_1, L'_1, L_2 are unconfounded, and the confounding of information on L_p is minimized. An intra-block estimate of L_p may be formed as

$$L_p = L_p \text{(crude)} + \tfrac{1}{3}(B_\mathrm{I} - B_\mathrm{II}), \qquad (9.9.5)$$

where B_I and B_II are totals of all blocks of the two types. This in fact is identical with L'_2, and the design is useless unless prior experience justifies neglect of L'_2. If the whole assay has $2K$ blocks, the components of the intra-block sum of squares are

Preparations: $3L_p^2/16K$,
Regression: $L_1^2/4K$,
Parallelism: $L_1'^2/4K$,
Quadratic: $L_2^2/12K$.

The contrast between block types can yield information on the difference of preparations; information from this source is usually too small to be worth recovery. The intra-block estimate of $(\bar{y}_T - \bar{y}_S)$ is $3L_p/8K$. Again the analysis is equivalent to the general regression procedure in § 9.5.

TABLE 9.9.4   (3, 3) design in $2K$ blocks of three (L_1 unconfounded and L_p minimally confounded)

Block type	S_1	S_2	S_3	T_1	T_2	T_3
I	×		×		×	
II		×		×		×

TABLE 9.9.5   (4, 4) design in $2K$ blocks of four (all contrasts unconfounded except L'_2)

Block type	S_1	S_2	S_3	S_4	T_1	T_2	T_3	T_4
I	×			×		×	×	
II		×	×		×			×

(iii) *(4, 4) designs*   For blocks of 2, the number of blocks should be a multiple of 4, equal numbers being assigned to the types $S_1, T_4; S_2, T_3; S_3, T_2; S_4, T_1$. This leaves L_p, L_1, L'_2, L_3 unconfounded, with L'_1, L_2, L'_3 completely confounded between blocks. For blocks of 4, a design having equal numbers of blocks of each of the two types in Table 9.9.5 has L'_2 completely confounded between block types but no other confounding. The complement of the design in blocks of 2 gives a design for blocks of 6 in sets of four blocks. The analyses should have no great difficulty for those who have mastered the (3, 3) design in blocks of 4. Designs in blocks of 3, 5 and 7, constructed on the analogy of Table 9.9.4, are not particularly important.

§ 9.10   INCOMPLETE BLOCK DESIGNS

The designs introduced above might reasonably be used also for $(0, k, k)$ slope ratio assays, though the contrasts needed in the analyses would be different. The confounding of $(1, k, k)$ assays is more troublesome because the total number of doses is odd. One useful procedure is to add a zero dose to each block of a $(0, k, k)$ design, so giving that dose greater replication than any of the others. Das and Kulkarni (1966) proposed modifying balanced incomplete block designs by augmenting them with zero doses. Kulshreshtha (1969) showed certain advantages for these, but later (1972) introduced a new type of design that he regarded as superior. His claim of higher precision is doubtless justified, although to some extent this comes from the effectively greater replication as his comparison is not based on equal total numbers of responses measured. Table 9.9.6 shows a Kulshreshtha design. Table 9.9.7, which is obtained by adding a zero

TABLE 9.9.6 $(1, 4, 4)$ design in $2K$ blocks of five (Kulshreshtha design)

Block type	C	S_1	S_2	S_3	S_4	T_1	T_2	T_3	T_4
I	X	X			X	X			X
II	X		X	X			X	X	

TABLE 9.9.7 $(1, 4, 4)$ design in $2K$ blocks of five (adapted from Table 9.9.5)

Block type	C	S_1	S_2	S_3	S_4	T_1	T_2	T_3	T_4
I	X	X			X		X	X	
II	X		X	X		X			X

dose to each block of Table 9.9.5, is almost certainly better in respect of validity tests without losing on precision; moreover, the principle underlying this design can be used for all k, whereas Kulshreshtha had to limit himself to k being even. If these designs were wanted often, the appropriate analyses in terms of contrasts could be developed by adaptation of the methods that the authors mentioned above have presented. For occasional use, a general regression approach is perhaps the best, especially as it is so easily obtained from standard computer programs.

Enough has been said to show the advantages of assay designs that take account of existing knowledge of the assay technique and of the preparations under examination, and that concentrate information on the most valuable information. For an assay where validity is highly suspect, the emphasis might have to be reversed in order to achieve power of tests at the price of low precision. Each new situation encountered calls for a careful choice of designs; competent statistical advice is at least as important at this stage of an investigation as it is for the analysis of results.

9.10   Blocks of unequal size

In agricultural experimentation, the number of units per block can usually be held constant. In animal experimentation, the available subjects may include potential blocks of several different sizes, as when each block is to consist of litter-mates. One possibility is to discard animals before the experiment begins, so as to reduce each block to the size of the smallest block, but this can be extravagant. Alternatively, several incomplete block designs of the types described in §§ 9.7 and 9.9 (one for each size of litter) could be combined into a single statistical analysis; in practice, the number of litters of each size would rarely suffice to give satisfactory symmetry in this way. The oestrone assay in Table 4.15.1 illustrates the difficulty. The design omitted each dose in turn from the litters of 3; it would have been symmetric if the number of these litters had been a multiple of 4, but inclusion of the last litter spoiled this. The number of possibilities is too great for detailed advice, but the general principle should be to approach as closely as possible to symmetry. Not only will the results then be as precise as possible, but the statistical analysis will be less laborious. 'Symmetry', however, is to be interpreted widely as relating either to the complete balance in § 9.7 or to the carefully selected confounding in § 9.9, according to the needs of the projected assay.

The assumption that intra-litter variance is independent of litter size, theoretically a requirement for the estimation of precision by the formulae of this book, is likely to be more nearly correct than would be the corresponding assumption in an agricultural field trial. Certainly one need seldom worry about using, say, blocks of 6, 5 and 4 in one experiment and assuming σ^2 to be the same for all.

10
Cross-over designs

10.1 Intra-subject estimation

In all the indirect assays of earlier chapters, only one response was measured on each subject: a dose was applied, the response was measured, and the subject was not used again. This may adversely influence precision, because natural variation between subjects is often responsible for much of the experimental variance. Many techniques that provide assay responses affect subjects so drastically that a second use of a subject might give biased assessment of response; indeed, some or all of the subjects may be killed in the course of obtaining one response. Some techniques can be applied repeatedly to one subject without appreciable disturbance of that subject's reactions. For example, insulin has commonly been assayed from the changes it induces in the blood-sugar concentration of rabbits (Burn *et al.*, 1950, Chapter V); provided that several days elapse between successive tests, one rabbit can be used a number of times. The responsiveness of subjects may change from one occasion of testing to another, and the statistical analysis must take account of this. Ideally an assay will be so designed that the precision of the potency estimate depends only upon the variance within subjects (that is to say, the variance between repeated tests on one subject), and is independent of the variance between subjects. Such an assay should be more precise than one in which each subject is used once only.

Arrangements in which each subject receives two or more different doses on successive occasions are known as *cross-over designs*. The simplest is that in which one group of subjects receives a dose of S followed by an approximately equipotent dose of T, while a second group receives the same doses in reverse order. Unless a form of the standard slope method (§ 3.15) is acceptable, the relative potency cannot be estimated. If a series of such trials is made, with a fixed dose of the standard and different doses of the test preparation, estimation becomes possible (Fieller *et al.*, 1939a, b; Fieller, 1940), but the rather specialized analysis will not be described here.

For the cross-over designs presented in this chapter, statistical independence of successive responses is assumed. The level of response may change from one occasion to another, as balance over occasions permits such effects to be eliminated. Non-independence can take such forms as additive residual effects of each dose on subsequent occasions, correlation between successive experimental error components, and autoregressive effects. Although these features may not greatly affect potency estimation, they can lead to heterogeneity of components of error and inflate some of the mean squares ordinarily included in error. Should there be reason to suspect their occurrence, a more cautious arrangement of the analysis of variance is needed; validity tests and limit calculations are usually possible under these more complicated models, but they sacrifice information if independence is complete (Finney, 1956).

10.2 The twin cross-over

The minimal requirement for a satisfactory parallel line assay is a (2, 2) design. Therefore the simplest useful type is the twin cross-over design (Smith *et al.*, 1944), which has sets of four subjects assigned to doses as in Table 10.2.1. The subjects should be numbered in random order, and the arrangement is repeated

TABLE 10.2.1 The twin cross-over design

Subject	Dose on Occasion 1	Occasion 2
I	S_1	T_2
II	S_2	T_1
III	T_1	S_2
IV	T_2	S_1

Repeat with each randomized set of four subjects.

on sets of four subjects. The difference in response for doses T_2, S_1, and also that for doses T_1, S_2, can be estimated from comparisons between tests on one subject. From the total of these is derived the average difference between high and low doses, or the regression coefficient; from their difference is derived the mean difference in response between the two preparations. The contrast for parallelism is confounded between subjects, so that the validity test has only the power of a non-cross-over design, but the gain in respect of the other two contrasts may be great. The balance over the two occasions eliminates any effect of a change in the general level of responsiveness, provided that the response on the second occasion is unaffected by what happened on the first.

The design of an experiment always determines the form of the statistical analysis: the twin cross-over cannot be analyzed in the same way as a simple (2, 2) point, for that would ignore the special constraints. The estimation of relative potency can be made according to any of the usual formulae, except that if any observations are missing special calculations will be needed, but the analysis of variance and the validity test require modification of the procedures in Chapters 4 and 5.

10.3 A cross-over assay of corticotrophin

Smith *et al.* (1944) have illustrated the calculations for a twin cross-over. The natural extension for (3, 3) assays is the triple cross-over, used extensively by Rerup (1958) in a series of papers on corticotrophin assay. Table 10.3.1 contains data from one of these (Rerup, 1959), in which 30 (see § 10.8) hypophysectomized male rats were divided into six sets of five for the sequences of treatment shown. On day 1, the left adrenals were removed and the decrease in ascorbic acid after intravenous injection of hormone was measured for each; on day 2, this was repeated for the right adrenals. The doses of S were 0·5, 1·0, 2·0

§ 10.3 CROSS-OVER DESIGNS

TABLE 10.3.1 Responses of 30 rats in a triple cross-over assay of corticotrophin (mg adrenal ascorbic acid / g adrenal weight).

Rat		Day 1		Day 2	'1 + 2'	'2 − 1'
I		3·41		3·62	7·03	0·21
II		3·80		3·30	7·10	−0·50
III	S_1	4·00	T_3	3·61	7·61	−0·39
IV		3·79		3·03	6·82	−0·76
V		3·72		3·08	6·80	−0·64
Totals		18·72		16·64	35·36	−2·08
VI		1·75		2·73	4·48	0·98
VII		3·81		3·67	7·48	−0·14
VIII	S_2	2·62	T_2	2·92	5·54	0·30
IX		3·57		3·18	6·75	−0·39
X		2·94		3·12	6·06	0·18
Totals		14·69		15·62	30·31	0·93
XI		3·48		3·89	7·37	0·41
XII		3·54		4·30	7·84	0·76
XIII	S_3	3·04	T_1	4·07	7·11	1·03
XIV		4·03		4·52	8·55	0·49
XV		2·94		4·11	7·05	1·17
Totals		17·03		20·89	37·92	3·86
XVI		4·09		3·15	7·24	−0·94
XVII		4·29		3·68	7·97	−0·61
XVIII	T_1	3·76	S_3	3·18	6·94	−0·58
XIX		3·34		2·95	6·29	−0·39
XX		4·33		3·98	8·31	−0·35
Totals		19·81		16·94	36·75	−2·87
XXI		4·11		4·13	8·24	0·02
XXII		4·01		3·59	7·60	−0·42
XXIII	T_2	4·20	S_2	3·77	7·97	−0·43
XXIV		3·76		4·01	7·77	0·25
XXV		2·93		2·36	5·29	−0·57
Totals		19·01		17·86	36·87	−1·15
XXVI		2·44		3·45	5·89	1·01
XXVII		3·33		3·61	6·94	0·28
XXVIII	T_3	3·71	S_1	3·68	7·39	−0·03
XXIX		3·43		3·98	7·41	0·55
XXX		3·23		3·68	6·91	0·45
Totals		16·14		18·40	34·54	2·26
Grand totals		105·40		106·35	211·75	0·95

milliunits of the U.S.P. reference standard, and those of T were 1·25, 2·5, 5·0 µg of the sample.

The design is closely related to the (3, 3) assay in blocks of 2 discussed in § 9.8. As will be demonstrated below, L_1' and L_2 are confounded between rats, while L_p, L_1 and L_2' are entirely based on intra-rat contrasts. A further five contrasts correspond to the interactions of the difference between days with each of these, measuring the extent to which a contrast changes from day 1 to day 2;

they necessarily follow the opposite pattern, $L'_1 D$ and $L_2 D$ being unconfounded. Additional columns in Table 10.3.1 contain totals and differences for the two days. From them, the first division of the analysis of variance is easily made. The sum of squares between rats is

$$(7{\cdot}03^2 + 7{\cdot}10^2 + \ldots + 6{\cdot}91^2 - 2 \times 747{\cdot}3010)/2 = 12{\cdot}2066.$$

The sum of squares within rats, obtainable by subtraction, can be independently calculated as

$$[(0{\cdot}21)^2 + (-0{\cdot}50)^2 + \ldots + (0{\cdot}45)^2]/2 = 5{\cdot}1619.$$

Table 10.3.2 contains dose totals for the days separately and for the combinations '1 + 2' and '2 − 1' again. From these in turn are formed (Table 10.3.3) the five standard contrasts, using the coefficients from Table 5.2.3; the contrasts

TABLE 10.3.2 Dose totals for Table 10.3.1

Day	S_1	S_2	S_3	T_1	T_2	T_3	Total
1	18·72	14·69	17·03	19·81	19·01	16·14	105·40
2	18·40	17·86	16·94	20·89	15·62	16·64	106·35
'1 + 2'	37·12	32·55	33·97	40·70	34·63	32·78	211·75
'2 − 1'	−0·32	3·17	−0·09	1·08	−3·39	0·50	0·95

TABLE 10.3.3 Calculation of contrasts from Table 10.3.2

Contrast	Day 1	Day 2	'1 + 2'	'2 − 1'
L_p	4·52	−0·05	4·47	−4·57
L_1	−5·36	−5·71	−11·07	−0·35
L'_1	−1·98	−2·79	−4·77	−0·81
L_2	4·30	5·91	10·21	1·61
L'_2	−8·44	6·67	−1·77	−15·13

formed from '1 + 2' are the complete dose contrasts for the assay, whereas those from '2 − 1' are the corresponding quantities for interactions with days. The confounding situation can be made clear by consideration of the sets of coefficients. For example, L_1 requires that coefficients $-1, 0, 1, -1, 0, 1$ be applied to each response for $S_1, S_2, S_3, T_1, T_2, T_3$ respectively. This is readily seen to be obtainable from the final column of differences in Table 10.3.1, as

$$-2{\cdot}08 - 3{\cdot}86 - 2{\cdot}87 - 2{\cdot}26 = -11{\cdot}07,$$

and it is therefore an intra-subject contrast. The interaction $L_1 D$ is formed from the same coefficients for responses on day 2 and those with signs reversed on day 1; it is obtainable in just the same way from the penultimate column

$$35{\cdot}36 - 37{\cdot}92 + 36{\cdot}75 - 34{\cdot}54 = -0{\cdot}35$$

and is an inter-subject contrast. Similarly, L_2 uses the coefficients $1, -2, 1, 1, -2, 1$, and is easily seen to be obtainable as

$$35{\cdot}36 - 2 \times 30{\cdot}31 + 37{\cdot}92 + 36{\cdot}75 - 2 \times 36{\cdot}87 + 34{\cdot}54 = 10{\cdot}21,$$

an inter-subject contrast. In this way, the contrasts in the two final columns of Table 10.3.3 can be numerically checked from contrasts between the sub-totals in the two final columns of Table 10.3.1, with simultaneous confirmation of the confounding pattern already stated. Rerup (1960) has described essentially the same form of analysis; some will find his graphical presentation helpful.

TABLE 10.3.4 Analysis of variance for Table 10.3.2

Adjustment for mean		747·3010	
Nature of variation	d.f.	Sum of squares	Mean square
Parallelism (L'_1)	1	0·5688	0·5688
Quadratic (L_2)	1	0·8687	0·8687
L_pD	1	0·3481	0·3481
L_1D	1	0·0031	0·0031
L'_2D	1	1·9076	1·9076
Error (inter-rat)	24	8·5103	0·35460
Between rats	29	12·2066	
Days (D)	1	0·0150	0·0150
Preparations (L_p)	1	0·3330	0·3330
Regression (L_1)	1	3·0636	3·0636
Difference of quadratics (L'_2)	1	0·0261	0·0261
L'_1D	1	0·0164	0·0164
L_2D	1	0·0216	0·0216
Error (intra-rat)	24	1·6862	0·07026
Total	59	17·3685	

By use of the standard divisors, the analysis of variance (Table 10.3.4) is constructed. Despite the anomaly of the total of responses for S_2 being less than that for S_3 (in a situation where increase in dose tends to reduce response), the deviation from parallelism is not serious. The only worries on validity are the difference between preparations and the interaction of L'_2 with the difference between days; both are statistically significant. Rerup has generally had satisfactory experience with this assay technique, but may have deliberately picked an awkward set of data to illustrate a methodological paper. Since the major indicators of invalidity show no alarming features, there seems to be no objection to using the data as illustration here (see also § 10.8).

As all information on L_p and L_1 comes from the intra-rat section of the analysis, equations (5.5.1) and (5.5.6) can be applied without modification. Hence

$$R = \frac{2 \cdot 0}{5 \cdot 0} \text{ antilog} \left[\frac{4 \times 0 \cdot 3010 \times 4 \cdot 47}{3 \times (-11 \cdot 07)} \right]$$

$$= 0 \cdot 4 \text{ antilog } \bar{1} \cdot 8379$$

$$= 0 \cdot 275.$$

From $N = 60$, $t = 2\cdot064$ with 24 d.f., and $s^2 = 0\cdot070\,26$,
$$Nt^2s^2 = 17\cdot959;$$
from equation (5.5.6),

$$R_L, R_U = 0\cdot4 \text{ antilog}\left[\frac{1\cdot2041 \times \{-49\cdot483 \pm \sqrt{5\cdot986 \times (39\cdot962 + 367\cdot635 - 35\cdot918)}\}}{367\cdot635 - 35\cdot918}\right]$$

$$= 0\cdot4 \text{ antilog}\left[\frac{1\cdot2041 \times (-49\cdot483 \pm 47\cdot169)}{331\cdot717}\right]$$

$$= 0\cdot4 \text{ antilog } \bar{1}\cdot6492,\ 0\cdot4 \text{ antilog } \bar{1}\cdot9916$$

$$= 0\cdot178,\ 0\cdot392.$$

Thus the test preparation is estimated to contain 0·275 IU corticotrophin/mg, with fiducial limits at 0·178, 0·392.

Had the 30 rats consisted of five litters of six, a natural modification of design would be to make each litter include one rat on each of the six dose sequences. This would have provided a randomized block structure on sequences and enabled 4 d.f. for litters to be isolated in the inter-rat analysis. Estimation of ρ would not be affected, but the power of the parallelism and other inter-rat validity tests should be increased.

10.4 Efficiency of cross-over designs

Comparison of inter- and intra-subject mean squares provides a helpful assessment of any gain from cross-overs. In the corticotrophin assay, if the variance within subjects between repeated responses to a fixed dose and the variance between doses are defined as σ_W^2, σ_B^2 respectively, then

(i) the intra-subject mean square is an estimate of σ_W^2,
(ii) the inter-subject mean square is an estimate of $\sigma_W^2 + 2\sigma_B^2$;

the second of these follows because contrasts between rats involve two responses per rat. Hence s_W^2, s_B^2, estimates of σ_W^2, σ_B^2, may be taken as

$$s_W^2 = 0\cdot0703,$$
$$s_B^2 = 0\cdot1422.$$

In an assay which did not use the cross-over principle, but instead based all conclusions on contrasts between rats, the variance per response would be $(\sigma_W^2 + \sigma_B^2)$, estimated by

$$s_W^2 + s_B^2 = 0\cdot2125,$$

a quantity 3·0 times greater than the intra-rat variance that applied to both L_p and L_1 in §10.3. Thus, in order to obtain a potency estimate whose fiducial range would be as narrow as that in the cross-over assay, a non-cross-over design based on three times as many responses would be needed: 30 rats with two tests

§ 10.5   CROSS-OVER DESIGNS

on each have led to a potency estimate as good as if 90 rats had been tested once each.

This convincingly demonstrates the superiority of the cross-over design. The gain, dependent upon the relative magnitudes of the two components of variance, will vary from one stock of subjects to another, being largest when the subjects are heterogeneous. A cross-over scarcely ever leads to a loss of precision, and will usually give some increase. Nevertheless, it has its defects. In order to ensure that the correct sequence of doses is followed for each subject, extra care in organization is needed; also, conclusions are delayed until the second day of the experiment.

10.5   Catalogue of cross-over designs

Cross-over designs have not yet been fully exploited in bioassay. The number of occasions need not be two. If it is equal to the number of doses, a Latin square ensures that on each occasion all doses are tested on equal numbers of subjects and that every subject receives each dose once. Tables 10.2.1 and 10.3.1 allocate doses to subjects in the incomplete block schemes proposed in § 9.9 for (2, 2) and (3, 3) assays in blocks of 2. Other designs in §§ 9.7 and 9.9 can be used as bases for cross-over designs, the number of occasions being always the number of doses per incomplete block; the balanced incomplete blocks lead to Youden squares (Youden, 1937, 1940). If the number of occasions exceeds the number of doses, a super-complete block scheme is required; for example, a Latin square design for a number of occasions equal to the number of doses can be augmented by an incomplete block scheme for the remaining occasions (Table 10.5.4; Cochran and Cox, 1957, Chapter 13).

The following catalogue of designs includes many of the types likely to be of practical use, and will help to clarify ideas. Statistical analysis presents few difficulties: it is always similar to that for the corresponding non-cross-over design except for isolation of inter-occasion components of the sum of squares (§ 10.6). Fisher and Yates (1963, Table XV) have provided adequate tables of Latin squares, and only the 4×4 is discussed here. No designs for which the minimum number of subjects exceeds 20 are listed.

TABLE 10.5.1   Cross-over design for a (2, 2) assay with four responses per subject

Subject	Dose on occasion:			
	1	2	3	4
I	S_1	T_2	T_1	S_2
II	S_2	T_1	T_2	S_1
III	T_1	S_2	S_1	T_2
IV	T_2	S_1	S_2	T_1

(i)   (2, 2) *designs*   If each subject can be used four times, and the investigator is prepared to wait for his results, a Latin square is ideal. The statistical analysis is simpler than for any of the incomplete block designs. For example, four subjects (numbered in random order) might be assigned to doses and occasions as in Table 10.5.1. All contrasts between doses are then intra-subject. If more than four subjects are to be used, a new Latin square for each set of four ought to be chosen at random from all possible squares (Fisher and Yates, 1963, Introduction). A design with $4K$ subjects would have its degrees of freedom partitioned as shown in Table 10.5.2. Bliss and Rose (1940) have given an example of this design used in the assay of a parathyroid extract (§ 10.6).

TABLE 10.5.2   Partition of the degrees of freedom in the analysis of variance for Table 10.5.1 with $4K$ subjects

Nature of variation	d.f.
Subjects	$4K - 1$
Occasions	3
Preparations	1
Regression	1
Parallelism	1
Error	$12K - 6$
Total	$16K - 1$

For twelve subjects, the design in Table 10.5.3 has advantages. It uses three Latin squares, which together include 12 of the 24 possible different orders of giving the four doses. Moreover, each dose on occasion 1 is followed by each of

TABLE 10.5.3   Cross-over design for a (2, 2) assay with 4 responses per subject, balanced for residual effects

Subject	Dose on occasion:			
	1	2	3	4
I	S_1	T_2	T_1	S_2
II	S_2	T_1	T_2	S_1
III	T_1	S_2	S_1	T_2
IV	T_2	S_1	S_2	T_1
V	S_1	S_2	T_2	T_1
VI	S_2	S_1	T_1	T_2
VII	T_1	T_2	S_2	S_1
VIII	T_2	T_1	S_1	S_2
IX	S_1	T_1	S_2	T_2
X	S_2	T_2	S_1	T_1
XI	T_1	S_1	T_2	S_2
XII	T_2	S_2	T_1	S_1

the others on occasion 2 (subjects I, V, IX have S_1 followed by T_2, S_2, T_1 respectively), and similarly for the other two pairs of successive occasions. With a little extra complexity of analysis, not only average differences between occasions and between subjects but also any residual (additive) influences of one dose on the response to the next may be eliminated from effect on the assay. This is important, for example, if a high dose on one day still has some effect on

§ 10.5   CROSS-OVER DESIGNS   211

the next day of testing, and reduces the apparent response to the new dose. Such designs are valuable for nutritional experiments, but the possibility of taking account of residual effects does not at present seem of great importance in biological assay. Further information on methods of construction and analysis has been given by Williams (1949, 1950), Patterson (1951, 1952), and Lucas (1956, 1957). Patterson (1950) drew attention to the danger that, when more than three occasions are used, the intra-subject error may not be homogeneous: with three occasions, intra-subject error can be regarded as coming from both linear and quadratic contrasts over occasions, and the first type may tend to be larger than the second. Although Lucas (1951) found this source of bias in error estimation not to be very important for dairy cattle, the situation could be different in bioassay with laboratory mammals. Finney (1956) discussed aspects of these designs peculiar to bioassay. Patterson and Lucas (1959) showed advantages in augmenting designs by an extra occasion repeating the treatments of the 'final' occasion, but this does not appear to have been tried in bioassay.

If each subject is to be used only twice, the design in Table 10.2.1 will usually be the best. A design based upon the balanced incomplete block design in § 9.7 requires a multiple of 12 subjects; it can be read from the first two occasions in Table 10.5.3. Myerscough and Schild (1958) adopted this design in an assay of ergometrine using human subjects. If each subject is to be used three times, designs are obtained by omitting the last occasion in Table 10.5.1. Omission of the last occasion in Table 10.5.3 leaves a design with three tests per subject balanced for residual effects. Colquhoun (1963) has described an assay of gastrin in which a single Youden square (without any balance for residuals) enabled four rats to provide a satisfactory potency estimate. The construction of super-complete block designs is illustrated by Table 10.5.4, which is compounded of Tables 10.2.1 and 10.5.1 to give a design with 6 tests on each subject.

TABLE 10.5.4   Cross-over design for a (2, 2) assay with 6 responses per subject

Subject	Dose on occasion:					
	1	2	3	4	5	6
I	S_1	T_2	T_1	S_2	T_2	S_1
II	S_2	T_1	T_2	S_1	T_1	S_2
III	T_1	S_2	S_1	T_2	S_2	T_1
IV	T_2	S_1	S_2	T_1	S_1	T_2

(ii) *(3, 3) designs* A (3, 3) assay with tests on two occasions was exemplified in § 10.3. A balanced incomplete block scheme can easily be constructed; 30 subjects use the 15 pairs of doses in Table 9.5.1 in both orders.

If each subject is to be used three times, a neatly confounded design is impossible. A design based on Table 9.9.4 gives the most intra-subject information on L_p and L_1; Table 10.5.5 shows the design constructed by forming two 3×3 Latin squares with the two types of block in Table 9.9.4. Table 10.5.6 shows the recommended confounding scheme in Table 9.9.3 incorporated into a design with four tests per subject. Balanced incomplete block (Youden square or generalized Youden square) designs for 3 or 4 tests per subject require at least 30 subjects. The only useful design with 5 responses per subject is the Youden square formed from any 6×6 Latin square by omission of the last column. The complete square gives a design for 6 responses per subject.

TABLE 10.5.5 Cross-over design for a (3, 3) assay with 3 responses per subject (L_1 orthogonal with subjects, L_p minimally confounded)

Subject	Dose on occasion:		
	1	2	3
I	S_1	S_3	T_2
II	S_3	T_2	S_1
III	T_2	S_1	S_3
IV	S_2	T_1	T_3
V	T_1	T_3	S_2
VI	T_3	S_2	T_1

TABLE 10.5.6 Cross-over design for a (3, 3) assay with 4 responses per subject (L_p and L_1 orthogonal with subjects)

Subject	Dose on occasion:			
	1	2	3	4
I	S_2	S_3	T_1	T_2
II	S_3	S_2	T_2	T_1
III	T_1	T_3	S_1	S_3
IV	T_3	T_1	S_3	S_1
V	T_2	S_1	T_3	S_2
VI	S_1	T_2	S_2	T_3

TABLE 10.5.7 Cross-over design for a (4, 4) assay with two responses per subject (L_p and L_1 orthogonal with subjects)

Subject	Dose on occasion:	
	1	2
I	S_1	T_4
II	S_2	T_3
III	S_3	T_2
IV	S_4	T_1
V	T_1	S_4
VI	T_2	S_3
VII	T_3	S_2
VIII	T_4	S_1

(iii) *(4, 4) designs* Designs for 2, 4 and 6 responses per subject, based on the confounded designs in § 9.9, are shown as Tables 10.5.7–10.5.9. Designs for three, five, or seven responses per subject cannot have very satisfactory confounding schemes. Those that use the first 3, 5 or 7 columns of Table 10.5.10

§ 10.5 CROSS-OVER DESIGNS

TABLE 10.5.8 Cross-over design for a (4, 4) assay with 4 responses per subject (L_p and L_1 orthogonal with subjects)

Subject	Dose on occasion:			
	1	2	3	4
I	S_1	T_2	S_4	T_3
II	S_2	S_3	T_4	T_1
III	S_3	S_2	T_1	T_4
IV	S_4	T_3	T_2	S_1
V	T_1	T_4	S_2	S_3
VI	T_2	S_1	T_3	S_4
VII	T_3	S_4	S_1	T_2
VIII	T_4	T_1	S_3	S_2

TABLE 10.5.9 Cross-over design for a (4, 4) assay with 6 responses per subject (L_p and L_1 orthogonal with subjects)

Subject	Dose on occasion:					
	1	2	3	4	5	6
I	S_1	S_2	S_3	T_2	T_3	T_4
II	S_2	S_3	T_1	T_3	T_2	S_4
III	S_3	T_2	T_3	S_4	T_1	S_2
IV	S_4	S_1	T_2	S_3	T_4	T_1
V	T_1	T_4	S_4	S_1	S_3	T_2
VI	T_2	T_3	T_4	S_2	S_1	S_3
VII	T_3	T_1	S_2	T_4	S_4	S_1
VIII	T_4	S_4	S_1	T_1	S_2	T_3

TABLE 10.5.10 Cross-over design for a (4, 4) assay with 8 responses per subject (To be used also for 3, 5 or 7 responses per subject)

Subject	Dose on occasion:							
	1	2	3	4	5	6	7	8
I	S_1	S_4	T_2	T_3	T_1	T_4	S_2	S_3
II	S_2	S_3	T_1	T_4	S_4	S_1	T_3	T_2
III	S_3	S_2	T_4	T_1	T_3	T_2	S_4	S_1
IV	S_4	S_1	T_3	T_2	S_2	S_3	T_1	T_4
V	T_1	T_4	S_2	S_3	S_1	S_4	T_2	T_3
VI	T_2	T_3	S_1	S_4	T_4	T_1	S_3	S_2
VII	T_3	T_2	S_4	S_1	S_3	S_2	T_4	T_1
VIII	T_4	T_1	S_3	S_2	T_2	T_3	S_1	S_4

may be about the best in respect of freedom of L_p and L_1 from confounding; this 8 × 8 Latin square is one of the many that might be used if each subject could be tested on 8 occasions. Balanced incomplete block designs all require many subjects, except for that with 7 responses per subject obtained by omission of the last column from any 8 × 8 Latin square.

(iv) *(1, 1, 1) designs* Deletion of the last column from any 3 × 3 Latin square gives a balanced incomplete block design with two responses per subject. This may be repeated for any number of sets of three subjects.

(v) *(1, 2, 2) designs* The balanced incomplete block designs listed in § 9.7 may be arranged to give designs for 2, 3 or 4 responses per subject. Tables 10.5.11 and 10.5.12 show the generalized Youden square designs for 2 and 3 responses; in a set of 10 subjects, each dose occurs twice on each occasion. For 4 responses per subject, the last column of any 5 × 5 Latin square may be deleted.

TABLE 10.5.11 Cross-over design (generalized Youden square) for a (1, 2, 2) assay with 2 responses per subject

Subject	Dose on occasion:	
	1	2
I	C	S_2
II	C	T_1
III	S_1	C
IV	S_1	T_2
V	S_2	S_1
VI	S_2	T_1
VII	T_1	S_1
VIII	T_1	T_2
IX	T_2	C
X	T_2	S_2

TABLE 10.5.12 Cross-over design (generalized Youden square) for a (1, 2, 2) assay with 3 responses per subject

Subject	Dose on occasion:		
	1	2	3
I	C	S_2	T_1
II	C	T_2	S_2
III	S_1	S_2	T_1
IV	S_1	T_1	C
V	S_2	C	S_1
VI	S_2	S_1	T_2
VII	T_1	S_1	T_2
VIII	T_1	T_2	S_2
IX	T_2	C	S_1
X	T_2	T_1	C

(vi) *(1, 3, 3) and (1, 4, 4) designs* The balanced incomplete block design in Table 9.7.3 is easily arranged as a Youden square, as is its complementary design, so giving (1, 3, 3) designs for 7 subjects on three and four occasions. For two or five occasions, at least 21 subjects are needed for a generalized Youden square. Table 10.5.13 shows a generalized Youden square, based on Table 9.7.5, for a (1, 4, 4) assay using 18 subjects on four occasions. The complementary balanced incomplete blocks can be arranged similarly for five occasions. No designs with a reasonable number of subjects are possible for two, three, six, or seven occasions. Latin squares provide designs for (1, 3, 3) on seven occasions and (1, 4, 4) on nine; omission of the final columns leaves designs for 6 and 8 occasions.

For $(1, k, k)$ assays above, only the balanced incomplete block types of design

TABLE 10.5.13 Cross-over design (generalized Youden square) for a (1, 4, 4) assay with 4 responses per subject

Subject	Dose on occasion:			
	1	2	3	4
I	C	S_1	S_3	T_3
II	C	S_2	T_1	T_2
III	S_1	S_2	S_3	T_2
IV	S_1	T_3	T_2	T_1
V	S_2	S_3	S_4	C
VI	S_2	T_1	S_1	T_4
VII	S_3	S_4	T_4	S_1
VIII	S_3	T_1	T_3	S_2
IX	S_4	T_2	T_3	C
X	S_4	T_2	T_4	S_2
XI	T_1	C	S_1	S_4
XII	T_1	T_4	C	S_3
XIII	T_2	S_4	T_1	S_3
XIV	T_2	T_4	C	S_1
XV	T_3	S_1	S_2	S_4
XVI	T_3	S_3	T_2	T_4
XVII	T_4	C	S_2	T_3
XVIII	T_4	T_3	S_4	T_1

have been described, as no useful confounding schemes are known. Though designs are tabulated as for symmetric assays, all based on balanced incomplete blocks are equally applicable to any unsymmetric set of doses, since all contrasts are made with the same precision; the symbols used for the doses of the symmetric assay should then be replaced by the symbols for the actual doses in random order. Designs based on the confounding schemes in § 9.9 lose their special properties of giving maximal precision to the important contrasts if used for unsymmetric sets of doses.

Each design tabulated in this chapter is shown in systematic order. For use, the allocation of subjects to the Roman numerals and the order of the columns should be randomized. If subjects are classified into litters (or other groups), and if the size of the litter is the same as the number of subjects required for a full replication of the design, inter-litter differences may be eliminated (so increasing the precision of inter-subject contrasts). For Youden square designs, this will increase the precision of the potency estimate, but for the others it will affect only the power of the validity tests. Methods of statistical analysis can be based upon standard methods for balanced incomplete block designs, as described in textbooks, or can use regression devices such as were described in § 9.5; illustrations are discussed in §§ 10.6 and 10.8.

10.6 A cross-over assay of a parathyroid extract

Bliss and Rose (1940) discussed the assay of a parathyroid extract with the design in Table 10.5.1 but using five Latin squares. As already noted, the analysis presents little difficulty. Bliss (1947a) used the responses on the first three of the four days of testing to illustrate the analysis of a (2, 2) assay with 3 responses per subject. That analysis will be repeated here, in rather simpler form.

Doses of 0·125 and 0·205 ml per kg of S and T were administered to 20 dogs, each dog receiving three of the four doses on occasions about ten days apart. The serum calcium of each dog was determined on the day after treatment, so giving 60 responses for analysis. Table 10.6.1 lists the sequences of doses and the responses, in four sets of five dogs, each set having the same set of doses though not necessarily in the same order; the five Latin squares, or rather the five Latin squares with final columns omitted, are composed of the first, second, . . . , fifth dogs from the groups.

TABLE 10.6.1 Serum calcium (mg/kg) of dogs on the day after injection with parathyroid extract

Set	Dog no.	Dose on			Serum Calcium on			Total for dog
		Day 1	Day 2	Day 3	Day 1	Day 2	Day 3	
I	342	T_1	T_2	S_2	147	154	148	449
	287	T_1	T_2	S_2	151	150	158	459
	264	T_2	S_2	T_1	144	138	144	426
	349	T_2	T_1	S_2	162	140	130	432
	247	T_2	S_2	T_1	158	160	150	468
II	256	T_2	T_1	S_1	158	143	148	449
	317	T_2	T_1	S_1	170	165	150	485
	288	T_1	S_1	T_2	136	153	172	461
	252	T_1	T_2	S_1	140	138	140	418
	308	T_1	S_1	T_2	130	134	138	402
III	280	S_1	S_2	T_2	138	170	160	468
	244	S_1	S_2	T_2	120	138	140	398
	273	S_2	T_2	S_1	146	154	140	440
	250	S_2	S_1	T_2	130	140	140	410
	271	S_2	T_2	S_1	152	162	150	464
IV	320	S_2	S_1	T_1	150	145	140	435
	309	S_2	S_1	T_1	150	140	146	436
	276	S_1	T_1	S_2	158	150	152	460
	278	S_1	S_2	T_1	132	160	149	441
	299	S_1	T_1	S_2	142	141	150	433
	Totals for days of testing				2914	2975	2945	8834

The analysis of variance may be constructed by the general method (§ 9.5), using inter-subject and intra-subject regressions on variates corresponding to L_p, L_1, L_1'. Table 10.6.2 contains dose totals for the four sets of subjects and the totals of all responses for these sets, symbolized by B_I, B_{II}, B_{III}, B_{IV}. Those who are experienced in the analysis of variance will realize almost intuitively that the same analysis will be obtained by forming an intra-subject estimate of the preparations difference:

$$L_p = L_p \text{ (crude)} - \tfrac{1}{3}(B_I + B_{II} - B_{III} - B_{IV}), \tag{10.6.1}$$

with a corresponding inter-subject estimate

$$L_p = \tfrac{1}{3}(B_I + B_{II} - B_{III} - B_{IV}), \tag{10.6.2}$$

§ 10.6　　　　　CROSS-OVER DESIGNS　　　　　217

and similar contrasts for L_1 and L_1'. The orthogonality conditions are satisfied, and the usual rules give the divisors. A still more direct procedure is shown in Table 10.6.3 where the three intra-subject contrasts are defined in terms of the dose totals in Table 10.6.2; as usual, the divisors are the sums of squares of coefficients. The inter-subject contrasts are most easily written as

$$\left. \begin{aligned} L_p &= B_{\mathrm{I}} + B_{\mathrm{II}} - B_{\mathrm{III}} - B_{\mathrm{IV}} = 64, \\ L_1 &= B_{\mathrm{I}} - B_{\mathrm{II}} + B_{\mathrm{III}} - B_{\mathrm{IV}} = -6, \\ L_1' &= -B_{\mathrm{I}} + B_{\mathrm{II}} + B_{\mathrm{III}} - B_{\mathrm{IV}} = -44, \end{aligned} \right\} \quad (10.6.3)$$

with divisor 60 for each (since every response enters with coefficient 1 or -1). The component for Latin squares is obtained from the five totals of four dogs; it is removed because of the possibility that each square was formed from four homogeneous animals.

TABLE 10.6.2 Dose and set totals for Table 10.6.1

Set	Total for				Total of set
	S_1	S_2	T_1	T_2	
I	—	734	732	768	2234
II	725	—	714	776	2215
III	688	736	—	756	2180
IV	717	762	726	—	2205
Total	2130	2232	2172	2300	8834

TABLE 10.6.3 Orthogonal coefficients for contrasts within subjects for Tables 10.6.1 and 10.6.2

Contrast	Set	S_1	S_2	T_1	T_2	Divisor	Sum
L_p	I	—	−2	1	1		
	II	−2	—	1	1		
	III	−1	−1	—	2		
	IV	−1	−1	2	—	120	133
L_1	I	—	1	−2	1		
	II	−1	—	−1	2		
	III	−2	1	—	1		
	IV	−1	2	−1	—	120	348
L_1'	I	—	−1	−1	2		
	II	1	—	−2	1		
	III	1	−2	—	1		
	IV	2	−1	−1	—	120	61

The analysis of variance, Table 10.6.4, discloses no sign of invalidity. The fact that the inter-dog error is four times the intra-dog error indicates a substantial gain in precision from the cross-over design. In the inter-dog section of the analysis, the regression component is smaller than the error mean square; indeed, L_1

has a small negative value. As g exceeds unity, no useful potency estimate can be obtained from inter-block contrasts. The inter-dog variation contributes so little information about the potency of the test preparation that an estimate based on intra-dog contrasts alone is likely to be preferable to a compound estimate.

TABLE 10.6.4   Analysis of variance for Table 10.6.1

Nature of variation	d.f.	Adjustment for mean 1 300 659·27 Sum of squares	Mean square
Latin squares	4	502·73	125·68
Preparations	1	68·27	68·27
Regression	1	0·60	0·60
Parallelism	1	32·27	32·27
Error (inter-dog)	12	2868·86	239·07
Between dogs	19	3472·73	
Days	2	93·03	46·52
Preparations	1	147·41	147·41
Regression	1	1009·20	1009·20
Parallelism	1	31·01	31·01
Error (intra-dog)	35	2027·35	57·92
Total	59	6780·73	

Inspection of Table 10.6.3 shows the intra-dog L_p to be a difference between 40 responses to T and 40 to S. Hence the mean difference per observation is estimated to be

$$\bar{y}_T - \bar{y}_S = \frac{L_p}{40} \tag{10.6.4}$$

$$= 3·325,$$

with

$$\text{Var}(\bar{y}_T - \bar{y}_S) = \frac{120 s_2^2}{(40)^2}$$

$$= \frac{3 s_2^2}{40}. \tag{10.6.5}$$

Similarly the regression coefficient, which is exactly the quantity that would be obtained as the intra-dog regression coefficient on a variate that takes the value -1 for every response to S_1 or T_1, and $+1$ for every response to S_2 or T_2, is

$$b = \frac{L_1}{80} \tag{10.6.6}$$

$$= 4·350,$$

with

$$\text{Var}(b) = \frac{120 s_2^2}{(80)^2}$$

$$= \frac{3 s_2^2}{160}. \tag{10.6.7}$$

Estimation of potency then proceeds in the usual manner. Thus

$$M = 0.7644,$$

$$g = 0.2365,$$

and
$$M_L, M_U = (0.7644 \pm 0.9276)/0.7635$$

$$= -0.2138, 2.2161.$$

The dose ratio between high and low doses of either preparation was 1·64, so that all log potency values must be multiplied by $\frac{1}{2}\log 1.64$, or 0·1074, in order to convert them to logarithms to base 10.

Hence, since the two preparations were given at equal nominal doses,

$$R = \text{antilog } 0.0821 = 1.208,$$

$$R_L = \text{antilog } \bar{1}.9770 = 0.948,$$

$$R_U = \text{antilog } 0.2380 = 1.730.$$

Thus 1 ml of T is estimated to have the potency of 1·208 ml of the standard parathyroid extract, with fiducial limits at 0·948 ml and 1·730 ml of S.

If the inter-dog analysis is combined with this, in the manner of § 9.5, the slight improvement in the precision of contrasts is rather more than offset by the lower value of b consequent upon the negative inter-dog L_1. The new value of R is 1·222, with limits a little more widely spaced at 0·955 and 1·783. When one section of the analysis shows so small a regression coefficient, it should be ignored and the estimation based entirely on the other.

One difference apparent between Table 10.3.4 and Table 10.6.4 is that the latter shows no components for interactions between days and doses. Such components could be extracted, but the effort is seldom worth while in routine assays with more than two test days.

10.7 Single subject assays

In cross-over assays, a few subjects are used repeatedly instead of a larger number once only. In an extreme type, every response is measured on one subject. For example, histamine may be assayed from contractions of an isolated strip of guinea-pig's gut, immersed in a water-bath with one end fixed and the other attached to a writing point that records a constant magnification of any contraction.

Responses to different doses, measured at short intervals of time, are used in the ordinary statistical analysis. The level of response may decline as the assay progresses; cross-over designs can aid the elimination of trend. Any designs in this chapter can be used with a succession of sequences previously allotted to different subjects. Thus a (2, 2) assay using 16 responses might be based on the Latin square in Table 10.5.1; if occasions were randomized to the order 4, 2, 3, 1 and subjects to the order II, I, IV, III, the gut would receive in succession the doses:

$S_1, T_1, T_2, S_2;\quad S_2, T_2, T_1, S_1;\quad T_1, S_1, S_2, T_2;\quad T_2, S_2, S_1, T_1.$

The standard statistical analysis would eliminate differences between sets of four responses and the average effect of order within sets of four, these corresponding to 'subjects' and 'occasions'. If changes in responsiveness within the short period of four tests were unimportant, a randomized block design could replace the Latin square, with a fresh randomization of order for each four, so increasing flexibility by removing restrictions on the number of replications. Schild (1942) described assays of this class; Smith and Vos (1943) and Noel (1945) published interesting examples. If one subject cannot be used often enough for the precision desired, two may be used and the results combined.

Reference was made in § 10.5 to designs specially adapted to balancing effects of preceding doses when there is reason to suspect that these may influence current responses. Analogous designs can be constructed (Sampford, 1957) for single subject assays. Consider the sequence

$S_2;\quad T_1, T_2, S_2, S_1;\quad S_2, T_2, T_1, S_1;\quad T_2, S_1, T_1, S_2.$

After a first conditioning dose, each dose of a (2, 2) scheme occurs once in each of three blocks; moreover, each dose is preceded once only by each *other* dose. Finney and Outhwaite (1956) described the modified statistical analysis that takes account of a simple model in which each expected response includes additive components for the current and immediately preceding dose, with the residual component proportional to the corresponding current one. Even for (2, 2) assays, this becomes complicated, and since little practical use has been made of the designs no details are presented here. For (3, 3) assays, slightly better balance may be achieved. The following two sequences might be used on two subjects with 19 responses measured on each, or they might be put end-to-end for one subject with 37 responses, omitting the first T_3 of (ii):

(i) $S_2;\quad S_2, T_1, T_3, S_1, T_2, S_3;\quad S_3, T_3, T_2, S_2, S_1, T_1;\quad T_1, S_2, S_3, T_2, S_1, T_3;$

(ii) $T_3;\quad T_3, S_2, T_2, T_1, S_3, S_1;\quad S_1, S_3, S_2, T_3, T_1, T_2;\quad T_2, T_3, S_3, T_1, S_1, S_2.$

Here each dose is preceded once by *every* dose, including itself.

The above designs are sometimes unsatisfactory because their completion needs a large number of responses per subject. Computation becomes troublesome if an accident or sudden deterioration of a piece of tissue forces abandonment before completion. A (3, 3) assay with measurement of 24 responses might be based upon Table 10.5.6, using in succession the dose sequences allotted to the six subjects in the table; if testing had to stop after 18 responses, the statistical analysis would be laborious. Designs based on complete or incomplete randomized blocks, and lacking the double restriction on order of dosing characteristic of true cross-overs, are more flexible. For them, changes in plan during the experiment need not be so harmful, and if blocks are small, intra-block changes in responsiveness will rarely be important.

Some workers, fearing sudden changes in responsiveness, have tried to match doses, adjusting the magnitudes of successive doses of the two preparations in an

§ 10.7 CROSS-OVER DESIGNS

endeavour to achieve equal responses. Evidently this is wasteful, as many responses contribute nothing to the final potency estimate. Moreover, satisfactory matching will be difficult if the responsiveness of the subject is changing. Vos (1943, 1950), objecting to cross-over designs because the responsiveness of a subject may not change at a constant rate, introduced a *constant standard* design. This had weaknesses because it provided no test of similarity, it lacked randomization, and the proposed statistical analysis using moving averages might bring trouble from serial correlations. Nevertheless, to change doses as knowledge of responsiveness accumulates may be helpful, especially when at the start little is known about the potency of T (Thompson, 1944, 1945). The same convenience may be gained from a design in randomized blocks of two that permits valid error estimation. Each successive (non-overlapping) pair of observations is randomized between S and T to give a dose sequence such as

$$S, T; \quad T, S; \quad T, S; \quad T, S; \quad S, T; \quad T, S; \quad S, T; \ldots$$

Here S could represent a fixed dose of the standard or one that was changed only after evidence of gross change in responsiveness of the subject; T, the dose of the test preparation, could be changed from block to block as information on potency accumulated. A linear regression of $(y_T - y_S)$ on the corresponding $(x_T - x_S)$, each difference being formed for every block of two, gives M as the intercept of the regression line on the x-axis. The T doses should be chosen with the aim of keeping the mean of $(y_T - y_S)$ close to zero, yet with as wide a range of values of x_T as linearity constraints permit.

Finney and Schild (1966) described a further development that permits validity tests. Suppose observations for the subject are divided into successive blocks of four, each block including S_1, S_2, T_1, T_2 in random order; the magnitudes of dose used for S_1, S_2, T_1, T_2 may be varied from block to block in accordance with increasing knowledge of potency and application of the principles of Chapter 6. Statistical analysis is most easily conducted in terms of the contrasts L_p, L_1, L_1' computed for each block separately. As usual, L_1 and L_1' serve to estimate the regression coefficient and to test parallelism respectively; L_p, however, not only estimates a numerator for M but also (in the most general conditions) contains further information on regression and parallelism.

Equations (3.7.6), (3.7.7) represented a simple mathematical model of expected responses. A more general form of these helps to clarify the Finney and Schild design. The equations for the two preparations may be written symmetrically as

$$\left. \begin{array}{l} Y_S = \gamma_i + \alpha_S + (\beta - \phi)x, \\ Y_T = \gamma_i + \alpha_T + (\beta + \phi)x, \end{array} \right\} \qquad (10.7.1)$$

where γ_i is a parameter for block i and α_S, α_T have their usual meanings; $\phi = 0$ corresponds to parallelism, with β as the common regression coefficient. In a valid assay, with $\phi = 0$,

$$\log \rho = (\alpha_T - \alpha_S)/\beta \qquad (10.7.2)$$

in the usual way. Now write x_S, $(x_S + d_S)$ for the log doses of S and x_T, $(x_T + d_T)$ similarly for T. In an ordinary (2, 2) assay, the doses would be constant from block to block and d_S, d_T would be equal. Here neither condition is essential, and an additional subscript i will be used to indicate block i. Further attention will be restricted to the case

$$d_{Si} = d_{Ti} = d_i, \tag{10.7.3}$$

without requiring d_i to be the same in all blocks; the more general assay without this constraint is permissible but seems less useful.

Simple manipulation of symbols shows that, for block i,

$$\mathrm{E}(L_{pi}) = 2(\alpha_T - \alpha_S) + 2\beta(x_{Ti} - x_{Si}) + 2\phi(x_{Ti} + x_{Si}), \tag{10.7.4}$$

$$\mathrm{E}(L_{1i}) = 2\beta d_i, \tag{10.7.5}$$

$$\mathrm{E}(L'_{1i}) = 2\phi d_i. \tag{10.7.6}$$

The variance of each such contrast is $4\sigma^2$. A first step in analysis is to form a weighted mean of values of $L'_{1i}/2d_i$ as an estimate of ϕ:

$$h = \Sigma L'_{1i} d_i / 2 \Sigma d_i^2, \tag{10.7.7}$$

the weights being reciprocals of variances. Moreover,

$$\mathrm{Var}(h) = \sigma^2 / \Sigma d_i^2; \tag{10.7.8}$$

summations are for all blocks.

If h does not deviate from 0 to an extent that raises doubts about validity, a linear regression of the L_{pi} on $2(x_{Ti} - x_{Si})$ estimates both $2(\alpha_T - \alpha_S)$ and β. Of course, a further estimate of β is obtainable as a weighted mean of the L_{1i}, this latter usually being the more precise. If there were uncertainty about validity, further information on ϕ could be obtained by including $(x_{Ti} + x_{Si})$ as a second variate in the L_{pi} regression, but this will seldom contribute much. In a valid assay, the potency estimate is obtained from equation (10.7.2), using a weighted combination of the two estimates of β.

The variance σ^2 can be estimated from four different components. If the assay has K blocks, $(K-2)$ d.f. come from the residual of the L_{pi} regression, $(K-1)$ d.f. from variation in L_{1i}/d_i, $(K-1)$ d.f. from variation in L_{2i}/d_i, and 1 d.f. from comparison of the two independent estimates of β. Finney and Schild illustrated the arithmetic in detail, in a paper that suffered many typographical errors. As usual, an analysis of variance helps the assembly of the various components. The combination of the two estimates of β requires weights inversely proportional to the variances of the estimates; the ratio of the weights is independent of σ^2. The composite estimate therefore is itself a linear contrast between the responses, and Fieller's theorem may be applied without trouble. A generalization without the condition (10.7.3) is permissible, though perhaps of less practical interest. Box and Hay (1953) described a similar assay that permitted both the elimination of a time trend and the use of a curvilinear regression of response on dose.

§ 10.8   CROSS-OVER DESIGNS   223

In practice, a constant d_i is often convenient, and this simplifies the weighting considerably. If $(x_{Ti} - x_{Si})$ is constant, no supplementary information on β is obtainable from the L_{pi} regression. If both d_i and $(x_{Ti} - x_{Si})$ are constant, the assay is of the standard (2, 2)-point pattern.

10.8 An exact analysis with missing subjects

In § 10.3, the data for the corticotrophin assay were oversimplified: rats number X and XXX in Table 10.3.1 either never existed or suffered accidental loss of their records. In order to provide a simple example, mean responses for the other four rats in each set were inserted, and the analysis in § 10.3 used these uncritically.

An exact analysis can be approached in three ways. One is to treat the four responses for rats X and XXX as missing values, and to estimate them by least squares. The values that simultaneously minimize both the inter-rat and the intra-rat sums of squares are easily proved to be those already inserted, the means for other rats with the same dose patterns. (This would not be so if a block structure, using 5 litters of 6 rats, had been imposed.) The major consequence is to reduce the degrees of freedom for each error by 2. An approximate analysis can be based upon this one modification. The potency estimate would be exactly as in § 10.3, but the increase of about 9 percent in error mean squares would reduce the fears of invalidity and increase the fiducial range. These statements can be made numerically exact, but to proceed further in this way is tedious.

An alternative is to follow § 4.23, for the inter-rat and intra-rat analyses separately. For the latter, variates x_1, x_2 are defined to correspond to L_p, L_1, and x_3, x_4, x_5 to correspond to $L'_2, L'_1 D, L_2 D$; thus x_5 for $L_2 D$ would take the values

	S_1	S_2	S_3	T_1	T_2	T_3
Day 1	-1	2	-1	-1	2	-1
Day 2	1	-2	1	1	-2	1

Multiple regression of the 28 'day 2 − day 1' differences on x_1, x_2, x_3, x_4, x_5 would then give tests of significance for three contrasts and estimates of L_p, L_1. A similar procedure with five other dummy variates would deal with the inter-rat analysis. As in § 4.23, a covariance between the regression coefficients on x_1, x_2 enters into Fieller's formula.

The third approach is to work directly with mean sums and differences of response for the six treatment sequences, and to build up appropriate estimates. This is particularly simple if fears of invalidity can be set aside. Here the approximate analysis involving only the changed degrees of freedom for errors in Table 10.3.4 suffices to dispose of invalidity. The tests are biased, but are usually good enough for practical purposes when less than 10 percent of the responses are missing. Further steps use only the final column of Table 10.3.1, with omission of rats X and XXX and consequential changes of subtotals. From variation within the 6 groups, the variance per response is estimated as s^2 with 22 d.f., where

$$22s^2 = \frac{1}{2}\left[\left(0\cdot21^2 + 0\cdot50^2 + \ldots - \frac{2\cdot08^2}{5}\right) + \ldots\right.$$
$$\left. + \left(1\cdot01^2 + 0\cdot28^2 + \ldots - \frac{1\cdot81^2}{4}\right)\right]$$
$$= 0\cdot2841 + 0\cdot5407 + 0\cdot2178 + 0\cdot1097 + 0\cdot2423 + 0\cdot2914$$
$$= 1\cdot6860.$$

This sum of squares corresponds with the intra-rat error sum of squares in Table 10.3.4, differing only because the entries for rats X and XXX in Table 10.3.1 were rounded to the nearest 0·01. Hence

$$s^2 = 0\cdot076\,64.$$

Table 10.8.1 shows mean differences between doses for each of the six dose sequences, together with variances expressed as multiples of s^2 (readily obtained by consideration of linear contrasts between 10 or 8 of the individual responses). In order to eliminate differences between days, equal weight must be given to the two orders in which each pair of doses was administered. Thus for $(T_3 - S_1)$ the estimated difference is

$$\tfrac{1}{2}(-0\cdot4160 - 0\cdot4525) = -0\cdot4342$$

with variance

$$\left(\frac{2s^2}{5} + \frac{s^2}{2}\right)\bigg/ 2^2 = 9s^2/40.$$

Similarly for $(T_2 - S_2)$ is obtained

$$\tfrac{1}{2}(0\cdot1875 + 0\cdot2300) = 0\cdot2088$$

with variance $9s^2/40$, and for $(T_1 - S_3)$

$$\tfrac{1}{2}(0\cdot7720 + 0\cdot5740) = 0\cdot6730$$

with variance $s^2/5$.

TABLE 10.8.1 Mean response differences, between days and within dose sequences

Rats	Dose difference	Mean	Variance
I–V	$T_3 - S_1$	−0·4160	$2s^2/5$
VI–IX	$T_2 - S_2$	0·1875	$s^2/2$
XI–XV	$T_1 - S_3$	0·7720	$2s^2/5$
XVI–XX	$T_1 - S_3$	0·5740	$2s^2/5$
XXI–XXV	$T_2 - S_2$	0·2300	$2s^2/5$
XXVI–XXIX	$T_3 - S_1$	−0·4525	$s^2/2$

A further mean of these three quantities estimates the mean difference between preparations:

§ 10.8 CROSS-OVER DESIGNS

$$a_T - a_S = \tfrac{1}{3}(-0.4342 + 0.2088 + 0.6730)$$

$$= 0.1492,$$

with

$$\text{Var}(a_T - a_S) = s^2 \left(\frac{9}{40} + \frac{9}{40} + \frac{1}{5}\right) \bigg/ 3^2$$

$$= 13s^2/180.$$

Similarly the regression coefficient is estimated by subtracting the $(T_1 - S_3)$ difference from the $(T_3 - S_1)$, and dividing by 4.

$$b = \frac{-0.4342 - 0.6730}{4}$$

$$= -0.2768$$

with

$$\text{Var}(b) = s^2 \left(\frac{9}{40} + \frac{1}{5}\right) \bigg/ 4^2$$

$$= 17s^2/640.$$

(The division by 4 is because the high and low doses of each preparation are coded as $+1$ and -1 when the appropriate log dose scale is used: cf. Chapter 5.) Moreover, the manner of forming $a_T - a_S$ and b shows them to have a covariance:

$$\text{Cov}(a_T - a_S, b) = s^2 \left(\frac{9}{40} - \frac{1}{5}\right) \bigg/ 3 \times 4$$

$$= s^2/480.$$

These steps have been presented in detail in order to show their logical development. They could easily be expressed in general algebraic form, but the resulting inelegant formulae are seldom needed and the approach from first principles seems more illuminating.

From here on, potency estimation and application of Fieller's theorem follow the standard pattern, remembering that the highest doses of the two preparations were 2·0 and 5·0 on their appropriate scales and that the ratio of consecutive doses was 2·0. Hence

$$R = \frac{2 \cdot 0}{5 \cdot 0} \text{ antilog } \frac{0 \cdot 1492 \times 0 \cdot 3010}{-0 \cdot 2768}$$

$$= 0 \cdot 4 \text{ antilog } \bar{1} \cdot 8378$$

$$= 0 \cdot 275.$$

Also, for 22 d.f., $t = 2·074$, and

$$g = \frac{(2 \cdot 074)^2 \times 17s^2}{640 \times (0 \cdot 2768)^2}$$

$$= 0 \cdot 1143.$$

Hence
$$M_L, M_U = \left[-0.5390 - \frac{640g}{17 \times 480} \pm \frac{2.074}{0.2768} \left[s^2 \left\{ \frac{13}{180} + \frac{2 \times 0.5390}{480} \right. \right. \right.$$
$$\left. \left. \left. + \frac{17 \times (0.5390)^2}{640} - 0.1143 \left(\frac{1}{180} - \frac{1}{17 \times 360} \right) \right\} \right]^{1/2} \right] \Big/ 0.8857$$
$$= [-0.5480 \pm 7.493\{s^2(0.072\,22 + 0.002\,25 + 0.007\,72$$
$$- 0.000\,62)\}^{1/2}]/0.8857$$
$$= [-0.5480 \pm 0.5924]/0.8857$$
$$= -1.288, 0.0501.$$

Therefore
$$R_L, R_U = 0.4 \text{ antilog } (-0.3010 \times 1.288), 0.4 \text{ antilog } (0.3010 \times 0.0501)$$
$$= 0.4 \text{ antilog } \bar{1}.6123, 0.4 \text{ antilog } 0.0151$$
$$= 0.164, 0.414.$$

In this assay, 4 responses out of 60 were missing. As a first guess, this might be expected to increase the fiducial range by about $\left(\frac{60}{56}\right)^{1/2}$, or a factor of less than 1·04 (on the log scale, but also approximately on the dose scale). Comparison with §10.3 shows the actual loss to be greater, a factor of 1·17 in the difference $R_U - R_L$, because of the increase in g and of asymmetry in the design.

11
Multiple assays

11.1 The economy of multiple assays

If several test preparations are to be assayed against one standard, each could be made the occasion of a separate small assay. Alternatively, each might be included in one *multiple assay*. For example, two or three doses of S and of each of three test preparations might be incorporated into a single experiment involving eight or twelve dose levels in all. A multiple assay may be more confusing to perform, and the danger of mistakes can be greater than in a simple assay; it will also have advantages, notably in its more economical use of subjects. If three test preparations are assayed separately against one standard, one-half of the total number of subjects will be assigned to S, and, as emphasized in Chapter 3, separate estimates of the regression coefficient will be needed in each assay. A multiple assay permits a more equal division of subjects between all the preparations, and the regression coefficient, estimated from the combined evidence of all subjects, will be more precisely determined.

The principles of assay design, and the general theory of experimental design for factors at two or more levels, must be applied to multiple assays. Contrasts between preparations form one factor in the design, now with several degrees of freedom. The statistical analysis introduces additional complexity but not new problems.

An investigator might reject the idea of a multiple assay, because it would require too much labour or space, or more subjects than are available at any one time. The advantage may then rest with separate assays, but the alternative of a series of small multiple assays, whose results will eventually be combined, should be considered. For example, suppose that three preparations are to be assayed against S; although a total of 120 subjects can be spared, perhaps not all are available simultaneously. If separate assays are performed, the best procedure would be to use 40 subjects in the assay of the first test preparation, 20 for S and 20 for the test preparation. Two further similar assays would be required for the other test preparations. Thus, in all, 60 subjects would be assigned to S and only 20 to each of the test preparations. A comprehensive multiple assay might assign 30 subjects to each of the four preparations in a single experiment. Unless the variance per response is increased on account of the complexity of the experiment, the precision of the potency estimates would certainly be improved. If a single large experiment is impossible, three small multiple assays, each of 40 subjects (ten for each preparation), could be used. Potency estimates, calculated separately for each assay, would subsequently be combined (Chapter 14) to give a single set of estimates based on 120 subjects.

Aspects of the general theory of statistical analysis have been discussed by

11.2 An assay of two tuberculins

Wadley (1949b) reported an assay of two tuberculins, A and B, against a standard, each being used at three levels of dose. The nine doses were allocated at random to nine sites on the skin of each of four guinea-pigs; after 24 hours, the diameter of the irritated area at each spot was taken as the response (Table 11.2.1).

TABLE 11.2.1 Responses in a multiple assay of tuberculins (diameters of irritated spots, in units of 0·25 mm)

Guinea-pig	Preparation (with concentration × 10^3)									Totals
	S			A			B			
	0·4	2·0	10·0	0·4	2·0	10·0	0·4	2·0	10·0	
I	36	52	64	45	40	65	33	44	70	449
II	41	48	62	38	42	65	36	57	63	452
III	44	48	100	45	62	57	33	54	78	521
IV	48	52	59	40	42	70	37	61	70	479
Totals	169	200	285	168	186	257	139	216	281	1901

The assay might be described as (3, 3, 3). The design is the obvious modification of a randomized block (3, 3) required to accommodate three preparations. The analysis of variance (Table 11.2.3) is very similar to those discussed in §§ 5.2 and 5.4. The total sum of squares divides into components for guinea-pigs, doses, and error, and the only novelty lies in the subdivision of the 8 d.f. between doses. There are now 2 d.f. between preparations. The linear and quadratic components are single degrees of freedom, found in the usual manner from the L_1 and L_2 contrasts totalled over all preparations. The parallelism component, or the difference in regressions for the three preparations, has 2 d.f., and is most easily obtained as the interaction Preparations × L_1.

The component for differences of quadratics is found similarly. A convenient computing scheme is shown in Table 11.2.2, where linear and quadratic contrasts are tabulated for each preparation and for the totals. The regression square is

$$347^2/24 = 5017,$$

and the parallelism component is

$$(116^2 + 89^2 + 142^2 - 8 \times 5017)/8 = 176.$$

The quadratic component and that for differences of quadratics are found in like manner, but with divisors of 72 and 24 respectively.

Table 11.2.3 shows no indications of invalidity. The estimation of potency should utilize all information from the assay, and not only those portions of it

§ 11.2 MULTIPLE ASSAYS

TABLE 11.2.2 Dose totals and contrasts for Table 11.2.1

Preparation	Low	Medium	High	Totals	L_1	L_2
S	169	200	285	654	116	54
A	168	186	257	611	89	53
B	139	216	281	636	142	−12
Totals	476	602	823	1901	347	95

TABLE 11.2.3 Analysis of variance for Table 11.2.1

Adjustment for mean		100 383	
Nature of variation	d.f.	Sum of squares	Mean square
Preparations	2	78	39
Regression	1	5017	5017
Quadratic	1	125	125
Parallelism	2	176	88
Differences of quadratics	2	119	60
Doses	8	5515	
Guinea-pigs	3	371	124
Error	24	1568	65·33
Total	35	7454	

that pertain to S and a particular test preparation. As in § 5.3, logarithms to base 5 lead to a dose metameter that takes values $-1, 0, 1$ for the three doses of each preparation. The average regression coefficient is then

$$b = \frac{347}{24}$$

$$= 14·46.$$

Also, for the first test preparation,

$$\bar{y}_A - \bar{y}_S = \frac{611 - 654}{12}$$

$$= -3·583.$$

Hence

$$M = -0·2478.$$

From the usual formula

$$\text{Var}(b) = s^2/24,$$

and s^2 has 24 d.f., so that

$$g = \frac{(2·064)^2 \times 65·33}{(14·46)^2 \times 24}$$

$$= 0·0555.$$

Equation (4.14.2) then gives for the fiducial limits of M

$$M_L, M_U = -\left[0{\cdot}2478 \pm \frac{2{\cdot}064}{14{\cdot}46}\left\{\left(0{\cdot}9445 \times \left(\frac{1}{12}+\frac{1}{12}\right)\right.\right.\right.$$
$$\left.\left.\left. + \frac{0{\cdot}2478^2}{24}\right) \times 65{\cdot}33\right\}^{1/2}\right]\bigg/0{\cdot}9445$$
$$= -0{\cdot}7509, 0{\cdot}2262.$$

Multiplication by $\log_{10} 5$ converts values of M to common logarithms, whence

$R = $ antilog $\bar{1}{\cdot}8268 = 0{\cdot}671$,

$R_L = $ antilog $\bar{1}{\cdot}4751 = 0{\cdot}299$,

$R_U = $ antilog $0{\cdot}1581 = 1{\cdot}439$.

Thus test preparation A is estimated to have a potency $0{\cdot}671$ times that of the standard, with fiducial limits at $0{\cdot}299$ and $1{\cdot}439$. Similar calculations for B give an estimate of potency of $0{\cdot}846$ times that of the standard with limits at $0{\cdot}384$ and $1{\cdot}830$.

Some may wonder why the data for A and S were not analyzed as a $(3, 3)$ assay, and those for B and S as a second assay of the same design. There are objections:

(i) Validity tests for either analysis have less power than those in Table 11.2.3, because of the fewer degrees of freedom for error (because the analyses are not independent of one another, no simple combination of tests is possible).

(ii) The regression coefficient in each analysis is less precisely estimated, having a variance $s^2/16$.

(iii) The quantity g in each analysis is larger, both because of the greater variance of b and because of the fewer degrees of freedom (15 instead of 24) for t.

(iv) In consequence of (ii) and (iii), the fiducial range for M is wider than in the composite analysis.

(v) Even though in many instances the effects of (i) to (iv) may be small, the labour of computing two separate analyses is certainly greater than is needed for one slightly more complex analysis.

Wadley (1948) mentioned a more complex multiple assay of tuberculins. Sixteen preparations were included, each at four levels, and the 64 doses were assigned to 64 sites on each of ten cows. Allowance for differences in responsiveness between regions of a cow was made by adoption of a balanced incomplete block design for the sixteen preparations, the four doses of a preparation being injected at neighbouring sites. Wadley stated that one such assay showed a 50 percent increase in precision relative to a randomized block design.

11.3 A multiple assay of vitamin B_{12}

Multiple assays have proved especially useful for microbiological assay techniques. A favourite device has been a large assay plate containing many small

§ 11.3   MULTIPLE ASSAYS   231

cavities, each of which can be filled with agar and used in standard fashion for measuring a zone of inhibition of bacterial growth. Positional effects on the plate and time-differences in the inoculations at the many sites can be controlled by appropriate constraints of design and by randomization.

Lees and Tootill (1955a) discussed the assay for which responses are shown in Table 11.3.1. Seven test preparations of vitamin B_{12} were assayed against a standard, using two doses of each preparation. For S and five other preparations, the doses were dilutions of 0·1, 0·2; for F, the dilutions were 0·05, 0·1, and for C they were 1/30, 1/15. Each dose was replicated four times, and the 64 sites on the plate were arranged in an 8×8 formation. A quasi-Latin square design enabled each preparation to appear once in each row and once in each column, four high doses and four low doses in each, in such a way that one parallelism contrast was confounded between rows and one between columns. The structure enables positional effects on the plate to be eliminated. This should go far to remove variation arising from any tendency to 'drift', that is to say a decrease in zone size from one side of the plate to the other, and variation associated with the edges of the plate; also, if solutions are applied consecutively, any time-trend will to a large extent be removed by the statistical analysis. An investigator may contend that an entirely objective order of applying the doses is impracticable. The statistician may be obliged to assent, but he should emphasize that all conclusions are then subject to assumptions about the unimportance of certain sources of variation (§ 4.19). Lees and Tootill described in detail how they conformed to the requirements of randomization. At each of the 64 sites, the diameter of the zone of inhibition of bacterial growth was measured.

TABLE 11.3.1   Responses in a multiple assay of seven vitamin B_{12} preparations
(Diameters of zones of inhibition, in mm $- 14·0$)

								Totals
A_2 4·0	D_2 3·6	E_1 1·9	B_2 3·6	F_1 2·3	S_1 2·5	G_1 2·3	C_2 4·3	24·5
G_2 3·8	E_2 4·4	D_1 0·5	S_2 4·2	C_1 1·5	B_1 1·5	A_1 1·2	F_2 4·3	21·4
C_2 4·3	B_2 3·0	S_1 2·6	D_2 2·8	G_1 1·6	E_1 1·7	F_1 1·6	A_2 3·6	21·2
B_1 1·0	C_1 1·0	F_2 4·0	A_1 1·3	E_2 3·3	G_2 3·0	S_2 3·8	D_1 0·9	18·3
E_1 1·6	G_1 1·5	A_2 3·3	F_1 1·9	B_2 3·0	C_2 3·6	D_2 3·3	S_1 2·2	20·4
S_1 1·9	F_1 1·9	C_2 3·4	G_1 1·0	D_2 2·8	A_2 3·4	B_2 2·9	E_1 1·0	18·3
D_1 0·9	A_1 1·0	G_2 3·4	C_1 1·0	S_2 3·8	F_2 3·4	E_2 3·4	B_1 0·6	17·5
F_2 3·6	S_2 3·8	B_1 0·3	E_2 3·5	A_1 1·0	D_1 0·3	C_1 1·0	G_2 3·4	16·9
21·1	20·2	19·4	19·3	19·3	19·4	19·5	20·3	158·5

The confounding structure should be apparent from inspection of rows 1, 3, 5, 6 as compared with rows 2, 4, 7, 8 and from columns 1, 3, 6, 8 as compared with columns 2, 4, 5, 7. The design was constructed to satisfy these constraints on rows and columns and the orders of rows and columns were then separately randomized. Thus Regression $\times (A+B+C+D-E-F-G-S)$ is confounded with rows and Regression $\times (A-B+C-D-E+F+G-S)$ with columns. Had there been concern to avoid total confounding, a design involving partial confounding of parallelism contrasts with rows and columns could have been adopted. Because of the severe constraints, the design used necessarily appears to have a markedly systematic pattern. An alternative would have been to take an 8×8 Latin square for the eight preparations and impose on it an arbitrary orthogonal partition into two sets of 32 cells such that each set consisted of four in each row, four in each column, and four of each preparation (Finney, 1945b, 1946c; orthogonal partitions of 8×8 Latin squares have not been investigated, but probably most can be so partitioned). The two sets of 32 would be used for the high and low doses. Such a design may seem simpler to understand than the quasi-Latin square; it is less satisfactory in use as it does not permit the isolation of parallelism constrasts. The quasi-Latin square is a particular instance of the orthogonal partition design, but it has much additional symmetry.

Table 11.3.1 includes row and column totals. Table 11.3.2 contains totals for preparations and doses, and differences between high and low doses. From the last line of Table 11.3.2, the contrasts for the two confounded degrees of freedom are

$$9 \cdot 8 + 9 \cdot 1 + 11 \cdot 1 + 9 \cdot 9 - 8 \cdot 4 - 7 \cdot 6 - 7 \cdot 2 - 6 \cdot 4 = 10 \cdot 3$$

and

$$9 \cdot 8 - 9 \cdot 1 + 11 \cdot 1 - 9 \cdot 9 - 8 \cdot 4 + 7 \cdot 6 + 7 \cdot 2 - 6 \cdot 4 = 1 \cdot 9.$$

The corresponding squares, $10 \cdot 3^2/64$ and $1 \cdot 9^2/64$, must be subtracted from the parallelism component of the analysis of variance. From Tables 11.3.1 and 11.3.2 this analysis can be rapidly computed. The steps are as follows, the number alongside each indicating where it fits into Table 11.3.3:

(1) $158 \cdot 5^2/64 = 392 \cdot 5352$

(2) $4 \cdot 0^2 + 3 \cdot 6^2 + 1 \cdot 9^2 + \ldots + 3 \cdot 4^2 - 392 \cdot 5352 = 93 \cdot 5948$

(3) $(24 \cdot 5^2 + 21 \cdot 4^2 + \ldots + 16 \cdot 9^2 - 8 \times 392 \cdot 5352)/8 = 5 \cdot 6460$

(4) $(21 \cdot 1^2 + 20 \cdot 2^2 + \ldots + 20 \cdot 3^2 - 8 \times 392 \cdot 5352)/8 = 0 \cdot 3760$

(5) $(24 \cdot 8^2 + 18 \cdot 8^2 + \ldots + 20 \cdot 0^2 - 8 \times 392 \cdot 5352)/8 = 9 \cdot 3336$

(6) $69 \cdot 5^2/64 = 75 \cdot 4727$

(7) $(6 \cdot 4^2 + 9 \cdot 8^2 + \ldots + 7 \cdot 2^2 - 8 \times 75 \cdot 4727)/8$

                                          $-10 \cdot 3^2/64 - 1 \cdot 9^2/64 = 0 \cdot 4620$

(8) $(2) - (3) - (4) - (5) - (6) - (7) = 2 \cdot 3045$

As in §11.2, the sum of squares for parallelism is formed as though it were a

§ 11.3   MULTIPLE ASSAYS   233

sum of squares for the interaction of two factors, preparations and regression; thus it measures the extent to which the regression coefficients for the separate preparations differ.

TABLE 11.3.2   Preparation and dose totals for Table 11.3.1

	S	A	B	C	D	E	F	G	Total
Low	9·2	4·5	3·4	4·5	2·6	6·2	7·7	6·4	44·5
High	15·6	14·3	12·5	15·6	12·5	14·6	15·3	13·6	114·0
Sum	24·8	18·8	15·9	20·1	15·1	20·8	23·0	20·0	158·5
Difference	6·4	9·8	9·1	11·1	9·9	8·4	7·6	7·2	69·5

TABLE 11.3.3   Analysis of variance for Table 11.3.1

1	Adjustment for mean		392·5352	
	Nature of variation	d.f.	Sum of squares	Mean square
3	Rows	7	5·6460	
4	Columns	7	0·3760	
5	Preparations	7	9·3336	1·3334
6	Regression	1	75·4727	75·4727
7	Parallelism	5	0·4620	0·0924
8	Error	36	2·3045	0·064014
2	Total	63	93·5948	

The preparations mean square is much greater than the error, so that unless the investigators have full confidence in linearity they should be a little worried about statistical invalidity. In fact, the vitamin B_{12} assay rests upon considerable experience of linearity, but Table 11.3.2 indicates that preparations B and D showed much lower potency than is implied by the doses chosen for them. The regression square is very satisfactorily large. The parallelism mean square is larger than error, but occasions no alarm. Although the columns mean square scarcely differs from error, that for rows is much larger. Had an unrestricted randomization of four replicates of each dose to the 64 sites been adopted, the error mean square would have been of the order of

$$(49 \times 0·064\,014 + 5·6460 + 0·3760)/63 = 0·1454.$$

This is more than twice the error mean square in Table 11.3.3: two assays with unrestricted randomization would be needed in order to approximate to the precision of the one assay under discussion.

In forming potency estimates and assessing their fiducial limits, simplified formulae for (2, 2) assays could be used, but some modifications would be necessary in order to take account of the multiple assay scheme. If an 8×8 square were to be used frequently for the simultaneous assay of seven preparations, formulae analogous to equations (5.4.1) and (5.4.5) might profitably be constructed, but in an isolated instance the surest way of avoiding mistakes is to return to first principles of calculation in terms of mean responses and regression coefficients. If doses are taken as $-1, 1$ on the logarithmic scale, the regression coefficient is estimated as

$$b = \Sigma S_{xy}/\Sigma S_{xx}$$
$$= 69.5/64$$
$$= 1.0859.$$

Also,
$$\text{Var}(b) = s^2/\Sigma S_{xx},$$

and for a probability of 0.95,

$$g = (2.028)^2 \times 0.064\,014/(1.0859^2 \times 64)$$
$$= 0.0035,$$

sufficiently small to be safely ignored. Standard formulae such as equations (4.14.1) and (4.14.2) enable the potency and its limits to be estimated for each preparation. For this purpose, $N_S = N_T = 8$ and $\Sigma S_{xx} = 64$; the dose ratio, 2·0, and the different dilutions used for the preparations must be introduced at the right points for converting M to R. For example, preparation D is estimated to have potency 0·679 relative to S, with limits at 0·623 and 0·738. Had the doses for D been chosen a little higher, the improvement in precision would have been small, but there would have been less reason to fear any invalidity on account of the difference in mean response. Since g is so small, essentially the same potency limits would have been obtained for each preparation from the approximation

$$\text{Var}(M) = \frac{s^2}{64b^2}(16 + M^2). \tag{11.3.1}$$

Had separate assays been used for each test preparation, at the same total cost only four responses could have been measured for the two preparations in one assay. The approximate variance would then have been

$$\text{Var}(M) = \frac{s^2}{8b^2}(4 + M^2). \tag{11.3.2}$$

Comparison of equations (11.3.1) and (11.3.2) shows that if s^2 were the same for both, use of seven separate assays would have increased Var(M) by a factor of about 2·0. This over-simplified statement may exaggerate the advantage of the multiple assay, as a design with only eight cells and one test preparation could have a more compact layout and might therefore have a smaller s^2.

11.4 Multiple slope ratio assays

The principles illustrated in §§ 11.2 and 11.3 may be applied to multiple slope ratio assays. For example, if three test preparations were tested simultaneously, a regression function

$$Y = a + b_S x_S + b_A x_A + b_B x_B + b_C x_C \tag{11.4.1}$$

would be estimated. The ratios b_A, b_B, b_C to b_S would estimate potencies. The scheme of analysis recommended in Chapter 7 may be extended to give validity tests and fiducial limits (Barraclough, 1955; Bliss, 1946b; Clarke, 1952).

11.5 Designs for multiple assays

Possible designs for multiple assays are very numerous. No attempt will be made to discuss and classify them in the same detail as for incomplete block and cross-over designs. To those who have understood Chapters 9 and 10, construction of designs for multiple assays should present no special difficulty. Guld *et al.* (1958) have reported an interesting example of tuberculin assay using large numbers of human subjects in a balanced incomplete block design with two responses per subject.

Equal numbers of doses for all preparations and regular spacing for the dose metameter are desirable features of these designs. Earlier recommendations on the choice and number of doses (Chapters 6 and 8) remain applicable. Although considerations of symmetry suggest that the number of responses measured per dose should be the same for all doses of all preparations, this is not ideal for precision: the advantage of assigning more subjects to S than to each of the test preparations is demonstrated in § 11.6, a modification rarely convenient in designs other than the completely randomized and randomized block types.

When the number of responses that can be measured in one block of subjects is as large as the total number of doses, a randomized block design should be used; in the assay of § 11.2, the block consisted of nine sites on one guinea-pig. When the number of responses per block is smaller than the total number of doses, the usual problems of incomplete blocks, balanced or confounded, arise. If equal precision on all dose contrasts is desired, a balanced incomplete block design must generally be chosen, and those in § 9.7 may be applied with appropriate change of nomenclature. Thus, if three test preparations are to be assayed against a standard, with two doses of each, Table 9.7.2 can be used to give an arrangement in blocks of four, the symbols for the eight doses in that table being replaced by those for the two levels of four preparations. The assay of tuberculins in § 11.2 could have been arranged for an experiment with only four tests per guinea-pig by use of Table 9.7.3. Designs for larger total numbers of doses may be taken from Cochran and Cox (1957, Tables 9.5 and 11.3). An investigator who wishes to increase the precision of estimation of the differences between preparations and the regression coefficient, at the expense of validity tests, may use a confounded arrangement analogous to those in § 9.9. Tables 11.5.1 and 11.5.2 are examples for three test preparations at two levels in blocks of four and for two test preparations at three levels in blocks of three respectively. The design for the assay in § 11.3 is related to that in Table 11.5.1.

Brownlee *et al.* (1948) described an assay of three streptomycin preparations using a quasi-Latin square in a manner similar to that of Lees and Tootill. The latter authors went on to present a 9×9 quasi-Latin square for two test preparations and a standard with three levels of each. In later papers (Lees and Tootill, 1955b, c), they have illustrated a 12×12 quasi-Latin square (see also

TABLE 11.5.1 Multiple assay of three test preparations at two dose levels in blocks of four (Preparations and regression contrasts unconfounded)

Block no.	S_1	S_2	A_1	A_2	B_1	B_2	C_1	C_2
I	X		X			X		X
II	X			X	X			X
III	X			X		X	X	
IV		X	X			X		X
V		X	X				X	X
VI		X		X	X		X	

TABLE 11.5.2 Multiple assay of two test preparations at three dose levels in blocks of three (Preparations and regression contrasts unconfounded)

Block no.	S_1	S_2	S_3	A_1	A_2	A_3	B_1	B_2	B_3
I	X			X					X
II	X				X			X	
III		X		X					X
IV		X				X		X	
V			X	X				X	
VI			X		X		X		

Harrison, Lees and Wood, 1951) and Youden and balanced Latin squares. Brownlee and Lapedes (1951) have compared several designs. Cross-over designs may be constructed in the manner of § 10.5. For example, a cross-over for three occasions per subject and 18 subjects is easily constructed from Table 11.5.2. Allott and O'Neill (1970) described a method of using seedling weights as responses for multiple assays of herbicide residues in soil.

11.6 The allocation of subjects to preparations

The recommendations on design in § 11.5 have been made with the assumption that, as for the assays discussed in §§ 11.2 and 11.3, the number of responses measured would be the same for each preparation. For a complicated design, such as that of the vitamin B_{12} assay, symmetry almost forces this equality; for simpler designs, the possibility of gaining in efficiency by unequal distribution must be borne in mind.

Consider the optimal allocation of a fixed total number of subjects, N, in a parallel line assay for which only one response per subject can be measured, with c test preparations to be assayed simultaneously. Unless greater precision is desired for some test preparations than for others, the same number of subjects, N_T, will be assigned to each, but this need not be equal to N_S, the number assigned to S. Whatever these numbers are,

$$N_S + cN_T = N. \tag{11.6.1}$$

For any one test preparation, provided that the doses have been well chosen and that g is small, Var(M) is approximately proportional to

$$\left(\frac{1}{N_S} + \frac{1}{N_T}\right).$$

Hence, for a fixed N, the variance will be a minimum if

$$N_S = c^{1/2} N_T \tag{11.6.2}$$

(Fieller, 1947, Brownlee and Lapedes, 1951), whence

$$\left.\begin{array}{l} N_S = \dfrac{N}{1 + c^{1/2}}, \\[2mm] N_T = \dfrac{N}{c^{1/2} + c}. \end{array}\right\} \tag{11.6.3}$$

The alternative of complete symmetry has

$$N_S = N_T = \frac{N}{1 + c}. \tag{11.6.4}$$

The recommended allocation of subjects reduces the value of $\left(\dfrac{1}{N_S} + \dfrac{1}{N_T}\right)$ by a factor

$$\frac{(1 + c^{1/2})^2}{2(1 + c)};$$

the efficiency of the asymmetric allocation relative to the symmetric is therefore

$$\text{Eff} = \frac{2(1 + c)}{(1 + c^{1/2})^2}. \tag{11.6.5}$$

When the choice of doses has been less successful, so that $(M - \bar{x}_S + \bar{x}_T)$ is large, $\left(\dfrac{1}{N_S} + \dfrac{1}{N_T}\right)$ plays a lesser role in determining the precision or the reliability of M, and the gain from allocating subjects in accordance with equation (11.6.3) is less. A complete discussion would be analogous to that in Chapter 6.

Table 11.6.1 shows some values of the efficiency function. Unless c is large, the gain is small, but even the 11 percent gain for an assay of four test preparations, when interpreted as a 9 percent saving of subjects for a specified precision, may be of economic importance, especially if large numbers of routine assays are affected. The very large values of c may be relevant to techniques such as those of radioligand assay (§16.6). The exact values of N_S, N_T given by equations (11.6.3) are seldom integers, but little is lost by using the nearest convenient integers. For example, for $c = 3, 4, 5,$ or 6, the rule of using $N_S = 2N_T$ will be good enough, giving 107, 111, 114, and 117 as percentage efficiencies.

If half the subjects were assigned to S and the remainder divided equally between the test preparations, so that

$$\left.\begin{array}{l} N_S = N/2, \\ N_T = N/2c, \end{array}\right\} \tag{11.6.6}$$

TABLE 11.6.1 Precision of a multiple parallel-line assay with optimal allocation of subjects, relative to symmetric distribution, as given by equation (11.6.5)

Number of test preparations (c)	Percent precision	Number of test preparations (c)	Percent precision
1	100	10	127
2	103	20	140
3	107	30	148
4	111	40	153
5	115	50	157
6	118	60	159
8	123	80	164
9	125	100	167

Var(M) would be the same as for the allocation in equation (11.6.4). This perhaps suggests that c separate assays would be as good as a multiple assay with a completely symmetric allocation of subjects. In reality, the multiple assay is better because all subjects that receive the standard preparation contribute to the one value of $\bar{y}_S$: the efficiency of the separate assays relative to a symmetric multiple assay is only

$$\text{Eff} = \frac{1+c}{2c}, \tag{11.6.7}$$

an expression that is little larger than 0·5 when c is moderately large. Moreover, in the multiple assay, all subjects help to estimate b.

If no block constraints are to be incorporated into the design, equations (11.6.3) are easily applied. Doses are chosen in the usual manner, and each dose of S has its number of subjects increased in the same proportion relative to the test preparations. For more complicated designs, perhaps the easiest method of construction is to regard the standard preparation as subdivided into two or more identical treatments. Thus, for a multiple assay of five test preparations, a fully symmetric design for seven preparations might be formed and S included twice; extra degrees of freedom for error arising from this duplication of one treatment could be neglected if calculation of the appropriate sum of squares proved awkward.

Discussion of the optimal allocation of a fixed total number of subjects in a slope ratio assay is more difficult, because the variance formulae are more complex. However, if N_0, N_S, N_T are the numbers of subjects assigned to the blanks, to S, and to each test preparation, respectively, so that

$$N_0 + N_S + cN_T = N, \tag{11.6.8}$$

the variances of b_S and of any b_T will be approximately proportional to $1/N_S$ and $1/N_T$. Under good conditions, for which g is small and h nearly unity, this means that the variance of R will be approximately proportional to

$$\left(\frac{1}{N_S} + \frac{1}{N_T}\right).$$

§ 11.6   MULTIPLE ASSAYS

If N_0 were held fixed, the optimal allocation would again have

$$N_S = c^{1/2} N_T.$$

The argument of Chapter 8 showed that there was no merit in making N_0 large, and a reasonable rule would seem to be to take it as equal to the number of subjects on any one dose of a test preparation. Hence

$$\left.\begin{aligned} N_0 &= \frac{N}{1 + k(c^{1/2} + c)}, \\ N_S &= \frac{Nkc^{1/2}}{1 + k(c^{1/2} + c)}, \\ N_T &= \frac{Nk}{1 + k(c^{1/2} + c)}, \end{aligned}\right\} \quad (11.6.9)$$

may be expected to be somewhere near the best possible. Of course, strict adherence to these formulae is not essential. For $c = 2$, a fully symmetric design will be almost as good as that recommended here; for $c = 3, 4, 5$, or 6, approximately double replication of S relative to each test preparation will be good.

12
Use of concomitant information

12.1 The combination of measurements

Earlier discussion of assays based upon quantitative responses has assumed that no doubt exists as to which uniquely defined numerical value should be taken as the response; it may be a single measurement, such as uterine weight, or it may be a compound of two or more measurements, such as a percentage reduction in blood sugar. In §§ 6.11 and 8.6 the problem of choosing between alternative measures of response and the possibility of combining measurements of different kinds into a single metameter were mentioned. This chapter and the next are concerned with the optimal construction of a composite metameter from measurements of two or more characteristics of the subjects.

One situation is that in which, after each application of a dose to a subject, any or all of several distinct measurements could have been affected by the dose. For example, the body weight of an animal and the weights or linear dimensions of individual organs might be measured. If one of these were relatively more affected by dose differences than the others, it would be chosen as *the* response; if two or more are comparable in their qualities as response metameters (perhaps having similar values of s/b), a combination of measurements may be appreciably better. Determination of an appropriate composite metameter, essentially a problem of discriminant functions, is discussed in Chapter 13.

A distinct situation is that of using *concomitant* measurements. Only one measurement that can properly be regarded as a response to the dose is available; others are correlated with it as properties of the subjects, but are not themselves affected by dose. Ideally these concomitants are measured before the randomization that allocates subjects to doses. For example, in the rabbit method for insulin assay, the initial blood sugar level cannot be affected by the dose of insulin that is to be given: any association between initial blood sugar and dose is due either to chance or to faulty randomization, and the latter explanation ought to be excluded by proper planning and conduct of the experiment. The final blood sugar level is affected by the dose of insulin, but may also be correlated with the initial value. The assayist is not compelled to follow the usual practice of using percentage change in blood sugar as his metameter. He would be quite in order to use the final value itself, or any function of the two, provided that the validity conditions are satisfied (Chapter 15). He will naturally wish to choose the function that will give him the most reliable estimate of ρ. In other assays, the response measured may be body weight or weight of a certain organ of the subject at the end of the experiment, and initial body weight may also have been recorded. Increase in body weight, or weight of the organ per unit initial body weight, is commonly assumed to be the ideal

§ 12.2   USE OF CONCOMITANT INFORMATION   241

metameter, yet it can lead to a less precise potency estimate than the unadjusted final weight.

If y is a response variate and z_1, z_2 are two concomitants, any new metameter formed as a function of y, z_1, z_2 can be considered for use in the assay. In practice, if y itself is judged to be a fairly satisfactory metameter, simple modifications of y are likely to be of interest, such as

$y/z_1,$

$y - z_1,$

$y - 3z_1 \log z_2,$

$y - \beta_1 z_1 - \beta_2 z_2$ (for any fixed β_1, β_2).

Since z_1, z_2 have only a random association with dose, the statistical validity of an assay is unaffected by the choice of these metameters. Indeed, for the last three functions and others like them, the expectation of any contrast between dose totals is exactly the same as for y alone, the randomization ensuring that the function of z_1, z_2 appearing in the contrast has expectation zero. Such simple but arbitrary compounds of available measurements have often been used as metameters. Examples were mentioned in the preceding paragraph. In a more complicated problem described by Fieller (1940), a series of similar insulin assays was used to determine a suitable function which was then adopted as the metameter for future assays.

In general, arbitrary adjustment for concomitant variates does not maximize precision. Even the use of weight gain is not always an ideal way of using the information provided by initial weights of subjects. Usually preferable is a technique that allows the data themselves to indicate the best metameter, though in a small assay the information may be too scanty for this to be satisfactory. The standard procedures of covariance analysis, as explained in many statistical texts, are assumed familiar to the reader; emphasis here is placed on the special points arising in the analysis of assay data. In practice, a choice cannot be made from all possible functions of the measurements, and some restriction to a class of functions must be accepted. The best linear function is usually good enough; selection from a larger class would be possible, at the expense of much heavier computations, but is seldom worth while. The phrase 'linear function' here includes linear functions of the logarithms or specified powers of the original measurements, where these seem more suitable.

12.2 Adjustment by proportionality

Among pharmacologists, a common method of adjusting for a concomitant variate is by using a ratio as the response metameter (cf. § 4.15). For example, if y is the weight of a particular organ at the end of an assay and z is the initial body weight of the animal, the metameter used might be mg organ weight per g body weight. To use *final* body weight rather than initial might seem more logical, but if body weight itself were affected by the assay stimulus this would tend to reduce the responsiveness of the metameter to dose changes. In practice, if the assay is of short duration, it probably matters little whether initial or final weight is used, though weights which cannot have been affected by dose are to be preferred. If z, y are measurements of the same character before and after the assay, as when y itself is final body weight, the metameter y/z will commonly lead to much the same conclusions as the increases, $(y - z)$.

12.3 Adjustment by covariance analysis

Although adjustment of responses by expressing them as proportions of the concomitant variate often markedly improves the precision of an assay, it is ideal only if among subjects at the same dose the response varies almost in proportion to the concomitant. Even with approximate proportionality, the adjustment will still be good, which probably explains its frequent successful employment. It can fail deplorably. Indeed, it can even reduce precision, as it will if the correlation between response and concomitant is negative. A further disadvantage of the method is that its application is restricted to a single concomitant variate, though in some assays simultaneous adjustment for two or more concomitants is desirable.

Bliss and Marks (1939a) emphasized that to assume proportionality is often unjustifiable, and stated the arguments in favour of adjustment by covariance analysis from the internal evidence of the data. They described calculations for parallel line assays, and illustrated them from a (2, 2) insulin assay in which the percentage fall in blood sugar of injected rabbits was adjusted for inequalities in the initial blood sugar. Bliss (1940a) discussed a similar analysis for an assay of vitamin D, in which ash content of the femur of rats was adjusted for weight of organic matter in the bones (a quantity unaffected by vitamin D). Bliss and Rose (1940) gave examples of assays for parathyroid extracts in which serum calcium of dogs was adjusted for initial body weight; these are chiefly of interest because incomplete block designs were used. Fieller *et al.* (1939b) and Fieller (1940), in a very full account of insulin assays using cross-over designs, discussed the adjustment of percentage fall in blood sugar by means of a covariance analysis on initial blood sugar. They used a single regression coefficient of response on concomitant, estimated from the combined evidence of many assays, to adjust the response for each assay. If the true regression coefficient does not vary from one assay to another, this adjustment will be more precise than any based on a single assay, and the complications introduced into the assessment of errors and fiducial limits will be less serious.

Finney (1947b) described a (3, 3) assay with covariance, following the general approach of this book but using a different computational pattern. In a personal communication (1957), Dr G.S. James pointed out the algebraic equivalence of the computations. Sakiz and Guillemin (1964) reported substantial gains from using covariance analysis in thyrotrophin assays. Leppäluoto (1972) provided further evidence of advantages in using covariance rather than the proportional change in ^{131}I uptake for thyrotrophin assays.

12.4 An example of covariance

Despite the reports of gains in precision from the use of covariance in bioassay, good published examples that quote individual responses are rare. One such is presented in § 12.5, but an experiment reported by Guillemin and Sakiz (1963) provides a better introduction. Not strictly a bioassay, this was concerned with comparing three conditions in their effects on the potency of one drug, but the computations are exactly as for an assay (Finney, 1971). Doses of $0.3\,\mu g$, $0.9\,\mu g$, $2.7\,\mu g$ of luteinizing hormone (LH) were injected into rats of three types: 'normal' (S), injected with prolactin 90 minutes previously (T), and hypophysectomized after prolactin treatment (U). The experiment will be regarded as an 'assay' of the effects of T and U on the potency of the hormone relative to the potency for S animals. The experiment was conducted on two days, 27 rats per day being assigned three to each of the nine combinations of dose with S, T, U. Because of slight differences in other conditions on the two days, each day will be regarded as a block of the experiment. For reasons explained by Guillemin and Sakiz, rats hypophysectomized but without prolactin have been omitted.

TABLE 12.4.1 Dose totals for LH 'assay' (totals for 6 rats)

Type of rat	Dose of LH (μg)	Ovary weight, z (μg)		Response, y (mg)	
S	0.3	569.8		357.5	
	0.9	619.1	1773.3	302.9	818.9
	2.7	584.4		158.5	
T	0.3	578.6		356.2	
	0.9	598.0	1671.2	289.3	752.0
	2.7	494.6		106.5	
U	0.3	525.1		334.0	
	0.9	574.6	1783.0	249.2	816.9
	2.7	683.3		233.7	
Total		5227.5		2387.8	
L_1 contrast		88.8		−549.0	

TABLE 12.4.2 Analysis of covariance for Table 12.4.1

Adjustments for means		506 051.0	231 152.3	105 585.0
Nature of variation	d.f.	S_{zz}	S_{yz}	S_{yy}
Blocks	1	54 232.4	14 412.9	3 830.4
Types	2	426.3	260.5	160.9
Regression (L_1)	1	219.0	−1 354.2	8 372.2
Quadratic (L_2)	1	201.7	186.4	172.3
Parallelism	2	2 472.4	1 537.7	962.1
Difference of quadratics	2	510.5	532.4	558.3
Treatments	8	3 829.9	1 162.8	10 225.8
Error	44	16 057.6	9 742.9	7 294.1
Total	53	74 119.9	25 318.6	21 350.3

The response for the assay was the ascorbic acid content of the left ovary of each rat. The weight of each ovary was also recorded; this is most unlikely to have been affected by the hormone and will be used as a concomitant. Table 12.4.1 summarizes totals of the two variates for the three doses of each preparation; values for individual rats are not available. From these totals and from the original records for individual rats, Table 12.4.2 was constructed. The analysis of sums of squares for the response (y) in the standard form for a bioassay has been repeated for the concomitant (z); an analysis of sums of products is made in exactly the same manner, except that every square of y (or z) or of a sub-total is replaced by the product of corresponding y and z individual values or sub-totals.

If z is ignored, the data on y can be used as in Chapter 11 to permit the 'assay' of T and U against S. A rapid scanning of the y analysis in Table 12.4.2 suffices to show that none of the mean squares for validity tests significantly exceeds

$$s^2 = 7294 \cdot 1/44 = 165 \cdot 77. \tag{12.4.1}$$

In the usual way

$$b = -549 \cdot 0/36 = -15 \cdot 25. \tag{12.4.2}$$

Hence for T,

$$M = -\frac{66 \cdot 9}{18} \bigg/ (-15 \cdot 25)$$

$$= 0 \cdot 2437 \tag{12.4.3}$$

and the usual calculations lead to 0·95 limits at $-0 \cdot 3311, 0 \cdot 8611$. Multiplication by 0·477 12 converts the logarithms from base 3 to base 10: the estimated potency of T relative to S is 1·307 with limits at 0·695, 2·575. Correspondingly for U, the potency estimate is 1·008 with limits at 0·527, 1·932.

Whereas the y analysis in Table 12.4.2 shows, as it ought, a highly significant regression mean square, no treatment mean square for z is large enough to suggest that ovary weight may have been influenced by the hormone. The covariance adjustment can therefore reasonably be undertaken. The error regression coefficient of y on z is

$$B = 9742 \cdot 9/16\ 057 \cdot 6$$

$$= 0 \cdot 6067. \tag{12.4.4}$$

The residual sum of squares for error is

$$7294 \cdot 1 - (9742 \cdot 9)^2/16\ 057 \cdot 6 = 1382 \cdot 6 \tag{12.4.5}$$

with $(44 - 1)$ degrees of freedom. The residual error mean square,

$$s^2 = 32 \cdot 153, \tag{12.4.6}$$

is so much smaller than that in equation (12.4.1) that the covariance adjustment is clearly attractive. Table 12.4.3 summarizes the tests.

A first step before estimating relative potency is to check assay validity. This

§ 12.4 USE OF CONCOMITANT INFORMATION 245

TABLE 12.4.3 Validity tests for LH 'assay' after adjustment of response for ovary weight

Nature of variation	d.f.	Sum of squares	Mean square
Types	2	1·7	0·8
L_2	1	20·1	20·1
Parallelism	2	6·1	3·0
Difference of curvatures	2	97·2	48·6
Residual error	43	1382·6	32·15

is achieved by taking each of four lines of Table 12.4.2 in turn, adding to the error line, and forming a reduced sum of squares exactly as in equation (12.4.5) above. For example, for curvature,

$$(7294\cdot1 + 172\cdot3) - (9742\cdot9 + 186\cdot4)^2/(16\,057\cdot6 + 201\cdot7) = 1402\cdot7. \tag{12.4.7}$$

The difference between this and the error residual is the appropriate sum of squares for L_2 after adjustment of y by linear regression on z. If any mean square here substantially exceeded the error mean square, the validity of assay calculations would be in doubt. The logic is essentially as for assay data without covariance. In fact, inspection shows that only one mean square exceeds the error and this not significantly. Assay calculations can therefore be continued using

$$y^* = y - Bz \tag{12.4.8}$$

where

$$\operatorname{Var}(B) = s^2/16\,057\cdot6; \tag{12.4.9}$$

the divisor is taken from the error line of Table 12.4.2, and the expression is best left in this form for the present.

Consider the potency calculations for T relative to S. From equation (12.4.8) and Table 12.4.2,

$$\begin{aligned}\bar{y}_T^* - \bar{y}_S^* &= L_p(y^*)/18 \\ &= \frac{L_p(y) - BL_p(z)}{18} \\ &= [-66\cdot9 - B(-102\cdot1)]/18 \\ &= -0\cdot2753,\end{aligned} \tag{12.4.10}$$

where L_p for the moment relates only to S and T. Also,

$$\operatorname{Var}(\bar{y}_T^* - \bar{y}_S^*) = s^2\left[\frac{1}{18} + \frac{1}{18} + \left(\frac{102\cdot1}{18}\right)^2 \cdot \frac{1}{16\,057\cdot6}\right]. \tag{12.4.11}$$

Similarly, the adjusted dose regression coefficient is

$$\begin{aligned}b^* &= L_1(y^*)/36 \\ &= \frac{L_1(y) - BL_1(z)}{36}\end{aligned} \tag{12.4.12}$$

$$= \frac{-549 \cdot 0 - 88 \cdot 8 B}{36}$$

$$= -16 \cdot 75,$$

with variance

$$\text{Var}(b^*) = s^2 \left[\frac{1}{36} + \left(\frac{88 \cdot 8}{36}\right)^2 \cdot \frac{1}{16\,057 \cdot 6} \right]. \tag{12.4.13}$$

Although the unadjusted 'preparations' and 'slope' contrasts for the assay are orthogonal, the regression adjustments that give rise to (12.4.10) and (12.4.12) both involve B and therefore introduce a covariance estimated by

$$\text{Cov}(\bar{y}_T^* - \bar{y}_S^*, b^*) = \left(\frac{-102 \cdot 1}{18}\right)\left(\frac{88 \cdot 8}{36}\right) \text{Var}(B). \tag{12.4.14}$$

Hence the estimate of logarithmic relative potency (base 3) is, from (12.4.10) and (12.4.12),

$$M = -0 \cdot 2753/(-16 \cdot 75)$$

$$= 0 \cdot 0164; \tag{12.4.15}$$

the variance–covariance matrix for numerator and denominator is, from (12.4.11), (12.4.13) and (12.4.14),

$$\begin{pmatrix} 3 \cdot 636\,99 & -0 \cdot 028\,01 \\ -0 \cdot 028\,01 & 0 \cdot 905\,33 \end{pmatrix}. \tag{12.4.16}$$

In this example omission of the terms arising from $\text{Var}(B)$ would not have made much difference, but in other circumstances they may be important. Calculations now continue, using Fieller's theorem in the usual way. Thus

$$g = \frac{2 \cdot 017^2 \times 0 \cdot 905\,33}{16 \cdot 752}$$

$$= 0 \cdot 0131,$$

and

$$M_L, M_U = -0 \cdot 2142, 0 \cdot 2483;$$

the limits are much narrower than those based on y alone. The estimated potency of T is 1·018, with limits at 0·790, 1·314. Similarly for U, the estimate is 1·029, with limits at 0·800 and 1·324. Guillemin and Sakiz reported almost identical values.

Here the gain from covariance is largely that due to reduction in s^2 by a factor of 5. As Bliss (1940a) found, for many sets of data the gain is less than might be hoped from consideration of s^2. Even though the regression on the concomitant variate is significant, the benefit of making allowance for it may be nullified by low precision in the estimation of the adjustment, and by the fact that the regression coefficient of response on dose chances to be lower than the regression coefficient when the concomitant variate is ignored. In such circumstances, no arbitrary adjustment, by proportionality or otherwise, is likely to prove any better. Provided that the regression on the concomitant is significant, adjustment can scarcely lead to any serious *loss* of precision, although there might be a small loss if the degrees of freedom for error were few.

The reader unfamiliar with the analysis of covariance may wonder at the complexity of the calculation for validity tests and estimation based on y^*; he may think that a simpler procedure would be to construct 54 adjusted responses from equation (12.4.8) and analyze these exactly as if they were not adjusted. He would obtain the same potency estimates, but the occurrence of B in every adjusted response introduces correlations that affect precision. Only if B were determined very precisely would this procedure be sufficiently exact.

The method is easily extended so as to adjust simultaneously for two or more distinct concomitant variates. The formulae become more complicated, but contain no intrinsic difficulties and follow the standard pattern of multiple linear regression. One useful application is adjustment for an unwanted classification in the subjects. For example, if Guillemin and Sakiz had been obliged to use two strains of rat and had not balanced their experiment in respect of them, a concomitant defined to have the value 0 for each rat of one strain, 1 for each rat of the other, would permit standard computations to remove any average effect of the strain difference.

The symmetry of design made statistical analysis fairly simple. Orthogonal coefficients could readily be used in subdivision of sums of squares between doses. Even with a design as unsymmetric as that in § 4.2, the method still applies; adjustment must then be to $(\bar{y}_T - \bar{y}_S)$ and to b, instead of to simple orthogonal contrasts. As long as the basic rules for handling variances and covariances of contrasts are remembered, no serious difficulty should arise, despite the tedious arithmetic.

12.5 An assay of a parathyroid extract

For an assay of more complex design, the computations inevitably become more laborious. An assay of a parathyroid extract (Bliss and Rose, 1940) illustrates the use of covariance analysis with a balanced incomplete block design. The response was the serum calcium level in a dog injected with a dose of S or T. In the belief that large dogs would respond less to a specified dose than would small dogs, the doses for a (2, 2) assay were chosen as 0·06 and 0·12 ml per kg body weight. Provided that the dogs used for the two preparations are selected from the same population, use of a fixed dose per unit body weight does not disturb the validity of the assay; as will be shown below, it did not remove all effects of inequalities in weight.

Each of 36 dogs was tested twice at an interval of two weeks. A balanced incomplete block cross-over design (§ 10.5) allowed every pair of doses to be assigned to six dogs, three in each of the two possible orders. In order to spread the labour over three days, the dogs were divided into blocks of 12. Table 12.5.1 shows the arrangement, the body weights of the dogs before injection, and the

TABLE 12.5.1 Initial body weight, z (in units of 0·1 kg) and final serum calcium, y (in units of 0·1 mg percent) in an assay of a parathyroid extract

Dog pair	Occasion 1	Occasion 2	Body weight, z						Final serum calcium, y					
			Block 1		Block 2		Block 3		Block 1		Block 2		Block 3	
			1	2	1	2	1	2	1	2	1	2	1	2
I	T_1	T_2	130	130	130	128	138	132	120	144	140	159	142	146
II	T_1	S_1	120	112	118	118	120	124	134	121	138	138	126	134
III	T_1	S_2	116	118	118	120	116	116	122	122	128	140	134	148
IV	T_2	S_1	118	120	114	114	108	106	134	121	148	140	141	130
V	T_2	S_2	102	116	112	110	104	118	124	142	134	136	130	140
VI	S_1	S_2	120	126	124	124	126	130	120	124	118	122	137	152
VII	T_2	T_1	88	90	96	96	90	90	132	128	136	120	142	138
VIII	S_1	T_1	114	120	86	88	108	112	110	125	129	129	130	133
IX	S_2	T_1	114	120	122	124	102	104	142	122	126	138	154	130
X	S_1	T_2	64	70	82	80	92	100	116	124	116	140	140	188
XI	S_2	T_2	110	118	124	124	162	156	142	148	130	152	144	141
XII	S_2	S_1	160	164	156	148	70	74	134	116	152	140	146	132

§ 12.5   USE OF CONCOMITANT INFORMATION   249

final serum calcium values. The close connexion with the twin cross-over design (§ 10.2) should be noted. The twin cross-over, using only four of the 12 dose sequences, would have given greater precision on the difference between mean responses to the two preparations and on the dose–response regression coefficient, and therefore greater precision on M, at the expense of a less powerful test of parallelism.

Bliss and Rose found the metameter 'increase in serum calcium' to estimate potency less precisely than the final values alone. Presumably differences between calcium levels for the same animal on different occasions were little affected by the initial status, and the use of increments instead of final values did more harm than good because it introduced irrelevant variation. The initial serum calcium values were not published, so that regression on that variate cannot be studied. This section is concerned only with estimation of ρ from the final serum calcium figures, and with the gain in precision resulting from a covariance analysis on body weight.

Bliss (1947a) described the calculations needed for an analysis of variance according to Yates's method for balanced incomplete blocks (§ 9.6), and these could be extended to the covariance; the method in § 9.5 is again more convenient. Tables 12.5.2 and 12.5.3 show sums and differences of pairs of measurements on each dog. In Table 12.5.2, the contrast between pairs of types I and VII and those of types VI and XII estimates the difference between preparations as an inter-dog contrast; the other types of pairs contribute no information on this difference. Analogously, in Table 12.5.3, the contrast between totals of types II, III, IV and V and of types VIII, IX, X and XI estimates the preparations difference as an intra-dog contrast. Thus Table 12.5.4 can be constructed to give sets of orthogonal contrasts for the inter-dog and intra-dog sections of the analysis. The coefficients are obtained exactly as in § 9.5 by addition or subtraction of the coefficients in Table 5.2.3 corresponding to each dose of a pair; in Table 12.5.4, a factor of 2 has been removed throughout. Use of the six mutually orthogonal contrasts in Table 12.5.4 is equivalent to forming regressions of the sums and differences in Tables 12.5.2 and 12.5.3 on three variates whose values are the coefficients in Table 5.2.3.

Table 12.5.5 is now easily completed. Total sums of squares and products are divided into portions between and within dogs. For example, from Table 12.5.2 the total sum of squares between dogs for z (35 d.f.) is

$$[(260)^2 + (232)^2 + (234)^2 + \ldots + (144)^2]/2 - (8214)^2/72 = 30\,573 \cdot 5.$$

From Table 12.5.3, the total sum of squares within dogs (36 d.f.) is

$$[0^2 + (8)^2 + (-2)^2 + \ldots + (-4)^2]/2 = 494 \cdot 0.$$

Sums of products are obtained by analogous computations on products of corresponding entries. The squares for preparations are

$$\frac{(-184)^2}{24} = 1410 \cdot 7$$

and

TABLE 12.5.2 Sums of pairs of measurements on each dog for Table 12.5.1

Dog pair	Doses	Block 1 z	Block 1 y	Block 2 z	Block 2 y	Block 3 z	Block 3 y	Totals z	Totals y
I	$T_1 + T_2$	260	264	258	299	270	288	788	851
II	$T_1 + S_1$	232	255	236	276	244	260	712	791
III	$T_1 + S_2$	234	244	238	268	232	282	704	794
IV	$T_2 + S_1$	238	255	228	288	214	271	680	814
V	$T_2 + S_2$	218	266	222	270	222	270	662	806
VI	$S_1 + S_2$	246	244	248	240	256	289	750	773
VII	$T_2 + T_1$	178	260	192	256	180	280	550	796
VIII	$S_1 + T_1$	234	235	174	258	220	263	628	756
IX	$S_2 + T_1$	234	264	246	264	206	284	686	812
X	$S_1 + T_2$	134	240	162	256	192	328	488	824
XI	$S_2 + T_2$	228	290	248	282	318	285	794	857
XII	$S_2 + S_1$	324	250	304	292	144	278	772	820
Totals		2760	3067	2756	3249	2698	3378	8214	9694

TABLE 12.5.3 Differences of pairs of measurements on each dog for Table 12.5.1

Dog pair	Doses	Block 1 z	Block 1 y	Block 2 z	Block 2 y	Block 3 z	Block 3 y	Totals z	Totals y
I	$T_1 - T_2$	0	−24	2	−19	6	−4	8	−47
II	$T_1 - S_1$	8	13	0	0	−4	−8	4	5
III	$T_1 - S_2$	−2	0	−2	−12	0	−14	−4	−26
IV	$T_2 - S_1$	−2	13	0	8	2	11	0	32
V	$T_2 - S_2$	−14	−18	2	−2	−14	−10	−26	−30
VI	$S_1 - S_2$	−6	−4	0	−4	−4	−15	−10	−23
VII	$T_2 - T_1$	−2	4	0	16	0	4	−2	24
VIII	$S_1 - T_1$	−6	−15	−2	0	−4	−3	−12	−18
IX	$S_2 - T_1$	−6	20	−2	−12	−2	24	−10	32
X	$S_1 - T_2$	−6	−8	2	−24	−8	−48	−12	−80
XI	$S_2 - T_2$	−8	−6	0	−22	6	3	−2	−25
XII	$S_2 - S_1$	−4	18	8	12	−4	14	0	44
Totals		−48	−7	8	−59	−26	−46	−66	−112

$$\frac{(10)^2}{48} = 2\cdot 1$$

for z, with similar expressions for y; the corresponding products are

$$\frac{(-184) \times 54}{24} = 414\cdot 0$$

and

$$\frac{10 \times 72}{48} = 15\cdot 0.$$

In the inter-dog analysis, a component for differences between blocks (2 d.f.) must be isolated. In the intra-dog analysis, allowance for a possible change in responsiveness between the first and second injections of each subject must be made. Since the dates differ for the three blocks, it seemed advisable to calculate

§ 12.5 USE OF CONCOMITANT INFORMATION 251

Table 12.5.4 Orthogonal coefficients for contrasts between and within dogs for Table 12.5.1

Dog pair	First dose + second dose			First dose − second dose		
	L_p	L_1	L_1'	L_p	L_1	L_1'
I	1	—	—	—	−1	−1
II	—	−1	—	1	—	−1
III	—	—	−1	1	−1	—
IV	—	—	1	1	1	—
V	—	1	—	1	—	1
VI	−1	—	—	—	−1	1
VII	1	—	—	—	1	1
VIII	—	−1	—	−1	—	1
IX	—	—	−1	−1	1	—
X	—	—	1	−1	−1	—
XI	—	1	—	−1	—	−1
XII	−1	—	—	—	1	−1
Divisor	24	24	24	48	48	48
Sum for x_1	−184	116	−222	10	6	−60
Sum for y	54	116	32	72	308	−24

the contrast for each block separately; hence the component for z entered in Table 12.5.5 as 'Dates' is

$$[(-48)^2 + (8)^2 + (-26)^2]/24 = 126 \cdot 8$$

with 3 d.f., instead of

$$\frac{(-66)^2}{72} = 60 \cdot 5$$

with 1 d.f. (the totals are taken from Table 12.5.3).

TABLE 12.5.5 Analysis of covariance for Table 12.5.1

Adjustments for means		937 080·5	1 105 923·8	1 305 189·4
Nature of variation	d.f.	S_{zz}	S_{zy}	S_{yy}
Blocks	2	100·3	−381·8	2 034·5
Preparations	1	1 410·7	−414·0	121·5
Regression	1	560·7	560·7	560·7
Parallelism	1	2 053·5	−296·0	42·7
Error (1)	30	26 448·3	2 216·3	3 805·2
Between dogs	35	30 573·5	1 685·2	6 564·6
Dates	3	126·8	44·2	235·2
Preparations	1	2·1	15·0	108·0
Regression	1	0·8	38·5	1 976·3
Parallelism	1	75·0	30·0	12·0
Error (2)	30	289·3	339·3	1 840·5
Total	71	31 067·5	2 152·2	10 736·6

As in § 9.5, the two portions of the analysis must be separately considered. Ignoring for the present any covariance adjustment, the inter-dog analysis has

$$s_1^2 = \frac{3805 \cdot 2}{30}$$

$$= 126 \cdot 8$$

as the variance per response. No sign of invalidity appears. Moreover, since L_p is a difference between two totals of 12 responses,

$$\bar{y}_T - \bar{y}_S = \frac{54}{12}$$

$$= 4 \cdot 500,$$

and

$$b = \frac{116}{24}$$

$$= 4 \cdot 833.$$

Hence

$$M = \frac{4 \cdot 500}{4 \cdot 833}$$

$$= 0 \cdot 9311.$$

Though the values of L_p and L_1 could be inserted in equation (5.4.1), in order to give R directly, the confounding makes equation (5.4.5) inapplicable and development from first principles is safer. From the way in which they have been formed,

$$\operatorname{Var}(\bar{y}_T - \bar{y}_S) = s_1^2/6,$$

and

$$\operatorname{Var}(b) = s_1^2/24.$$

The contrasts are orthogonal, and the covariance is therefore zero. Also,

$$g = \frac{(2 \cdot 042)^2 \times 126 \cdot 8}{24 \times (4 \cdot 833)^2}$$

$$= 0 \cdot 9432,$$

a value near to unity because, at probability 0·05, b is only just statistically significant. So large a value of g must give exceedingly wide fiducial limits to M; in fact,

$$M_L, M_U = \left[0 \cdot 9311 \pm \frac{2 \cdot 042}{4 \cdot 833} \left\{ s_1^2 \left(\frac{0 \cdot 0568}{6} + \frac{0 \cdot 9311^2}{24} \right) \right\}^{1/2} \right] \Big/ 0 \cdot 0568$$

$$= -1 \cdot 49, 34 \cdot 28.$$

Multiplication by 0·1505 converts to logarithms to base 10, and therefore

§ 12.5   USE OF CONCOMITANT INFORMATION   253

$$R = \text{antilog } 0\cdot140 = 1\cdot38,$$
$$R_L = \text{antilog } \bar{1}\cdot776 = 0\cdot60,$$
$$R_U = \text{antilog } 5\cdot159 = 144\,000.$$

The upper limit is so high as to make the result useless; conclusions from a variance formula for M would have been entirely wrong.

The error regression of y on z for the inter-dog analysis is not significant (Table 12.5.6). The mean square for the regression is so little greater than the residual that covariance cannot improve the precision of potency estimation.

TABLE 12.5.6   Error regression analysis for both sections of Table 12.5.5

Nature of variation	d.f.	Sum of squares	Mean square
Regression on z	1	185·7	185·7
Residual	29	3619·5	124·8
Error (1)	30	3805·2	
Regression on z	1	397·9	397·9
Residual	29	1442·6	49·74
Error (2)	30	1840·5	

For the intra-dog analysis, before any covariance adjustment is tried,

$$s_2^2 = \frac{1840\cdot5}{30}$$
$$= 61\cdot35.$$

Again no invalidity appears. Also,

$$\bar{y}_T - \bar{y}_S = \frac{72}{24}$$
$$= 3\cdot000,$$

and
$$b = \frac{308}{48}$$
$$= 6\cdot417.$$

Hence
$$M = 0\cdot4675.$$

From the divisors in Table 12.5.4,
$$\text{Var}(\bar{y}_T - \bar{y}_S) = s_2^2/12$$
$$\text{Var}(b) = s_2^2/48.$$

Since s_2^2 is about half s_1^2, these two variances are about one quarter of the corresponding inter-dog values. Also

$$g = \frac{(2 \cdot 042)^2 \times 61 \cdot 35}{48 \times (6 \cdot 417)^2}$$

$$= 0 \cdot 1294.$$

Calculations of the usual type give

$$R = \text{antilog } 0 \cdot 0704 = 1 \cdot 176,$$
$$R_L = \text{antilog } \bar{1} \cdot 9612 = 0 \cdot 915,$$
$$R_U = \text{antilog } 0 \cdot 2005 = 1 \cdot 587.$$

The value of R agrees well with the very imprecise inter-dog estimate.

The intra-dog error regression analysis (Table 12.5.6) shows that adjustment for inequalities in z effects a significant reduction in s_2^2. Some improvement in fiducial limits is to be expected from a covariance adjustment: the percentage narrowing of the limits is likely to be about half the percentage reduction in s_2^2, or possibly a little greater because of the effect on g. Before proceeding with the adjustment, however, validity needs to be re-examined; Table 12.5.7, constructed exactly as was Table 12.4.3, shows no cause for concern.

TABLE 12.5.7   Validity tests for parathyroid assay after adjustment of response for body weight, in intra-dog analysis

Nature of variation	d.f.	Sum of squares	Mean square
L_p	1	75·1	75·1
L'_1	1	35·5	35·5
Residual error	29	1442·6	49·74

Now the error regression of y on z gives

$$B = \frac{339 \cdot 3}{289 \cdot 3}$$

$$= 1 \cdot 173,$$

an estimate of the increase in serum calcium per unit increase in body weight. Hence, for the adjusted contrasts,

$$\bar{y}_T^* - \bar{y}_S^* = \frac{72 - 1 \cdot 173 \times 10}{24}$$

$$= 2 \cdot 511,$$

and

$$b^* = \frac{308 - 1 \cdot 173 \times 6}{48}$$

$$= 6 \cdot 270.$$

Therefore

§ 12.5 USE OF CONCOMITANT INFORMATION

$$M^* = \frac{2\cdot 511}{6\cdot 270}$$
$$= 0\cdot 4005.$$

Again,

$$\text{Var}(\bar{y}_T^* - \bar{y}_S^*) = s_2^2 \left[\frac{1}{12} + \left(\frac{10}{24}\right)^2 \cdot \frac{1}{289\cdot 3} \right]$$
$$= 0\cdot 083\,93\,s_2^2,$$

$$\text{Var}(b^*) = s_2^2 \left[\frac{1}{48} + \left(\frac{6}{48}\right)^2 \cdot \frac{1}{289\cdot 3} \right]$$
$$= 0\cdot 020\,89\,s_2^2,$$

and

$$\text{Cov}\{(\bar{y}_T^* - \bar{y}_S^*), b^*\} = s_2^2 \left(\frac{10}{24}\right) \cdot \left(\frac{6}{48}\right) \cdot \frac{1}{289\cdot 3}$$
$$= 0\cdot 000\,18\,s_2^2,$$

where s_2^2 is now the residual variance per response,

$$s_2^2 = 49\cdot 74.$$

Since s_2^2 has only 29 d.f., $t = 2\cdot 045$, and

$$g = 0\cdot 1105.$$

Calculations like those of § 12.4 then give

$$R = \text{antilog } 0\cdot 0603 = 1\cdot 149,$$
$$R_L = \text{antilog } \bar{1}\cdot 9590 = 0\cdot 910,$$
$$R_U = \text{antilog } 0\cdot 1762 = 1\cdot 500.$$

Results so far obtained may be summarized as:

	Inter-dog	Intra-dog
$\bar{y}_T - \bar{y}_S$	4·500	2·511
$\text{Var}(\bar{y}_T - \bar{y}_S)$	21·133	4·175
b	4·833	6·270
$\text{Var}(b)$	5·283	1·039
$\text{Cov}\{(\bar{y}_T - \bar{y}_S), b\}$	0	0·009

Despite the inadequacy of the inter-dog information used above, it may still contribute something to a composite potency estimate. But for the covariance adjustment in the intra-dog analysis, the second analogue of Fieller's theorem would apply as in § 9.5. Unfortunately, the extra terms in $\text{Var}(\bar{y}_T^* - \bar{y}_S^*)$ and $\text{Var}(b^*)$, and the appearance of a non-zero covariance between this numerator and denominator, destroy the exact conditions. However, for these data, the two variances are almost exactly in the 4:1 ratio that was obtained for the inter-dog analysis, and the covariances of numerator and denominator are so small as not to matter in this context. To a first-order approximation that should be adequate with as many degrees of freedom as there are here, the Fieller formulae can be applied in an obvious way.

The weighted means are

$$\bar{y}_T - \bar{y}_S = \frac{4\cdot175 \times 4\cdot500 + 21\cdot133 \times 2\cdot511}{4\cdot175 + 21\cdot133}$$

$$= 2\cdot839,$$

and similarly

$$b = 6\cdot033.$$

Consequently

$$M = \frac{2\cdot839}{6\cdot033}$$

$$= 0\cdot4706.$$

Next can be formed

$$\text{Var}(\bar{y}_T - \bar{y}_S) = \left(\frac{4\cdot175}{25\cdot308}\right)^2 \times 21\cdot133 + \left(\frac{21\cdot133}{25\cdot308}\right)^2 \times 4\cdot175 = 3\cdot486,$$

$$\text{Var}(b) = \left(\frac{4\cdot175}{25\cdot308}\right)^2 \times 5\cdot283 + \left(\frac{21\cdot133}{25\cdot308}\right)^2 \times 1\cdot039 = 0\cdot868,$$

$$\text{Cov}\{(\bar{y}_T - \bar{y}_S), b\} = \left(\frac{21\cdot133}{25\cdot308}\right)^2 \times 0\cdot009 = 0\cdot006.$$

Equation (4.13.8) may be written approximately here as

$$\tan\theta = \frac{s_2^2}{2s_1^2}$$

$$= 0\cdot196,$$

and therefore

$$\theta = 24°.$$

The 0·05 level of d for this angle and 29, 30 d.f. is about 2·04. Consequently, equation (4.13.23) gives

$$M_L, M_U = \left[0\cdot4706 - \frac{0\cdot0992 \times 0\cdot006}{0\cdot868} \pm \frac{2\cdot04}{6\cdot033}\{3\cdot486 - 2\times0\cdot4706\times0\cdot006 + (0\cdot4706)^2 \times 0\cdot868 - 0\cdot0992 \times 3\cdot486\}^{1/2}\right]\bigg/0\cdot9008$$

§ 12.6   USE OF CONCOMITANT INFORMATION   257

$$= [0 \cdot 4699 \pm 0 \cdot 6167]/0 \cdot 9008,$$
$$= -0 \cdot 163, 1 \cdot 206,$$

whence

$$R = \text{antilog } 0 \cdot 0708 = 1 \cdot 18,$$
$$R_L = \text{antilog } \bar{1} \cdot 9755 = 0 \cdot 95,$$
$$R_U = \text{antilog } 0 \cdot 1815 = 1 \cdot 52.$$

Table 12.5.8 summarizes the four estimations of potency for the parathyroid extract. The inter-dog estimate alone is clearly useless. Adjustment of the intra-dog estimate for variations in body weight narrows the fiducial range by about 13 percent, and combination of this estimate with the inter-dog information effects a further 3 percent reduction.

TABLE 12.5.8 Comparison of estimates and fiducial limits in the assay of a parathyroid extract

	Inter-dog	Intra-dog		Combined
		No adjustment	Covariance	
Estimated potency of test preparation relative to standard	1·38	1·18	1·15	1·18
g	0·943	0·129	0·110	0·099
Limits	0·60–144 000	0·91–1·59	0·91–1·50	0·95–1·52
Limits as % of estimate	43–104 000	77–135	79–130	81–129

12.6   Validity tests from the concomitant

In any covariance analysis, the analysis of variance on the concomitant variate gives valuable information. If that variate was measured before the experiment began and if randomization over doses was properly conducted, no real effects can be associated with dose contrasts for the concomitant; any evidence to the contrary casts doubt upon the genuineness of the randomization, and therefore upon the validity of any potency estimation. If the concomitant was measured after doses or other experimental treatments had been administered, statistical significance of a contrast may point to a real effect. Adjustment of responses by covariance may then remove some of the effects that are directly relevant to the assay and lead to poorer estimation of potency; in such circumstances, the methods of Chapter 13 may be more appropriate. The requirement that the z analysis shall not show dose effects is not an assumption on which the mathematics of covariance depends but is a condition important to the logic of interpretation.

The error mean square for z from Table 12.4.2 is about 365, and the mean square for parallelism is just significantly greater. Since no other mean square is even as large as 365, and the general mean square for treatments (479 with

8 d.f.) is not large, no great alarm is created. Had the tests given stronger warnings of treatment effects on z, the obligation would rest with the investigators, not the statistician, to decide what should be done. Obviously the belief that weight of ovaries is unaffected by differences in hormone treatment would have needed re-examination. Guillemin and Sakiz imply that rats were allocated at random to treatment and dose. If they knew the rules to have been followed strictly, the fear of trouble from non-random or non-independent allocation could be dismissed; much rests upon certainty by the investigators, for it is not impossible that an instruction for randomization is interpreted by a technical assistant as permitting haphazard or even systematic selection from cages. In the parathyroid assay (§ 12.5), the intra-dog analysis also has a significant parallelism mean square, and the situation could be similarly discussed.

Thus the analysis of one or more concomitant variates can throw light on the validity of a bioassay; the various tests relate to the fundamental logical validity rather than to questions of the statistical model.

12.7 The economics of covariance adjustment

Inspection of Table 12.5.8 may suggest that adjustment of potency estimates by means of covariance analysis gives a small return for a great amount of extra computation. That the labour is considerable cannot be denied, but in many circumstances the gain from it is greater than could be obtained in any other way for the same trouble and labour. In the parathyroid assay, the covariance adjustment narrowed the fiducial range for the potency estimate by about 13 percent. The alternative of achieving the same precision from a larger assay without covariance would require about 30 percent more observations, so that an experiment using serum calcium determinations alone would need an extra block of 12 dogs. For an easily measured concomitant such as body weight, the advantage obviously lies with the saving of subjects; for a concomitant whose measurement was itself laborious, the assayist might prefer to be more prodigal in his use of subjects and to omit the covariance (cf. § 2.9). No general advice should be based upon a single experience, for the improvement effected by covariance will vary from one type of assay to another. Even within one class of assays, it will vary with the stock of animals and the details of assay procedure. Nevertheless, now that computation has become so inexpensive, if a concomitant can be fairly easily obtained a covariance analysis is almost always worth trial.

One further aspect of covariance deserves mention. In the assay discussed in § 12.5, dose was defined as quantity per kg body weight of dog. This common practice is itself intended to make allowance for dependence of responsiveness upon size; it may be compared with the expression of dose in a direct assay as quantity per unit body weight (§ 2.7), and with the expression of response as proportion of body weight (§ 4.15). If adjustment for inequalities in the weights of subjects is to be made by covariance analysis, the giving of doses proportional to weights is unnecessary, although a reasonable uniformity of weight is desirable in order to avoid risk of appreciable non-linearity of the weight–response regression. In any evaluation of the economics of covariance analysis for such an assay, therefore, the saving of labour associated with separate calculation of each dose must be credited to the covariance procedure (Jerne and Wood, 1949).

13
Composite responses

13.1 Discriminants

The other class of problem (§ 12.1) involving measurements of several variates on each subject must now be discussed. After every application of the stimulus, two or more different response variates are measured; relative potency might be estimated from any one variate, and a combined estimate might be formed by averaging the several values of M or R. In general, the separate estimates would differ in precision, so that some kind of weighted average would be wanted. The precision of the combined estimate could not be simply assessed, however, since different response variates from one subject are likely to be correlated.

A better procedure is to consider a composite response metameter y^*, defined as a function of the several response measures $y_1, y_2, y_3, \ldots$

$$y^* = y^*(y_1, y_2, y_3, \ldots). \tag{13.1.1}$$

The ideal function will be one leading to an estimate of ρ substantially more precise than if any single response variate were used. Some limitation of the class of metameter functions to be considered is inevitable, and here only linear functions

$$y^* = h_1 y_1 + h_2 y_2 + h_3 y_3 + \ldots \tag{13.1.2}$$

will be used (Bliss and Bartels, 1946). For assay purposes, all multiples of y^* are equivalent and an arbitrary standardization of $h_1 = 1$ can be introduced.

For alternative response metameters in a parallel line assay, if the doses have been well chosen and b is large relative to its standard error, by the approximation in equation (6.11.1)

$$\mathrm{Var}(M) \propto \frac{s^2}{b^2}. \tag{13.1.3}$$

Hence a reasonable method for choosing the h_i in equation (13.1.2) would be to determine them by the condition that, in the analysis of y^*, s^2/b^2 shall be a minimum.

The non-applicability of a variance formula for M, important though this can be, need not inhibit the use of such a working rule for guidance. If g is not negligible, the process will not minimize the fiducial range for M exactly, but, since s^2/b^2 is almost always the major influence on precision, it should still give results fairly close to the best possible. Complications arise with assay designs which have two separate error variances (§§ 9.3–9.7), but these will not be considered here.

The minimizing of s^2/b^2 is the same as the maximizing of b^2/s^2, a process which amounts to the selection of y^* so as to maximize the ratio of b to its standard error. This is in fact the determination of a *discriminant function* (Fisher, 1970) for the contrast representing the regression of response on dose. Theory shows this to be formally equivalent to calculation of a multiple linear regression of x, the dose metameter, on $y_1, y_2, y_3, \ldots$; the partial regression coefficients are optimal values for $h_1, h_2, h_3 \ldots$. For imagine an analysis of covariance table to be set up for x and the y_i, in structure like the conventional analysis of variance for the assay, except that the single degree of freedom for regression is left in with the error. The analysis of covariance for x and y^* is then obtainable from the relations

$$\left. \begin{array}{l} S_{xy^*} = \sum_i h_i S_{xy_i} \\ S_{y^*y^*} = \sum_i h_i^2 S_{y_i y_i} + 2\sum_i \sum_j h_i h_j S_{y_i y_j}. \end{array} \right\} \quad (13.1.4)$$

The values of the h_i are to be chosen so as to maximize the proportion of the error component of $S_{y^*y^*}$ removed by a linear regression on x. This requirement is equivalent to maximizing the error correlation coefficient between x and y^*, which in turn is equivalent to maximizing the proportion of the error component of S_{xx} removed by linear regression on y^*. In its final version, the condition is exactly that for determining the partial regression coefficients of x on the y_i.

This use of multiple linear regression is purely formal. The values of x cannot be regarded as normally distributed about the regression function; indeed, since they are selected by the experimenter, they are not 'distributed' in any true sense. Nevertheless, for a test of significance of whether or not certain of the y_i improve discrimination, the usual regression method is valid. As for a simple regression or correlation coefficient, the significance test requires only that either x or the y_i shall have normally distributed errors. For assessment of the precision of any coefficient h_i, once it has been shown to deviate significantly from zero, the regression analogy is not applicable.

Discriminant functions seem likely to be less useful in slope ratio assays, because the situations in which the slope ratio model has been found appropriate lend themselves less readily to the making of multiple response measurements. For alternative response metameters in a slope ratio assay, equation (7.6.7) shows that, approximately,

$$\text{Var}(R) \propto \frac{s^2}{b_S^2}, \quad (13.1.5)$$

provided that the value of R is not much altered by the choice of metameter. Hence the coefficients in equation (13.1.2) might be determined so as to minimize s^2/b_S^2. The analysis would be more complicated than for a parallel line assay, because of the non-orthogonality of b_S and b_T, though the same principles apply; it is not discussed further here.

13.2 An assay analysis using a discriminant function

The discriminant function technique has been employed in many fields of statistical inquiry, though it has been little used in biological assay. Ginsburg (1951) presented results of a (2, 2) assay based on the antidiuretic activity of vasopressin; percentage water excretion of rats in three consecutive periods constituted three response variates. Unfortunately the assay included only 12 subjects, and the variances were too great for any determination of a discriminant, or indeed for any satisfactory potency estimation ($g = 0.60, 0.37, 0.49$ for the three variates). The published analysis used the data as though they were single responses from 36 subjects.

An artificial example has therefore been constructed. This is based upon the prolactin assay used by Finney (1947b) in a discussion of covariance, but it bears no relation to a genuine discriminant situation. Table 13.2.1 purports to show data for two response variates from a (3, 3) assay; y_1 in fact is the genuine response for the assay (crop-gland weight of a pigeon) and y_2 has been manufactured from an irrelevant variate. The y_1 data show signs of a positive association between response and variance, but this is ignored here.

TABLE 13.2.1   Artificial data for a (3, 3) assay with two response variates
(a) The variate y_1 (g)

Dose of standard preparation (IU)			Dose of test preparation (mg)		
1·25	2·50	5·00	0·125	0·250	0·500
3·8	5·3	8·5	2·8	4·8	6·0
3·9	10·2	14·4	6·5	4·7	13·0
4·8	8·1	5·4	3·5	5·4	8·3
6·2	7·5	8·5	3·6	7·4	6·0
18·7	31·1	36·8	16·4	22·3	33·3

(b) The variate y_2 (%)

Dose of standard preparation (IU)			Dose of test preparation (mg)		
1·25	2·50	5·00	0·125	0·250	0·500
51	49	47	55	50	45
55	53	51	55	53	52
46	46	39	54	50	50
51	51	41	56	52	53
203	199	178	220	205	200

Taking $-1, 0, 1$ as the values of x, the dose metameter, the various contrasts are:

	x	y_1	y_2
L_p	0	$-14·6$	45
L_1	16	35·0	-45
L_1'	0	$-1·2$	5
L_2	0	$-1·6$	-7
L_2'	0	11·8	27

Table 13.2.2 is the analysis of covariance prepared for examining the regression

§ 13.2 COMPOSITE RESPONSES

TABLE 13.2.2 Analysis of covariance for Table 13.2.1

Adjustments for means		1048·08	7963·04	60 501·04	0	0	0
Nature of variation	d.f.	$S_{y_1 y_1}$	$S_{y_1 y_2}$	$S_{y_2 y_2}$	$S_{y_1 x}$	$S_{y_2 x}$	S_{xx}
Preparations	1	8·88	−27·38	84·38	0	0	0
Regression	1	76·56	−98·44	126·56	35·0	−45·0	16
Parallelism	1	0·09	−0·38	1·56	0	0	0
Quadratic	1	0·05	0·23	1·02	0	0	0
Difference of quadratics	1	2·91	6·64	15·19	0	0	0
Error	18	103·81	74·99	205·25	0	0	0
Total	23	192·30	−44·34	433·96	35·0	−45·0	16
Error + regression	19	180·37	−23·45	331·81	35·0	−45·0	16

of x on y_1, y_2. The method of construction of L_1 makes clear that its values for x, y_1, y_2 are $S_{xx}, S_{y_1 x}$, and $S_{y_2 x}$ respectively.

The coefficients of the discriminant function are proportional to the linear regression coefficients of x on y_1, y_2, after elimination of the irrelevant variation associated with L_p, L'_1, L_2 and L'_2. Hence they may be found from equations based upon (Error + Regression):

$$180·37 h_1 - 23·45 h_2 = 35·0,$$
$$-23·45 h_1 + 331·81 h_2 = -45·0,$$

and are

$$h_1 = 0·178\ 050,$$
$$h_2 = -0·123\ 036.$$

In continuation of the regression analogy, the sum of squares accounted for by the regression of x on y_1, y_2 is calculated as

$$35·0 h_1 - 45·0 h_2 = 11·77,$$

as compared with

$$35·0^2 / 180·37 = 6·79$$

for y_1 alone. The significance of the improvement in discrimination between doses attributable to the use of y_2 as well as y_1 is then tested as shown in Table 13.2.3; the usual normality assumptions for y_1, y_2 ensure validity of the significance tests. Clearly y_2 improves discrimination, and a similar test demonstrates that inclusion of y_1 is an improvement on the use of y_2 alone.

Only the ratios of the h_i are really of importance, and for arithmetical purposes the coefficient of y_1 is conveniently reduced to unity, so as to give

$$y^* = y_1 - 0·6910 y_2. \tag{13.2.1}$$

Equation (13.2.1) could be approximated by

TABLE 13.2.3 Error regression analysis for a discriminant calculated from Table 13.2.2

Nature of variation	d.f.	Sum of squares	Mean square
Regression on y_1 alone	1	6·79	
Additional for y_2	1	4·98	4·98
Regression on y_1, y_2	2	11·77	
Regression on y_2 alone	1	6·10	
Additional for y_1	1	5·67	5·67
Regression on y_1, y_2	2	11·77	
Residual	17	4·23	0·249
Total	19	16·00	

$$y^* = y_1 - 0·7y_2$$

with negligible loss in discrimination, but for this example the full version is retained. By the second of equations (13.1.4), each component of the analysis of variance of y^*, Table 13.2.2, may be calculated from

$$S_{y^*y^*} = S_{y_1 y_1} - 1·382 S_{y_1 y_2} + 0·477\,481\,S_{y_2 y_2}. \tag{13.2.2}$$

The dose components may alternatively be derived from

$$L_p(y^*) = L_p(y_1) - 0·6910\,L_p(y_2)$$
$$= -45·70$$
$$L_1(y^*) = 66·10,$$
$$L_1'(y^*) = -4·66,$$
$$L_2(y^*) = 3·24,$$
$$L_2'(y^*) = -6·86.$$

The regression component is assigned two degrees of freedom instead of one, since two variates have been used in the determination of the discriminator. The justification may not be immediately obvious, but it leads to exactly the same variance ratio for a test of significance as comes from the 2 d.f. for regression in Table 13.2.3. The extra degree of freedom is removed from the error line of Table 13.2.4.

TABLE 13.2.4 Analysis of variance for the discriminant function, y^*

Nature of variation	d.f.	Sum of squares	Mean square
Preparations	1	87·01	87·01
Regression	2	273·03	136·52
Parallelism	1	1·36	1·36
Quadratic	1	0·22	0·22
Difference of quadratics	1	0·99	0·99
Error	17	98·17	5·775
Total	23	460·78	

The test of significance for the regression component suggested by Table 13.2.4 is exact, but other tests of significance from this table and the assessment of fiducial limits that follows are only approximate; the fault lies in assuming the h_i to be known exactly, whereas they are only estimates. Table 13.2.4 gives no indication of invalidity, except for the difference between preparations. From the dose contrasts in the metameter y^*,

$$M = -\frac{45 \cdot 70}{12} \bigg/ \frac{66 \cdot 10}{16}$$

$$= -3 \cdot 808/4 \cdot 131$$

$$= -0 \cdot 9218.$$

If the usual formulae are accepted as adequate approximations,

$$g = \frac{(2 \cdot 110)^2 \times 5 \cdot 775}{(4 \cdot 131)^2 \times 16}$$

$$= 0 \cdot 0942,$$

and

$$M_L, M_U$$

$$= -\left[0 \cdot 9218 \pm \frac{2 \cdot 110}{4 \cdot 131} \times \left\{ 5 \cdot 775 \times \left(\frac{0 \cdot 9059}{6} + \frac{(-0 \cdot 9218)^2}{16} \right) \right\}^{1/2} \right] \bigg/ 0 \cdot 9058$$

$$= -[0 \cdot 9218 \pm 0 \cdot 5545]/0 \cdot 9058$$

$$= -1 \cdot 6298, -0 \cdot 4055.$$

Hence

$$R = 10 \text{ antilog } \bar{1} \cdot 7225 = 5 \cdot 28,$$

$$R_L = 10 \text{ antilog } \bar{1} \cdot 5094 = 3 \cdot 28,$$

$$R_U = 10 \text{ antilog } \bar{1} \cdot 8779 = 7 \cdot 55.$$

As is to be expected, the limits are considerably narrower than those calculated from y_1 alone, namely 2·36 and 13·3: the interval is narrowed from 35–196 percent to 62–143 percent. Nevertheless, the only fully justifiable statement from the present analysis is that the potency of T is estimated to be 5·28 IU per mg. The true limits of error presumably should be a little wider than the current approximations of 3·28 IU per mg and 7·55 IU per mg.

13.3 Exact multivariate analysis

Rao (1954) analyzed similar data using exact theory of multivariate statistics. His very clear paper depends largely on a general result for multivariate statistical theory (Rao, 1965). Suppose that an analysis of covariance for p variates $y_1, y_2, \ldots, y_p$ has error sums of squares and products represented by w_{ij} for $1 \leq i \leq p$, $1 \leq j \leq p$ with f degrees of freedom, and that u_{ij} are corresponding quantities from a line of the analysis with 1 d.f. Then

$$F = \frac{f-p+1}{p}\left(\frac{\|w_{ij}+u_{ij}\|}{\|w_{ij}\|} - 1\right) \qquad (13.3.1)$$

follows the standard F or variance ratio distribution with p, $(f-p+1)$ d.f.; here $\|w_{ij}\|$ symbolizes the symmetric determinant formed by the sums of squares and products w_{ij}. The theory of course depends upon underlying normality of error distributions. Bennett (1964) has proposed a similar method.

This may be applied to the various lines of Table 13.2.2 in turn, with $p=2$. As a composite test of the preparations difference, after addition of the Preparations and Error lines to form $(w_{ij}+u_{ij})$,

$$F = \frac{18-2+1}{2}\left(\left\|\begin{matrix}112{\cdot}69 & 47{\cdot}61 \\ 47{\cdot}61 & 289{\cdot}63\end{matrix}\right\| \middle/ \left\|\begin{matrix}103{\cdot}83 & 74{\cdot}99 \\ 74{\cdot}99 & 205{\cdot}25\end{matrix}\right\| - 1\right)$$

$$= 8{\cdot}5 \times \left(\frac{30\,371{\cdot}69}{15\,683{\cdot}50} - 1\right)$$

$$= 16{\cdot}46, \text{ with } 2, 17 \text{ d.f.};$$

evidently the data show a marked difference between mean responses to the two preparations, but for this illustrative example no further account will be taken of it. Similar calculations for regression, the values of $(w_{ij}+u_{ij})$ being taken from the final line of Table 13.2.2, give

$$F = 23{\cdot}64 \text{ with } 2, 17 \text{ d.f.},$$

leaving no doubt of the statistical significance of the dependence of response on dose. The same formula applied to parallelism, quadratic curvature, and difference of quadratics gives values of $0{\cdot}13$, $0{\cdot}04$, and $0{\cdot}64$ for F, none of which contains any suggestion of significance.

Rao also used an alternative to (13.3.1); if the u_{ij} are based on 2 d.f., then

$$F = \frac{f-p+1}{p}\left(\frac{\|w_{ij}+u_{ij}\|^{1/2}}{\|w_{ij}\|^{1/2}} - 1\right) \qquad (13.3.2)$$

follows the F distribution with $2p$, $2(f-p+1)$ d.f. This does not generalize to greater numbers of degrees of freedom in any simple manner, though approximations are available (Rao, 1965). As a single test of deviations from linearity:

$$F = \frac{18-2+1}{2}\left(\left\|\begin{matrix}106{\cdot}77 & 81{\cdot}86 \\ 81{\cdot}86 & 221{\cdot}46\end{matrix}\right\|^{1/2} \middle/ (15\,683{\cdot}50)^{1/2} - 1\right)$$

$$= 8{\cdot}5 \times (1{\cdot}080\,39^{1/2} - 1)$$

$$= 0{\cdot}34, \text{ with } 4, 34 \text{ d.f.}$$

By a procedure analogous to Fieller's theorem, limits to a composite estimate of the logarithm of relative potency can be obtained. For suppose that μ is the true value. Then

$$\bar{y}_{1T} - \bar{y}_{1S} - \mu b_1, \quad \bar{y}_{2T} - \bar{y}_{2S} - \mu b_2$$

(the numerical suffix identifying the variate) are normally distributed with means zero; their variances and covariance are proportional to the residual variances and covariance per response, the factor of proportionality being

$$\frac{1}{12} + \frac{1}{12} + \frac{\mu^2}{16},$$

the relation to the structure and replication of the experiment being obvious. Consequently, a possible set of values for u_{ij} is given by

$$u_{ij} = (\bar{y}_{iT} - \bar{y}_{iS} - \mu b_i)(\bar{y}_{jT} - \bar{y}_{jS} - \mu b_j) \Big/ \left(\frac{1}{6} + \frac{\mu^2}{16}\right). \tag{13.3.3}$$

These quantities can be used in a procedure essentially that described in a different context in § 14.3, which section should be read for a fuller explanation.

If the u_{ij} defined by (13.3.3) are substituted into the right-hand side of equation (13.3.1), the behaviour of F can be studied as a function of μ. In the example, the expression can be reduced to

$$F = 8 \cdot 5 \times \left[205 \cdot 25 \left(\frac{14 \cdot 6}{12} + \frac{35 \cdot 0\mu}{16}\right)^2 + 2 \times 74 \cdot 99 \left(\frac{14 \cdot 6}{12} + \frac{35 \cdot 0\mu}{16}\right)\left(\frac{45 \cdot 0}{12} + \frac{45 \cdot 0\mu}{16}\right) \right.$$
$$\left. + 103 \cdot 81 \left(\frac{45 \cdot 0}{12} + \frac{45 \cdot 0\mu}{16}\right)^2 \right] \Big/ \left[(205 \cdot 25 \times 103 \cdot 81 - 74 \cdot 99)^2 \left(\frac{1}{6} + \frac{\mu^2}{16}\right) \right],$$
$$\tag{13.3.4}$$

but of course if p were greater the function would be more complicated. This F can show the same kind of complexities of behaviour that are discussed for $J(\mu)$ in § 14.3, with multiple maxima and minima, but the particular numerical values for equation (13.3.4) are quite reasonable. The minimum of the function will usually occur near the value of μ already found by the discriminant method in § 13.2. If f were large, pF at the minimum would be approximately a $\chi^2_{[p-1]}$ statistic for testing whether the potency estimates from the p variates are in agreement; here presumably $pF/(p-1)$ is better regarded as a variance ratio with $(p-1)$, $(f-p+1)$ d.f. For equation (13.3.4), the minimum has been explored by tabulation and interpolation. The minimum occurs at

$$\hat{\mu} = -0 \cdot 9345,$$

and the test statistic is 0·65 with 1, 17 d.f. which gives no suggestion of a discrepancy between y_1 and y_2.

Again as in § 14.3, $pF - \text{Min}(pF)$ for large f would be a $\chi^2_{[1]}$ statistic that could be used to set limits to μ. Here one may use a variance ratio with 1, $(f-p+1)$ d.f. as a better approximation. The 0·95 probability level for a variance ratio with 1, 17 d.f. is 4·45, and therefore the 0·95 limits to μ will be set at those values for which $2F - 0 \cdot 65$ attains the value 4·45, that is to say at the values of μ giving $F = 2 \cdot 55$. Again by interpolation in a tabulation of F, the function is seen to be less than 2·55 for all μ in $-1 \cdot 653 \leq \mu \leq -0 \cdot 414$, and these are therefore the required limits.

The standard conversion of scales now gives

$$R = 10 \text{ antilog } \bar{1}\cdot719 = 5\cdot24,$$
$$R_L = 10 \text{ antilog } \bar{1}\cdot502 = 3\cdot18,$$
$$R_U = 10 \text{ antilog } \bar{1}\cdot875 = 7\cdot50.$$

These are very close indeed to the results obtained in § 13.2. If so small an assay can give such accuracy in the earlier approximation, faith in the approximation may reasonably be encouraged. The position might be much less satisfactory if the variates agreed less well in their separate estimations of potency.

13.4 Discriminants and concomitants

The types of problem discussed in Chapters 12 and 13 might occur simultaneously. If data on concomitant variates, themselves unaffected by the stimulus, are available, as well as two or more response variates, the ideal response metameter will be a discriminant adjusted by covariance. The statistical technique is a little more complicated than the ordinary discriminant function process, but does not involve essentially new methods. Cochran and Bliss (1948) discussed an example which was almost a biological assay, although unfortunately its more interesting features from this point of view were ignored.

13.5 The economics of discriminant analysis

Like the adjustment of assay data by covariance, the use of a discriminant function as a response metameter increases the computational labour. Whether or not this is worth while can be assessed only by comparison of the cost of measuring the additional variates and computing the discriminant with that of obtaining equal precision from a larger number of subjects with measurement of only one variate (cf. § 12.7).

Too few uses of discriminant analysis in assays have been published for general conclusions to be drawn. Liddle *et al.* (1955) described a technique for aldosterone assay in which either a potassium response or a sodium response could be used for estimating potency from parallel regression lines. The values of s/b were 0·37 and 0·48 for these responses. An index formed as a linear combination of the two responses gave 0·24 for s/b; this suggests that the number of administrations of doses required in order to achieve a stated precision could be reduced to one-half or one-quarter of that needed when account is taken of only one response. Munson and Sheps (1958) found log (comb weight/body weight) a better metameter than log (comb weight) in a chick assay for androgens. They considered alternatives, but the simple ratio was almost as good as a discriminant function calculated for each assay separately.

14
Combination of estimates

14.1 Weighted means

If a test preparation is assayed by several analytic dilution assays, whether all of the same design and technique or using entirely different techniques, the estimates must agree except for differences attributable to sampling error. Hence Fisher (1949) took the disagreement between two potency estimates for a tuberculin preparation, based respectively on cows and on guinea-pigs, as evidence of a qualitative difference between S and T. If agreement is satisfactory, a composite estimate of ρ that summarizes all available information is desirable. Even for comparative assays, if good agreement between several estimates is found, a composite figure will naturally be required. In the formation of the combined estimate, most importance will be attached to the individual values that are most precise or reliable, and some type of weighted mean will therefore be wanted.

If the g for each assay is small and the variance per response is either a known σ^2 (as will often be true for assays based on a quantal response: see Chapters 17, 18) or an estimated s^2 with a moderately large number of degrees of freedom, a good procedure is to take a weighted mean of the individual estimates, using weights inversely proportional to the variances. For parallel line assays the calculations should be made on M, and for slope ratio assays on R, but otherwise the two types are treated identically. Either M or R can be assigned limits by the use of a variance formula, and $\mathrm{Var}(M)$ or $\mathrm{Var}(R)$ is not subject to serious sampling errors.

Suppose that the separate values of M, the log potency in a series of parallel line assays of the same preparation, are $M_1, M_2, M_3, \ldots, M_p$. Write the weight for assay i as the reciprocal of the variance:

$$\omega_i = 1/\mathrm{Var}(M_i). \tag{14.1.1}$$

The weighted mean, $\bar{M}$, is then defined by

$$\bar{M} = \frac{\Sigma \omega_i M_i}{\Sigma \omega_i} \tag{14.1.2}$$

with variance

$$\mathrm{Var}(\bar{M}) = 1/\Sigma \omega_i, \tag{14.1.3}$$

where Σ denotes summation over all the assays.

In determining fiducial limits for $\bar{M}$, several cases should be distinguished:
(i) If the ω_i are true weights, based upon known variances per response, σ_i^2, normal deviates can be used in assigning limits to $\bar{M}$ with greater

justification than for the individual M_i. Hence the 0·95 limits may be taken as

$$\begin{aligned} \bar{M}_L &= \bar{M} - 1\cdot960\,\{\mathrm{Var}(\bar{M})\}^{1/2}, \\ \bar{M}_U &= \bar{M} + 1\cdot960\,\{\mathrm{Var}(\bar{M})\}^{1/2}. \end{aligned} \qquad (14.1.4)$$

Moreover, reference of

$$\chi^2_{[p-1]} = \Sigma\omega_i M_i^2 - (\Sigma\omega_i M_i)^2/\Sigma\omega_i \qquad (14.1.5)$$

to Appendix Table IV tests the homogeneity of the p estimates. If the separate M_i came from valid assays but this test discloses differences between them, the hypothesis that all were analytic dilution assays is contradicted, though they may still be valid comparative assays.

(ii) If the p assays are based upon the same type of response, and the values of s_i^2, the estimated variances per response, are satisfactorily homogeneous (Bartlett's test, § 3.11, enables this point to be examined, though it is inclined to give a large χ^2 for reasons other than heterogeneity), a pooled variance should be formed:

$$\bar{s}^2 = \frac{\Sigma f_i s_i^2}{\Sigma f_i}, \qquad (14.1.6)$$

where f_i is the number of degrees of freedom for s_i^2. The values of $\mathrm{Var}(M_i)$ may then be recomputed, using $\bar{s}^2$ in place of s_i^2, so as to give more reliable values. In calculating $\bar{M}$, the factor $1/\bar{s}^2$ may be omitted from each ω_i provided that at the end it is reinserted as a multiplier of $\mathrm{Var}(\bar{M})$. The fiducial limits are

$$\begin{aligned} \bar{M}_L &= \bar{M} - t\,\{\mathrm{Var}(\bar{M})\}^{1/2}, \\ \bar{M}_U &= \bar{M} + t\,\{\mathrm{Var}(\bar{M})\}^{1/2}, \end{aligned} \qquad (14.1.7)$$

where t has Σf_i degrees of freedom. Since f_i is likely to be fairly large, equations (14.1.4) will usually be close approximations to equations (14.1.7) for the 0·95 limits. Similarly, equation (14.1.5) may be used to give an approximate test of homogeneity of the M_i; more exactly, the expression on the right of equation (14.1.5) may be divided by $(p-1)$ and regarded as a variance ratio with $(p-1, \Sigma f_i)$ degrees of freedom (Appendix Table II).

(iii) The instructions given under (i) and (ii) are exact except for the dependence on the smallness of g for each assay and the consequent legitimacy of using $\mathrm{Var}(M)$. When the s_i^2 cannot be pooled, but each is based upon many degrees of freedom and each g is small, calculation as for case (i) is unlikely to be seriously misleading. Probably data for which each f_i is at least 20 can be safely treated in this manner. If the numbers of degrees of freedom are small, a complete solution can be given only when $p = 2$. The problem then calls for the Behrens distribution (§ 4.13), which may be used both for testing the significance of the difference between M_1 and

§ 14.2

M_2 and for assessing the precision of $\bar{M}$. When $p > 2$, equations (14.1.1)–(14.1.3) may still be used, but no tables are available for the determination of fiducial limits. A value of t, with degrees of freedom fewer than Σf_i but more than the smallest number for any s_i^2, should be used as an approximation in equations (14.1.7).

14.2 Illustrative calculations

In recent years, many international drug standards have been studied by collaborative programmes of assay extending over several laboratories and countries. The *Bulletin of the World Health Organization* contains numerous reports of these projects. One such (Lightbown et al., 1965) reported 49 assays for comparing two standard oleandomycin preparations. As has become customary, the paper lists the relative potency and the value of ω_i for every assay; in this series, all were parallel line assays, though various techniques were used.

In its first two columns, Table 14.2.1 shows the results recorded for the four assays from Laboratory 6, and various arithmetical processes will be illustrated on them. First of all, the values of M and its standard error have been reconstructed, the latter solely for comparison with the eventual SE of the mean. Then a column of the products ωM is added. By equation (14.1.2),

$$\bar{M} = -102 \cdot 9602 / 6470$$
$$= -0 \cdot 0159.$$

Also, by equation (14.1.5),

$$\chi^2_{[3]} = 15 \cdot 3867 \times 0 \cdot 019 + 16 \cdot 7140 \times 0 \cdot 0137 + \ldots - (102 \cdot 9602)^2 / 6470$$
$$= 4 \cdot 11.$$

This small χ^2 arouses no fears of disagreement between the assays. Hence

$$\text{Var}(\bar{M}) = 1/6470$$
$$= (0 \cdot 0124)^2.$$

From equations (14.1.4),

$$M_L, M_U = -0 \cdot 0159 - 1 \cdot 960 \times 0 \cdot 0124, \quad -0 \cdot 0159 + 1 \cdot 960 \times 0 \cdot 0124$$
$$= \bar{1} \cdot 9598, 0 \cdot 0084.$$

TABLE 14.2.1 Summary of potencies for four assays of oleandomycin

Assay	R	ω	M	SE(M)	ωM
1	0·973	1293	−0·0119	0·0278	−15·3867
2	1·032	1220	0·0137	0·0286	16·7140
3	0·880	1722	−0·0555	0·0241	−95·5710
4	0·991	2235	−0·0039	0·0211	−8·7165
Total	—	6470	—	—	−102·9602

And therefore finally the estimated potency is 0·964 with 0·95 limits at 0·912, 1·020.

The four assays were all of the same design, eight replicate blocks of a (3, 3) assay, and consequently each provided an estimate s_i^2 with 35 d.f. Had a pooled s^2 been used in forming the variances of the separate M_i, the calculations would have been essentially the same. They could be made a little more exact by using a variance ratio with 3, 140 d.f. instead of χ^2 to test the heterogeneity of the M_i; t with 140 d.f. should replace 1·960.

If M_3, M_4 were the only potency estimates available, and s_3^2, s_4^2 were too unequal to be pooled, recourse would be had to the Behrens distribution. The test of agreement would be

$$d = \frac{-0·0039 + 0·0555}{\left(\dfrac{1}{1722} + \dfrac{1}{2235}\right)^{1/2}}$$

$$= 1·61,$$

with
$$\tan^2\theta = 2235/1722 = 1·298$$

and 35, 35 d.f. (i.e. f_3, f_4 d.f.). From a fuller version of Appendix Table III, the 0·05 significance level for d, at $\theta = 49°$, is 2·025, so that M_3, M_4 are not significantly different.

The weighted mean estimate is

$$\bar{M} = -\frac{1722 \times 0·0555 + 2235 \times 0·0039}{1722 + 2235}$$

$$= -0·0264,$$

with
$$\text{Var}(\bar{M}) = 1/(1722 + 2235) = 0·0159^2.$$

Now for θ the ratio to be used is $\text{Var}(M_4)/\text{Var}(M_3)$, the reciprocal of the previous one, with f_3, f_4 d.f. (§ 4.13). Here, because $f_3 = f_4$, it makes no real difference, but

$$\tan^2\theta = 1722/2235 = 0·770.$$

Hence $\theta = 41°$, and the deviate for 35, 35 d.f. is again 2·025. The usual calculations lead to 0·941 as a potency estimate, with limits at 0·874, 1·143.

No complete solution can be given to the problem of combining three or more estimates when the variances from the separate assays are heterogeneous. When the numbers of degrees of freedom are all equal and fairly large, the approximation of a pooled s^2 still should not be far wrong. For four assays with very different s_i^2 based upon 8, 21, 12, 6 d.f., the situation might be very different. The Behrens theory that underlies § 4.13 could be generalized from 2 to p independent estimates, but tables analogous to Appendix Table III would need to be very extensive. Without them, only lower and upper bounds to the d-deviate for a specified probability can be stated. Clearly it cannot be smaller than the t-deviate for Σf_i degrees of freedom. Also, it will not often be greater

§ 14.3 COMBINATION OF ESTIMATES

than the largest d-deviate among those for all possible pairs of estimates; whether in somewhat unusual circumstances this extreme can be exceeded is not known.

14.3 The combination of parallel line assays

The suggestions in § 14.1 depend upon g being negligible. In parallel line assays, g is often too large to be ignored, and methods for combining estimates not subject to this restriction are therefore needed. The only prospects for a practicable general method seem to involve restriction to assays that share the same variance per response, σ^2. Such a restriction is not unreasonable for assays using the same technique and subjects from similar sources, but it can scarcely apply to assays of the same test preparation that have used different techniques or different species of animals. Bartlett's test can be used to examine the hypothesis that several independent variance estimates correspond to the same σ^2; the test is well known to be sensitive to non-normality of distribution, and the user should not be too ready to discard a belief in homogeneity.

Previous editions of this book suggested several approximate procedures. Armitage, Bennett and Finney (1976) have described a method that should supersede these earlier proposals: it is no more laborious to apply, it is more informative, and it is a natural generalization of Fieller's theorem. The essentials of the new method, except for an important logical development of the distinction between a validity test and estimation, can be found in several earlier papers (Armitage, 1970; Armitage and Bennett, 1974; Armitage, Bailey, Petrie, Annable and Stack-Dunne, 1974; Bennett, 1962, 1963a, b).

Suppose that assay i in a series of p assays to be combined has as its regression lines

$$\left. \begin{array}{l} Y_{Si} = \alpha_{Si} + \beta_i x, \\ Y_{Ti} = \alpha_{Ti} + \beta_i x, \end{array} \right\} \quad (14.3.1)$$

so that the log potency ratio is

$$\mu_i = (\alpha_{Ti} - \alpha_{Si})/\beta_i. \quad (14.3.2)$$

From calculations within each assay, the estimated regressions are formed in the usual way as

$$\left. \begin{array}{l} Y_{Si} = a_{Si} + b_i x, \\ Y_{Ti} = a_{Ti} + b_i x, \end{array} \right\} \quad (14.3.3)$$

and the estimate of μ_i is

$$M_i = d_i/b_i \quad (14.3.4)$$

where

$$d_i = a_{Ti} - a_{Si}. \quad (14.3.5)$$

If all assays have the same σ^2, the standard formulae for variances and covariance can be written

$$\left.\begin{aligned}\text{Var}(d_i) &= \sigma^2 v_{11i},\\ \text{Var}(b_i) &= \sigma^2 v_{22i},\\ \text{Cov}(d_i, b_i) &= \sigma^2 v_{12i},\end{aligned}\right\} \quad (14.3.6)$$

where the multipliers of σ^2 are functions only of the design of the assay, the numbers of replicates, and the doses. For example, in a completely randomized design with N_S, N_T subjects for S, T respectively,

$$\left.\begin{aligned}v_{11} &= \frac{1}{N_S} + \frac{1}{N_T} + \frac{(\bar{x}_T - \bar{x}_S)^2}{\Sigma S_{xx}},\\ v_{22} &= 1/\Sigma S_{xx},\\ v_{12} &= -\frac{(\bar{x}_T - \bar{x}_S)}{\Sigma S_{xx}},\end{aligned}\right\} \quad (14.3.7)$$

(the subscript i will be omitted in various formulae where it is obvious that a particular one of the p assays is being discussed). Very often assays will all have used the same doses, and symmetry will ensure that all have the same $\bar{x}_S, \bar{x}_T$, in which case one can work in terms of

$$M_i + (\bar{x}_T - \bar{x}_S) = (\bar{y}_{Ti} - \bar{y}_{Si})/b_i \quad (14.3.8)$$

which simplifies (14.3.7) by removing the third term of v_{11} and making $v_{12} = 0$. On the other hand, complexities of design such as incomplete block schemes or the kind of accidental loss of subjects discussed in §§ 4.21–4.23 will complicate (14.3.7), and in particular will make v_{12} take a less trivial form. The general specification of (14.3.1)–(14.3.6) includes the combining of any mixture of parallel line designs. The estimated variance s^2, obtained by pooling error sums of squares, will usually have a large value of f, its degrees of freedom. The constancy of σ^2 is essential to the *theory* of what follows; *in practice*, a pooled s^2 may still give reasonable results even if variances are heterogeneous, provided that s_i is not noticeably correlated with M_i.

Under the usual assumptions of normality, the b_i, d_i and s^2 are sufficient statistics for the parameters β_i, μ_i and σ^2. The log likelihood can be written

$$L = K(\sigma^2) - \frac{1}{2\sigma^2} \sum_{i=1}^{p} \{v_{11}(b-\beta)^2 - 2v_{12}(b-\beta)(d-\beta\mu) + v_{22}(d-\beta\mu)^2\}$$

$$\div (v_{11}v_{22} - v_{12}^2), \quad (14.3.9)$$

the i subscript having been omitted from every symbol within the summation; $K(\sigma^2)$ is a function of σ^2 and of individual responses that does not involve the β_i, δ_i. Evidently the unrestricted maximum of L will have $\beta_i = b_i$, $\mu_i = d_i/b_i$ and, as is well known, apart from a factor that slightly biases the estimation, s^2 is the maximum likelihood estimator of σ^2. With the usual minor modification that departs from strict maximization but removes bias and conforms to standard distribution theory for linear models, the unrestricted maximum of L may be written

§ 14.3 COMBINATION OF ESTIMATES

$$L_0 = K(s^2). \tag{14.3.10}$$

Now the conditions of bioassay require that each of the assays shall be estimating the same true potency. This implies the constraints expressed by

$$\mu_i = \mu, \quad i = 1, 2, \ldots, p, \tag{14.3.11}$$

where μ is the log potency applicable throughout. Under this condition, (14.3.9) can be rewritten

$$L = K(\sigma^2) - \frac{1}{2\sigma^2} \sum_{i=1}^{p} \frac{(b\mu - d)^2}{v_{22}\mu^2 - 2v_{12}\mu + v_{11}}$$

$$- \frac{1}{2\sigma^2} \sum_{i=1}^{p} \frac{\{\beta(v_{22}\mu^2 - 2v_{12}\mu + v_{11}) - \mu(v_{22}d - v_{12}b) + (v_{12}d - v_{11}b)\}^2}{(v_{22}\mu^2 - 2v_{12}\mu + v_{11})(v_{11}v_{22} - v_{12}^2)}; \tag{14.3.12}$$

the additional subscript i is to be understood as added to every symbol except σ and μ. For any fixed value of μ, the second summation can be minimized (and therefore L maximized) by making each term zero and estimating

$$\hat{\beta}_i = \frac{\mu(v_{22i}d_i - v_{12i}b_i) - (v_{12i}d_i - v_{11i}b_i)}{v_{22i}\mu^2 - 2v_{12i}\mu + v_{11i}}. \tag{14.3.13}$$

The additional condition for maximizing with respect to μ is most easily obtained from differentiation of (14.3.9), and after some reduction finding

$$\hat{\mu} = \sum \frac{\hat{\beta}\{v_{22}d - v_{12}(b - \hat{\beta})\}}{v_{11}v_{22} - v_{12}^2} \Big/ \sum \frac{v_{22}\hat{\beta}^2}{v_{11}v_{22} - v_{12}^2}. \tag{14.3.14}$$

Equations (14.3.13), (14.3.14) can be iterated in order to find the set of values for $\hat{\mu}$ and the $\hat{\beta}_i$.

The maximum of L under the condition (14.3.11) for a specified μ is obtained by substitution of $\hat{\beta}_i$ from (14.3.13) into (14.3.12). With the appropriate s^2, the maximum is

$$L_p = K(s^2) - \frac{1}{2s^2} \sum_{i=1}^{p} \frac{(b\mu - d)^2}{v_{22}\mu^2 - 2v_{12}\mu + v_{11}}. \tag{14.3.15}$$

Maximization with respect to μ, thus relaxing one of the p constraints implicit in (14.3.11), leads to

$$L_{p-1} = K(s^2) - \frac{1}{2s^2} \sum_{i=1}^{p} \frac{(b\hat{\mu} - d)^2}{v_{22}\hat{\mu}^2 - 2v_{12}\hat{\mu} + v_{11}}, \tag{14.3.16}$$

with $\hat{\mu}$ as determined by the iteration described above. The subscripts on L in (14.3.10), (14.3.15), (14.3.16) represent the numbers of constraints imposed. In each formula, s^2 ought to be estimated from the corresponding maximization, but in practice s^2 can be taken as the residual mean square (with f degrees of freedom) defined earlier.

General statistical theory shows that, for sufficiently large f,

$$2(L_0 - L_{p-1}) \quad \text{and} \quad 2(L_{p-1} - L_p)$$

are asymptotically distributed as $\chi^2_{[p-1]}$ and $\chi^2_{[1]}$ respectively. The first provides a significance test on the evidence against equality of the μ_i, the second provides a significance test on the deviation of $\hat{\mu}$ from any specified value. For smaller f, replacement of χ^2 tests by corresponding variance ratio tests should be a better approximation.

Further discussion is simplified by introducing the function

$$J(\mu) = \sum_{i=1}^{p} \frac{(b_i\mu - d_i)^2}{v_{22i}\mu^2 - 2v_{12i}\mu + v_{11i}}, \tag{14.3.17}$$

in which each term is the expression appropriate to the derivation of Fieller limits from the evidence of a single assay (§ 4.12). Then $\hat{\mu}$ is the value of μ that minimizes $J(\mu)$, and

$$F = J(\hat{\mu})/(p-1)s^2 \tag{14.3.18}$$

can be tested against the variance ratio distribution with $(p-1)$, f degrees of freedom; a significantly large F indicates that the hypothesis of a common μ for the p assays is untenable, and thus controverts the validity of the set of assays as estimating a single relative potency. If the hypothesis of a common μ is deemed acceptable, the statistic

$$[J(\mu) - J(\hat{\mu})]/s^2 \tag{14.3.19}$$

can be referred to the variance ratio distribution with 1, f degrees of freedom; all values of μ for which it is less than the tabular F at a chosen probability are to be regarded as acceptable at that probability.

Very commonly, the iteration of equations (14.3.13), (14.3.14) leads to the unique minimum of $J(\mu)$. When the separate M_i are in good agreement, this procedure is adequate for determining $\hat{\mu}$. More generally, however, $J(\mu)$ can have many distinct minima, and unless iteration begins close to $\hat{\mu}$ it will not necessarily lead to the *minimum minimorum*. The possibility of multiple minima suggests that a more thorough study of $J(\mu)$ is desirable.

Expansion of $J(\mu)$ in inverse powers of μ shows that for large μ (positive or negative) the function tends to a limit

$$J_\infty = \sum \frac{b_i^2}{v_{22i}}, \tag{14.3.20}$$

and that the approach to this asymptotic value will usually be from opposite sides at the two extremes. As Armitage *et al.* (1976) have shown, artificial examples can readily be constructed in which $J(\mu)$ has p minima, one corresponding to each constituent assay, though not necessarily occurring close to the M_i; Dr W. Knight (personal communication) has pointed out the possibility of having up to $(2p-1)$ minima. Close agreement between assays causes minima to coalesce. Each pair of consecutive minima must be separated by a maximum, and maxima and minima can occur at any magnitudes of $J(\mu)$, whether above or below J_∞. The only safe method for determining the actual number of minima seems to be to tabulate or graph $J(\mu)$ over a wide range of μ, a simple piece of computation for a function of a single unknown. From an interactive computer terminal, one can rapidly locate minima approximately and then subtabulate at finer intervals in order to secure any desired numerical accuracy. In this way, the test of agreement between the p assays represented by equation (14.3.18) can be made. Having found $J(\hat{\mu})$, (14.3.19) shows the acceptable range of values for μ to consist of all values for which

$$J(\mu) \leq J(\hat{\mu}) + s^2 F(1, f), \tag{14.3.21}$$

where $F(1, f)$ is the tabular value of F with the stated degrees of freedom at the chosen probability. From a tabulation or graph of $J(\mu)$, the limits to the range can be read. If $J(\mu)$ has several minima, this range of values may consist of two or more distinct segments. For example, $J(\hat{\mu}) + s^2 F(1, f)$ may exceed two minima of $J(\mu)$ but be less than a maximum that separates them. Moreover, if

$$J_\infty < J(\hat{\mu}) + s^2 F(1, f)$$

one segment must be open-ended, in the manner that has been described for Fieller's theorem when $g > 1$.

Armitage *et al.* discussed these possibilities in considerable detail, and gave various numerical examples. In practice, good data will rarely produce the more bizarre forms of multiple segments. The behaviour of $J(\mu)$ is not dependent upon σ^2 and s^2. Indeed, if all assays are of the same design but possibly differ in replication, so that $v_{11}:v_{12}:v_{22}$ is independent of i, the function $J(\mu)$ condenses so as to be essentially of the same form as for $p = 1$: it will have only one maximum and one minimum, though the minimum will not be at zero unless all M_i are equal. When the mathematical form of $J(\mu)$ is such as to have several minima, whether or not the set of values acceptable at a specified probability constitutes more than one segment depends upon s^2. The intersection with a horizontal as defined by (14.3.21) will, for some values of s^2, take every form that the pattern of $J(\mu)$ makes possible. Nevertheless, data giving a reasonably small value to F in (14.3.18) will rarely show the multi-segment phenomenon at levels of probability likely to be considered interesting. More usually, the occurrence of a probability set forming several segments will correspond with evidence against a common value of μ sufficiently strong to cause *either* rejection of the whole hypothesis *or* recognition that some of the constituent assays are untrustworthy and possibly invalid.

A simple example is illuminating. Consider a $(2, 2)$ assay with n subjects at each of the doses and an interval of $1 \cdot 0$ between log doses of each preparation. Then

$$v_{11} = v_{22} = 1/n, \quad v_{12} = 0.$$

Suppose that two independent assays of the same test preparation give the same b and also

$$d_1 = d, \quad d_2 = -d.$$
Then
$$M_1 = d/b, \quad M_2 = -d/b.$$
Working in terms of true variances (replacement of σ^2 by s^2 makes no essential difference to the discussion that follows), for each assay
$$g = t^2\sigma^2/nb^2.$$
If this g is sufficiently small, a weighted mean would be used as in § 14.1 and would lead to the estimate
$$\bar{M} = 0$$
with
$$\text{Var}(\bar{M}) = \sigma^2(d^2 + b^2)/2nb^4.$$
The method of the present section involves
$$J(\mu) = 2n\left(b^2 + \frac{d^2 - b^2}{\mu^2 + 1}\right),$$
for which
$$J_\infty = 2nb^2.$$
If $d^2 > b^2$, $J(\mu) > J_\infty$ for all μ; there is then no true minimum for $J(\mu)$ and the evidence appears to suggest an indefinitely large value, positive or negative, for $\hat{\mu}$. In accordance with the theory outlined above, the test of agreement between the assays is given by
$$\chi^2_{[1]} = 2nb^2/\sigma^2,$$
and the set of values of μ acceptable at a stated probability is determined by
$$\frac{2n(d^2 - b^2)}{\mu^2 + 1} \leqslant \chi^2 \sigma^2,$$
where χ^2 is the tabular value for the probability. More reasonably, if $d^2 < b^2$, the minimizing value of μ is 0, for which
$$J(\hat{\mu}) = 2nd^2.$$
The test of agreement between the assays is then given by
$$\chi^2_{[1]} = 2nd^2/\sigma^2,$$
and the set of acceptable values of μ is determined by
$$\frac{2n\mu^2(b^2 - d^2)}{\mu^2 + 1} \leqslant \chi^2 \sigma^2.$$

The limits so defined are always wider than the approximation based on $\text{Var}(\bar{M})$, and may be very much wider.

In practice, the two assays would scarcely be regarded as suitable for combining unless either

§ 14.4 COMBINATION OF ESTIMATES

$$d^2 < 2\cdot 5\sigma^2/n$$

or knowledge external to the assays makes certain that $\mu_1 = \mu_2$. Moreover, any single assay in which the biological system has been sufficiently responsive to dose changes to give a good estimation (and to have g small) will have

$$b^2 > 25\sigma^2/n.$$

Use of $\mathrm{Var}(\bar{M})$ in place of 0·95 limits based upon $J(\mu)$ then underestimates the width of the interval by at most 5 percent. On the other hand, if b were as small as 3 times its SE, so that $b^2 = 9\sigma^2/n$, this underestimation could be almost 20 percent.

Both in this example and in the general situation, the problems disappear under the stronger hypotheses that all β_i are equal as well as all μ_i. When all v_{12} are zero, this leads to

$$\hat{\mu} = \bar{d}/\bar{b},$$

where $\bar{d}, \bar{b}$ are weighted means; non-zero values of v_{12} complicate the formula without making it intrinsically more difficult. The example of the pair of (2, 2) assays gives

$$\bar{d} = 0, \quad \bar{b} = b,$$

and a straightforward application of Fieller's theorem for the limits.

Armitage *et al.* (1974) compared alternative methods of combining bioassay results. They showed that $\bar{M}$, as defined by equation (14.1.2), will usually be a biased estimator of μ, and obtained an approximate formula for the bias. They then used a computer to simulate large numbers of results from assays with known parameters, on which they studied the behaviour of $\bar{M}$ and of $\hat{\mu}$. Their findings verified the bias in $\bar{M}$, the magnitude of which agreed well with the formula; this bias is small, but not negligible when L_p deviates much from zero in some assays. The maximum likelihood estimate, $\hat{\mu}$, appeared remarkably free from bias, but a variance for it based upon general asymptotic theory often seriously underestimated the true uncertainty, especially when s^2 was based on few degrees of freedom. When estimates of individually low precision or with few degrees of freedom are combined, inadequacies in the results are scarcely surprising; the good performance of maximum likelihood and computations that today are relatively inexpensive should encourage the general adoption of the method described here.

14.4 An example of combining cross-over assays

The method of § 14.3 can be illustrated by four assays of the same preparation of insulin (Smith *et al.*, 1944). All had twin cross-over designs (Table 10.2.1) with 3 rabbits on each of the four sequences of doses. The doses of S were related to the sensitivity of individual rabbits, but were based on dilutions of standard insulin at 1 and 2 IU/ml. The doses of T were prepared so as to be equally potent on the provisional assumption that T contained 22 IU insulin per mg. The response was the percentage reduction in blood sugar with an agreed adjustment for initial blood sugar. In the first assay, one rabbit was lost and analysis had to proceed as in § 10.8; details need not be given here.

TABLE 14.4.1 Summary of four assays of the same test preparation of insulin

Assay no.	d	b	v_{11}	v_{22}	v_{12}	d.f.	Error sum of squares
1	0·833	3·500	3/16	3/64	−1/96	7	70·67
2	−3·942	3·262	1/6	1/24	0	8	151·87
3	−0·432	5·584	1/6	1/24	0	8	267·32
4	2·150	5·759	1/6	1/24	0	8	325·40

Table 14.4.1 summarizes the essential information for the four assays, in the notation of § 14.3. The separate error mean squares may look very different, but equation (3.11.1) gives

$$\chi^2_{[3]} = 3\cdot71;$$

use of a pooled variance

$$s^2 = 815\cdot26/31$$
$$= 26\cdot299 \text{ with 31 d.f.}$$

need not be questioned. The first two assays have somewhat smaller regression coefficients than the second two, but not sufficiently so as to forbid an assumption of a constant β. In general, however, it seems safer to regard different assays as having potentially different β_i.

From the table and equation (14.3.17), the function for minimization is easily constructed:

$$J(\mu) = 192(3\cdot500\mu - 0\cdot833)^2/(9\mu^2 + 4\mu + 36) + 24(3\cdot262\mu + 3\cdot942)^2/(\mu^2 + 4)$$
$$+ 24(5\cdot584\mu + 0\cdot432)^2/(\mu^2 + 4) + 24(5\cdot759\mu - 2\cdot150)^2/(\mu^2 + 4).$$

A simple computer program enables this to be tabulated over the range $\mu = -1\cdot0$ to $\mu = 1\cdot0$ at intervals of 0·1, and successive subtabulations rapidly locate the minimizing value of μ to any reasonable numerical accuracy. A more sophisticated optimizing program could be used, but is barely worth introducing for minimizing a function of a single unknown. Indeed the direct tabulation has the advantage that it will also detect secondary minima if these exist (for these data they do not). The minimum occurs at

$$\hat{\mu} = -0\cdot0032,$$

for which

$$J(\hat{\mu}) = 125\cdot79.$$

From equation (14.3.18),

$$F = 125 \cdot 79/(3 \times 26 \cdot 299)$$
$$= 1 \cdot 59,$$

with (3, 31) d.f., showing no significant difference between the potency estimates for the four assays.

Now the tabular value of F for (1, 31) d.f. is 4·16, and therefore 0·95 limits to μ are set by

$$J(\mu) = 125 \cdot 79 + 26 \cdot 299 \times 4 \cdot 16$$
$$= 235 \cdot 18.$$

Further tabulations of $J(\mu)$ locate these as $-0 \cdot 489, 0 \cdot 486$. Hence the combined estimate of relative potency (in IU/mg) is

$$R = 22 \cdot 0 \text{ antilog} (-0 \cdot 0032 \times 0 \cdot 1505)$$
$$= 22 \cdot 0,$$

with limits at 18·6 and 26·0 IU/mg.

In this example, $J(\mu)$ lies almost entirely below its asymptote of

$$J_\infty = 2061 \cdot 04.$$

At a high negative value of μ, however, the function crosses the asymptote, rises slightly above J_∞, and then approaches the asymptote from above. The conclusions are largely as in Bennett's earlier discussion (1962) of the same assays, except for the modifications in theory introduced by Armitage *et al.* (1976).

14.5 Replication within blocks

A topic that may conveniently be examined here concerns assays in which doses are replicated within all or some of the blocks of subjects. For example, if litters of eight subjects were available for a (2, 2) assay such as that in § 4.15, the natural course would be to assign two subjects from each litter to each dose. How should this affect the statistical analysis? For the laboratory animals in general use, litters large enough to allow more than two subjects per dose are uncommon, but blocks based upon some other classification of subjects can be larger.

If an assay employs litter-mate control and has one subject per dose from each litter, the component of the analysis of variance commonly termed Error (cf. Table 4.16.1) is in reality composed of Doses × Litters interactions. When replication within blocks is adopted, the analysis of variance will have a component for intra-litter error independent of these interactions. An experimenter who wished only to test the significance of the separate dose contrasts would inquire whether a contrast was large in comparison with its variations from litter to litter; as is right and customary in the analysis of experiments, he would use the interaction with litters for his estimate of error. Examination of the mathematical model appropriate to bioassay shows that the situation here is somewhat different.

The right procedure can be guessed from an alternative approach. Each litter may be regarded as constituting a self-contained assay. When litters allow only two subjects per dose, analysis for each separately is unattractive because the estimates would be of very low precision. Nevertheless, they could be formed: if they were, the intra-litter variance would be used without further thought. Provided that the intra-litter variances from the several litters could be pooled, the method of combining independent estimates of potency (§ 14.3) could be used. The identity of design, however, permits an alternative and simpler approach. Interactions between dose differences and litters play no part in the assessment of error, and are relevant only to tests of the agreement between litters in respect of the magnitudes of separate contrasts.

The argument is clarified by study of the structure of the analysis of variance for a (3, 3) point assay. Suppose that K litters each contain $6r$ subjects, so that every litter provides r subjects per dose. The natural mathematical model to adopt will be one that allows for litter differences in general level of response and also in the regression coefficients yet maintains a fixed log potency. Write β_i for the regression coefficient in litter i. The difference in expected response between doses of S, T with equal dose metameters must therefore be $\beta_i \log \rho$. Consequently, with the symbols and conventions of Chapter 5 and again writing μ for $\log \rho$, the deviations of expected responses in litter i from the mean response for this litter are:

Dose	x	Expected response
S_1	-1	$-\tfrac{1}{2}\beta_i(\mu + 2)$
S_2	0	$-\tfrac{1}{2}\beta_i\mu$
S_3	1	$-\tfrac{1}{2}\beta_i(\mu - 2)$
T_1	-1	$\tfrac{1}{2}\beta_i(\mu - 2)$
T_2	0	$\tfrac{1}{2}\beta_i\mu$
T_3	1	$\tfrac{1}{2}\beta_i(\mu + 2).$

The difference between any observed response and its expectation according to this model will be ϵ, a random error that will as usual be supposed normally distributed; ϵ has mean zero and its variance, σ^2, will be assumed to be the same for all litters.

Essentially the same model would be used if the experiment were intended only for examining differences between the effects of six different treatments. The one modification is the restriction in the parametric representation of expected responses brought about by the nature of a parallel line assay, in consequence of which two parameters instead of five suffice to specify the expected responses. Formation of the expectations of the mean squares in the analysis of variance is a familiar task in the study of experimental designs. Table 14.5.1 contains the results, derivation of which is well illustrated by consideration of preparations contrasts. Within litter i, one may use the expected responses to form

$$L_p(i) = 3r\beta_i\mu + \text{sums and differences of } 6r \text{ values of } \epsilon. \tag{14.5.1}$$

TABLE 14.5.1 Expectations of mean squares in the analysis of variance for a (3, 3)-point assay with K litters of $6r$ subjects each

Nature of variation	d.f.	Expected mean square
Litters	$K-1$	—
Preparations (L_p)	1	$\sigma^2 + \frac{3}{2}rK\bar{\beta}^2\mu^2$
Regression (L_1)	1	$\sigma^2 + 4rK\bar{\beta}^2$
Parallelism (L_1')	1	σ^2
Quadratic (L_2)	1	σ^2
Difference of quadratics (L_2')	1	σ^2
$L_p \times$ Litters	$K-1$	$\sigma^2 + \frac{3}{2}rV_\beta\mu^2$
$L_1 \times$ Litters	$K-1$	$\sigma^2 + 4rV_\beta$
$L_1' \times$ Litters	$K-1$	σ^2
$L_2 \times$ Litters	$K-1$	σ^2
$L_2' \times$ Litters	$K-1$	σ^2
Error	$6K(r-1)$	σ^2
Total	$6rK-1$	—

By summation over all litters

$$L_p = 3r(\Sigma\beta_i)\mu + \text{sums and differences of } 6rK \text{ values of } \epsilon, \qquad (14.5.2)$$

where Σ represents summation over the K litters. Hence

$$E[L_p(i)^2] = 9r^2\beta_i^2\mu^2 + 6r\sigma^2 \qquad (14.5.3)$$

and

$$E[L_p^2] = 9r^2(\Sigma\beta_i)^2\mu^2 + 6rK\sigma^2. \qquad (14.5.4)$$

Remembering that the sums of squares for L_p and $L_p \times$ Litters are respectively

$$L_p^2/6rK$$

and

$$\frac{\Sigma L_p(i)^2}{6r} - \frac{L_p^2}{6rK},$$

the expected mean squares in Table 14.5.1 follow, where the notation

$$\bar{\beta} = \Sigma\beta_i/K \qquad (14.5.5)$$

and

$$V_\beta = \Sigma(\beta_i - \bar{\beta})^2/(K-1) \qquad (14.5.6)$$

has been introduced. The formulae for L_1 and $L_1 \times$ Litters are derived similarly. A variant on the model, which does not make any essential difference to the discussion of potency estimation, is to regard the litters as randomly selected from a population in which the β_i are distributed about a mean B with variance Ψ^2. The corresponding changes in Table 14.5.1 are that V_β must be replaced by Ψ^2 and $\bar{\beta}^2$ by $(B^2 + \Psi^2/K)$.

The table shows that, unless all β_i are necessarily equal, the five constituents of the Doses × Litters interaction are not homogeneous; Preparations × Litters and Regression × Litters have expectations that are unequal and greater than the other three, and therefore should not be pooled. An estimate of σ^2, with $3(2rK - K - 1)$ d.f., can be formed by pooling $L_1' \times$ Litters, $L_2 \times$ Litters, and $L_2' \times$ Litters with the true error.

As an adjunct to the testing of validity (§ 4.7), a test of significance of the difference between mean responses to the two preparations may be both interesting and important. For the test, the mean square for L_p should be compared with either the pooled error or with the mean square for $L_p \times$ Litters, according to which model has been adopted for the β_i, using Appendix Table II. A similar comparison for L_1 tests the significance of the deviation from zero of the mean regression coefficient, a test of little interest since no assay would be undertaken without strong prior evidence of regression. The essential validity tests are those for L'_1, L_2, and perhaps L'_2; these might be compared each with the corresponding interaction mean square, or all with the pooled mean square.

From equation (14.5.2),

$$E(L_p) = 3rK\bar{\beta}\mu, \qquad (14.5.7)$$

and similarly

$$E(L_1) = 4rK\bar{\beta}. \qquad (14.5.8)$$

Hence for a valid assay, with dose scales so chosen that the values of x are the actual log doses, $\log \rho$ is estimated by

$$M = 4L_p/3L_1. \qquad (14.5.9)$$

Exactly as in § 4.12, Fieller's theorem can be applied to the expression

$$4L_p - 3L_1\mu,$$

the variance of which is derived from

$$\mathrm{Var}(L_p) = 6rK\sigma^2, \qquad (14.5.10)$$

and

$$\mathrm{Var}(L_1) = 4rK\sigma^2, \qquad (14.5.11)$$

$$\mathrm{Cov}(L_p, L_1) = 0. \qquad (14.5.12)$$

Although the model that regards the β_i as randomly selected from a population increases $\mathrm{Var}(L_p)$ and $\mathrm{Var}(L_1)$ by multiples of Ψ^2, it also increases the covariance by a compensating amount, so that for either specification

$$\mathrm{Var}(4L_p - 3L_1\mu) = 12rK(8 + 3\mu^2)\sigma^2. \qquad (14.5.13)$$

Thus the fiducial limits are the roots of

$$(4L_p - 3L_1\mu)^2 = 12rK(8 + 3\mu^2)s^2t^2, \qquad (14.5.14)$$

which equation readily leads to equation (5.5.6) with $N = 6rK$. The value of s^2 might be the intra-litter estimate with $6K(r-1)$ degrees of freedom, or, as recommended above, the pooled estimate with $3(2rK - K - 1)$ degrees of freedom. The choice should be made on grounds of general policy before the experiment is done, and not simply as the smaller of the alternatives.

Table 14.5.1 indicates that information on μ, additional to that obtained from the ratio of the L_p and L_1 totals, is contained in the mean square for $L_p \times$ Litters. In fact, an estimate could be constructed from the mean squares for

§ 14.5 COMBINATION OF ESTIMATES

$L_p \times$ Litters, $L_1 \times$ Litters, and Error, but the precision seems likely to be too low to justify development of the appropriate theory.

The theory discussed above for a $(3,3)$ assay is easily modified to suit any parallel line or slope ratio assay in which the blocks are large enough to provide two or more subjects for each dose; even the use of blocks of different sizes, as when some litters of eight subjects and others of four are available for a $(2,2)$ assay, does not alter it. In all such assays, the fiducial limits to the relative potency should be based upon an estimate of intra-litter variance (which may include some Doses × Litters interactions with the same expected mean squares) and not on the mean square from $L_p \times$ Litters and $L_1 \times$ Litters (Finney and Wood, 1951).

What are the implications when $r = 1$, that is to say when only one subject per dose can be taken from each block? Because the analysis provides no direct estimate of intra-block variance, the usual practice is to describe as 'error' the pooled Doses × Litters interactions. Provided that the inter-block variance of the regression coefficients is zero or negligible, this will be satisfactory, but if V_β or Ψ^2 is large it will lead to over-estimation of variance and thus to fiducial limits that are too widely spaced. The difficulty could be overcome by removing the $L_p \times$ Litters and $L_1 \times$ Litters components from the estimate of error, but this would often leave too few degrees of freedom for a satisfactory estimate. Although V_β is likely to be small in many assays, the possibility must be borne in mind that the familiar methods of analysis for assays such as the oestrone assay in § 4.18 or the penicillin assay in § 5.4 can underestimate the precision of the estimate for relative potency. The consequences for incomplete block designs are similar in principle but more complicated in arithmetic; they will not be discussed here, but are reasonably straightforward for confounded designs (§ 9.9) and very awkward for balanced incomplete blocks (§ 9.7).

In their $(2, 2)$ point penicillin assays, Knudsen and Randall (1945) found greater inter-plate variance for L_1 than for L_p, and concluded that the numerator and denominator of M must be assigned different variances for calculating limits. As shown in § 6.7, failure to randomize does not satisfactorily explain the situation. An alternative worth considering is that the plates differed in values of β, so that the expected mean squares for $L_p \times$ Plates and $L_1 \times$ Plates would differ from one another and from that for $L_1' \times$ Plates. The correct procedure then would not be that adopted from Knudsen and Randall, but rather to use the mean square from $L_1' \times$ Plates as s^2.

Bliss (1952a, b) strongly criticized the model suggested here, because in assays that he examined he found no evidence of a non-zero V_β. The present model includes the simpler as a special case that is doubtless the most important in practice, but Bliss's contention that to estimate σ^2 by pooling of L_p and L_1 as well as of other contrasts will allow appropriately for non-linearity and even for imperfect parallelism seems to lack justification.

The alternative procedure of regarding each litter or block as a small self-contained assay, and then combining estimates as in § 14.3, ought in theory to estimate potency more precisely but scarcely makes practical sense. It differs from what has just been described by attaching greater weight to those litters in which the estimated regression coefficient is large, instead of simply adding values of L_p. It cannot bring appreciable advantages unless the β_i are clearly different and are well-estimated, not a very likely situation to be encountered when a block contains only two or three replicates.

14.6 Two assays of vitamin D_3

Table 14.6.1 contains data from two assays of vitamin D_3 in the same oil, each using the line test scores as responses; the original scores (to the nearest half-unit) have been multiplied by two for arithmetical convenience. These (2, 2) assays used eight rats from each of six litters, two per litter for each dose. In the first assay, the doses of T were chosen on the basis of an assumed potency of 4/3 IU/mg, and in the second this was changed to 16/15 IU/mg. The dose ratio, D, was 2 in both.

The analyses of variance follow the standard pattern, except that the pairs of litter-mates on each dose permit direct estimation of an intra-litter variance as described in § 14.5. For the first assay,

$$L_p = -45 - 63 + 45 + 61$$
$$= -2,$$

and the square for preparations is therefore

$$\frac{(-2)^2}{48} = 0\cdot0833.$$

A component for the interaction of preparations and litters is obtained by forming the L_p contrast for each litter, whence by the usual rules:

$$[(-2)^2 + (7)^2 + (3)^2 + (-6)^2 + (-3)^2 + (-1)^2 - 8 \times 0\cdot0833]/8 = 13\cdot4167.$$

The other litter interactions are similarly formed. Just as the sum of the components for preparations, regression, and parallelism is checked by agreement with the sum of squares of deviation between doses, so the sum of the litter interactions is checked by agreement with 17·7500, the complete Doses × Litters sum of squares. As a further check, the error (within litters) may be formed independently, from differences between pairs of responses:

$$[(4-3)^2 + (4-2)^2 + (2-3)^2 + \ldots + (7-6)^2]/2 = 20\cdot0000.$$

The two analyses of variance appear in Table 14.6.2.

As concluded in § 14.5, the estimate of error variance should be taken from the intra-litter sum of squares augmented by interactions of dose contrasts other than L_p and L_1 with litters, so giving for each assay a value of s^2 with 29 d.f. Comparison of L_p × Litters and L_1 × Litters with this error does not give any clear indication of variation in the regression coefficient from litter to litter. The large mean square for L_p × Litters in Assay 1 is worrying, but the smallness of the L_1 × Litters mean square in the same analysis suggests that chance is

§ 14.6 COMBINATION OF ESTIMATES

TABLE 14.6.1 Line test scores (multiplied by 2) for two assays of the same preparation of vitamin D_3

(a) Assay 1

Litter	Standard preparation		Test preparation		Total
	0·4 IU	0·8 IU	0·3 mg	0·6 mg	
I	3, 4 } 7	4, 6 } 10	4, 2 } 6	4, 5 } 9	32
II	2, 4 } 6	4, 5 } 9	5, 5 } 10	5, 7 } 12	37
III	3, 2 } 5	3, 4 } 7	3, 4 } 7	4, 4 } 8	27
IV	4, 4 } 8	5, 7 } 12	2, 3 } 5	4, 5 } 9	34
V	4, 5 } 9	6, 7 } 13	4, 5 } 9	5, 5 } 10	41
VI	4, 6 } 10	5, 7 } 12	4, 4 } 8	6, 7 } 13	43
Total	45	63	45	61	214

(b) Assay 2

Litter	Standard preparation		Test preparation		Total
	0·4 IU	0·8 IU	0·375 mg	0·75 mg	
I	3, 3 } 6	4, 7 } 11	4, 6 } 10	5, 7 } 12	39
II	2, 2 } 4	4, 4 } 8	4, 5 } 9	5, 6 } 11	32
III	2, 3 } 5	5, 5 } 10	4, 4 } 8	5, 5 } 10	33
IV	3, 3 } 6	4, 5 } 9	4, 5 } 9	5, 7 } 12	36
V	3, 3 } 6	4, 4 } 8	3, 2 } 5	6, 5 } 11	30
VI	3, 4 } 7	4, 5 } 9	3, 6 } 9	4, 7 } 11	36
Total	34	55	50	67	206

TABLE 14.6.2 Analyses of variance for Table 14.6.1

(a) Assay 1

Adjustment for mean		954·0833	
Nature of variation	d.f.	Sum of squares	Mean square
Litters	5	21·9167	4·3833
Preparations	1	0·0833	0·0833
Regression	1	24·0833	24·0833
Parallelism	1	0·0833	0·0833
$L_p \times$ Litters	5	13·4167	2·6833
$L_1 \times$ Litters	5	1·9167	0·3833
$L_1' \times$ Litters	5	2·4167	} 0·7730
Error (intra-litter)	24	20·0000	
Total	47	83·9167	

(b) Assay 2

Adjustment for mean		884·0833	
Nature of variation	d.f.	Sum of squares	Mean square
Litters	5	6·6667	1·3333
Preparations	1	16·3333	16·3333
Regression	1	30·0833	30·0833
Parallelism	1	0·3333	0·3333
$L_p \times$ Litters	5	2·9167	0·5833
$L_1 \times$ Litters	5	1·1667	0·2333
$L_1' \times$ Litters	5	4·4167	} 0·9799
Error (intra-litter)	24	24·0000	
Total	47	85·9167	

responsible. The small suspicion of statistical invalidity may be disregarded for the purpose of illustrating computations. Assay 2 has a regrettably large mean square for preparations, indicating that the change in assumed potency made for this assay was for the worse, but no signs of non-parallelism appear.

In order that the formulae of § 14.3 may be applied, $(\bar{x}_T - \bar{x}_S)$ should be expressed on the same metametric scale. Both assays had a dose ratio of 2, and therefore logarithms to base $2^{1/2}$ are convenient; all common logarithms are therefore divided by $\frac{1}{2}\log_{10} 2$. No further comment on estimation from the assays separately is needed; the results appear in Table 14.6.3. The two values of s^2 differ little, and the pooled variance

$$s^2 = 0.876\ 44 \text{ with } 58 \text{ d.f.}$$

can be adopted. In terms of logarithms to base $2^{1/2}$,

$$b_1 = 0.708\ 33,$$
$$b_2 = 0.791\ 67,$$

and

§ 14.6   COMBINATION OF ESTIMATES

$$d_1 = -0.083\ 33 + 0.830\ 07 b_1$$
$$= 0.504\ 63,$$
$$d_2 = 1.314\ 09.$$

Construction of the variance of $(b\mu - d)$ requires care, and the expression must be written

$$b(\mu + \bar{x}_T - \bar{x}_S) - (\bar{y}_T - \bar{y}_S)$$

in order to display the orthogonal or statistically independent parts. This shows that the simplest form of (14.3.17) is

$$J(\mu) = 24(0.708\ 33\mu - 0.504\ 63)^2/[(\mu - 0.830\ 07)^2 + 4]$$
$$+ 24(0.791\ 67\mu - 1.314\ 09)^2/[(\mu - 0.186\ 22)^2 + 4].$$

As in § 14.4, tabulation of $J(\mu)$ locates the minimum at

$$\hat{\mu} = 1.2233$$

and

$$J(\hat{\mu}) = 2.6427.$$

The test of agreement between the assays is given by

$$\chi^2_{[1]} = 3.015.$$

There is no sign of disagreement and $J(\mu)$ has only the single minimum. The limits are the values of μ that make

$$J(\mu) = 2.6427 + 0.876\ 44 \times 2.0017^2$$
$$= 6.1545;$$

further tabulations give the limits as 0.723, 1.831. Multiplication by $\frac{1}{2}\log_{10} 2$ and taking of antilogarithms leads to the values summarized in Table 14.6.3.

TABLE 14.6.3   Comparison of estimates and fiducial limits in two assays of vitamin D_3

Method	Potency (IU per mg)		Limits as percent of estimate
	Estimate	0·95 Limits	
Assay 1	1·280	0·968–1·672	76–131
Assay 2	1·778	1·356–2·738	76–154
Combined	1·528	1·285–1·886	84–123

The stronger hypothesis that $\beta_1 = \beta_2$ would require a different maximization of the full log-likelihood function in (14.3.9). The only parameters to be estimated now would be β, the product of $\beta\mu$, and σ^2, and the resulting estimate for μ itself would have somewhat narrower limits than those found above for $\hat{\mu}$. With two assays as alike as these, the estimate will not be very different from

$$(d_1 + d_2)/(b_1 + b_2),$$

the limits to which could easily be found by Fieller's theorem. Full calculation would not be difficult, but this model seems unlikely to be appropriate often.

14.7 Combination of slope ratio estimates

Because slope ratio assays commonly have small values of g, the simple weighting method of § 14.1 often suffices for the combination of estimates. It remains applicable if parallel line and slope ratio estimates are to be combined into a single figure.

When g is not small, the method of § 14.3 can be applied with little modification. For assay i, the estimate is

$$R_i = b_{Ti}/b_{Si}. \tag{14.7.1}$$

If σ^2 is constant for all assays, variances and covariances can be written

$$\left. \begin{array}{l} \operatorname{Var}(b_{Si}) = \sigma^2 v_{Si}, \\ \operatorname{Var}(b_{Ti}) = \sigma^2 v_{Ti}, \\ \operatorname{Cov}(b_{Si}, b_{Ti}) = \sigma^2 c_i. \end{array} \right\} \tag{14.7.2}$$

Apart from changes of symbolism, these are exactly like equations (14.3.4) and (14.3.6). Consequently the log likelihood can be written as in (14.3.9):

$$L = K(\sigma^2) - \frac{1}{2\sigma^2} \sum_{i=1}^{b} \{v_S(b_T - \beta_T)^2 - 2c(b_S - \beta_S)(b_T - \beta_T) \\ + v_T(b_S - \beta_S)^2\}/(v_S v_T - c^2) \tag{14.7.3}$$

and further development is almost exactly as in § 14.3. Bennett (1963a) first made suggestions along these lines.

14.8 Antiserum activity

Another form of combined estimation arises in connexion with the neutralization of hormones by antisera (Robyn, Diczfalusy and Finney, 1968). Suppose that a hormone H can be neutralized by an antiserum AS. Define an antiunit (AU) of AS to be such that 1·0 AU neutralizes the specific biological activity of 1·0 IU (international unit) of H. Under a condition that will be termed *simple additivity*, it follows that a dose consisting of z_H IU of H with z_S ml of AS acts exactly as a dose $(z_H - \theta z_S)$ IU of H alone, where θ is the neutralizing potency of the antiserum expressed in AU/ml. An equivalent statement is that a composite dose

$$\frac{z}{\gamma} \text{ IU of H}, \quad \frac{z(1-\gamma)}{\theta \gamma} \text{ ml of AS} \tag{14.8.1}$$

should, for any z and for any γ in the range $0 < \gamma \leqslant 1\cdot 0$, act exactly as z IU of H alone; as z varies, the composite doses retain the property that a fraction γ of the hormone present is unneutralized.

In order to estimate θ, a mixture of H and AS can be assayed against H alone. Suppose that T_0 is a preliminary guess at θ. For any arbitrary choice of γ, doses of a mixture corresponding to that specified by (14.8.1) may be formed as

$$\frac{z}{\gamma} \text{ IU of H,} \quad \frac{z(1-\gamma)}{T_0\gamma} \text{ ml of AS,} \qquad (14.8.2)$$

with suitable choices of z. This mixture can be assayed against H using any appropriate assay design; for example, a (3, 3) assay could use three values of z in the ratio 1:2:4 for H and for the mixture. The potency of the mixture relative to H alone is

$$\rho = \frac{1}{\gamma} - \frac{\theta(1-\gamma)}{T_0\gamma}$$

$$= 1 + \left(1 - \frac{\theta}{T_0}\right)\frac{1-\gamma}{\gamma}. \qquad (14.8.3)$$

Consequently, if R is the estimate of ρ from the assay, the equation

$$R = 1 + \left(1 - \frac{T}{T_0}\right)\frac{1-\gamma}{\gamma} \qquad (14.8.4)$$

defines T, an estimate of θ.

Equation (14.8.3) shows that
(i) if $\theta = T_0$, then $\rho = 1$;
(ii) if $\gamma = 1$, of course there is no AS in the mixture and $\rho = 1$;
(iii) if γ is close to 1, deviation of T_0 from θ will have less effect in making ρ different from unity than if γ is small.

However, the interest lies in θ and not in ρ. From (14.8.4),

$$T = T_0(1 - R\gamma)/(1 - \gamma). \qquad (14.8.5)$$

An alternative formula that displays more clearly the departure from the guessed value is

$$T = T_0 + \frac{T_0\gamma}{1-\gamma}(1 - R). \qquad (14.8.6)$$

Note that $R = 1$ corresponds to $T = T_0$. If analysis of the assay leads to R_L, R_U as fiducial limits for ρ, limits for θ are calculated as

$$\left.\begin{array}{l} T_L = T_0(1 - R_U\gamma)/(1 - \gamma), \\ T_U = T_0(1 - R_L\gamma)/(1 - \gamma), \end{array}\right\} \qquad (14.8.7)$$

the upper limit for ρ leading to the lower limit for θ. The relation between R_L, R_U and T_L, T_U can cause the fiducial range for θ to be quite different from that for ρ. The fraction of the content of H neutralized in each composite dose is

$$\frac{T(1-\gamma)}{T_0} = 1 - R\gamma. \qquad (14.8.8)$$

A single assay can thus estimate the potency of AS. Equation (14.8.5) indicates that, when γ is small and neutralization almost complete, the value of R has little effect on T and imprecision in R will therefore not matter greatly. On the other hand, when γ is large, even small errors in R have serious consequences for T. The purely statistical argument here must be modified because of the relatively greater importance of technical errors when γ is small; accurate measurement of dilutions of H and AS becomes vital. The ideal plan of assay will use the smallest γ that avoids serious risk of trouble from technical errors. In practice, to take γ about 0·5 seems satisfactory.

Robyn and Diczfalusy (1968a) described an assay of the neutralizing potency of antigonadotrophic serum, for which a pilot study had indicated a potency of about 1280 AU/ml. They used a standard (3, 3) parallel line assay with 5 litters of 6 immature male rats each, the response being weight of the total accessory reproductive organs. As standards, they had doses of 1·0, 2·0, 4·0 IU of human chorionic gonadotrophin (HCG); the 'test preparation' comprised doses of 2·0 IU + 1/1280 ml, 4·0 IU + 1/640 ml, 8·0 IU + 1/320 ml of HCG and antiserum. If the guessed potency were correct, the test doses would have $\gamma = 0\cdot5$ and the assay would indicate a relative potency that agreed with a theoretical value of 1·0. In the event, standard calculations for a (3, 3) assay led to

$$R = 0\cdot772,$$

$$R_L = 0\cdot647, \quad R_U = 0\cdot911.$$

Hence from (14.8.5)

$$T = 1280\,(1 - 0\cdot772 \times 0\cdot5)/0\cdot5$$
$$= 1570,$$

and from (14.8.7)

$$T_L = 1280\,(1 - 0\cdot911 \times 0\cdot5)/0\cdot5$$
$$= 1390,$$

$$T_U = 1730.$$

The assay gave no indications of invalidity. The final estimate of antiserum potency is appreciably higher than that used in designing the assay, but internal evidence casts no doubts on the legitimacy of the assertion of a potency of 1570 AU/ml with limits at 1390 AU/ml, 1730 AU/ml. Note that, if the limits were reported as percentages of the estimates (a practice with some assayists), they would appear more widely spaced for R than for T; for these data, $(T_U - T_L)/T$ is less than $(R_U - R_L)/R$. No simple rule connects the quantities, and with other data the position may be reversed. When γ is large, even an apparently very precise estimation of ρ may be unsatisfactory after conversion to estimation of θ.

In the same paper, Robyn and Diczfalusy collated results from many similar assays of HCG neutralizing potency, and also of the potency of antigonadotrophic sera in neutralizing the activity of luteinizing hormone. Subsequently (1968b) these authors reported assays of the potency of antisera in neutralizing human follicle stimulating hormone (FSH). They explored both the validity and the limitations of these assay procedures.

What a single assay cannot do is to put to the test the assumption of simple additivity, a *sine qua non* for validity of the whole procedure. For this purpose, experiments must be conducted with several values of γ, either as a multiple assay against the standard or as separate assays with different γ. The values of T can then be examined for consistency with one another. A simple procedure is to plot T against γ and to judge whether deviations from an average value show any systematic trend. Formal statistical test is more troublesome, especially if all values of γ have been used in one assay, so producing correlations between values of T based upon the same data for responses to H alone.

Suppose that a multiple assay has been conducted, with H alone as the standard preparation and series of doses corresponding to two or more different values of γ as test preparations. Then, from (14.8.3), the regression of response on dose for all the data can be expressed as

$$Y = \alpha + \beta \log \left\{ 1 + \frac{(1-\gamma)(T_0 - \theta)}{\gamma T_0} \right\} + \beta x, \tag{14.8.9}$$

where x is the logarithm of the content of H in the composite dose. In the manner of Chapter 11, a relative potency can be estimated for each series; several distinct (but not independent) estimates of θ are then obtainable by comparison with the standard regression for which $\gamma = 1 \cdot 0$. Optimal estimation of θ requires a more complicated analysis. The sum of squares of deviations of observed responses from Y as defined by (14.8.9), the summation being over every dose included, must be minimized with respect to the three parameters α, β, θ. The difference between the residual sum of squares for this analysis and that for the analysis as a multiple assay is a sum of squares for testing the departure from equality of values of θ estimated separately for each dose series, that is to say for testing agreement with the condition of additivity of hormone and antiserum doses.

Robyn (1969) and Petrusz, Robyn, Diczfalusy and Finney (1970) presented experimental evidence that mixtures of HCG and anti-HCG, and also of FSH and anti-FSH, behave in conformity with the predictions of additivity for moderate values of γ, though other factors complicate the situation for $\gamma \leq 0 \cdot 1$. Romani *et al.* (1974) have compared independent estimates of neutralizing potencies of anti-HCG sera from assays at different levels of neutralization. Since most of their assays were satisfactorily precise, they were able to use variances in a rather simple manner. Taking first the usual asymptotic variance, $\text{Var}(M)$, for a log potency estimate, they used the further approximation

$$\text{Var}(R) = R^2 \text{Var}(M) \tag{14.8.10}$$

for the variance of R. Then from equation (14.8.5)

$$\text{Var}(T) = T_0^2 \gamma^2 \text{Var}(R)/(1-\gamma)^2. \tag{14.8.11}$$

They used equations (14.1.1)–(14.1.7), with T_i and $\text{Var}(T_i)$ replacing M_i and

Var(M_i) to examine weighted mean potency estimates $\bar{T}$, and to apply χ^2 tests for heterogeneity of individual estimates. The method is statistically unsophisticated; its use was advised by the present author as computationally simple and adequate for the mass of data available. The chief weakness, use of Var(M) and Var(R), seems likely to exaggerate the value of χ^2 because of neglect of g; as conclusions were that almost always agreement between estimates was good even when the same antiserum was assayed at about 50 percent and about 90 percent neutralization, there seems little cause to worry further. What has been presented here is a simplified account of the biological problems, with emphasis on statistical considerations; the original papers should be consulted for fuller information.

14.9 Neutralization of antiserum

The series of studies that underlay § 14.8 disclosed the existence of gonadotrophin preparations that have immunological but little or no biological activity (Robyn and Diczfalusy, 1968b; Robyn, Petrusz and Diczfalusy, 1969; Petrusz, 1969). In order to measure the potency of these, the concept of additivity between hormone and antiserum has been extended. The anti-antiunit (AAU) of antiserum neutralizing factor (ANF) is so defined that 1·0 AAU combines with and neutralizes the effect of 1·0 AU of AS; if present in excess relative to the AS, ANF does not augment the potency of H. A dose consisting of z_H, z_S, z_N respectively in IU of H, ml of AS, and mg of the hormone containing ANF will in general act as a dose

$$(z_H - \theta_S z_S + \theta_S \theta_N z_N) \text{ IU of H alone;}$$

if this quantity is negative, the dose acts as though it were zero, and if ANF is in excess (i.e. $z_S - \theta_N z_N < 0$) the dose acts as z_H IU of H alone. Just as θ_S is the neutralizing potency of the antiserum in AU per ml, so $\theta_S \theta_N$ is the content of ANF in the hormone expressed in AAU per mg.

Corresponding to (14.8.1), three equivalent doses can be stated:

IU of H	ml of AS	mg of ANF	
z	0	0	
$\dfrac{z}{\gamma_S}$	$\dfrac{z(1-\gamma_S)}{\theta_S \gamma_S}$	0	(14.9.1)
$\dfrac{z}{\gamma_S}$	$\dfrac{z(1-\gamma_S)}{\theta_S \gamma_S \gamma_N}$	$\dfrac{z(1-\gamma_S)(1-\gamma_N)}{\theta_S \theta_N \gamma_S \gamma_N}$	

for any z, and any γ_S, γ_N with $0 < \gamma_S \leq 1\cdot 0$, $0 < \gamma_N \leq 1\cdot 0$. As z is varied, the dose remains such that a fraction γ_N of the antiserum is left unneutralized by ANF, and this in turn leaves a fraction γ_S of H unneutralized.

Estimation of θ_N now requires combination of two assays. Suppose that T_{S0}, T_{N0} are preliminary guesses at θ_S, θ_N. Use these with arbitrarily chosen γ_S, γ_N to choose sequences of doses guessed to be equivalent according to the formulae of the previous paragraph. In one assay compare the dose combinations that have

§ 14.9   COMBINATION OF ESTIMATES

$\gamma_S = \gamma_N = 1\cdot 0$ and $\gamma_S < 1\cdot 0$, $\gamma_N = 1\cdot 0$; exactly as in § 14.8, obtain an estimate of relative potency, R_S, from which, by equation (14.8.4),

$$R_S = 1 + \left(1 - \frac{T_S}{T_{S0}}\right) \frac{1 - \gamma_S}{\gamma_S} \tag{14.9.2}$$

determines T_S, the estimate of θ_S. In the second assay compare the dose combinations that have $\gamma_S < 1\cdot 0$, $\gamma_N = 1\cdot 0$ and $\gamma_S < 1\cdot 0$, $\gamma_N < 1\cdot 0$. Here the relative potency, R_N, is related to the estimate, T_N, of θ_N by the equation

$$R_N = \left[1 - \frac{T_S(1 - \gamma_S)}{T_{S0}\gamma_N}\left(1 - \frac{T_N(1 - \gamma_N)}{T_{N0}}\right)\right] \Big/ \left[1 - \frac{T_S}{T_{S0}}(1 - \gamma_S)\right] \tag{14.9.3}$$

By substitution from equation (14.9.2) and rearrangement of terms, this gives

$$T_S T_N = \frac{T_{S0} T_{N0}}{(1 - \gamma_S)(1 - \gamma_N)} [\gamma_S \gamma_N R_S R_N - \gamma_S R_S + 1 - \gamma_N] \tag{14.9.4}$$

as an equation for $T_S T_N$, the estimate of the required potency $\theta_S \theta_N$.

Since $T_{S0} T_{N0}$ is the preliminary guess for the potency, a form of (14.9.4) analogous to (14.8.6) is interesting:

$$T_S T_N = T_{S0} T_{N0} + \frac{T_{S0} T_{N0}}{(1 - \gamma_S)(1 - \gamma_N)} [\gamma_S(1 - R_S) - \gamma_S \gamma_N(1 - R_S R_N)]. \tag{14.9.5}$$

There is little chance of obtaining any exact theory of fiducial limits for the estimation of $\theta_S \theta_N$: equation (14.9.4) involves two distinct relative potencies and is non-linear in them. Perhaps the best that can be done is to use an approximate variance,

$$\operatorname{Var}(T_S T_N) = \left[\frac{T_{S0} T_{N0} \gamma_S}{(1 - \gamma_S)(1 - \gamma_N)}\right]^2 \left[(1 - \gamma_N R_N)^2 \operatorname{Var}(R_S) + \gamma_N^2 R_S^2 \operatorname{Var}(R_N)\right], \tag{14.9.6}$$

to calculate approximate limits. If it were possible to make the first assay very precise, or to estimate θ_S by combining evidence from several reasonably good assays, limits could be determined much as in § 14.8. In (14.9.4), R_S would be replaced by a numerical value calculated from (14.9.2) with T_S assumed known; replacement of R_N by corresponding upper and lower limits would then give limits for $\theta_S \theta_N$.

From a statistical point of view, alternative experimental procedures can be envisaged, though they may be less practicable experimentally. For example, the first assay might be as described above, but the second might compare dose combinations with $\gamma_S = \gamma_N = 1\cdot 0$ and $\gamma_S < 1\cdot 0$, $\gamma_N < 1\cdot 0$, so leading to a direct estimate for $R_S R_N$. The required estimate of ANF potency would still be obtainable from (14.9.4) or (14.9.5), but its variance would be somewhat simpler: equation (14.9.6) would be modified in an obvious manner. Again, the two assays might be collapsed into a single multiple assay of each of the two types of composite dose against H alone. The estimates R_S and $R_S R_N$ would now be correlated, each using the same responses for H alone, and consequently a covariance term would be introduced into equation (14.9.6).

The extended form of the additivity rule cannot be tested by the simple patterns of experimentation described. Even if additivity has already been established for hormone and antiserum, at least as an adequate approximation, validity of the whole system of calculations can be examined only by trial of several values of γ_N. The general equation analogous to equation (14.8.9) is

$$Y = \alpha + \beta \log \left\{ 1 + \frac{(1-\gamma_S)(T_{S0}-\theta_S)}{\gamma_S T_{S0}} - \frac{\theta_S(1-\gamma_S)(1-\gamma_N)(T_{N0}-\theta_N)}{\gamma_S \gamma_N T_{S0} T_{N0}} \right\} + \beta x,$$

(14.9.8)

where again x is the logarithm of the content of H in the composite dose. Here also, general non-linear least squares procedures could be used to examine whether an experiment that used various γ_S, γ_N would have results consistent with unique values for the parameters θ_S, θ_N.

Petrusz, Diczfalusy and Finney (1971a, b) have described this theory, with slightly different notation. They illustrated the calculations from studies of human chorionic and human menopausal gonadotrophins. Tests with various γ_S, γ_N enabled them to examine the behaviour of the estimated ANF potency under various conditions of neutralization. Incubation of antiserum with the ANF at 37°C showed satisfactorily constant potency, but incubation at 4°C showed very clear trends. The full least squares analysis outlined above was not undertaken, as cruder methods left the issue in no doubt. The reasons for these phenomena, discussed in Petrusz et al. (1971b), lie beyond the scope of the present work. Although some workers have expressed doubts about the behaviour of ANF, Qazi et al. (1974) have produced extensive evidence from assays *in vitro* that the phenomena described above do occur.

15
Validity and the choice of metameters

15.1 Fundamental validity

Questions relating to the validity of assays (especially statistical validity) must now be examined in detail. Some of the ideas of this chapter have been presented earlier (Chapters 3, 4 and 6), but their reconsideration should enable the reader to appreciate more fully their importance to the whole practice of assay. This chapter relates only to analytic dilution assays, since the whole problem of what is being estimated, and consequently what is the validity of the estimate, is much less clearly defined for purely comparative assays. The discussion in §§ 15.1 to 15.5 owes much to an important paper by Jerne and Wood (1949). Papers by Gaddum (1950), Miles (1952) and Schild (1950) should also be read, with particular attention to Miles's remarks on heterogeneous materials.

The fundamental condition for assay validity is that expressed by the condition of similarity (§ 3.3): the dose–response curves for the standard and test preparations are identical except for replacement of z by ρz, where z measures the actual amount of the dose. Unless the condition of similarity is satisfied, the test preparation must not be regarded simply as a dilution of the standard preparation in an inert diluent, and the assay is therefore fundamentally invalid. If the true response curves could be determined exactly, an exact check on this condition could be applied. Even a small discrepancy between the forms of the two curves would prove invalidity, and, however valuable the dose–response relations might be as expressions of the reactions of the subjects to the two preparations, they would not be a legitimate basis for an analytic dilution assay. A point sometimes forgotten is that similarity requires identity of form in the response curves over the complete range of doses, not merely over some restricted portion for which they are linear. In particular, if the S response curve asymptotically approaches a limiting value, that for T must have the same asymptote.

The necessity for insistence on similarity has been challenged by authors (Kendall *et al.*, 1963; Cornfield, 1964) who have suggested the practical usefulness of a potency functionally dependent upon level of response as a parameter for a situation where response regressions are not parallel. Finney (1965) and Cornfield (1967) have commented further on this idea, which has won little acceptance and will not be considered further here.

In practice, a response curve is not known exactly and must be estimated from experimental data. The curves so estimated for the two preparations may fail to be identical, not only because errors of estimation (deriving from the natural variability of responses) obscure the truth but because the true curves do not conform to the condition of similarity. The situation is complicated by the fact that estimation presupposes a mathematical model for the response curve, and even this is usually not known exactly. For example, the expected response might be assumed to be a linear function of a specified dose metameter, or a cubic polynomial, or a logistic function; estimation then involves determination of the best numerical values for the unknown parameters. Choice of a model that is adequate only over a limited range of doses may lead to an apparent deviation from similarity. An attempt might be made to represent the regression of response on log dose by a straight line, when in reality the relation is logistic but effectively linear over a moderate range of doses. If the doses used for S lie within the linear portion of the curve, but those chosen for T happen to be a little too high, the best-fitting straight line for T will tend to have a lesser slope than the corresponding line for S. A test of parallelism applied as a test of similarity may show a significant deviation, yet the true curves might conform perfectly to the condition of similarity. This is an additional reason for the use of symmetric designs, with doses of T guessed as well as possible to be equivalent to corresponding doses of S (§§ 6.8, 8.2), since the two response curves are thereby examined over corresponding portions. In a valid parallel line assay, any deviation from parallelism must then be the result of chance fluctuations in the responses, and a significance test is solely a test for similarity.

No test of statistical significance can give an absolute decision on this question, but it can lead to objective and standardized evaluation of the evidence. If the assayist were very anxious not to reject any valid assays, he might test deviation from parallelism at a probability of 0·01. He would then fail to reject assays which deviated only slightly and he would still reject 1 in 100 of perfectly good assays. If, on the other hand, he were anxious to reject all suspicious data, he might choose a probability level of, say, 0·1 for his significance test. He would reject 1 in 10 of good assays and would still fail to reject assays deviating from similarity so slightly as not to be detectable without more extensive evidence (cf. § 4.10). The responsibility rests with him: on the basis of past experience with like material and the evidence of the present data, he must decide whether or not he believes the assay to be valid. Statistical analysis cannot enable him to evade his responsibility, but can ensure that all relevant evidence is presented in a clear, objective manner.

For simplicity of discussion, a representation of the dose–response relation by a linear regression on log dose will be assumed; analogous arguments could be given for a power dose metameter (§ 7.1). On the hypothesis of similarity, if the design is symmetric and the difference between mean responses to the two preparations is not significant, the data for both preparations should relate to the same portion of the response curve. Any significant deviation from parallelism is then likely to be the result of fundamental invalidity, such as would require rejection of the whole assay. The investigator who accepts an assay in the face of this evidence must realize that he is asserting the observed deviation from parallelism to be in reality the result of chance, despite the low probability: he claims that the 1 in 20 or the 1 in 100 chance has occurred. Deviations from linearity, on the other hand, may merely be signs that the statistical analysis is invalid because the wrong metameters are being used. A significant difference between preparations does not of itself call for rejection of the assay, but is a danger signal. Just as a valid assay might appear to give non-parallel response curves on account of unsatisfactory choice of doses, so an invalid assay might appear to give two parallel lines if tested over ranges of dose for which levels of response differ. Consequently, if a significant difference between preparations is found, the evidence on non-linearity should be examined with especial care; the need for this is one strong reason for preferring assays with at least three points on each line.

15.2 The validity of (2, 2) assays

Some have claimed that in a (2, 2)-point assay the parallelism contrast may also be regarded as providing a significance test for deviations from linearity. Unfortunately, the need for distinguishing the two types of invalidity is greatest in circumstances which make the contrast most susceptible to disturbances from both (i.e. when a poor choice of doses has produced a large contrast for 'Preparations'). In an assay with three or more doses of each preparation, a significant difference between preparations, accompanied by either significant non-linearity or significant deviation from parallelism, would be a strong indication that the assay was statistically or fundamentally invalid. Simultaneous non-parallelism of the best-fitting straight lines and non-linearity are in theory not inconsistent with similarity, since they might be due to a non-linear regression with a poor choice of doses (Fig. 4.10.3, page 79); nevertheless, they hint at grave dangers, and an assay in which they occur should be treated with great reserve. When three or more doses of each preparation have been tested, the investigator has some chance of distinguishing the possibilities, and is moderately safe from the danger that two different sources of invalidity compensate for one another in their effects on a particular test criterion. Without *a priori* knowledge that the effective stimuli in the two preparations are qualitatively identical, the employment of (2, 2) designs is sheer obscurantism. The many merits of the (2, 2) scheme can be enjoyed only when experiments have shown that, for the group of materials under study, there is little danger of invalidity, at least of the fundamental kind that controverts the condition of similarity.

15.3 Statistical validity

For strict validity, evaluation of data from an analytic dilution assay (or indeed of data from a comparative assay) almost inevitably requires assumptions additional to the condition of similarity. Whether failure of the data to satisfy these assumptions perfectly affects the validity of conclusions to an extent that is of practical importance remains to be seen. The question of truth or falsity of assumptions which are theoretically requisite for the applicability of bioassay techniques has been widely discussed, especially by Jerne and Wood (1949). The difficulties are not peculiar to biological assay. Most of the statistical techniques commonly applied to biological data involve assumptions about the nature of distributions; an investigator is seldom, if ever, certain that the assumptions are correct, and he ought therefore to consider how far deviations from them destroy the practical validity of methods.[*] Use of distribution-free techniques might seem to be the obvious way of avoiding this difficulty, but these neglect completely all relevant knowledge. In practice, the investigator may have good reason to believe that a distribution is unimodal and roughly symmetric with finite moments, yet he may not be able to describe it more exactly; a distribution-free analysis fails to utilize this rather vague information. The problem, moral and philosophical as well as mathematical and statistical, is not unique to biological assay.

The usual statistical analysis of an assay involves expression of the relation between dose and response in terms of a linear regression. The assumptions in the simple type of assay, with several doses of each preparation tested on a homogeneous set of subjects, will now be listed. The specification would need to be more complicated if litter-mate or other block restrictions were present, or if it were possible to make several tests on the same subject, but the alterations would not affect principles. If z and u are the dose and response as actually measured, the implicit assumptions are:

(i) Errors of measurement in the magnitude of the dose, z, are negligible. For the test preparation, of course, the factor ρ is unknown, and only the ratios between values of z are assumed to be exactly known.

(ii) For any fixed z, values of the response u are random observations from a frequency distribution

$$f(u)\,du, \tag{15.3.1}$$

in which z plays the part of a parameter.

(iii) There exists at least one pair of one-to-one transformations

$$\left.\begin{array}{l} x = x(z), \\ y = y(u), \end{array}\right\} \tag{15.3.2}$$

uniquely defining x and y as functions of z and u respectively, such that y has a linear regression on x:

$$\begin{aligned} Y = \mathrm{E}(y) &= \textstyle\int_{-\infty}^{\infty} y f(u)\,du \\ &= \alpha + \beta x \end{aligned} \tag{15.3.3}$$

(x and y are known as the dose and response metameters).

---

[*] For example, employment of the t-test for examining the significance of a difference between two means assumes that the variations of each set of observations about their true means are normally distributed, and that the two variances are equal. The test is often used when one of these conditions is manifestly untrue (if the observations are essentially positive measurements, they cannot be normally distributed), with a confident belief that conclusions will be near enough to the truth for practical purposes. To give logical justification for this confidence is difficult: it rests in part upon the central limit theorem, in part upon empirical investigation and experience, and in part upon ignorance of any satisfactory alternative course of action. Most applications of statistical theory to 'real' data involve assuming that a particular mathematical model adequately represents the factors determining the data. The logic and validity of this step still need more consideration by statisticians.

§ 15.4   VALIDITY AND THE CHOICE OF METAMETERS   301

(iv) For any fixed x (or z), the response metameter y is normally distributed about its expectation.

(v) For any fixed x, the variance of y,

$$\text{Var}(y) = \text{E}\left[(y - \alpha - \beta x)^2\right], \qquad (15.3.4)$$

is either
- (a) constant.
- or (b) a known function of Y, the expectation of y,
- or (c) expressible as a function of Y with the aid of a finite number of additional parameters.

(vi) Explicit formulation of a pair of transformations, satisfying (iii), (iv) and (v), is either known completely or known in terms of a finite number of additional parameters. There is no *a priori* reason why some parameters should not occur in both transformations, but the dose transformation does not involve the response, and vice versa.

(vii) All the standard theory of mathematical statistics is accepted as applicable to entities that satisfy the strict conditions laid down in the relevant theorems.

(viii) If due attention is paid to the inexactness of correspondence between mathematical entities and measurements of material bodies, inferences of practical value may be drawn from the application of techniques developed in mathematical statistics to the data of biological experimentation and observation.

All these assumptions are at present intended as exact statements of fact. How far they must be modified, because a scientific measurement is not the same thing as a mathematical concept, remains to be discussed.

15.4 Comments on statistical validity

The assumptions listed in § 15.3 call for comment:—

(i) This is needed because standard regression techniques assume that the dose measurement is free from error. The condition might be relaxed to allow for a specific distribution of errors of measurement of z, but these will usually be so small in comparison with the errors of response measurements that they can be ignored. Any such relaxation would require a modified regression theory. 'There are a few assays in which the error of measuring the dose is inavoidably large; for example, when the dose is measured as numbers of bacteria injected, the magnitude of each dose is dependent upon a plate count of bacteria and is thus subject to considerable error. The magnitude of the error in x should then be estimated separately if possible and the computations modified accordingly' (Jerne and Wood, 1949).

(ii) Randomness can be ensured only by a suitably randomized assignment of subjects to doses. Any departure from strict randomization may result in the two preparations, or different doses of either preparation, being tested on non-comparable subjects (§§ 4.21, 6.7).

(iii) Even if the exact mathematical specification of the frequency distribution of responses were known, transformation of z into x and of u into y so as to satisfy (iii) may be impossible. Alternatively, several pairs of transformations exactly satisfying the requirements may exist. Moreover, whether or not any such transformations exist, other transformations not exactly correct but adequate within the limits of experimental error can easily be devised. For example if

$$\left. \begin{array}{l} x = \log z, \\ y = u, \end{array} \right\} \quad (15.4.1)$$

give a linear regression, or at least no serious deviation from linearity according to the evidence of the data, then alternatives such as

$$\left. \begin{array}{l} x = \log z, \\ y = u^{0.9}, \end{array} \right\} \quad (15.4.2)$$

or

$$\left. \begin{array}{l} x = \log (z + 0.014), \\ y = (u - 0.12)^{1.05}, \end{array} \right\} \quad (15.4.3)$$

or

$$\left. \begin{array}{l} x = 3 \log z + z^{0.01}, \\ y = 17u^{0.95} + 4u^{0.15} - \log u, \end{array} \right\} \quad (15.4.4)$$

are likely to accord with any reasonable amount of data just as satisfactorily.

Theoretical support for a particular pair of transformations is rare; usually empirical evidence is all that is available, though of course experience of similar work will be taken into account. No technique can find the 'best' metameters, unless choice is restricted to families that use a finite number of additional parameters. For example, the best metameters of the forms

$$\left. \begin{array}{l} x = z^i \\ y = (u - C)^\lambda \end{array} \right\} \quad (15.4.5)$$

might be sought, where i, C, λ are also to be estimated from the data. Such a procedure would be impracticable for routine use, and, unless the data were very extensive, would be unlikely to give estimates of all the parameters with reasonable precision.

Despite its apparent crudeness, the usual practice is to assume that some very simple choice of metameters will suffice, for example

$$\left. \begin{array}{l} x = z, \\ y = u \end{array} \right\} \quad (15.4.6)$$

in slope ratio assays, and (15.4.1) in parallel line assays, with occasional uses of others such as

$$\left.\begin{array}{l} x = \log z, \\ y = \log u \end{array}\right\} \tag{15.4.7}$$

for special purposes (cf. § 16.2). In the absence of theoretical reasons for choosing a particular metametric transformation, nothing is gained by a complicated rule if a simple one accords just as well with the data. An assayist who has doubts about the adequacy of the metameters that he proposes to use must design his experiments so as to permit the making of the appropriate validity tests; in a well-conducted and apparently valid assay, the conclusions drawn are unlikely to be seriously wrong solely because of the choice of metameters (§ 15.6).

(iv) Strictly interpreted, this is a severe restriction on the y metameter. For a wide class of frequency distributions, however, the central limit theorem (Cramér 1946; Kendall and Stuart, 1977) shows the distribution of a linear function of independent observations to be more closely approximated by a normal distribution than is that of individual observations. Many sampling studies and experiments have verified the rapid trend to normality of distribution for combinations of even very few observations from distributions that are themselves far from normal. Now $(\bar{y}_T - \bar{y}_S)$ and b, or b_T and b_S, the pairs of quantities whose ratio leads to potency estimates, are linear functions of the observations. Provided that there are twenty or more responses in all (and few assays will be attempted with less), the ill consequences of non-normality seem unlikely to be serious. This is indeed fortunate, for no test of the significance of deviations from normality is sufficiently sensitive to be of practical value when applied to small numbers of observations spread over several dose groups.

Though the central limit theorem gives a reasonable assurance that the distributions of linear functions of responses are approximately normal, it does not ensure that the variances assessed in the ordinary manner behave as would variance estimates from a truly normal distribution. Moreover, for regression coefficients the approach to normality may be to some extent upset by the unequal weighting of the several responses. Geary (1947) showed that tests of significance based upon the t distribution can be seriously misleading if the data are not normally distributed; Jerne and Wood (1949) pointed out that, if the true distribution is skew, 'fiducial limits' to a potency estimate calculated as though the measured responses were normally distributed may be very different from the correct values. In the absence of detailed information about the nature of any non-normality, nothing further can be done. The danger to the statistical validity of assays must not be forgotten, but should not be exaggerated: § 15.6 shows that gross non-normality may often fail to affect the fiducial limits to an appreciable extent, unless accompanied by other and even more obvious signs of invalidity.

(v) (a) In the absence of strong contrary evidence, the response metameter is usually assumed to have a homoscedastic regression. The assumption is unlikely to be exactly true, but may be effectively true over a limited range of expected responses. In a symmetric design with a good choice of doses, even though the assumption be false, it will lead to results almost exactly the same as would a properly weighted analysis. If the assumptions relating to linearity and similarity are exactly true, the assumption of constant variance will never invalidate the estimation of potency. The only possible harm is that the data are not used in the most efficient manner possible; by making allowance for dependence of the variance on Y, an estimate having higher precision might be obtained.

(v) (b) In assays based on quantal responses (Chapter 17), a dependence of variance on Y is generally assumed. Such a dependence is also occasionally appropriate for quantitative responses, and may legitimately be used as described in § 16.3 if the approximate form of the dependence is known. For example

$$\mathrm{Var}(y) = KY, \tag{15.4.8}$$

or

$$\mathrm{Var}(y) = KY^2 \tag{15.4.9}$$

might be taken. If the change in $\mathrm{Var}(y)$ as Y changes is very noticeable, a rough guess at the form of dependence is likely to be better than ignoring it completely.

(v) (c) When several responses have been measured for each dose, from tests at that dose on a number of different subjects or on the same subject at different times, empirical evidence on the form of any relationship between $\mathrm{Var}(y)$ and Y is available. If a general formula involving additional parameters, for example

$$\mathrm{Var}(Y) = VY^J \tag{15.4.10}$$

or

$$\mathrm{Var}(Y) = V_1 Y + V_2 Y^2, \tag{15.4.11}$$

were postulated, estimates of these parameters might be formed from the data. An exact statistical theory would be complicated, but an approximate analysis would probably be fairly satisfactory. (cf. § 16.7).

Fieller (1947) expressed the opinion that to take account of heteroscedasticity, by a transformation for equalizing variances or otherwise, is of primary importance in the analysis of assays. Undoubtedly any criterion for the heterogeneity of variance will be sensitive to certain types of change in the response metameter, as is exemplified in § 15.6. In general the effects of failure to allow perfectly for the relation between $\mathrm{Var}(y)$ and Y are of the second order: provided that the main assumptions of the analysis are correct, the estimation of ρ will be valid unless the variance function is grossly wrong. Use of the wrong expression for Var(y) is equivalent to attaching a wrong weight to different responses in the assay, not making the best possible use of the data, and so reducing the precision of the estimate. Experience of the weighting of data suggests that the loss is likely to be trivial as long as approximately the right relative values are given to the variances (Tukey, 1948). On the other hand, application of a stringent criterion for homoscedasticity could lead to rejection of a metameter, even though the corresponding potency estimate would be satisfactorily close to that based upon an apparently better metameter.

(vi) In practice, only very simple transformations are used.
(vii) If this were not believed, the subject of mathematical statistics would not exist!
(viii) If this were not believed, books on statistical science would not be written!

If all the assumptions stated were exactly satisfied, the standard processes of calculation, of which many examples have been given in other chapters, would lead to fiducial limits, in the sense of the orthodox theory of statistical estimation. Then R_L is the lowest value for the unknown parameter ρ that is not contradicted by a significance test based on the observed responses, and R_U is the highest value for ρ that is not contradicted by a significance test (§ 2.4). Both significance tests use single-tail probabilities, in this book conventionally 0·025, so that in asserting that the true ρ lies between R_L and R_U, the fiducial probability of being correct is 0·95.

15.5 The objectivity of statistical analysis

The assayist rarely has the knowledge implied by the conditions of statistical validity. Without them, statistical analysis of his assay cannot be exactly correct, and he might therefore seem to be debarred from the use of statistical science. Such criticism is not confined to bioassay, but can be levelled at almost any application of statistical methods to biological data that represents variation by a theoretical frequency distribution. Few would refuse to use theorems of pure geometry in the applied science of surveying, on the grounds that the 'straight lines' there encountered do not have the ideal properties required by Euclid's definition. Nor should statistical methods be dismissed from the evaluation of scientific data because the entities observed are not precisely the same as those used in theories of mathematical statistics (cf. § 2.5). A more practical though possibly not easily justified procedure is to devise a theory with which the observations usually agree, and then to examine each experiment or series of observations for evidence that the data deviate from theory so much as to invalidate a statistical analysis.

Any assay involves four classes of parameter:—

(i) the relative potency, ρ;

(ii)   the parameters α, β of the linear regression, as defined in (iii) of §15.3;
(iii) parameters defining the metametric transformations;
(iv)   parameters occurring in the expression of Var(y) as a function of Y.

Data from one assay will not give adequate information for the estimation of all these, especially as even the algebraic models containing the last two classes of parameter are unlikely to be known. Conclusions drawn from a single assay of a new type, perhaps based upon very simple assumptions about parameters, such as

$$\left.\begin{aligned} x &= \log z, \\ y &= u, \\ \text{and} \quad \text{Var}(y) &= \text{constant}, \end{aligned}\right\} \quad (15.5.1)$$

are necessarily tentative. The situation alters when data from a number of similar assays have accumulated, or when a planned investigation of the regression equation (§ 3.5) has preceded any assays. After they have been examined in relation to all the evidence, the assumptions can be modified as necessary. As the volume of data grows, more powerful tests of linearity, normality, homoscedasticity, and the rest become possible. Thus the difficulties are evaded by admitting the impracticability, or even impossibility, of determining for every assay separately the ideal dose and response metameters. Instead, the evidence of preliminary investigations, and of past assays of the same type, are used to guide the future choice of metameters. These future assays are then so planned as to make them provide:

(i)   tests of fundamental validity, essentially tests of deviation from the condition of similarity;
(ii)   tests of statistical validity, in terms of the metameters adopted. Chief of these is the test of linearity of regression, since the central limit theorem gives reasonable confidence that only serious non-normality will upset the conclusions, and non-constancy of variance is seldom of first-order importance. Of course, the data must be tested for evidence of non-normality or heteroscedasticity.

The design of an assay should permit both sets of tests to have adequate power. In each assay performed, the appropriate tests are made. In the absence of strong evidence against either fundamental or statistical validity, the assay is accepted as valid: the calculations are completed just as though all assumptions were known to be exactly satisfied. Invalidity in respect of the test for similarity must lead to rejection of an assay, unless it seems properly explicable as an alternative manifestation of statistical invalidity. If an assay shows only statistical invalidity, the analysis may be modified using other metameters, or decision between acceptance and rejection may be deferred until further information has accumulated.

When an assay is in use for routine estimation, strong reasons for belief in its fundamental validity should have been established so that the objectivity of statistical analysis can be secured. The metameters have been decided: the whim of the statistician does not influence conclusions. Validity tests are not undertaken to demonstrate the validity of assumptions peculiar to one assay; they are confirmation that no abnormal behaviour of the subjects, or other disturbance from unknown causes, has upset either the fundamental or the statistical validity. The fiducial limits eventually stated may not be exactly the same as would be obtained if the exact transformations appropriate to the data were known, but, as the experimenter has no hope of ever possessing this knowledge, that is no cause for despair. For example, one set of data might give $R = 1 \cdot 12$ with limits at $0 \cdot 95, 1 \cdot 32$ if the responses as measured were assumed to have a linear regression on log dose, and $R = 1 \cdot 08$ with limits at $0 \cdot 92, 1 \cdot 27$ if the logarithm of the response and log dose were used. The existence of two or more different estimates of a parameter from a particular set of data, depending upon the estimation procedure chosen, is not in itself unusual; if the limits are interpreted as conventional guides to the reliability of the estimates, rather than as satisfying the theory of fiducial inference in the strictest mathematical sense, the occurrence of slightly different values according to the form of analysis ceases to be objectionable. When a scheme of analysis is decided objectively, and not after inspection of the data from a particular assay, the subjective judgement of the statistician no longer determines the conclusions. The statement that ρ lies between $0 \cdot 95$ and $1 \cdot 32$ may not be the best possible statement of this character that might be made from the assay, but it possesses the advantage of being objectively determinable, and the assayist may be sure that a large proportion of such statements will be correct. As a guide to the practical action based on the assay, the limits should be trustworthy and objective. This attitude is no less justifiable than in any branch of biometry in which some form of distribution is arbitrarily assumed.

Empirical investigation suggests that choice of metameters will not affect the fiducial limits to an extent that would influence the action to be taken on the results of an assay, unless it is so extreme as to make statistical invalidity, in one or more of its forms, unmistakably clear (§ 15.6). In any programme of assays for a particular substance or group of substances, unless the choice of metameters is in no doubt, continuous records of validity tests should be kept (by control charts or otherwise), and the form of statistical analysis should be modified if the accumulated evidence makes this seem desirable (§ 5.10). Until evidence to the contrary is apparent, the simpler of two alternative metametric transformations may be preferred.

15.6 Alternative metametric transformations

Not uncommonly the responses recorded for an assay will suggest that some of the conditions for statistical validity are violated. An occasional very large response may hint at skewness of the distribution of y for fixed x, or variance estimates within dose groups may appear heterogeneous. Undoubtedly the evidence for invalidity can be strong: what then should be done? Though some would assert that either the responses should be transformed or a distribution-free method of analysis should be adopted, the experienced statistician may suspect that little harm would come of using the simple metameter $y = u$, especially if the design and choice of doses are of the quality advised in Chapters 6 and 8.

In an attempt to study the effects of alternative metametric transformations on inference from assays, I reported (Finney, 1949c) the consequences of using the metameter

$$y = u^\lambda \tag{15.6.1}$$

on responses for two assays. The results supported the conclusion that a wide range of values of λ was tolerable so far as validity tests were concerned, and that within this range the potency estimate and its fiducial limits were near enough to constancy for practical purposes.

At that time, 13 analyses of a single assay (at intervals of 0·5 in λ) required a lot of arithmetic labour. To-day, such computations are trivial and a greater range of transformations can easily be tried. A valuable family of metameters is that proposed by Box and Cox (1964):

$$y = [(u - C)^\lambda - 1]/\lambda; \tag{15.6.2}$$

as early as 1903, Kapteyn had suggested essentially the same transformation in his work on growth. For any C smaller than the minimum response, equation (15.6.2) represents a scale of transformations continuous over all positive and negative λ, with the limiting form

$$y = \ln(u - C) \tag{15.6.3}$$

as $\lambda \to 0$. I have analyzed several assays with a large number of different metameters defined by the pairs C, λ. Although obviously details of results vary with changes in C, the effects are surprisingly small unless C approaches the magnitude of the minimal observed response, in which case variances naturally behave somewhat oddly. The emphasis here is therefore on metameters defined by (15.6.2) and (15.6.3) with $C = 0$.

Table 15.6.1 records 48 responses from a (3, 3) assay of vitamin D reported by Gridgeman (1951). The natural way to analyze these would be without

TABLE 15.6.1 Responses in an assay of vitamin D in an oil (logarithms of tarsal–metatarsal distances)

Doses per chick per day					
S (standard units)			T (mg)		
1·200	2·000	3·333	1·552	2·587	4·312
1·20	1·16	0·66	1·50	0·47	1·24
1·48	0·91	0·93	1·44	1·25	0·80
1·29	0·94	0·99	1·60	1·08	0·72
1·54	0·93	1·10	1·50	1·29	0·84
1·67	1·00	0·81	1·51	1·21	0·87
1·21	1·10	0·63	1·55	0·93	1·07
1·40	1·06	1·03	1·53	1·21	0·90
1·42	1·03	1·02	1·39	1·17	1·00

§ 15.6   VALIDITY AND THE CHOICE OF METAMETERS   309

transformation, that is to say with $\lambda = 1\cdot 0$, though on this scale there is evidence of curvature of regression. Table 15.6.2 summarizes information on a number of validity tests. The column headed 'z' contains the z statistic ($= \frac{1}{2}$ natural logarithm of ratio of variances) for the ratio of variances within dose-groups obtained by pooling the two sets of 7 d.f. for the highest doses and the two sets of 7 d.f. for the lowest doses. Significance levels for z are shown at the foot of the column: large values, whether positive or negative, indicate heteroscedasticity. Other validity tests might be presented in terms of variance ratios with 1, 42 d.f. from the analysis of variance. The square roots of these, alternatively obtained as ratios of contrast values to their standard errors, are of course t statistics with 42 d.f. These values of t, including that for regression as well as those for the other four contrasts, appear in the remaining columns, and at the foot are shown significance levels for t with 42 d.f. It is evident that values of λ less than $-1\cdot 0$ are undesirable because of heteroscedasticity and because the regression coefficient is imprecise (to an extent that may give $g > 1\cdot 0$). Over most of the whole range of λ studied, only curvature of regression (L_2) gives cause for alarm; however, curvature is quite marked for all positive λ, and at $\lambda = 1\cdot 0$ is clearly significant. On the evidence of this one assay, a logarithmic metameter seems a good choice.

TABLE 15.6.2   Validity tests for vitamin D assay in Table 15.6.3

Metameter index, λ	Fisher's z	L_p	L_1	L_1'	L_2	L_2'
$-5\cdot 0$	3·46	$-0\cdot 66$	$-1\cdot 08$	0·24	1·05	1·41
$-4\cdot 0$	2·89	$-0\cdot 52$	$-1\cdot 66$	0·26	0·98	1·33
$-3\cdot 0$	2·33	$-0\cdot 31$	$-2\cdot 51$	0·24	0·95	1·17
$-2\cdot 0$	1·80	$-0\cdot 01$	$-3\cdot 73$	0·17	1·05	0·94
$-1\cdot 0$	1·28	0·38	$-5\cdot 33$	0·01	1·34	0·65
0·0	0·79	0·85	$-7\cdot 20$	$-0\cdot 25$	1·88	0·33
0·2	0·69	0·95	$-7\cdot 58$	$-0\cdot 31$	2·02	0·28
0·4	0·60	1·05	$-7\cdot 96$	$-0\cdot 37$	2·17	0·22
0·6	0·51	1·15	$-8\cdot 33$	$-0\cdot 43$	2·32	0·17
0·8	0·41	1·24	$-8\cdot 69$	$-0\cdot 50$	2·48	0·12
1·0	0·32	1·34	$-9\cdot 04$	$-0\cdot 57$	2·64	0·07
2·0	$-0\cdot 12$	1·73	$-10\cdot 38$	$-0\cdot 89$	3·43	$-0\cdot 08$
3·0	$-0\cdot 54$	1·93	$-10\cdot 90$	$-1\cdot 12$	4·01	$-0\cdot 10$
4·0	$-0\cdot 93$	1·94	$-10\cdot 65$	$-1\cdot 25$	4·30	$-0\cdot 03$
5·0	$-1\cdot 31$	1·83	$-9\cdot 95$	$-1\cdot 27$	4·33	0·06
Probabilities						
0·10	0·35				1·68	
0·05	0·45				2·02	
0·01	0·65				2·70	

Fig 15.6.1 presents, on a logarithmic scale in order to show extremes more adequately, the behaviour of the potency estimate and of 0·95 and 0·99 limits calculated for every value of λ. Note particularly that for large negative λ, where $g > 1\cdot 0$, the limits are exclusive. The interesting indication from Fig. 15.6.1 is that conclusions on potency are insensitive to the choice of metameter over a

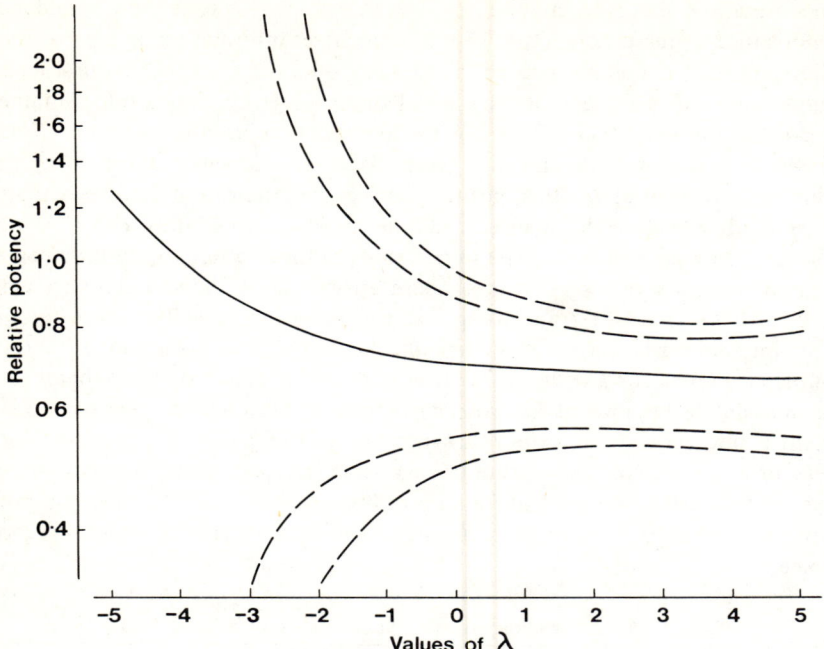

Fig. 15.6.1 Potency estimates and fiducial limits for the assay of vitamin D, Table 15.6.1 (0·95 and 0·99 limits are shown; for large negative values of λ, the limits become exclusive and have not been shown.)

wider range than the combined evidence of validity tests would permit the statistician to accept. Thus an uncritical analysis with $\lambda = 1\cdot 0$, ignoring all fears about curvature, would give a potency estimate of 0·683, with 0·95 limits at 0·560, 0·824. On the internal evidence of the data, $\lambda = 0\cdot 0$ is much better for avoiding curvature; with it, the estimate is 0·701, with limits 0·544, 0·888. However theoretically interesting, the difference is scarcely such as to lead to any qualitatively different conclusions or actions. A square-root transformation of course almost bisects the intervals between the findings with $\lambda = 1\cdot 0$ and $\lambda = 0\cdot 0$. Even a statistician stubbornly determined to push his analysis to extremes could scarcely justify going above $\lambda = 1\cdot 5$ at which $R = 0\cdot 678$ (limits 0·564, 0·805) or below $\lambda = -1\cdot 5$ at which $R = 0\cdot 749$ (limits 0·488, 1·133). Except for a notable widening of limits for this last strange choice of metameter, the practical conclusions remain fairly stable.

Table 15.6.3 contains responses from an assay of prolactin. With them, the same procedures have been followed. Table 15.6.4 shows z with 6, 6 d.f. for the ratio that compares within-dose variances at high and low doses (omitting all use of the middle dose of each preparation), and also the various values of t with 18 d.f. The only indications of statistical invalidity are those suggestive of heteroscedasticity, and these are absent for $-1\cdot 5 \leqslant \lambda \leqslant 0\cdot 7$; the lack of any suggestion

§ 15.6 VALIDITY AND THE CHOICE OF METAMETERS

TABLE 15.6.3 Responses in an assay of prolactin (crop-gland weights in g)

Doses per pigeon					
	S (IU)			T (mg)	
1·25	2·50	5·00	0·125	0·250	0·500
3·8	5·3	8·5	2·8	4·8	6·0
3·9	10·2	14·4	6·5	4·7	13·0
4·8	8·1	5·4	3·5	5·4	8·3
6·2	7·5	8·5	3·6	7·4	6·0

TABLE 15.6.4 Validity tests for prolactin assay in Table 15.6.3

Metameter index, λ	Fisher's z	L_p	L_1	L'_1	L_2	L'_2
−5·0	−3·02	−1·47	2·87	1·60	−1·42	−0·71
−4·0	−2·35	−1·55	3·34	1·55	−1·52	−0·56
−3·0	−1·69	−1·64	3·86	1·41	−1·56	−0·29
−2·0	−1·05	−1·69	4·33	1·12	−1·47	0·08
−1·0	−0·41	−1·66	4·53	0·69	−1·15	0·47
0·0	0·24	−1·50	4·27	0·23	−0·63	0·72
0·2	0·38	−1·46	4·17	0·14	−0·52	0·74
0·4	0·51	−1·40	4·05	0·07	−0·41	0·74
0·6	0·65	−1·35	3·92	0·00	−0·30	0·74
0·8	0·79	−1·30	3·78	−0·07	−0·20	0·73
1·0	0·93	−1·24	3·64	−0·12	−0·10	0·71
2·0	1·64	−0·99	2·99	−0·32	0·30	0·54
3·0	2·38	−0·80	2·50	−0·41	0·54	0·33
4·0	3·14	−0·68	2·17	−0·46	0·69	0·14
5·0	3·92	−0·62	1·96	−0·50	0·77	0·00
Probabilities						
0·10	0·56			1·73		
0·05	0·73			2·10		
0·01	1·07			2·88		

of non-parallelism or curvature over the whole range of λ studied is very striking.

Again the potency estimate is little affected by the choice of metameter, 6·52 IU per mg at $\lambda = -1·5$, changing slowly to 6·80 IU per mg at $\lambda = 1·0$. The latter value of λ is a little high for variance homogeneity, and if the choice had to rest on these data alone the square root or logarithmic metameter might be preferred. Fiducial limits are wide, but despite large values of g they do not change to an extent that is of great practical importance over the reasonable range of λ. Fig. 15.6.2 shows the behaviour of the estimate and its limits, and indicates clearly how the limits expand dramatically when the metametric transformation becomes extreme to the point of folly.

As an example of similar procedures with a slope ratio assay, a study has been made of the riboflavin assay for which the responses are shown in Table 7.10.2. The same family of metameters was used, and the analysis in § 7.10 was repeated for each metameter. Table 15.6.5 summarizes the validity tests. Here the z for testing homoscedasticity has been calculated for the ratio of the error mean

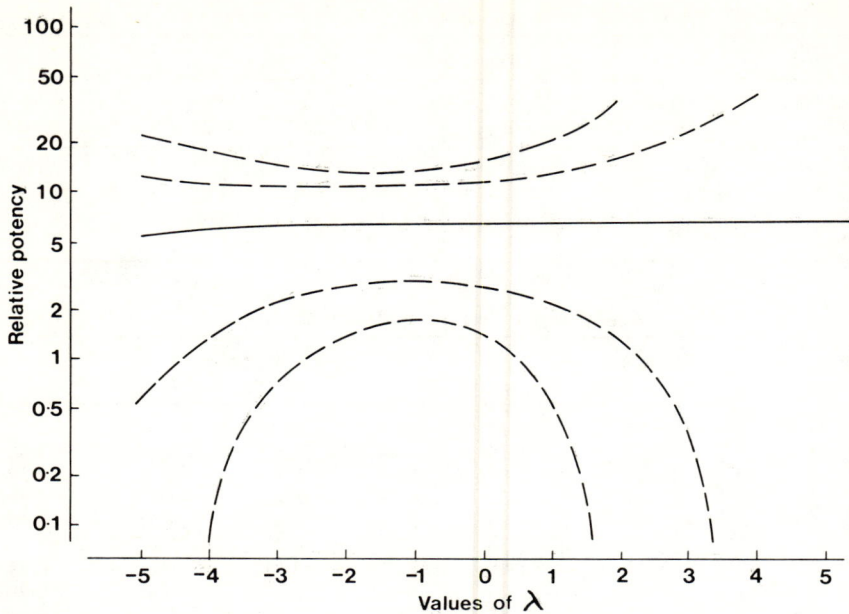

Fig. 15.6.2 Potency estimates and fiducial limits for the assay of prolactin, Table 15.6.3. (0·95 and 0·99 limits are shown; for large values of λ, positive or negative, the limits become exclusive and have not been shown.)

TABLE 15.6.5 Validity tests for riboflavin assay in Table 7.10.2

Metameter index, λ	Fisher's z	\multicolumn{4}{c}{Value of t for}			
		b_S	b_T	L_B	L_I
−5·0	−7·62	7·2	7·1	−5·79	0·10
−4·0	−6·41	9·0	8·8	−6·89	0·21
−3·0	−5·16	11·9	11·4	−8·32	0·41
−2·0	−3·88	17·1	15·8	−9·95	0·74
−1·0	−2·58	26·6	22·9	−10·89	1·08
0·0	−1·25	39·8	31·1	−8·19	0·63
0·2	−0·98	42·1	32·1	−6·92	0·33
0·4	−0·72	44·0	32·6	−5·43	−0·05
0·6	−0·45	45·3	32·8	−3·80	−0·50
0·8	−0·18	46·0	32·4	−2·09	−1·01
1·0	0·09	46·2	31·6	−0·39	−1·54
2·0	1·43	41·1	24·3	6·09	−4·03
3·0	2·78	33·1	16·8	8·43	−5·49
4·0	4·13	26·5	11·6	8·43	−6·00
5·0	5·48	21·6	8·2	7·65	−5·99
Probabilities					
0·10	0·83, −0·60			1·75	
0·05	1·10, −0·78			2·13	
0·01	1·66, −1·14			2·95	

square pooled for the two high dose groups to that for the blanks, with 6, 3 d.f.; the upper and lower significance levels are shown at the foot of the column. Evidently heteroscedasticity appears very strongly outside the range $0.0 < \lambda < 2.0$. The estimate of β_S is always large enough to keep g small, but the validity test for blanks is much more sensitive to changes in λ than was any test for parallel line assays. Indeed, trouble from L_B seems indicated unless $0.8 < \lambda < 1.3$, whereas no invalidity on account of non-intersection appears as long as $\lambda < 1.2$. Thus only a metameter with λ close to 1.0 seems acceptable.

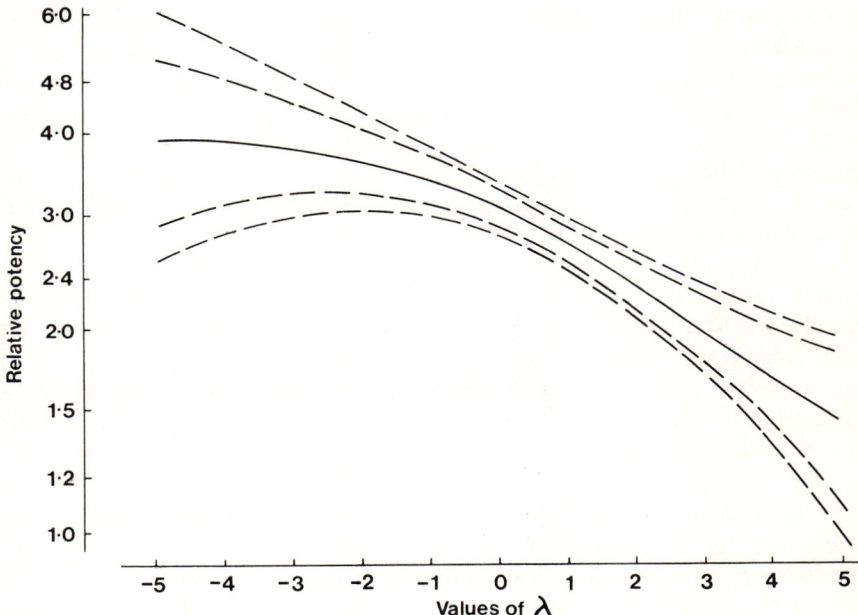

Fig. 15.6.3 Potency estimates and fiducial limits for the assay of riboflavin, Table 7.10.2 (0.95 and 0.99 limits are shown; for $-5.0 < \lambda < 5.0$, the limits are always inclusive.)

Fig. 15.6.3 shows the behaviour of the estimate and fiducial limits. Not surprisingly, the potency estimate changes with λ to a greater extent than it did in the other assays. The difference between $2.82\,\mu g/g$ (limits 2.66, 2.97) with $\lambda = 0.8$ and $2.62\,\mu g/g$ (limits 2.47, 2.78) with $\lambda = 1.3$ is only 7 percent, but for an assay as precise as is obtainable from a good microbiological technique this may not be negligible. The validity tests point so strongly to $\lambda = 1.0$ that the conclusions in § 7.10 seem adequately confirmed.

15.7 Further notes on metameters

The investigations reported in §15.6, a corollary to the work of Jerne and Wood (1949), are a warning to assayists against uncritical acceptance of a standard pattern of computation without thought of the assumptions involved.

Although there is no suggestion that similar calculations should be undertaken as part of the statistical analysis of every assay, they may on occasion help the interpretation of an awkward set of data. However little any alternative might alter the inferences from a particular assay, the theoretical implications of the actual choice must not be neglected.

In general, one should not expect to determine the metameter for an assay from the internal evidence of the assay. The choice of metameter for a class of assays should be based on the combined evidence of a number of independent experiments. Within the family of metameters represented by equation (15.6.2), often the one best satisfying the requirements of validity comes close to minimizing the width of the fiducial interval, but that is not in itself a criterion for a good metameter. This very useful family is not the only possibility. Another would be a transformation that recognizes the existence of an upper and a lower limit to possible values of the response and the consequent sigmoidal regression of u on x, as in equation (3.8.1) and in § 16.6. One of the very few publications that compares different metameters for the same data is that of Winder (1950).

To some extent related to the problem of metameter choice is that of the rejection of individual responses. The United States Pharmacopeia has commended statistical tests for the identification and rejection of markedly aberrant responses. Most statisticians are reluctant to reject a datum solely because it appears not to conform to the pattern of the remaining data: if the model postulated for a frequency distribution is somewhat wrong, the biases resulting from rejection may be more serious than would follow retention of the datum. If other evidence that an experimental or observational mistake occurred accords with a purely statistical test, rejection may be justifiable. In the absence of such evidence, rejection is analogous to a transformation determined by the other data; the same reserve as already urged is advisable in all consideration of using a value calculated from the remainder of the data in place of a response that is seriously discrepant from the general pattern.

15.8 Response metameters and covariance analysis

Earlier chapters have referred to attempts to reduce the variance of potency estimates by dosing each subject in proportion to its body weight (§ 4.15), or by expressing responses as so much per unit of body weight (§ 12.6). The latter is of course another way of defining a response metameter. An assayist will usually try to choose animal subjects all of about the same size, but if he is unable to do so adjustment of response in this manner may be helpful. There is, however, no guarantee that responses will on average increase in proportion to weight: they might be unaffected by weight over a wide range, or they might change much less markedly than weight, in which case the metameter described would overadjust and could increase variance instead of decreasing it. A covariance analysis (Chapter 12) is a better way of taking account of a concomitant variate. It allows the data themselves to determine the optimal linear function to be used as a metameter, including the possibility that no adjustment for the concomitant is desirable, and it is exceptional in providing statistical procedures that validly take account of the selection of the metameter.

Of course, this argument is not restricted to body weight. Any other measure of general body size could be used, or indeed any other measurement that is not itself affected by the applied stimulus.

16
General transformations and radioligand assays

16.1 Curvature and scedasticity

All the statistical methods and numerical examples discussed in Chapters 4 to 14 have involved a homoscedastic linear regression of response on either dose or log dose. These two major types of dose–response relation are not the only important possibilities. Some assays exhibit undoubted curvature or heteroscedasticity of the dose–response relation on either of the simple metameter systems mentioned; curvature at least must almost always be encountered if doses are sufficiently extreme.

As examples in § 15.6 illustrated, small changes in the metametric transformations need not seriously affect either the estimate of potency or its fiducial limits. The conclusion was that a single assay could not be expected to determine its own metameters. Preliminary investigations and experience of similar assays should indicate transformations likely to be sufficiently good; for a future assay, these metameters may be postulated, but tests for detecting serious discrepancies are still necessary. After dose and response metameters that linearize the regression have been found, a further problem concerns the variance of responses. The temptingly convenient assumption of constancy for the variance of individual values of y simplifies subsequent calculations. Even if untrue, it will not seriously bias the potency estimate from a symmetric assay with well-chosen doses, though it may invalidate the assessment of fiducial limits. In order that the mathematical model may correspond more closely to biological reality, a dependence of the variance of y upon Y, the expected response, may sometimes have to be considered. For many bioassay techniques, responses have constant variance on the original scale of measurement u: the variance of any non-linear transform of u is therefore dependent upon U (or upon Y). Quite generally, the method in § 3.13 may be used to give the maximum likelihood estimate of potency, provided that the functional dependence of $\text{Var}(y)$ on Y can be expressed. In practice, the assayist rarely knows that any one formulation of this dependence is preferable to all alternatives. Preliminary investigations for the standard preparation ought to have given some indication, however, and a reasonable approximation will be good enough for most purposes (§ 16.8).

16.2 An assay of *dl*-tryptophan

Table 16.2.1 contains data from an assay of *dl*-tryptophan, using *Lactobacillus arabinosus* as the test subject (W.F.J. Cuthbertson; private communication). In the microbiological assay of many amino-acids, the regression of response on dose has been found to be linearized by logarithmic transformation of both variates (Wood, 1946b, 1947a). Metameters defined by

§ 16.2   GENERAL TRANSFORMATIONS AND RADIOLIGAND ASSAYS   317

$$\left.\begin{array}{l} x = \log z, \\ y = \log u, \end{array}\right\} \quad (16.2.1)$$

therefore seemed worth trial. Table 16.2.2 is a metametric version of Table 16.2.1. For arithmetical convenience, 2 has been added to the dose metameters for the test preparation, a modification equivalent to expressing the doses in units of 0·01 ml. The design of the assay is open to criticism, but this is reserved for § 16.4.

TABLE 16.2.1   Responses in an assay of *dl*-tryptophan (ml N/10 NaOH)

\multicolumn{8}{c}{Dose of standard preparation (μg per tube)}							
2·0	4·0	6·0	8·0	10·0	12·0	14·0	16·0	18·0
2·2	3·7	5·6	7·1	8·5	9·8	11·3	12·2	13·2
2·2	3·8	5·6	7·3	8·5	9·8	11·3	12·2	13·6
2·2	4·0	5·6	7·6	8·7	—	11·4	12·4	—

\multicolumn{7}{c}{Dose of test preparation (ml per tube)}						
0·05	0·10	0·15	0·20	0·30	0·40	0·50
1·2	1·8	2·8	3·4	5·3	6·7	8·2
1·3	2·0	3·0	3·6	5·6	7·0	8·4

TABLE 16.2.2   Metametric transformations of the data in Table 16.2.1

\multicolumn{9}{c}{Dose metameter for standard preparation}								
0·301	0·602	0·778	0·903	1·000	1·079	1·146	1·204	1·255
0·342	0·568	0·748	0·851	0·929	0·991	1·053	1·086	1·121
0·342	0·580	0·748	0·863	0·929	0·991	1·053	1·086	1·134
0·342	0·602	0·748	0·881	0·940	—	1·057	1·093	—

\multicolumn{7}{c}{Dose metameter for test preparation}						
0·699	1·000	1·176	1·301	1·477	1·602	1·699
0·079	0·255	0·447	0·531	0·724	0·826	0·914
0·114	0·301	0·477	0·556	0·748	0·845	0·924

Dose metameter for test preparation calculated from doses measured in units of 0·01 ml.

This section summarizes an unweighted analysis, ignoring any heteroscedasticity in the regression of y on x. When the response metameters in Table 16.2.2 are plotted against the dose metameters (Fig 16.2.1), the regressions appear to be nearly linear over the range of doses. An impression that the variance of y is greater at low doses than at high will be ignored for the present. Parallel lines

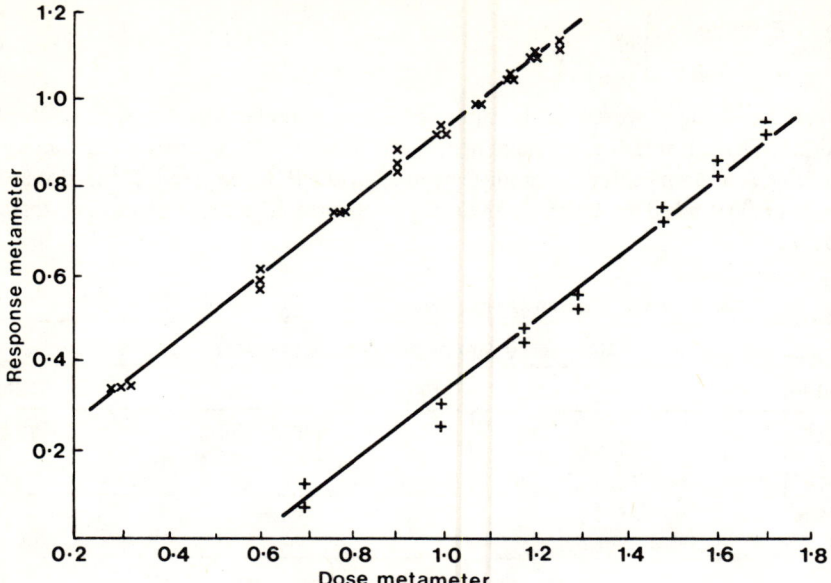

Fig. 16.2.1 Linear regressions for the assay of *dl*-tryptophan, Tables 16.2.1, 16.2.2
× : Response metameters for standard preparation
+ : Response metameters for test preparation

The straight lines are those mentioned in § 16.2 as drawn by eye, but the calculated lines (16.3.14) are almost identical with them.

drawn by eye in Fig. 16.2.1 indicate that (on the *x*-scales used) *M* is about −0·72; these lines will be used in the computations of § 16.3.

The calculations for an unweighted linear regression analysis are similar to those for Table 4.4.1, though the unequal spacing of the dose metameter makes them more laborious. For reasons that will be clear later, analyses of variance have been made separately for the two preparations, these being later combined into a single analysis of variance on conventional lines. The two analyses, Table 16.2.3, should give no difficulty. In each, the total sum of squares is divided into components between and within doses. For *S*,

$$S_{xx} = 2{\cdot}192\,386,$$
$$S_{xy} = 1{\cdot}828\,998;$$

subtraction of the square for regression,

$$S_{xy}^2/S_{xx} = 1{\cdot}525\,842,$$

from the sum of squares between doses leaves the component for linearity.

A disquieting feature of Table 16.2.3 is that the mean square for deviations from linearity for *S* is significant relative to that within doses and for *T* is almost significant. Fig. 16.2.1 exhibits no consistent trend of deviations such as would

§ 16.2 GENERAL TRANSFORMATIONS AND RADIOLIGAND ASSAYS

TABLE 16.2.3   Separate unweighted analyses of variance for Table 16.2.2

(a) *Standard preparation*

Adjustment for mean		17·771 283	
Nature of variation	d.f.	Sum of squares	Mean square
Regression	1	1·525 842	
Linearity	7	0·002 327	0·000 332
Between doses	8	1·528 169	
Within doses	16	0·001 260	0·000 079
Total	24	1·529 429	

(b) *Test preparation*

Adjustment for mean		4·280 220	
Nature of variation	d.f.	Sum of squares	Mean square
Regression	1	1·071 422	
Linearity	5	0·007 858	0·001 572
Between doses	6	1·079 280	
Within doses	7	0·002 951	0·000 422
Total	13	1·082 231	

occur if the regression were in reality curved. Inadequate randomization, an alternative possible explanation, is a danger to which attention has already been drawn in §§ 4.19, 4.20, 6.7 and 7.5: the tubes for each dose may have been filled consecutively, incubated in adjacent positions, or in some other manner subjected to conditions producing a correlation of responses. If the order of the doses was properly randomized at every stage, the mean square for deviations from linearity may be used as s^2 in the subsequent analysis; failure to include any element of randomization would violate a cardinal principle of unbiased experimentation and would make the data of questionable value. The error and linearity mean squares for T in Table 16.2.3 are considerably larger than the corresponding mean squares for S. This is a consequence of the neglect of the dependence of $\text{Var}(y)$ on Y: in reality, the variance of individual values of y about their regression line decreases as Y increases, and the general level of response is lower for T than for S.

Table 16.2.4 combines the two analyses. For the regression component,

$$(1 \cdot 828\ 998 + 1 \cdot 263\ 721)^2 / (2 \cdot 192\ 386 + 1 \cdot 490\ 534) = (3 \cdot 092\ 719)^2 / 3 \cdot 682\ 920$$

$$= 2 \cdot 597\ 100.$$

The difference between this and the sum of the corresponding squares for the preparations separately is the component for parallelism. Table 16.2.4 shows no evidence of lack of parallelism, but emphasizes the need to use the mean square for deviations from linearity,

$$s^2 = 0.000\,848\,8,$$

as the estimated variance per response. The choice of doses has made L_p highly significant, but if the linearity of the regression be accepted, no serious worry arises.

TABLE 16.2.4 Combined unweighted analysis of variance for Table 16.2.2

Adjustment for mean		21·295 763	
Nature of variation	d.f.	Sum of squares	Mean square
Preparations	1	0·755 740	
Regression	1	2·597 100	
Parallelism	1	0·000 164	0·000 164
Linearity	12	0·010 185	0·000 848 8
Between doses	15	3·363 189	
Within doses	23	0·004 211	0·000 183 1
Total	38	3·367 400	

Now

$$b = \frac{3.092\,719}{3.682\,920}$$

$$= 0.839\,75,$$

and therefore

$$g = \frac{(2.179)^2 \times 0.000\,848\,8}{(0.8397)^2 \times 3.6829}$$

$$= 0.0016.$$

In the usual manner,

$$M = -0.7259,$$

and by Fieller's theorem

$$M_L, M_U = -0.7558, -0.6983.$$

From these values, the relative potency is estimated as $18.80\,\mu g$ dl-tryptophan per ml, with fiducial limits at $17.56\,\mu g$ per ml and $20.03\,\mu g$ per ml.

16.3 Weighted analysis of the assay of dl-tryptophan

Inspection of Table 16.2.1 suggests that the variance of u is nearly constant, whereas Table 16.2.2 and Fig. 16.2.1 show that the variance of $\log u$ decreases with increasing U. The general method described in § 3.13 therefore seems preferable to the unweighted analysis in § 16.2. Suppose that

$$\text{Var}(u) = \sigma^2, \tag{16.3.1}$$

§ 16.3 GENERAL TRANSFORMATIONS AND RADIOLIGAND ASSAYS

a constant, is the variance of u about its expectation U. The metametric transformation, equation (3.7.5), may be written (now using natural logarithms)

$$Y = f^{-1}(U) = \ln U, \qquad (16.3.2)$$

whence

$$U = f(Y) = e^Y. \qquad (16.3.3)$$

Since

$$f'(Y) = U, \qquad (16.3.4)$$

equations (3.13.6) and (3.13.13) become

$$w = U^2 \qquad (16.3.5)$$

and

$$y = Y - 1 + \frac{u}{U} \qquad (16.3.6)$$

for the weighting coefficient and working response respectively. In desk computation, logarithms to base 10 will be preferred; if equation (16.3.2) is so interpreted, and

$$\epsilon = \ln 10 = 2\cdot302\ 585, \qquad (16.3.7)$$

then

$$w = \epsilon^2 U^2 = 5\cdot301\ 90 U^2 \qquad (16.3.8)$$

and

$$y = Y - 0\cdot434\ 29 + 0\cdot434\ 29 \left(\frac{u}{U}\right). \qquad (16.3.9)$$

The weighting coefficient could be taken as U^2; the factor ϵ^2 must then be introduced later, as a multiplier of the total weight and of all the sums of squares and products, before the calculation of any variances or fiducial limits. For ease of comparison with § 16.2, however, equation (16.3.8) is used here.

The first computational step is the estimation of σ^2, the supposedly constant variance of the actual responses. This is a mean square for u, not y. Although subsequent calculations are made with the aid of y, they are brought to the u-scale by the weighting coefficient, so that weighted sums of squares of y are in the same units as σ^2 and its estimate. Sums of squares within dose groups in Table 16.2.1 amount to

$$\left.\begin{array}{l} S_{uu} = 0\cdot313\ 33, \text{ with 16 d.f.,} \\ S_{uu} = 0\cdot175\ 00, \text{ with 7 d.f.} \end{array}\right\} \qquad (16.3.10)$$

for S and T respectively. Neither preparation shows any dependence of variance on U. In contrast to Table 16.2.3, the two mean squares are nearly equal. A pooled estimate is

$$s^2 = 0\cdot021\ 23, \text{ with 23 d.f.} \qquad (16.3.11)$$

The main computing table, Table 16.3.1, is similar to that familiar for assays based on quantal responses and using the probit transformation (§§ 18.1 and

and 18.2). Systematic completion of the table minimizes its complexities. Here desk computation is described (cf. §16.5), with details of the filling of each column of Table 16.3.1:

- (i) Insert x, the dose metameter, from Table 16.2.2.
- (ii) Insert n, the number of responses measured for each dose.
- (iii) Insert the mean value of u, the measured response for each dose, as obtained from Table 16.2.1. The SE's of these means, $s/\sqrt{n}$, being of the order of 0·1, justify at most two places of decimals.
- (iv) On a diagram (Fig. 16.2.1) representing $\log u$ plotted against x, draw by eye two regression lines constrained to be parallel; because the weighting coefficient increases as U increases, regard deviations from the line as less serious for small than for large values of x. Then, either directly from the diagram or with the aid of a regression coefficient measured on the diagram, enter Y, the expected value for $\log u$, at each dose. In Fig. 16.2.1, the regression coefficient was measured as 0·82, and successive values of Y were constructed from this.
- (v) Insert $U = $ antilog Y for each dose.
- (vi) Calculate nw from equation (16.3.8); three or four digits in weights are adequate for most purposes, and nw is here taken to the nearest integer.
- (vii) Calculate y, the working response, from equation (16.3.9). The variance of any y is approximately σ^2/nw. Thus, even with nw at its maximum of 2430 and σ^2 estimated by s^2 in equation (16.3.11), the standard error of y would be of the order of 0·003: the data warrant at most three decimals in y.
- (viii), (ix) Enter individual products nwx, nwy, retaining all decimals in order to assist checking.
- (x) Reserve a final column for new calculations of expected response metameters.
- (xi) Form totals of nw, nwx and nwy for each preparation.
- (xii) Use equation (3.13.18) to give weighted sums of squares and products of deviations for each preparation, as shown at the foot of the table, and add corresponding pairs.

Steps (v)–(xii) constitute one cycle of the iteration. Before the potency estimate is obtained, the need for further cycles must be considered. The weighted means, by equations (3.13.16) and (3.13.17), are

$$\left.\begin{array}{ll} \bar{x}_S = 1\cdot1013, & \bar{x}_T = 1\cdot5451, \\ \bar{y}_S = 1\cdot0101, & \bar{y}_T = 0\cdot7845. \end{array}\right\} \tag{16.3.12}$$

Also,

$$b = 288\cdot6888/352\cdot2020$$
$$= 0\cdot8197, \tag{16.3.13}$$

§16.3 GENERAL TRANSFORMATIONS AND RADIOLIGAND ASSAYS

TABLE 16.3.1 Computations for the weighted analysis of the assay of *dl*-tryptophan

x	n	u	Y	U	nw	y	nwx	nwy	Y_S, Y_T
Standard preparation									
0·301	3	2·20	0·352	2·249	80	0·343	24·080	27·440	0·354
0·602	3	3·83	0·599	3·972	251	0·583	151·102	146·333	0·601
0·778	3	5·60	0·743	5·534	487	0·748	378·886	364·276	0·745
0·903	3	7·33	0·846	7·015	783	0·866	707·049	678·078	0·848
1·000	3	8·57	0·925	8·414	1 126	0·933	1 126·000	1 050·558	0·927
1·079	2	9·80	0·990	9·772	1 013	0·991	1 093·027	1 003·883	0·992
1·146	3	11·33	1·045	11·092	1 957	1·054	2 242·722	2 062·678	1·047
1·204	3	12·27	1·092	12·359	2 430	1·089	2 925·720	2 646·270	1·094
1·255	2	13·40	1·134	13·614	1 965	1·127	2 466·075	2 214·555	1·136
					10 092		11 114·661	10 194·071	
Test preparation									
0·699	2	1·25	0·091	1·233	16	0·097	11·184	1·552	0·091
1·000	2	1·90	0·338	2·178	50	0·283	50·000	14·150	0·338
1·176	2	2·90	0·482	3·034	98	0·463	115·248	45·374	0·482
1·301	2	3·50	0·585	3·846	157	0·546	204·257	85·722	0·584
1·477	2	5·45	0·729	5·358	304	0·736	449·008	223·744	0·729
1·602	2	6·85	0·831	6·777	487	0·836	780·174	407·132	0·831
1·699	2	8·30	0·911	8·147	704	0·919	1 196·096	646·976	0·911
					1 816		2 805·967	1 424·650	

	S_{xx}	S_{xy}	S_{yy}
Standard preparation:	12 524·4766	11 455·0154	10 481·0908
	12 240·9522	11 227·0752	10 297·1744
	283·5244	227·9402	183·9164
Test preparation:	4 404·2783	2 262·0266	1 171·5762
	4 335·6007	2 201·2780	1 117·6364
	68·6776	60·7486	53·9368
Total	352·2020	288·6888	237·8562

very close to 0·82, the value used in the first cycle. In the usual manner, the two regression equations are evaluated as

$$Y_S = 0·1074 + 0·8197x, \atop Y_T = -0·4820 + 0·8197x.\} \quad (16.3.14)$$

The final column in Table 16.3.1 contains the value of Y_S or Y_T calculated for each dose; the T entries are almost identical with those that began the first cycle, the S entries are 0·002 higher. Had the differences been appreciable, a second cycle would have been performed with the values of Y_S, Y_T replacing Y. Iteration would have continued until all differences became negligible.

The S_{yy} values at the foot of Table 16.3.1 are weighted sums of squares between doses, with 8 d.f. for S and 6 d.f. for T. The method of determining weights, with insertion of the factor ϵ^2, ensures that these are comparable with the sums of squares within doses in equations (16.3.10). As usual, $(S_{xy})^2/S_{xx}$ is the portion of the sum of squares between doses attributable to the linear

TABLE 16.3.2 Separate weighted analyses of variance for Tables 16.2.1 and 16.2.2

(a) *Standard preparation*

Nature of variation	d.f.	Sum of squares	Mean square
Regression	1	183·2531	
Linearity	7	0·6633	0·0948
Between doses	8	183·9164	
Within doses	16	0·3133	0·0196

(b) *Test preparation*

Nature of variation	d.f.	Sum of squares	Mean square
Regression	1	53·7350	
Linearity	5	0·2048	0·0410
Between doses	6	53·9398	
Within doses	7	0·1750	0·0250

regression of y on x. Table 16.3.2 may be compared with Table 16.2.3; total sums of squares are never wanted and need not be calculated.

Again the deviations from linearity, at least for S, are too great to be explained by intra-dose variation. On the assumption that the variation between doses is validly estimated after a proper randomization, a pooled mean square for deviations from linearity must be taken as s^2, the estimate of σ^2, instead of that in equation (16.3.11). The mean squares for the two preparations separately differ appreciably, but, as a consequence of the weighting of the data, the difference is much less marked than in Table 16.2.3. A rigorous analysis would require separate consideration of intra-dose and inter-dose variations, taking account of the different numbers of tubes per dose; this will not be attempted here, as it cannot be of much importance for these data. Now therefore

$$s^2 = 0.07234, \text{ with 12 d.f.} \tag{16.3.15}$$

In the combined analysis of variance, Table 16.3.3, the component for a single regression coefficient is

$$\frac{(\Sigma S_{xy})^2}{\Sigma S_{xx}} = 236.6290,$$

and that for deviations from parallelism is obtained by subtraction of this quantity from the sum of corresponding components in Table 16.3.2. The component for preparations is similarly found as

$$\Sigma \frac{(Snwy)^2}{Snw} - \frac{(\Sigma Snwy)^2}{\Sigma Snw} = 10\,297 \cdot 1744 + 1117 \cdot 6364 - 11\,336 \cdot 4694$$

$$= 78 \cdot 3414.$$

§ 16.3 GENERAL TRANSFORMATIONS AND RADIOLIGAND ASSAYS

TABLE 16.3.3 Combined weighted analysis of variance for Tables 16.2.1, 16.2.2

Nature of variation	d.f.	Sum of squares	Mean square
Preparations	1	78·3414	
Regression	1	236·6290	
Parallelism	1	0·3591	0·359 1
Linearity	12	0·8681	0·072 34
Between doses	15	316·1976	
Within doses	23	0·4883	0·021 23

The relatively greater weight given to high responses here, as compared with Table 16.2.4, makes the square for the deviations from parallelism just significantly greater than s^2; the T regression line is steeper than that for S. Taken in conjunction with the large difference between preparations, this casts serious doubt upon the validity of the assay. On internal evidence, the condition of similarity seems not to be satisfied, although some slight non-linearity at high doses may be responsible. For the sake of illustrating the method, the calculations are concluded as though the assay were valid.

From equations (16.3.12) and (16.3.13),

$$M = -0.4438 - \frac{0.2256}{0.8197}$$

$$= -0.7190.$$

Also, by adaptation of equations (3.13.21) and (3.13.22),

$$\text{Var}(\bar{y}_S - \bar{y}_T) = s^2 \sum \frac{1}{Snw} \quad (16.3.16)$$

and

$$\text{Var}(b) = s^2 / \sum S_{xx}, \quad (16.3.17)$$

where the variance per unit weight must be taken from equation (16.3.15). Hence

$$g = \frac{(2.179)^2 \times 0.072\ 34}{(0.8197)^2 \times 352.2020}$$

$$= 0.0015.$$

Fieller's theorem gives

$$M_L, M_U$$

$$= -0.4438 - \left[0.2752 \pm \frac{2.179}{0.8197} \left\{ 0.072\ 34 \times \left(0.9985 \times \left(\frac{1}{10\ 092} + \frac{1}{1816} \right) \right. \right. \right.$$

$$\left. \left. \left. + \frac{(0.2752)^2}{352.2020} \right) \right\}^{1/2} \right] \bigg/ 0.9985$$

$$= -0.7404, -0.6984.$$

The potency of T is therefore estimated to be $19.10\ \mu g$ tryptophan per ml, with fiducial limits at 18·18 μg per ml and 20·03 μg per ml. These limits are 95·2 percent and 104·9 percent of the estimate, as compared with limits at 93·4 percent and 106·5 percent of the estimate in the unweighted analysis.

The more laborious computations of this section have not much altered the estimate of potency in § 16.2, but the more rational interpretation of the relation of variance to response has improved the precision. This illustrates the general experience that, almost always in a valid assay and often even when (as here) validity is in doubt, a wrong choice of weighting will seldom give a misleading estimate of potency. Of course,

$$\text{Var}(u) = \text{constant}$$

is not necessarily true in any exact sense, but it is certainly nearer the truth than the assertion

$$\text{Var}(\log u) = \text{constant}$$

implicit in § 16.2; further refinement would have little effect, even if the data were sufficient to suggest any.

When the dependence of $\text{Var}(y)$ upon Y is as pronounced as in the tryptophan assay, an analysis that takes no account of differential weighting is bound to involve serious loss of precision. Assayists must recognize that to obtain the most precise conclusions sometimes necessitates laborious computation (but see § 16.5). Here the weighted analysis has also disclosed evidence of invalidity: if the assay is invalid, this needs to be known and ought not to be concealed by imperfect analysis. The non-parallelism may indicate fundamental invalidity. Alternatively, it may be due only to an unsuccessful choice of doses having brought the highest doses of S or the lowest doses of T outside the region of linearity, although Fig. 16.2.1 gives little support to this view.

16.4 Criticism of the design of the tryptophan assay

Though not directly appropriate to this chapter, discussion of the design of the assay of dl-tryptophan seems desirable. No critic can put himself in the state of knowledge and state of mind of the investigator who planned the assay 30 years earlier. If the assay was exploratory, rather than one of a routine series, he may have known little about the range of linearity of regression. Indeed, he may have had no idea what metameters might linearize the regression (Wood's proposals for this were about contemporary with the assay), and may have chosen a wide range of doses in the expectation that he would later discard some. The analysis in this chapter perhaps makes the best of the data; criticism of the design may be relevant to future similar assays without being condemnation of past practices.

The design was far from optimal (Chapter 6) and the narrow limits of error might have been made much narrower by a different choice of doses. The following points should be noted:—

(i) Allocation of unequal numbers of subjects to preparations had its usual bad consequences, especially in respect of $\text{Var}(\bar{y}_S - \bar{y}_T)$.

(ii) The dose-range for T was too narrow; a better guess at the relative potency might have led to choice of higher doses for T.

(iii) A wider dose-range could have reduced $(M - \bar{x}_S + \bar{x}_T)$ by making the mean responses to the preparations almost equal. In § 16.3, this quantity made a large contribution to the width of the fiducial range for M.

(iv) The non-linearity and non-parallelism reported in § 16.3 must raise anxiety about flaws in randomization as an alternative to fundamental invalidity or a by-product of non-linearity. Strict attention to randomization should make the intra-dose mean square an estimate of σ^2, the variance per response. That variance might not be as low as the value 0·021 in Table 16.3.3, but ought to be less than the value 0·072 used in § 16.3.

(v) A further improvement would be to reduce the number of doses tested, so as to give a (3, 3) or a (4, 4)-point symmetric design. Regular spacing of doses on the logarithmic scale, over a range as wide for T as for S, is desirable. In this way ΣS_{xx} would be increased, g decreased and possibly even Snw increased. The degrees of freedom for the intra-dose s^2 would be increased at the expense of those for deviations from linearity, so that the change would not be ideal if for any reason imperfect randomization were inevitable.

The changes suggested in (i), (ii), (iii) would narrow the fiducial range to an extent that can be roughly estimated. Equal division of subjects, and a dose-range for T about equal to that for S, would give Snw and S_{xx} for each preparation about 78 percent of that found in § 16.3 for S. Hence ΣS_{xx} would be increased to about 440, and g would be decreased to about 0·0012; if $(M - \bar{x}_S + \bar{x}_T)$ were negligible, the semi-fiducial range for M would become approximately

$$\frac{2 \cdot 18}{0 \cdot 82 \times 0 \cdot 999} \times \left\{ 0 \cdot 072 \times 0 \cdot 999 \times \frac{2}{7870} \right\}^{1/2}$$
$$= 0 \cdot 0114,$$

instead of the previous 0·0210. Such changes almost cut the fiducial range by half, and so improve the precision of R to the same extent as would a fourfold increase in subjects on the old design. Inadequate knowledge on which to base the choice of doses would reduce the gain in precision. The considerations advanced in (iv) and (v), however, ought at least to compensate; they might themselves give still bigger advantages, as well as making computation much simpler.

16.5 Least squares by computer

The analysis illustrated in § 16.3 could be undertaken as there described if not wanted very often. If this or a statistical technique much like it were to become part of a routine assay process, a special computer program would be written to perform the estimation by minimizing the sum of squares. Without this facility, a general program for weighted linear or non-linear least squares can be used. All that is necessary should be the writing of a short routine specifying the features peculiar to the model. Here the function to be minimized is

$$S(u - e^Y)^2 \qquad (16.5.1)$$

where

$$\left.\begin{aligned} Y_S &= \alpha_S + \beta x, \\ Y_T &= \alpha_T + \beta x, \end{aligned}\right\} \qquad (16.5.2)$$

for the two preparations; summation is over the 25 responses for S and the 14 for T. Note that no weight is inserted because, as stated by equation (16.3.1), u has a constant variance. With the aid of the program MINISQUARE, discussed more extensively in §16.13, the equations

$$\left.\begin{aligned} Y_S &= 0\cdot1076 + 0\cdot8194x, \\ Y_T &= -0\cdot4816 + 0\cdot8194x \end{aligned}\right\} \qquad (16.5.3)$$

were obtained. These correspond to equations (16.3.14); obviously the difference is trivial, arising solely from the failure to pursue iteration of either series of calculations to sufficient numerical exactness. With appropriate extensions, the program could complete the analysis of variance in Tables 16.3.2 and 16.3.3.

An alternative procedure, not exactly equivalent but closely related to the same model and equally amenable to computer processing, is to minimize a weighted sum of squares of $\ln u$:

$$Sw(\ln u - Y)^2, \qquad (16.5.4)$$

where Y is as before and

$$w = e^{2Y}. \qquad (16.5.5)$$

The estimated regression lines are

$$\left.\begin{aligned} Y_S &= 0\cdot1050 + 0\cdot8215x, \\ Y_T &= -0\cdot4862 + 0\cdot8215x, \end{aligned}\right\} \qquad (16.5.6)$$

not materially different from equations (16.5.3). However, minimization of (16.5.4) can be more troublesome than that of (16.5.1) because the expression tends to zero if the parameters are so chosen as to make every value of Y large and negative. An ill-planned iteration may move towards this extreme instead of to the desired finite minimizing conditions.

In either form of analysis, the potency estimation can be completed as in §16.3. If the model were wanted regularly, all the necessary arithmetic would be incorporated into the computer program and the assayist would no longer be aware that a fully weighted analysis involved the laborious calculations of §16.3.

16.6 Radioligand assay

Radioligand assays (RLA) are widely used as research tools and for routine clinical estimations of antigens; they have the attraction of estimating potency from very small quantities of materials and with reputedly high precision. They include radioimmunoassays (RIA) in which antigens are labelled with

§16.6 GENERAL TRANSFORMATIONS AND RADIOLIGAND ASSAYS

radio-isotopes, and immunoradiometric assays (IRMA) in which antibodies are labelled. The literature is immense; no attempt is made here to give a full historical survey even from a statistical viewpoint, or to discuss laboratory procedure. Rodbard (1974), Rodbard and Frazier (1975) and Rodbard and Hutt (1974) gave simple accounts of schemes of statistical analysis that also lead into the literature. Although RLA are perhaps not strictly bioassays, since they do not depend upon responses measured in living organisms or tissues, they are so similar in structure that they need consideration from the viewpoint of bioassay.

Many early users of radioligand assay procedures were so impressed by the precision and by the smoothness of response curves constructed almost freehand from the data, that they ignored the requirements of validity emphasized in earlier chapters. In particular, they commonly neglected the merits of simultaneous trial of standard and test preparations, statistical tests of validity, and even estimation of error. Rodbard and Frazier commented: 'Unfortunately, many persons still utilize graphical methods alone, or linear interpolation between adjacent points on the dose–response curve. These methods do not provide estimates of the precision of unknowns, are subject to erratic behaviour and to subjective biases, and forfeit important information about the assay system.' If experimentation is well controlled, the response curve may look very satisfactorily smooth yet not well fitted by any obvious family of functions. Even where some mathematically expressed curve seems likely to fit satisfactorily, the process of estimating its parameters according to recognized statistical principles may have intimidated those who did not have access to, or were unaccustomed to using, good computing facilities. Consequently, the temptation to rely on graphical or interpolatory methods has been strong.

If a suitable family of non-linear regression functions can be identified, estimation of the particular parameters from a set of data should now be free from great difficulty. As in §16.3, iterative computation can be based upon a computer implementation of the general method of §3.13 with the formulae appropriate to the particular curves. Other ways of maximizing a log likelihood or minimizing a weighted sum of squares (§16.5) may prove more satisfactory.

For RLA, the response measured at a dose is the radiation count in a fixed time. Among practitioners, the usual symbolism is that B and F denote bound and free counts for the labelled ligand at any dose, and $T (= B + F)$ is the total count. At zero dose, B_0 is used for the bound count, and at 'infinite' dose the non-specific count is N (though B_∞ would be a better symbol for a count in the presence of an unlimited amount of the ligand); direct observations are possible at both extremes. Many proposals for statistical analysis have involved using as response variables such functions as

$$\frac{B-N}{T-N}, \text{ the fraction bound,} \qquad (16.6.1)$$

$$\frac{F}{T-N}, \text{ the fraction free,} \qquad (16.6.2)$$

$\dfrac{B-N}{B_0-N}$, the 'bound/initial bound' ratio, (16.6.3)

$\dfrac{B-N}{F}$, the 'bound/free' ratio, (16.6.4)

and their reciprocals, where B_0 and N now represent either mean counts at extreme doses or values extrapolated from a smooth curve. Plotted against dose or logarithm of dose, any of them will show an approximately linear regression over some range of doses, but will almost certainly be disconcertingly non-linear if the range is extended.

Some workers truncate the data so as to restrict analysis to a linear segment, scarcely a satisfactory general practice. Rodbard and Cooper (1970), noting that the ratio (16.6.3) must have a regression curve that runs between asymptotes at 0·0 and 1·0, suggested that a logit transform would give a more nearly linear dose regression. In this and subsequent papers, Rodbard and his colleagues discussed how to fit a weighted regression of $\ln\{(B-N)/(B_0-B)\}$ on log dose. Though the model proves suitable for many assays, no implication that it should apply to all radioligand assays is intended. After examining various representations of response curves, Midgley, Niswender and Rebar (1969) also concluded that the logit transform is the most useful; they proposed estimation by weighted least squares as advocated in § 16.9 but without the iteration that takes account of expected weights. The model is related to a general theory described by Ekins and Newman (1970). As with quantal response (§ 17.9), experimental data do not readily discriminate between different sigmoid equations, and reasonable alternatives will commonly lead to essentially the same conclusions on relative potency. Arrigucci *et al.* (1973) and Harding *et al.* (1973) discussed entirely different formulations that may sometimes be useful.

16.7 The RLA sigmoids

The statistician recognizes that to try to base a system of analysis on derived functions such as (16.6.1)–(16.6.4) is tactically unwise, because values of the functions are not independent. Far better is the strategy of first expressing the expectation of individual responses by a mathematical model in terms of dose and the parameters. Rodbard has represented the expected bound count for RLA by a logistic type of curve that in his notation is

$$y = d + \dfrac{a-d}{1+(X/c)^b},$$ (16.7.1)

where X is dose and a, b, c, d are parameters (cf. Burger *et al.*, 1973); he takes $b > 0$ as a convention, since a reversal of this sign simply interchanges a and d. If $b = 1\cdot 0$, this regression function becomes a rectangular hyperbola (Leclerq *et al.*, 1971). Representation of dose logarithmically gives the more familiar logistic form, and leads to an analysis that seems appropriate to many RLA. In (16.7.1), a and d correspond to B_0 and N respectively, and c is a dose for which the expected count is half-way between the two limits, sometimes termed the

§ 16.7 GENERAL TRANSFORMATIONS AND RADIOLIGAND ASSAYS

midrange or the ID50. Here the approach will be that of § 3.13. In order to preserve analogies, terminology and notation must be changed; differences from Rodbard are not important to the conclusions (Finney, 1976b).

Suppose that $U = E(u|x)$ is the expected count at a dose, and define D and C as the expectations of the counts at extremes of dose, with $D > C$; for RIA, D, C correspond to B_0, N respectively, and for IRMA the correspondence is reversed. The regression curve of bound count on x must be asymptotic to D, C at the extremes. It can be written

$$U = C + (D - C) F(x), \qquad (16.7.2)$$

where $F(x)$ is a function that changes smoothly from 1 to 0 for RIA (or 0 to 1 for IRMA) as x changes from $-\infty$ to $+\infty$. Of the many different sigmoidal curves likely to agree adequately with experimental data, experience of bioassay suggests

$$F(x) = \frac{1}{1 + e^{-2Y}} \qquad (16.7.3)$$

and

$$F(x) = \int_{-\infty}^{Y} \frac{1}{\sqrt{2\pi}} e^{-\frac{1}{2}t^2} dt, \qquad (16.7.4)$$

where

$$Y = \alpha + \beta x, \qquad (16.7.5)$$

as especially suitable. To any logistic sigmoid of the form (16.7.3) can be found a corresponding Gaussian of form (16.7.4) that has slightly different α, β and that is scarcely distinguishable in shape, except for behaviour close to the asymptotes (c.f. § 17.14). Rodbard used the logistic, though in accordance with American practice he favoured Y in place of $2Y$; the convention here makes corresponding pairs of α, β for (16.7.3) and (16.7.4) more nearly equal. When a comprehensive computer program is used, internal formulae must be consistent with the Y or with the $2Y$ convention, but the assayist need not be aware which definition is being employed. Rodbard's sign convention is also different: here $\beta < 0$ for RIA, $\beta > 0$ for IRMA. A particular version of the model, with $C = 0$, was mentioned in § 3.8. Although other sigmoids are also considered in Chapter 17, unless biochemical or biophysical theory eventually points to some alternative the practical choice for RLA seems to lie between the logistic and the normal. The logistic has won acceptance among many practitioners, and is perhaps to be preferred on general grounds; it is adopted for subsequent sections, but no statistical analysis is likely to discriminate between the two in respect of goodness of fit to data.

Hereafter attention is restricted to RIA, as the changes required for IRMA are too obvious to need repeated comment; the notation is always that introduced above, except where explicit reference is made to Rodbard. Theory and a numerical example (§ 16.10) will be based on the logistic, but experience of comparing (16.7.3), (16.7.4) shows clearly that only very unusual data will give appreciably different potency estimates when analyzed in the two ways (cf. Finney, Holt and Sheffield, 1975). From equations (16.7.2) and (16.7.3)

$$\frac{U-\overset{N}{C}}{\underset{B_0}{D}-\underset{N}{C}} = \frac{1}{1+e^{-2Y}}. \tag{16.7.6}$$

Thus the proportionate position of U between its asymptotes is a logistic function of Y. The metameter

$$Y = \tfrac{1}{2}\ln\left(\frac{U-C}{D-U}\right), \tag{16.7.7}$$

is termed the *logit* of $\{(U-C)/(D-C)\}$.

At this point, a summary of correspondences between notations may be useful. The changes advocated here are not devised unthinkingly; they are an attempt to harmonize RLA practices with that customary in bioassay and used throughout this book. The most important differences are summarized in Table 16.7.1. Note the different meanings of Y in columns 3 and 4 and the retention of Rodbard's logit definition in column 3. Because

$$\begin{aligned}
\text{logit}\left(\frac{Y-d}{a-d}\right) &= -\text{logit}\left[1-\left(\frac{Y-d}{a-d}\right)\right] \\
&= -\text{logit}\left(\frac{a-Y}{a-d}\right),
\end{aligned} \tag{16.7.8}$$

conventions can be reversed by working in terms of the complement of $\{(Y-d)/(a-d)\}$; this would not affect potency estimates, but could confuse the pattern of signs or interchange the roles of C and D (Rodbard's d and a). Hereafter the symbols in column 4 are taken as standard.

In terms of expectations, any of the representations on the last four lines of Table 16.7.1 may be used, and various ways of displaying them graphically as straight lines or as simple curves are effectively equivalent. Users of radioligand assays, however, have often fitted equations to empirical functions of the counts. For example, if N, B_0 and B refer to actual counts or to means over replicates, an empirical logit or the function

$$\ln\left(\frac{B-N}{B_0-B}\right) \tag{16.7.9}$$

might be calculated for the B at each dose. When plotted against log dose, these quantities may be so close to linearity that a straight line can be drawn by eye to correspond to Rodbard's $b(\log c - \log X)$. Alternatively an unweighted or weighted linear regression equation may be calculated as a step towards estimating his b and c. Because N, B_0 and B are all subject to experimental error, this is not a satisfactory general method; near the ends of the dose scale an occasional observation may have $B > B_0$ or $B < N$. The difficulty is not removed merely by omitting awkward counts: these may be providing evidence that the observed B_0 is smaller than its unknown expectation, or that the observed N is greater

§16.7 GENERAL TRANSFORMATIONS AND RADIOLIGAND ASSAYS

TABLE 16.7.1 Comparison of RIA notations

Name	Usual RIA practice	Rodbard	This book
Dose	X	X or x	z
$\log_{10}$ dose	—	—	x
Lower limit	N	d	C
Upper limit	B_0	a	D
Midrange or ID50	—	c	$10^{-\alpha/\beta}$
'Slope'	—	b	-2β
Bound count	B	Y	U
Bound/initial bound	$\dfrac{B - N}{B_0 - N}$ (*)	$= d + \dfrac{a-d}{1+(X/c)^b}$	$= C + \dfrac{D-C}{1+\exp(-2Y)}$
—	$\dfrac{B - N}{B_0 - B}$	$\dfrac{Y-d}{a-d} = \dfrac{1}{1+(X/c)^b}$	$\dfrac{U-C}{D-C} = \dfrac{1}{1+\exp(-2Y)}$
—	—	$\dfrac{Y-d}{a-Y} = (X/c)^{-b}$	$\dfrac{U-C}{D-U} = \exp(2Y)$
—	—	$\operatorname{logit}\left(\dfrac{Y-d}{a-d}\right) = b(\log c - \log X)$	$Y = \alpha + \beta x$

(*)In much RIA writing, this expression is by convention abbreviated to 'B/B_0', a well-established but potentially misleading usage.

than its unknown expectation, and estimation should take account of that evidence. In most assays, the exact placement of B_0 and N scarcely affects estimates of relative potency based upon the middle region of the response curve, but can be important if all counts on one test preparation are towards the upper or the lower end of the scale. Arbitrary rejection of some data (perhaps even of counts at less extreme doses where $B_0 > B > N$ but B approaches one of the limits) leaves a regrettable uncertainty and subjectivity attaching to individual analyses.

The difficulty can be overcome by the general approach of § 3.13, much as is done for quantal responses in Chapters 17 and 18, introducing working logits and iterating towards an optimal regression equation (Rodbard and Lewald, 1970). A natural development from (16.7.2) is to allow also for errors of estimation in C and D, and to estimate the four parameters simultaneously. Healy (1972) introduced this direct approach to estimation of the parameters by a general least squares computer program; Rodbard and Hutt (1974) discussed further details and generalized from Healy's assumption of a constant variance per count. Among representations of the same relation between doses and expectations of count, choice of the apparently simplest may not be a good start to estimation. Reference back to single independent counts avoids the complications in the correlated error structure of functions such as those in (16.6.1)–(16.6.4). The function (16.7.2), in combination with (16.7.3) and (16.7.5), represents the expectation of each observed count in terms of dose, where dose is regarded as infinite and zero for counts pertaining to C and D respectively. Before the process can be completed, attention must be given to the frequency distribution of observed counts at a dose.

Nothing said here is intended to imply that the models described in this section are adequate for all RLA. Rodbard (1974) has mentioned some circumstances in which simple linearization techniques break down.

16.8 The variance function

The deviation of an RIA count from its expectation, $(u - U)$, is an error that is compounded of many contributions and that almost certainly depends upon the expectation. If the Poissonian error of counts were all that mattered, the variance of u would be equal to U. Rodbard and Cooper (1970) and Rodbard (1971) discussed sources of error, and identified a complex structure. Their formulae for variances are important to full understanding of the assays, but are not themselves suitable for practical use. Many discrete distributions can be adequately approximated by the normal when means are large, and Rodbard found no evidence against approximate normality for the error distribution here.

Rodbard and Cooper proposed that a quadratic or other polynomial function of count or other response metameter could be adequate to represent the variance. They found the same kind of function, with different coefficients, to be applicable to various response metameters. Rodbard and Hutt (1974) developed the idea further. When attention is restricted to raw counts, the function $\phi(U)$ (introduced in § 3.10 to represent $\text{Var}(u)$ at the dose where $\text{E}(u) = U$), must vanish for $U = 0$, and Rodbard's quadratic may be written

§ 16.8 GENERAL TRANSFORMATIONS AND RADIOLIGAND ASSAYS

$$\phi(U) = V_1 U + V_2 U^2. \qquad (16.8.1)$$

For IRMA, Rodbard and Hutt found

$$\phi(U) = V_2 U^2 \qquad (16.8.2)$$

to be adequate.

A variance function is required primarily so that its reciprocal can be used as a weight for observations, and exactness in relative weights is seldom important to the outcome of a statistical analysis (which is fortunate because of the impossibility of ever being certain about the function). Although the variance parameters V_1 and V_2 could be estimated as part of a general statistical analysis, such a procedure is not ideal because a single assay is unlikely to contain adequate information on them. A better practice is to estimate V_2/V_1 from a large body of data and then, subject to occasional checks, to assume this ratio constant. Equation (16.8.1) has weaknesses, notably in that it may behave oddly for small U, and it cannot truly be regarded as intermediate between proportionality to U and proportionality to U^2. A more adaptable alternative (Finney, 1976b) is

$$\phi(U) = VU^J, \qquad (16.8.3)$$

where the index J takes account of the rate at which variance increases with count. Equations (16.8.1) and (16.8.3) can be nearly indistinguishable over a wide range of U; (16.8.3) can also accommodate more extreme patterns of variance by taking J outside the range 1·0 to 2·0, including a constant variance ($J = 0·0$) or a variance decreasing as U increases. Like that of V_2/V_1, estimation of J from a large body of data is desirable (§ 16.10).

Healy's simultaneous estimation of the four parameters assumed raw counts to have equal variances. Others have made the more extreme assumption that equal weights could be attached to all logits; if doses are chosen so as to keep counts well away from the asymptotes, even this can be adequate. Methods that do not take full account of the variances of counts are not necessarily wholly wrong, but they often encourage constraints on choice of doses that are not otherwise desirable. For their originators they may perform well; others may employ them uncritically in routine analyses and neglect limitations on validity. No laboratory can guarantee unfailingly to produce good data, or data always suitable for analyses that neglect the error structure. For example, fourfold replication of counts at doses corresponding to the two asymptotes may suffice for a comprehensive analysis; an approximation that takes C and D to be equal to mean counts at extremes of dose could be misleading without more replication. Today, computation to a standard pattern is so cheap that routine analyses should employ a comprehensive computer program (§ 16.9). Rodbard and Hutt (1974) have devised one full program, and Rodbard (1974) has commented on many aspects of the computations. Cook (1975) reviewed various available computer programs. An important feature of a good program is that it should allow a user readily to modify the variance function $\phi(U)$, perhaps by incorporating a family of functions that conform to his own experience.

16.9 Estimation and computation for RLA

The equation for the response curve was written in the form (16.7.2) because C and D have analogies with the response rates at zero dose and at infinite dose in an assay based on quantal responses (§17.16). No quantal response rate can lie outside the range 0·0 to 1·0, but an observed value can be less than C or greater than D. In RLA, a count has an absolute lower bound of zero but has no upper bound; a particular observed count can lie outside the asymptotic values C and D purely as a result of experimental error, though large deviations should be rare unless caused by gross errors in technique.

Before potency estimation is discussed, statistical techniques for data from a single response curve need consideration. One possibility is to estimate α, β, C and D by minimizing the weighted sum of squares

$$\Sigma \, [(u - U)^2/\phi(U)], \tag{16.9.1}$$

where summation is over all relevant counts and the appropriate form of U, equation (16.7.2), is to be inserted. If u can be assumed normally distributed about U, a relatively simple alternative is to maximize the likelihood, which is equivalent to maximizing

$$-\tfrac{1}{2} \Sigma \log [\phi(U)] - \tfrac{1}{2} \Sigma \, [(u - U)^2/\phi(U)]. \tag{16.9.2}$$

As is well known, the two methods of estimation are asymptotically equivalent, essentially because in the neighbourhood of the extremum the first term of (16.9.2) is 'constant to a higher order' than the second term. One practical advantage of least squares is that a general factor of proportionality in the variance function – V_1 in (16.8.1), V in (16.8.3) – is a divisor of the whole expression (16.9.1) and is therefore estimated by a residual mean square, whereas in (16.9.2) this factor enters differently into the first term and must therefore be included in the full process of optimization. For the present, weighted least squares will be chosen, but maximum likelihood and another alternative are compared with it in §16.11.

The classical method of numerical optimization for a non-linear function such as (16.9.1) has been illustrated in §16.3, being an adaptation of the general theory in §3.13. Especially when many parameters are involved, other approaches may make better use of modern computing facilities. The program MINISQUARE, mentioned in §16.5, has been used for the example in §16.10; both this and a related program, MAXLIKE, use simplex procedures (Nelder and Mead, 1965) to aid early stages of iteration and also simplify the arithmetic by using finite differences to estimate differential coefficients. Because the main burden of calculation can now be carried in this way, details of steps analogous to those of §16.3 need not be displayed here.

In potency estimation, the line

$$Y = \alpha + \beta x \tag{16.9.3}$$

plays its familiar parallel line role. It is estimated by, say,

§16.9 GENERAL TRANSFORMATIONS AND RADIOLIGAND ASSAYS

$$Y = a + bx \tag{16.9.4}$$

(a, b correspond to α, β and are not Rodbard's symbols). In order to estimate the dose for any specified fraction bound P, the value of Y corresponding to P must be determined from

$$P = \frac{1}{1 + e^{-2Y}} \tag{16.9.5}$$

or

$$Y = \tfrac{1}{2} \ln \left(\frac{P}{1-P} \right), \tag{16.9.6}$$

as is evident from (16.7.2) and (16.7.3). The numerical value of Y is inserted in (16.9.4) and x is calculated. In particular, for $P = 0.5$ and $Y = 0.0$,

$$m = -a/b \tag{16.9.7}$$

is the estimated log dose, and the corresponding absolute dose estimates Rodbard's parameter c. Fieller's theorem allows fiducial limits to m or another such x to be calculated from the variances and covariance of a and b.

Rodbard's parametrization leads directly to estimation of dose for 50 percent bound, though Fieller's theorem is still needed in calculating limits to a ratio of two values of c used as estimating a relative potency. For good RLA data, the sampling distribution of such a direct estimate of the ID50 should be close to symmetry and perhaps normality. However, if particular data estimate Rodbard's b (or the present β) as only a few times greater than its standard error, the sampling distribution may be severely skewed, as is intuitively apparent from (16.7.1), exactly the conditions that make Fieller's theorem essential and approximate standard errors inadequate to the calculation of limits to m. A general method that avoids risks from this source is desirable. Assayists once needed to seek ways of avoiding the laborious calculations inherent in the use of an estimator such as m; today the extra arithmetic is no more than a few lines in a program, never seen by the routine user and making no appreciable difference to computer time.

For an assay, observations must be made simultaneously on doses of S and T. The condition of similarity readily shows that β, C and D must be the same for both. Thus in terms of Y the situation is akin to that of simple parallel line assay: the two preparations have versions of (16.9.3) differing only in α, and

$$M = (a_T - a_S)/b \tag{16.9.8}$$

estimates the logarithm of relative potency. The procedure for estimating the parameters is extended so as to estimate $\alpha_S, \alpha_T, \beta, C, D$ simultaneously from the data, with asymptotic variances and covariances as by-products. The general form of Fieller's theorem puts limits on $\log \rho$.

A common RLA practice is to test enough doses of S for good estimation of the response curve but to include only a single dose of T (perhaps with replicate determinations of count at this dose). Hence β, C and D are estimated solely from the S data; the estimate of α_T is the value of a_T that, in combination with the other parameters, puts the curve for T through the mean count at the one dose. Many different test preparations may be assayed relative to the same data on S, with just the one dose for each T. In principle, this is open to the same objections as it would be in a simple parallel line assay (cf. §§ 3.14, 3.16): belief in a common β cannot be tested. If the observed counts for T are close to the estimates of C or of D, the small change in response that corresponds to any stated dose change means low precision for potency estimations. If the mean count for a particular test preparation exceeds D as estimated from the S data alone, the potency must be estimated as zero, but the counts encourage some upward revision of the estimate of D. Similarly, if the mean count for T is less than C, the potency estimate is infinite but the estimate of C needs to be revised downwards. Such data clearly indicate that the dose of T was ill-chosen.

Nevertheless, the statistician should refrain from unthinking condemnation of the single dose design. The requirement of a common β is a condition for specificity of the assay, necessary but by no means sufficient to secure validity. In the development of a new RIA technique and in pilot studies, inclusion of several doses of one or more test preparations seems essential (Midgley, 1966). When a technique becomes adopted as a routine, other steps will have been incorporated in order to give full confidence in specificity, and often understanding will be such that doses can be well-chosen. An additional consideration, encouraging single dose designs both for sound and for illegitimate reasons, is the need for routine assay of large numbers of samples for urgent clinical purposes. Insistence on multi-dose designs reduces the number of different test preparations that can be incorporated in one assay (commonly 50 or more), and may encounter the difficulty that under clinical conditions some samples for assays are severely limited in size. In some RIA, trial of more than one dose may be impossible because no suitable diluent can be found. Decision on these matters should rest with investigators who know the quality of their data, the closeness of agreement with a response curve that they can achieve, and, above all, the success with which they can use existing knowledge to choose a dose.

Whatever may be done with routine assays, in a RIA for research purposes the practice of including at least two doses of each test preparation should obtain. Assayists should also keep in mind the possibility that, even when most test preparations must be restricted to one dose, a few might still be allowed a second dose as a contribution to quality control on the assay system. There is no reason why a computer program should not deal easily with such a flexible design. At the stage of statistical analysis, a further feature that may require incorporation is the possibility to discard from definitive analysis preparations that show anomalous values of β in a preliminary computation. Good programming must serve the needs of assayists, so that simplicity of computation need never be a consideration when the number of doses per preparation is being decided. The statistician should seek systematization of computation, and availability of programs that estimate all parameters simultaneously and speedily, as a stimulus to use of designs in accord with general principles of good experimental design wherever they are practicable.

Banting (1974) adopted an analogous model for bioassay of herbicides, his response being fresh weight of herbage. He proceeded by analogy with quantal response techniques, and was able to obtain results that aided the comparison of different chemical compounds. A method closer to those of §§ 16.6–16.9 but with the constraint of $C = 0$ would be more consistent with general principles; the variance function would need study. Nevertheless, Banting's analyses were an imaginative approach to the interpretation of herbicide data, and they illustrate well the potential for assay techniques of the kind discussed in this chapter.

16.10 Assay of oestradiol

Results from a series of RIA of oestradiol, follicle-stimulating hormone (FSH) and luteinizing hormone (LH), kindly supplied by Professor S. Z. Cekan, will be used to illustrate the method of statistical analysis (Finney, 1976b). Counts ranged from under 50 to over 20 000. From replicate counts at a dose, sometimes as many as 6 or 8 and sometimes only 2 or 3 being available, means and variance estimates were calculated for 345 dose groups in all. Despite enormous scatter because of the few degrees of freedom, a rough plotting of variance against the mean showed clearly the marked increase in Var(u) as U increases. To calculate a quadratic regression, or a linear regression for a doubly logarithmic plot, would be improper since both variates are subject to error, though the insensitivity of subsequent procedures to the value of V_2/V_1 or of J suggests that refinement may be unimportant. Inspection of the logarithmic diagram suggested that equation (16.8.2) would be a reasonable approximation to the variance function, since it is equivalent to (16.8.3) with $J = 2 \cdot 0$, though $J = 1 \cdot 5$ looked better. Although the FSH and LH counts were in general greater than those for oestradiol, the data do not indicate heterogeneity in J attributable to the hormones or to the contrast between standard and test preparations. My colleague Miss P. Phillips set up a full maximum likelihood estimation appropriate to equation (16.8.3) and the approximation of normal distribution for counts; this confirmed the homogeneity of the various sets of data and gave $J = 1 \cdot 3$ as about the maximum likelihood estimate, but can tell nothing about bias. Equation (16.8.1) can be discussed and estimated similarly.

TABLE 16.10.1 Counts recorded for the standard curve in a radioimmunoassay of oestradiol

Dose (units of 10 pg)	Counts				Means
0	1627	1567	1720	1660	1681
	1704	1689	1759	1722	
0·625	1182	1291	1294	1312	1270
1·25	1029	1112	986	1074	1050
2·5	702	784	733	777	749
5·0	486	485	501	460	483
10·0	307	277	285	275	286
20·0	196	164	193	182	184
non-specific	25	39	51	38	38

Each of the assays of oestradiol, conducted as described by Aso *et al.* (1975), had four-fold replication for direct estimation of the non-specific count, eight-fold replication at zero dose, and four-fold replication at 6 doses of the standard preparation, rising by factors of 2 from 6·25 pg to 200·0 pg. The response curve shown by these 36 counts was to be used for estimating the potency of 40 or more test preparations, each tested in duplicate at a dose of 0·2 ml plasma. Table 16.10.1 shows the data for one of the standard curves.

As with other assays, the statistical analysis will be most satisfactory if the logarithmic scale for doses is adjusted so that the weighted mean dose is near to zero. All the original measures of dose were divided by 10, so giving a unit of 10 pg. This has no algebraic consequence, but it increases numerical accuracy by enabling compensating changes in $\text{Var}(a)$ and $\text{Cov}(a, b)$ to reduce loss of leading digits in subtraction. Table 16.10.2 shows estimates obtained with weighting proportional to the reciprocal of $\phi(U)$ with $J = 2·0, 1·5, 1·0$, and $0·0$ in equation (16.8.3). As in §16.5, the computations used the program MINISQUARE which produces a matrix of asymptotic variances and covariances essentially as in §3.13.

TABLE 16.10.2 Estimated parameters for the standard response curve from Table 16.10.1 [using equation (16.8.3) and weighted least squares]

Estimator	$J = 2·0$	$J = 1·5$	$J = 1·0$	$J = 0·0$
a	0·3167 ± 0·0553	0·3252 ± 0·0326	0·3299 ± 0·0249	0·3318 ± 0·0237
b	−1·1604 ± 0·0458	−1·1689 ± 0·0319	−1·1747 ± 0·0297	−1·1747 ± 0·0460
$\hat{C}$	40·55 ± 2·00	40·11 ± 2·50	39·83 ± 4·26	38·36 ± 18·69
$\hat{D}$	1688·30 ± 54·72	1683·83 ± 28·19	1681·46 ± 19·23	1679·80 ± 14·55

Because each test preparation was used only at a single dose, estimation was akin to a standard curve procedure (§3.14), with the complication that the additional parameters C and D affect the metametric transformation of responses before these are related to the regression estimated for S. The amount of calculation involved in the method of §§16.7 and 16.9 can be reduced by a different but equivalent approach: M is the difference between X, the log dose of T, and the log dose of S at which the count indicated by the fitted S curve equals the mean count observed for T. Variance calculations, though more complicated, require little programming once the formulae have been developed.

The formulae are most easily obtained by working in terms of the logit transformation, though the same results can be obtained in various ways. Suppose that for T at a dose with logarithm X, n independent counts have mean $\bar{u}$. Then, taking $a, b, \hat{C}, \hat{D}$ as estimated from the S data, the logit is calculated as [*]

---

[*] Those who follow the other convention on the definition of the logit will omit the factor $\frac{1}{2}$ in (16.10.1), and in consequence should omit the factors 4, 2 and 2 from the left-hand sides of (16.10.4), (16.10.5) and (16.10.6). Their values of α, β and the estimates a, b of these parameters will be double those obtained here, but C, D are unaltered. All other formulae are self-compensating. When logit calculations are embodied entirely within a computer program, the programmer's definition is immaterial to the user.

§16.10 GENERAL TRANSFORMATIONS AND RADIOLIGAND ASSAYS

$$y = \tfrac{1}{2} \ln \left[\frac{\bar{u} - \hat{C}}{\hat{D} - \bar{u}} \right], \tag{16.10.1}$$

whence the usual formula for log potency is

$$M = -X + (y - a)/b. \tag{16.10.2}$$

The variance of this y must involve $\hat{C}, \hat{D}$ and their variances as well as the variance of $\bar{u}$. The last is estimated by

$$\phi(\bar{u}) = V\bar{u}^J/n, \tag{16.10.3}$$

where the multiplier V is taken as the residual mean square in the least squares calculations for the standard curve. Well-known procedures for obtaining a first-order approximation to the variance of the function in (16.10.1) give

$$4 \operatorname{Var}(y) = \frac{\operatorname{Var}(\bar{u})(\hat{D} - \hat{C})^2}{(\bar{u} - \hat{C})^2(\hat{D} - \bar{u})^2} + \frac{\operatorname{Var}(\hat{C})}{(\bar{u} - \hat{C})^2} + \frac{2 \operatorname{Cov}(\hat{C}, \hat{D})}{(\bar{u} - \hat{C})(\hat{D} - \bar{u})} + \frac{\operatorname{Var}(\hat{D})}{(\hat{D} - \bar{u})^2}. \tag{16.10.4}$$

Similarly

$$2 \operatorname{Cov}(y, a) = -\frac{\operatorname{Cov}(a, \hat{C})}{\bar{u} - \hat{C}} - \frac{\operatorname{Cov}(a, \hat{D})}{\hat{D} - \bar{u}}, \tag{16.10.5}$$

$$2 \operatorname{Cov}(y, b) = -\frac{\operatorname{Cov}(b, \hat{C})}{\bar{u} - \hat{C}} - \frac{\operatorname{Cov}(b, \hat{D})}{\hat{D} - \bar{u}}. \tag{16.10.6}$$

In manual calculation, the economy of neglecting the covariances and all but the first term of the variance will seldom do much harm. The full formulae of the present section are easily programmed for a computer; unnecessary approximations should not be introduced into a good program (§ 16.11). In relation to the large investment now made in RLA, the effort of systematically organizing calculations for limits of error of potency estimates is trivial. The precision of these assays is often high, but not – as some have assumed – so high that errors of estimation can properly be overlooked. From these constituents can be formed:

$$\operatorname{Var}(y - a) = \operatorname{Var}(y) - 2 \operatorname{Cov}(y, a) + \operatorname{Var}(a), \tag{16.10.7}$$

$$\operatorname{Cov}(y - a, b) = \operatorname{Cov}(y, b) - \operatorname{Cov}(a, b). \tag{16.10.8}$$

All these formulae are obtainable from equation (16.10.3) and the covariance matrix arising in the estimation of $a, b, \hat{C}, \hat{D}$. Together with $\operatorname{Var}(b)$, equations (16.10.7) and (16.10.8) enable Fieller's theorem to be applied to the ratio $(y - a)/b$, so leading to limits for M.

Four oestradiol test preparations, representative of various regions of the response curve, have been chosen to illustrate calculations based on the values of J used for Table 16.10.2; the results are summarized in Table 16.10.3. As is to be expected, the fiducial limits are relatively wider towards either extreme of dose, and almost explode for preparation 9. Most preparations gave mean counts in the range 200–1200, corresponding to potencies between the extremes of dose used for S (Table 16.10.1). Preparations 9 and 10 show the behaviour of

$$M = 2 \cdot 3374$$

with limits at 2·2710, 2·4034. These correspond exactly with the results summarized in Table 16.10.3, an estimate of 217 pg/ml, with limits at 187 pg/ml and 253 pg/ml.

Although $\phi(U)$ has been taken to be given by (16.8.3), the alternative of (16.8.1) would not have made any appreciable difference, especially when J approximately 2·0 is compared with a very small value of V_1/V_2.

Use of equation (16.10.1) is impossible if $\bar{u}$ chances to be less than $\hat{C}$ or greater than $\hat{D}$. For well-chosen doses in a well-conducted assay, this is unlikely, but it could happen where counts were unusually variable or where misjudgement of the potency of T caused choice of a dose giving very high or very low counts. A mean count outside the range of $\hat{C}$ and $\hat{D}$ would suggest that errors of estimation in the standard curve have led to $\hat{C}$ being too great or $\hat{D}$ being too small. If one test preparation in a set of 50 behaves in this manner, the best policy is to write it off as having a very high or a very low potency and then, if possible, to include it at one or more different doses in a new assay; the current information is too poor to permit any reasonably precise estimate of potency (§ 16.9).

16.11 Alternative principles of estimation

Earlier sections have been presented in terms of minimization of the weighted sum of squares, expression (16.9.1). The alternative of maximizing the log likelihood, expression (16.9.2), has been asserted to be almost equivalent. Another possibility needs consideration, that of avoiding explicit use of the variance function by transformation of responses; this is possible with (16.8.3) but not with (16.8.1).

Statistical texts advise that if observations for which the variance is proportional to the expectation are transformed by taking the square root, and observations for which the standard deviation is proportional to the expectation are transformed logarithmically, the variance will be stabilized. A generalization (Finney, 1976b) for equation (16.8.3) is that to a first-order approximation $u^{1-J/2}$ will have constant variance (if J tends to 2·0, the transformation effectively approaches $\ln u$ as a limit). Though earlier discussion has assumed a normal distribution of u about U, the only evidence is that Poisson-distributed counts rapidly approach normality for large expectation. The alternative of an assumed normality of distribution of $u^{1-J/2}$ about $U^{1-J/2}$ with constant variance deserves trial. This requires minimization of

$$\Sigma (u^{1-J/2} - U^{1-J/2})^2. \tag{16.11.1}$$

Table 16.11.1 summarizes, for $J = 1 \cdot 5$ and for the standard preparation in the oestradiol assay, estimates of parameters by maximum likelihood and by transformation to constant variance (here $u^{0 \cdot 25}$). As is evident, differences from Table 16.10.2 are small. Similar calculations for other values of J fully support the conclusion that even for estimating α, β, C, D the three principles are effectively equivalent. Results comparable with Table 16.10.3 have not been computed for maximum likelihood or transformed u, but they would be virtually indistinguishable. Computing time was no shorter for transformed u than for weighted least squares, any time saved on weighting being lost on repeated transformation of U. Maximum likelihood was slower because one more parameter had to be estimated.

TABLE 16.11.1 Estimated parameters for the standard response curve from Table 16.10.1 [using equation (16.8.3), $J = 1\cdot 5$, and two methods of estimation]

Estimator	Maximum likelihood	Transformed u
a	$0\cdot3248 \pm 0\cdot0305$	$0\cdot3246 \pm 0\cdot0335$
b	$-1\cdot1693 \pm 0\cdot0298$	$-1\cdot1646 \pm 0\cdot0328$
$\hat{C}$	$39\cdot69 \pm 2\cdot35$	$37\cdot57 \pm 2\cdot43$
$\hat{D}$	$1681\cdot11 \pm 26\cdot55$	$1681\cdot72 \pm 29\cdot08$

All iterations converged satisfactorily for $0 \leqslant J \leqslant 2\cdot 0$. Need for $J < 0$ is almost inconceivable. Out of curiosity, $J = 2\cdot 5$ was tried for the oestradiol data: convergence seemed rather sensitive to good choice of starting values, especially for maximum likelihood and least squares, but this could be because so large a value of J was strongly in conflict with the data rather than because of any general properties. In practice, $J > 2\cdot 0$ seems unlikely.

16.12 Design for RLA

General comments about experimental design for parallel line and quantal response assays in Chapters 6 and 19 are applicable to RLA with the logit model. The consequences of a variance monotonically dependent upon expected count are not easily specified but must not be ignored. Counts at zero and infinite doses must be well replicated in order to estimate C and D precisely. On the other hand, most of the information on ρ comes from middle regions of the response curve, where the regression of count on log dose is not far from linearity. Hence replication at extremes of dose needs to be sufficient to ensure convergence of iterations but not excessive at the expense of more informative intermediate doses. Refinement of RLA techniques may one day permit doses to be chosen from a wide but effectively linear section of the curve; without the need to estimate C and D, resources could then be efficiently redeployed. Presumably the many long-established parallel-line assays in which the regression of response (e.g. blood sugar percent, uterine weight) on log dose is linear within a usable range must in reality have upper and lower asymptotes at extremes.

Except as a very temporary expediency, the decision whether to have several doses of each test preparation (§ 16.9) should never rest upon ease of subsequent statistical analysis. The duty of statisticians, and of the programs they write, is always to analyze data from investigations planned to give maximum information, not to demand that the form of data be planned to suit their convenience.

16.13 Programs for RLA

Especially when RLA are employed for clinical purposes, large numbers may be conducted regularly as a laboratory routine, usually with many test preparations in each. Statistical analysis by desk calculator is laborious (even though approximations and subjective adjustments simplify the arithmetic), the more so if potency estimates are to be given limits of error. Analysis by computer must today be regarded as the standard. In various centres, minicomputers have been programmed to handle the work, but commonly they lack the capacity or the output facilities desirable for full study and interpretation of large routine assays. Many proposals for major computer programs have been made (Cook, 1975), notably in a series of papers by Rodbard and his colleagues. Rodbard's program (1974), closely related to the methods described above but differing in some respects, goes far towards providing a general facility. Because of steadily improving understanding, changes in experimental practice, and developments of design, no program yet written is likely to represent the final word.

Experience in many fields has taught that a computer program for extensive routine use must

(i) accept input readily in the form (or forms) in which data are naturally recorded and stored;
(ii) provide output that is compact, well arranged and documented, easily understood, and as far as possible suitable for passing immediately to the user of the results;
(iii) be versatile as well as arithmetically accurate, with particular attention to any features of the data (whether known in advance or detected only during analysis) that differ from the standard pattern, by providing where appropriate either a modified analysis or rejection with helpful comments;
(iv) in all respects be designed for use by non-statisticians, without fear that gross error may result from uncritical acceptance.

The labour of writing and documenting a general program lies mainly in the input, internal organization, and output, and not in complications of formulae. Such steps as taking equation (16.10.4) in full instead of only its first term also have negligible effects on computing time.

There are more positive reasons for calculating limits of error from the full procedure described in §§ 16.9 and 16.10. Even under good conditions (the oestradiol assay is an example) an occasional test preparation may give very high or very low counts: less experienced laboratories may use a standard program uncritically, lacking the statistical expertise to tell where special care is needed. Also, the logistic and other sigmoidal response curves model under discussion may in the future be found appropriate to other assays that have intrinsically greater variances and that cannot so readily tolerate a neglect of various terms. Proper attention to all the terms and to the full formulation of Fieller's theorem is a considerable protection against unguarded inference in any of these circumstances. The approximation may give pleasingly narrow but totally misleading limits of error; the full computations, on the other hand, may point to the truth

§16.13 GENERAL TRANSFORMATIONS AND RADIOLIGAND ASSAYS

by "explosive" behaviour, excessively wide limits indicating a breakdown of the estimation process that can then be examined.

This chapter does not present any new general program, but it comments on widely applicable principles. Though written with respect to curves that can be included under equation (16.7.2), with a variance functionally dependent on the expected count, no claim is made that every RLA can be so analyzed.

Although the logistic regression function is implicit in the analysis, logit values need not be explicit to the user of a program. To start the iterative phase of a program such as MINISQUARE, the only requirement is a set of rough values for the parameters, and 'steps' associated with each so as to define an initial range for search. Starting values can be taken from a rough sketch of the curve. Alternatively, C and D may be guessed as equal to the highest and lowest counts recorded, β is commonly near to $-2 \cdot 3$ for RIA, near to $2 \cdot 3$ for IRMA[*], and α can be taken as $-\beta x_m$, where x_m is the logarithm (to base 10) of the dose for the median value of the recorded counts. The step to be associated with each parameter as a start for the iteration might be roughly 10 percent of the value taken for the parameter, with a minimum of $0 \cdot 1$ for α. A good program should converge even from a bad start and ultimately attain estimates that minimize the weighted sum of squares of deviations of u from U. Experience of an assay system, however, should lead to improvement in the tentative suggestions made here. The minimum itself leads to an estimate of the residual variance per unit weight with degrees of freedom 4 less than the number of counts. In the usual way, the inverse matrix of second differential coefficients leads to a matrix of multipliers for the residual variance that gives variances and covariances for the parameters. Although the second differential coefficients can be estimated numerically from second differences, a definitive program ought to include evaluation of the true mathematical functions. Healy (1972) described and illustrated what is essentially this system of computation, simplified by rejecting outliers and then assuming $\text{Var}(u)$ to be constant.

The method in §16.10 and the argument above need revision when the assay uses more than one dose of each test preparation. This improvement in assay design makes possible additional tests of validity, though it may be prevented by other considerations (§16.9). As noted earlier, an assay must have the same β, C, D for all preparations, but each preparation has its own α; an assay that includes p test preparations has in all $(p + 4)$ parameters for the response curves. The residual sum of squares can be compared with the smaller minimum achievable when each preparation is allotted its own parameters, so giving a validity test of a familiar kind for the homogeneity of β, C and D. If $p \leq 6$, simultaneous optimization in respect of the parameters is practicable without any major change in program. For larger p, the requirement for inverting $(p + 4) \times (p + 4)$ matrices might seem excessive, but fortunately they have a pattern that permits convenient computations. Many optimization programs require inversion as part of

---

[*] Corresponding to $-1 \cdot 0$, $1 \cdot 0$ if x were taken as the natural logarithm of dose; §16.7 commented on this as a special model, and more generally it provides a good start.

Newton–Raphson or similar calculations in each cycle of iteration. The simplex procedure at the heart of MINISQUARE does not itself use matrix inversion, though the program as a whole also requires some Newton–Raphson type optimizations. The parameters C and D are usually more easily estimated than the others, and a two-phase alternation of iteration can be effective. Provisional estimates of C and D are held fixed during an iteration for β and the α's that can be arranged much as the classical probit or logit calculations with quantal responses (§§ 18.1, 18.2); next the estimates of C and D are revised with β and the α's fixed, and alternation continues as long as necessary.

The ideal program structure depends upon the design of assay and the quality of data for which it is used. If only one dose per test preparation is customary, the right approach is simple, and even with several doses no great problem should be encountered for small p. Routine analysis of assays with two or more doses per preparation is not only inherently more expensive in computing but also requires new features. Unless laboratory conditions are well controlled and materials assayed are thoroughly familiar, an occasional preparation may show strong evidence that it differs from the standard and most others in one or more of the parameters β, C, D. Some such indications may be dismissed as due to random variation in counts, but extreme instances demand exclusion from the analysis lest conclusions on other parameters be distorted. A comprehensive program therefore should facilitate preliminary screening of the data, so that gross anomalies can be automatically excluded and others slightly less extreme drawn to the attention of the assayist for decision before definitive analysis begins.

Other suggestions have been made. For example, the p test preparations can be put into p separate assay calculations each with the same data for S. This is simple but clumsy, and it fails to make full use of all information on β, C, D. It is adopted by many people, but with modern computing facilities it is unlikely to save time and does not merit acceptance without fuller study. Similar devices for evading the problem of a truly comprehensive program are perhaps valuable temporary expedients: analysis as p separate assays is usually preferable to reliance on freehand graphical methods and neglect of limit calculations, adopted because alternatives are thought to be not available or too difficult.

As in many other fields, the need for a comprehensive program, with many safeguards built into it, is greatest where statistical expertise is least. If uncritical acceptance of what emerges from the computer is prevented by expert scrutiny, a program of relatively unsophisticated structure and content may be tolerable. If a laboratory has great biological competence but no staff with statistical skills, RLA provides scope for dangerous misunderstandings that may be aggravated by inadequate programs. The extent to which RLA is used surely justifies substantial investment in developing one or more large and comprehensive programs; the planning of these requires statistical wisdom, programming skill, and of course very close association with those who practise RLA. I have recently written a program RADIMM that is useful for research, but that has primarily served to teach me that, for routine assays, the main programming problems lie in input, scrutiny of data, and quality control.

17
Quantal responses and the tolerance distribution

17.1 The use of quantal responses

For some stimulus–subject systems, measurement of a response attributable to the action of the stimulus is impossible or impracticable; all that can be done is to record whether or not the subject manifests a certain reaction. The quantal response so used can be death or any other easily recognizable change in the subject. Thus, insecticides may be assayed by assigning batches of insects to doses of S and T and then analyzing the relation between death-rate and dose. Other examples are the assay of fungicides by means of spore germination rates, the assay of insulin by the mouse convulsion method, the assay of oestrogenic hormones by techniques dependent upon the induction of oestrus in test subjects, and the assay of vitamins by their success in curing deficiency symptoms. Many types of quantal response used for assays involve irreversible change in the subjects that respond, so that each subject can be used once only: an insect that has died after spraying, or a spore that has germinated, cannot be used again. Even subjects that have failed to respond may have been so affected by the stimulus that thereafter they react differently from others not previously exposed to the stimulus. Some responses, however, may imply no permanent effect on the subjects. Increased precision is then to be expected from a design that allows several doses to be tried on each subject. For the present, attention will be restricted to assays in which each subject is used once.

Quantal response assays have an important relation to direct assays. The *tolerance* of any one subject is defined as the dose that would just suffice to produce the characteristic response. To any lesser dose the subject will fail to respond, to any greater dose it will respond. The tolerance for any one subject may vary with time, but the experimenter can discuss only the value that it has at the instant of his test. In a direct assay, the tolerance is measured for each subject, and the estimation of potency is a comparison of mean tolerances. In a quantal response assay, all that can be done with each subject is to apply a selected dose, to observe whether or not the response occurs, and so to record whether the tolerance was less or greater than the dose; the objective is still to make inferences about mean tolerances. To measure directly the amount of an insecticide that will kill an aphid, or the amount of oestradiol that will just inhibit pregnancy in a rat, is impossible, yet assays based on these responses are similar in purpose to the 'cat' method for digitalis assay (§§ 2.2, 2.3). Information on whether the tolerance of each of a set of subjects is greater or less than a specified dose (not necessarily the same dose for every subject) is clearly less valuable than information on direct measurements of tolerance for the same parameters. This represents the distribution of individual log tolerances in terms of two parameters, one for location and one for scale. As usual,

$$x = \log z \qquad (17.4.2)$$

is the dose metameter, and

$$\int_{-\infty}^{\infty} f(\theta) \, d\theta = 1. \qquad (17.4.3)$$

Either or both tails of the distribution may have finite limits, but for convenience of general theory these may be regarded as contained within infinite limits. The normal tolerance distribution

$$dP = \frac{1}{\sigma\sqrt{2\pi}} \exp\left\{-\frac{(x-\mu)^2}{2\sigma^2}\right\} dx \qquad (17.4.4)$$

is one example of (17.4.1), having its mean

$$\mu = -\frac{\alpha}{\beta} \qquad (17.4.5)$$

and its standard deviation

$$\sigma = \frac{1}{\beta}. \qquad (17.4.6)$$

The probability that a dose whose measure on the metametric scale is x will cause the characteristic response in a subject chosen at random is the probability that the tolerance of the subject is less than x, or

$$P = \int_{-\infty}^{x} \beta f(\alpha + \beta x) \, dx. \qquad (17.4.7)$$

In relation to the specified density function, define Y, the *equivalent deviate* of P, by

$$P = \int_{-\infty}^{Y} f(\theta) \, d\theta. \qquad (17.4.8)$$

A value of P or of Y determines the other uniquely. The relation between them can be written as a numerical table that gives a metametric transformation of the response rate P. From (17.4.7) and (17.4.8),

$$Y = \alpha + \beta x, \qquad (17.4.9)$$

which equation completely characterizes the tolerance distribution within the agreed family.

The procedure for estimating α and β may be regarded as finding a linear relation between x and the Y-transform of the probability of response; the probability is itself estimated by p, the proportion of subjects observed to respond to a particular dose, and from these empirical values an equation

$$Y = a + bx \qquad (17.4.10)$$

may be calculated as an estimate of (17.4.9). With the aid of an iterative process like that of § 3.13, the computations can be made very similar to those for a

§17.5   QUANTAL RESPONSES AND THE TOLERANCE DISTRIBUTION

linear regression. Any other quantity relating to the tolerance distribution (such as the mean tolerance or the median effective dose) may easily be estimated by using a, b in place of α, β. The particular forms of distribution considered in this chapter are symmetric, so that mean and median coincide, but that need not be so. For the normal tolerance distribution (17.4.4), equation (17.4.8) gives Y as the *normal equivalent deviate* (N.E.D.) of P (Gaddum, 1933).

In an assay, each preparation tested will have corresponding to it an equation like (17.4.9). From the considerations advanced in §2.5 as pertaining generally to dilution assays, or from the condition of similarity (§3.3), the values of β must be identical. Hence the 'regression' lines given by (17.4.9) must be parallel, except for deviations attributable to sampling variation. The log relative potency is again estimated by the horizontal distance between two parallel lines. In fact, equation (4.11.4) will apply, though the estimation of the lines and the assessment of precision introduce new features. In this chapter, only the estimation of equation (17.4.9) for a single preparation will be discussed; the simple extension of theory needed for assay purposes can properly appear with the instructions for assay computations (§18.1).

17.5 Estimation of the two parameters

Suppose that n subjects receive a dose whose logarithm is x, under conditions that make the reactions of each subject independent of all others, and r subjects respond. The *binomial distribution* states the probability of this result relative to all possible results with n subjects:

$$P(r) = \binom{n}{r} P^r Q^{n-r}, \qquad (17.5.1)$$

where $P(=1-Q)$ is defined by equation (17.4.7) as a function of x involving the parameters α, β, and

$$\binom{n}{r} = \frac{n!}{r!(n-r)!}. \qquad (17.5.2)$$

Typical data come from independent trials at a series of ν doses, so that to the value x_i of the dose metameter correspond n_i, r_i ($i = 1, 2, \ldots, \nu$). The method of maximum likelihood is one of several mathematical techniques that may be adopted in order to develop a computational procedure (§§3.13, 20.3). Writing S for summation over i, the logarithm of the likelihood becomes

$$L = \text{constant} + S\{\ln P(r_i)\}$$
$$= \text{constant} + S(r_i \ln P_i) + S\{(n_i - r_i) \ln (1 - P_i)\}, \qquad (17.5.3)$$

where

$$P_i = \int_{-\infty}^{x_i} \beta f(\alpha + \beta x)\, dx$$
$$= \int_{-\infty}^{\alpha + \beta x_i} f(\theta)\, d\theta. \qquad (17.5.4)$$

Maximization of L can be organized as an iterative process by easy adaptation of the theory in § 3.13. This classical approach has been described in detail in many places (Finney, 1971d), and need not be set out again here.

The method is easily summarized. For each dose, define an *empirical response rate*

$$p = r/n. \tag{17.5.5}$$

Use the equivalent deviate transforms of p (ignoring momentarily any doses for which $p = 0 \cdot 0$ or $p = 1 \cdot 0$) to give a linear equation

$$Y = a_1 + b_1 x; \tag{17.5.6}$$

a rough sketch or inspection of a table of equivalent deviates and values of x suffices here. Using Y as determined by (17.5.6) for each dose, find the *weighting coefficient*

$$w = \frac{Z^2}{PQ}, \tag{17.5.7}$$

where P is determined by (17.4.8),

$$Q = 1 - P, \tag{17.5.8}$$

and

$$Z = f(Y). \tag{17.5.9}$$

Also find the *working equivalent deviate*

$$\left.\begin{aligned} y &= Y - \frac{P-p}{Z} \\ &= Y + \frac{Q-q}{Z}. \end{aligned}\right\} \tag{17.5.10}$$

Now use all the data (including any instances of $p = 0 \cdot 0$ or $p = 1 \cdot 0$) in a calculation of a linear regression of y on x for the ν doses, with weight nw attached to each observation. The result is a new regression equation

$$Y = a_2 + b_2 x \tag{17.5.11}$$

which replaces (17.5.6). The process from (17.5.6) onwards can be repeated, and the iterations will often converge rapidly. Full instructions for such computations are presented in § 18.1. For the present, chief emphasis is placed on the general principles.

No further progress can be made in general terms. Application of the method to actual data requires that a form be assumed for the function $f(\theta)$: the distribution (17.4.1) must be replaced by the normal distribution (17.4.4), or by another specific distributional form. In §§ 17.8–17.13 several of these will be discussed, but something must first be said about validity tests and about variances of the estimates of parameters. No confusion should arise from omitting the subscript that identifies the cycle of iteration, because in the calculations of one cycle all steps employ only the approximations obtained in the immediately preceding cycle.

17.6 The test of homogeneity

If an experiment has tested subjects at each of ν doses, the investigator will usually wish to assess the significance of evidence against the model represented by equation (17.4.7). A χ^2 test that compares observed frequencies with their expectations is the natural procedure to adopt. At any dose,

$$E(r) = nP. \tag{17.6.1}$$

Since two parameters have been estimated,

$$\chi^2_{[\nu-2]} = S \frac{n\{r - E(r)\}^2}{E(r)\{n - E(r)\}}; \tag{17.6.2}$$

the summation extends over all doses, and χ^2 has $(\nu - 2)$ degrees of freedom (Appendix Table IV; Fisher and Yates, 1963, Table IV). A more convenient formula is

$$\chi^2_{[\nu-2]} = S_{yy} - S^2_{xy}/S_{xx}, \tag{17.6.3}$$

where the symbols on the right represent the weighted sums of squares and products used in the regression calculations. Equations (17.6.2) and (17.6.3) give the same result if iteration has been continued well towards the limit. For most purposes, (17.6.3) is close enough to the truth at the last cycle calculated; since it is so easily evaluated, it will be used unless a result on the borderline of significance indicates a need for more detailed computation.

An alternative homogeneity test can be based on equation (17.5.3). If a model could be found that would fit the observations perfectly at every dose, the value of L would be

$$L_0 = S(r \ln p) + S\{(n - r) \ln (1 - p)\}, \tag{17.6.4}$$

apart from a constant the value of which does not affect what follows. Writing $\hat{P}$ to represent values of P computed from the maximum likelihood estimates a and b, the value of L with only 2 d.f. taken up by the fitting of the model is

$$L_{\nu-2} = S(r \ln \hat{P}) + S\{(n - r) \ln (1 - \hat{P})\}. \tag{17.6.5}$$

General statistical theory states that

$$2(L_0 - L_{\nu-2}) \tag{17.6.6}$$

will also behave asymptotically as χ^2 with $(\nu - 2)$ d.f. Indeed in the limit for large values of the n_i the expressions in (17.6.2) and (17.6.6) approach equality. There is some evidence that when the n_i are not large, (17.6.6) is less well approximated by the true χ^2 distribution than is (17.6.2), but firm knowledge is lacking. On the other hand, if the type of computation for maximizing likelihood that is described in § 18.5 be adopted, the expression (17.6.6) is evaluated as a matter of course and the test is therefore very convenient. In practice the two tests can be regarded as almost equivalent. In the analyses of the three examples in § 17.14, the two forms of homogeneity test were compared, and the numerical values for (17.6.2) and (17.6.6) were found to agree well throughout.

Reference of χ^2 to the tabulated distribution (Appendix Table IV) presupposes that none of the expected frequencies is very small. If $E(r)$ or $\{n - E(r)\}$ is less than 5 for any dose, χ^2 should be computed by equation (17.6.2) after combining observed and expected frequencies for this and the next dose; the number of degrees of freedom is then reduced by 1. This has been discussed elsewhere (Finney, 1971, §4.6). The effect of small expectations is to increase the dispersion of χ^2 values calculated without grouping of doses, so that data having no heterogeneity show very large or very small values too frequently. A large χ^2 should always be examined closely in order to see whether a spurious appearance of significance is produced by excessive contributions from classes with small expectations. A small χ^2 is safe, in the sense that small expectations are unlikely to cause the concealment of a genuinely significant deviation from chance.

A significantly large value of χ^2 would indicate the failure of the form assumed for the tolerance distribution. This is a test of statistical validity (§15.3), corresponding closely to the test of deviations from linearity in an assay based upon quantitative responses. If *a priori* the investigator believes in the validity of the statistical model, and this χ^2 is not significant, he may regard it as confirmation of his right to proceed with calculations of a standard type (§15.5). The action to be taken when χ^2 is significant will be discussed later (§18.1).

17.7 Variances of estimates

If the data show no heterogeneity (i.e. χ^2 is non-significant), the variances of the estimates of the parameters may be found by formulae that are generalizations of the formulae for unweighted regressions. The variance of $\bar{y}$ is

$$\text{Var}(\bar{y}) = 1/Snw, \tag{17.7.1}$$

and that of b is

$$\text{Var}(b) = 1/Snw(x - \bar{x})^2$$

$$= 1/S_{xx}, \tag{17.7.2}$$

both expressions that arise naturally in the standard calculations that §17.5 outlined; as in simple unweighted regression, $\bar{y}$ and b are uncorrelated. Since

$$a = \bar{y} - b\bar{x},$$

$$\text{Var}(a) = \frac{1}{Snw} + \frac{\bar{x}^2}{S_{xx}}. \tag{17.7.3}$$

Fieller's theorem is readily applied to an estimate of log ED50:

$$m = \bar{x} - \bar{y}/b. \tag{17.7.4}$$

The above equations are the same as for ordinary linear regression, except that the observations are differentially weighted and σ^2, the variance per observation, is replaced by unity. The variances are usually regarded as expected values, not empirical estimates, and in the determination of fiducial limits the normal distribution replaces the *t* distribution. No rigorous theory supports this practice. More strictly, the variances used are compounded from estimated binomial variances. In almost all assays, the total number of subjects is so large that any attempt to assign degrees of freedom on the basis of variation within doses would still lead to a *t* distribution well approximated by the normal.

17.8 Normal sigmoid

In § 2.5, the assumption that the distribution of log tolerances is normal was recommended as a working hypothesis for direct assays. The argument is neither stronger nor less strong than in any other branch of biometry. The true distribution may not be normal, but, in the absence of evidence favouring a specific alternative, the fact that the normal accords fairly well with observation and is mathematically tractable makes the hypothesis of normality appealing; the central limit theorem (Cramér, 1946, § 17.4; Kendall and Stuart, 1977) gives reason for hoping that conclusions based on the normal assumption will be close to the truth when means of several observations are involved. If the log tolerances of cats used in the assay of digitalis by a direct technique can be regarded as normally distributed, then a hypothesis of normality for the log tolerance of aphids in an indirect assay of derris, or for the log tolerance of rats in the assay of a hormonal contraceptive, seems permissible unless controverting evidence is found. Positive evidence for normality is rare, but considerations such as those of § 2.5 and § 15.4 justify this step. To adopt different hypotheses about the tolerance distribution according to whether assays are direct or quantal is irrational unless either the quantal assays are in some essential respects different or expediency carries greater weight than consistency.

The normal distribution (17.4.4) may be written

$$dP = \frac{\beta}{\sqrt{2\pi}} \exp\left\{-\frac{(\alpha + \beta x)^2}{2}\right\} dx. \tag{17.8.1}$$

Hence, by equation (17.5.9),

$$Z = \frac{1}{\sqrt{2\pi}} e^{-\frac{1}{2}Y^2}. \tag{17.8.2}$$

From equations (17.5.6), (17.5.7) and (17.5.10), *w* and *y* are calculated, and the iterative process can begin.

The idea of using the normal deviate in order to represent quantal response data by a linear regression has a long history (Finney, 1971, § 3.6). It appears to have originated with Fechner (1860) and, with improvements by other workers, it was used by psychometrists for fifty years before it was either rediscovered or adapted for the study of dose–response relationships in biology. Its first systematic use for biological assay was by Gaddum (1933), who developed his method of normal equivalent deviates in a manner very similar to that used later by Bliss (1935a, b; 1938) with probits. In a paper far in advance of its time, Thomson (1919) had outlined the essential feature of maximum likelihood estimation as applied to quantal response data. Bliss, however, was the first to present the full method within the context of bioassay. He introduced (Bliss, 1934a, b) the *probit* of P, the normal equivalent deviate increased by 5·0, with the object of making the occurrence of negative values very rare. Thus the metametric transformation is taken to be

$$P = \int_{-\infty}^{Y-5} \frac{1}{\sqrt{2\pi}} e^{-\frac{1}{2}\theta^2} \, d\theta. \tag{17.8.3}$$

The advantage of this is slight, for the avoidance of negative values is bought at the price of more cumbrous tables. Biologists today are perhaps less disconcerted by negative numbers than they were in 1935! Nevertheless, the use of probits has become so well established that to change to the N.E.D. for the classical method of computation scarcely seems desirable now. The basic tables for the iterative calculations (Appendix Tables V, VI, VII) are therefore given in terms of probits. Finney (1971) has discussed the probit transformation and many of its applications in considerable detail. Equations (17.5.7) and (17.5.10) apply without modification or simplification.

17.9 Logistic sigmoid

Berkson (1944) advocated the logistic function as an alternative to the normal sigmoid for the representation of the regression of a quantal response rate on dose. He agreed that 'In view of the wide use of the normal curve to represent the distribution of biologic traits and also because of direct experimental evidence of the normal distribution of susceptibility, it is to be conceded that the integral of the normal curve recommends itself'. He then stated that 'However, the logistic function is very near to the integrated normal curve, it applies to a wide range of physicochemical phenomena and therefore may have a better theoretic basis than the integrated normal curve. Moreover there are reasons for believing it to be easier to handle statistically'. The argument that the logistic curve may be more appropriate to assays than the normal is here a little obscure. When the reason for unlike behaviour of similarly treated subjects is primarily their intrinsic differences in susceptibility, specification in terms of a frequency distribution of individual tolerances is natural, and in the absence of evidence for any alternative, the assumption of normality seems the most natural. As shown below, use of the logistic implies a tolerance distribution defined by equation (17.9.1), a distribution that seems less likely to correspond to reality.

On the other hand, not all assays are of this kind, and stimulus–subject reactions of a very different nature may be used. Bliss and Packard (1941) studied the relation between the dose of X-rays to eggs of *Drosophila melanogaster* and the mutation rate. The probability of a mutation increases with increasing dosage, but presumably because of an increased chance that the appropriate gene-locus is 'hit' rather than because individual flies differ in susceptibility to mutation. The titration of sera by means of their effectiveness in producing haemolysis of red blood cells is another situation in which the physicochemical model of the reaction may be more reasonable than a tolerance distribution as the foundation of a statistical method. If such reactions are used for biological

§ 17.9 QUANTAL RESPONSES AND THE TOLERANCE DISTRIBUTION

assays, the logistic or autocatalytic function

$$P = \frac{1}{1 + e^{-(\alpha + \beta x)}} \qquad (17.9.1)$$

may well be appropriate. Indeed, equation (17.9.1) has been developed theoretically for representing phenomena such as haemolysis and population growth. Proposals to use the equation in bioassay once aroused controversy, but the choice should depend upon the nature of the biological reactions in use.

Berkson's metametric transformation, corresponding to equation (17.9.1), may be written

$$P = \frac{e^Y}{1 + e^Y}, \qquad (17.9.2)$$

or

$$Y = \ln(P/Q); \qquad (17.9.3)$$

he termed Y the *logit* of P. Essentially the same transformation, differing only by a factor of 2, was proposed by Fisher and Yates in the first edition of their *Statistical Tables* and independently by Wilson and Worcester (1943a, b, d), Worcester and Wilson (1943).

Although the logistic transformation is particularly suited to assays in which a frequency distribution of individual tolerances is inappropriate, it can also be generated by a tolerance distribution in the ordinary manner. The mathematical structure of the analysis is the same for both situations, and expression of the methods and formulae in terms of the tolerance distribution enables the computational terminology to be kept unaltered. In the notation of § 17.4, the basic density function

$$f(\theta) = \tfrac{1}{2} \operatorname{sech}^2 \theta \qquad (17.9.4)$$

corresponds to

$$P = \tfrac{1}{2}(1 + \tanh Y)$$
$$= \frac{e^{2Y}}{1 + e^{2Y}} \qquad (17.9.5)$$

which is the metametric transformation for the Wilson and Worcester form of the logistic transformation. From equation (17.9.5),

$$Y = \tanh^{-1}(2P - 1)$$
$$= \tfrac{1}{2} \ln(P/Q). \qquad (17.9.6)$$

Since

$$Z = \tfrac{1}{2} \operatorname{sech}^2 Y = 2PQ, \qquad (17.9.7)$$

the weighting coefficient is

$$w = 4PQ \qquad (17.9.8)$$

and the working equivalent deviate

$$y = Y + \frac{1}{2P} - \frac{q}{2PQ}. \tag{17.9.9}$$

Berkson's transform is twice as large. This does not make any difference of principle, but the form here has the advantage that the transform is very similar in magnitude to the N.E.D. Addition of 5·0 allows the logit to be made comparable with the probit, with the same merits and demerits.

17.10 Wilson–Worcester sigmoid

In a general discussion of two-parameter tolerance distributions, Wilson and Worcester (1943c) proposed a method of estimation essentially equivalent to that of § 17.5. They suggested several possible distributions, including the normal and the logistic, and introduced one for which

$$f(\theta) = \tfrac{1}{2}(1 + \theta^2)^{-3/2}. \tag{17.10.1}$$

The corresponding metameter is defined by

$$P = \tfrac{1}{2}\left\{1 + \frac{Y}{(1 + Y^2)^{1/2}}\right\}, \tag{17.10.2}$$

or

$$Y = \frac{2P - 1}{2(PQ)^{1/2}}. \tag{17.10.3}$$

For this sigmoid,

$$w = 16 P^2 Q^2, \tag{17.10.4}$$

and the working equivalent deviate does not simplify. The authors claimed no special merits for this, beyond its being a mathematically tractable member of the family of two-parameter sigmoids. It has some interest in comparisons with the more plausible sigmoids in §§ 17.8 and 17.9.

17.11 Cauchy–Urban sigmoid

An earlier proposal as an alternative to the normal distribution was due to Urban (1909, 1910): for the study of quantal responses in psychometry, he used

$$f(\theta) = \frac{1}{\pi(1 + \theta^2)}. \tag{17.11.1}$$

The distribution, commonly known as the Cauchy distribution, is well known to statisticians as a simple example of a distribution that has no mean, variance, or higher moments. The corresponding metameter is defined by

$$P = \tfrac{1}{2} + \pi^{-1} \tan^{-1} Y, \tag{17.11.2}$$

or

$$Y = \cot(\pi Q). \tag{17.11.3}$$

This sigmoid also has no special merits, and indeed its long 'tails' (representing relatively high probabilities for very extreme tolerances) makes it somewhat unrealistic. The weighting coefficient has the cumbrous form

$$w = 1/\{\pi(\cot^{-1} Y)(\pi - \cot^{-1} Y)(1 + Y^2)^2\}. \tag{17.11.4}$$

17.12 Angle sigmoid

In other contexts, and especially for standard analysis of variance, data expressed as percentages or proportions are often transformed to a variate

$$\sin^{-1} \sqrt{\text{proportion}}.$$

This has the property of making the variance of the transform into a constant. Knudsen and Curtis (1947) proposed to take advantage of this for the analysis of assays. Suppose that the tolerance distribution is defined by

$$f(\theta) = \begin{cases} \sin 2\theta & 0 \leq \theta \leq \frac{\pi}{2} \\ 0 & \theta < 0, \ \theta > \frac{\pi}{2} \end{cases}. \tag{17.12.1}$$

Then

$$P = \sin^2 Y \tag{17.12.2}$$

gives the metametric transformation, and since

$$Z = \sin 2Y,$$

equation (17.5.7) becomes

$$w = 4. \tag{17.12.3}$$

The working equivalent deviate is

$$y = Y + \tfrac{1}{2} \cot Y - q \csc 2Y. \tag{17.12.4}$$

The distribution represented by (17.12.1) is unlikely to be in truth the tolerance distribution for any of the dose–response relationships encountered in biological assay. It may be a reasonable approximation, provided that the data avoid extremes of dose at which the finite limits to the distribution, $-\alpha/\beta$ and $\left(\frac{\pi}{2} - \alpha\right)/\beta$, become important. For hand calculation, without or with a desk calculator, the constancy of the weighting coefficient can be a considerable convenience and saving of time, though extremes of dose may demand tiresome modifications in order to maximize the likelihood with due regard to the limits of $f(\theta)$. With the advent of computers, the advantages are trivial and are more than counterbalanced by the awkwardness of discontinuity at the limits.

17.13 'Linear' sigmoid

The simplest of all assumptions that might be made about the tolerance distribution is that the regression of proportion responding on x is linear: the tolerance distribution is rectangular, and is expressed by

$$f(\theta) = \begin{cases} 1 & 0 \leq \theta \leq 1 \\ 0 & \theta < 0, \quad \theta > 1 \end{cases} \quad (17.13.1)$$

in the notation of §17.4. Then

$$P = Y, \quad (17.13.2)$$

and equations (17.5.7) and (17.5.10) reduce to the simple forms

$$w = \frac{1}{PQ} \quad (17.13.3)$$

and

$$y = p. \quad (17.13.4)$$

The working equivalent deviate is identical with the empirical response rate, and the analysis consists in fitting a linear regression of p on x. This is not an unweighted regression, such as was often used in early attempts to analyze quantal response data, but uses a weighting coefficient dependent upon P, the expected response rate.

The assumption of a rectangular distribution will not bear close examination. The exact limitation of range implies the existence of a dose $-\alpha/\beta$ below which no subject can respond, and a dose $(1-\alpha)/\beta$ above which all subjects must respond. Estimation necessarily attaches excessive importance to data for doses that happen to give very small or very large response rates. As for the angle transformation, the discontinuity of behaviour at the limits is awkward in any full maximum likelihood calculations. The dependence of w on P is so marked that unweighted regressions could be very misleading, and this apparently simple 'transformation' in reality has no merits for bioassay. Nevertheless, for data from a well-designed assay using reasonably large numbers of subjects and no very extreme doses, conclusions will often agree closely with those based on more logical alternatives.

17.14 Comparisons between transformations

In §§17.8–17.13 comparable formulae have been presented for six tolerance distributions and their corresponding transformations. Doubtless others can be devised. Each transformation can be made the basis of computations such as are described in §17.6 and illustrated in §18.2, except for simple (though tedious) modifications when the distribution has finite limits. Although the normal and the logistic may each be appropriate for some types of data, the others have little theoretical interest. The constant weighting coefficient for the angle has been a great practical attraction, but today's computing facilities have reduced its importance.

A first step to an understanding of how far the choice of transformations influences conclusions is to examine the similarity of the transformations themselves. Fig. 17.14.1 shows how x and P are related for four tolerance distributions standardized so as to have zero mean and unit variance. Because of their infinite variances, the Wilson–Worcester and Cauchy–Urban transformations cannot be

§ 17.14 QUANTAL RESPONSES AND THE TOLERANCE DISTRIBUTION

Fig. 17.14.1 Relation of P to x for tolerance distributions with zero mean and unit variance
    A: Logistic C: Angle
    B: Normal D: Rectangular

standardized in this way. For the normal, logistic, angle, and rectangular distributions, the relations between x and P are similar over a wide range, and between response rates of 0·05 and 0·95 only intensive experimentation could discriminate between them. Indeed, all but the rectangular are very nearly the same between 0·02 and 0·98; in practice, the rectangular has little to recommend it, for it has no theoretical merits and, unless the correct weighting procedure be abandoned, it does not save much computation. When the probit of P is plotted against x, a

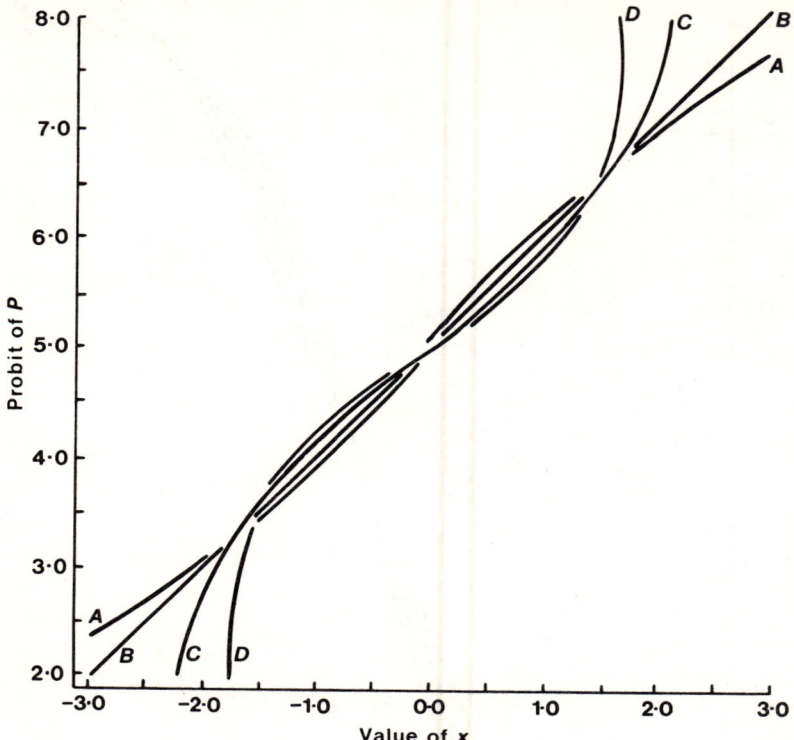

Fig. 17.14.2 Relation of probit of P to x for tolerance distributions with zero mean and unit variance

    A: Logistic C: Angle
    B: Normal D: Rectangular

straight line will be obtained only if the tolerance distribution is normal. Fig. 17.14.2 shows this line, and also the curves that arise if the distribution is in reality one of the other three. Inspection shows clearly the difficulty of detecting non-normality of distribution by a search for curvature in the probit diagram. Even if the true distribution were rectangular, curvature could not be detected between responses of 0·05 and 0·95 without considerably larger numbers of subjects than an assay generally employs; the logistic is scarcely distinguishable from the normal between response rates of 0·01 and 0·99.

In practice, discrimination between normal, logistic, and angle specifications of the dose–response relation is unlikely to be possible, and no doubt other transformations could be suggested that are also indistinguishable. This apparent difficulty in reality removes the major problem of deciding on a transformation for a particular set of data. If data cannot discriminate between alternatives, conclusions are not likely to be seriously influenced by the choice of hypothesis. An argument like that of § 15.5 may then be invoked to defend a choice made primarily on the score of convenience.

§ 17.14   QUANTAL RESPONSES AND THE TOLERANCE DISTRIBUTION   365

In illustration, three assays have been analyzed by use of each of the transformations now under discussion. The first is the insulin assay for which data are given in Table 18.2.1. This includes many doses of each preparation and, as shown by Fig. 18.2.1, the linearity on the probit scale is good; hence it might be expected to be fairly sensitive to changes in the metameter. Table 17.14.1 summarizes the validity tests, potency estimate, and fiducial limits for each transformation, iteration having been continued until close approximation to the

TABLE 17.14.1   Comparison between alternative response metameters for the analysis of Table 18.2.1

Metameter:	Probit	Logit	Wilson–Worcester	Cauchy–Urban	Angle	Linear
$\chi^2_{[10]}$ (linearity)	5·41	6·35	8·51	11·67	3·85	1·64[*]
$\chi^2_{[1]}$ (parallelism)	0·29	0·38	0·56	0·75	0·16	0·01
R (IU/mg)	13·37	13·35	13·32	13·35	13·44	13·27
R_L (IU/mg)	11·07	11·00	10·88	10·79	11·26	10·96
R_U (IU/mg)	16·08	16·13	16·28	16·52	15·99	15·73

[*] 9 d.f. only, because one dose is estimated to be on the horizontal segment of the curve.

solution of the maximum likelihood equations was attained. The numerical values differ, but the practical conclusions are essentially the same for each transformation: the assay gives no evidence of invalidity, and the potency of the test preparation is estimated to be about 13·4 IU/mg and almost certainly lies between 11·1 IU/mg and 16·1 IU/mg.

Tables 17.14.2 and 17.14.3 contain data for two other assays. Tables 17.14.4 and 17.14.5 summarize analyses in a manner similar to that of Table 17.14.1. Again the alternative metameters agree in their evidence on assay validity, both statistical and fundamental; also, for all practical purposes, the estimates of potency and their fiducial limits are the same by the three transformations.

TABLE 17.14.2   Data for an assay of insulin by the mouse convulsion method (Hemmingsen, 1933)

Standard preparation			Test preparation		
Dose (0·001 IU)	No. of mice	No. convulsed	Dose (mg)	No. of mice	No. convulsed
4·0	12	1	0·267	12	1
6·0	24	8	0·4	24	16
9·0	24	15	0·6	24	22
13·5	10	8	0·9	10	10

Somewhat surprisingly, the two least plausible analyses (angle and linear) do not show themselves as obviously inappropriate. In two of the assays, the best fit placed one dose in the horizontal region of the response curve; that is to say, one dose was estimated to be such that response was either impossible or certain to occur. Apart from this slight anomaly, there is no indication that the fit of the

TABLE 17.14.3 Data for an assay of pethidine against morphine as analgesics in mice (Grewal, 1952)

Morphine			Pethidine		
Dose mg/kg	No. of mice	No. responding	Dose mg/kg	No. of mice	No. responding
1·5	103	19	5·0	60	13
3·0	120	53	7·5	85	27
6·0	123	83	10·0	60	32
			15·0	90	55
			20·0	60	44

TABLE 17.14.4 Comparison between alternative response metameters for the analysis of Table 17.14.2

Metameter:	Probit	Logit	Wilson–Worcester	Cauchy–Urban	Angle	Linear
$\chi^2_{[4]}$ (linearity)	1·43	1·26	1·02	1·19	1·75(*)	2·30(*)
$\chi^2_{[1]}$ (parallelism)	2·79	2·93	3·29	3·74	(†)	2·56
R (IU/g)	21·82	21·73	21·62	21·75	22·01	22·50
R_L (IU/g)	17·73	17·64	17·46	17·24	17·91	18·29
R_U (IU/g)	28·26	28·11	28·13	29·15	28·67	29·49

(*) 3 d.f. only, because one dose is estimated to be on the horizontal segment of the curve.
(†) Not available, because simultaneous fitting of parallel lines does not place any dose on the horizontal segment.

TABLE 17.14.5 Comparison between alternative response metameters for the analysis of Table 17.14.3

Metameter:	Probit	Logit	Wilson–Worcester	Cauchy–Urban	Angle	Linear
$\chi^2_{[4]}$ (linearity)	1·93	2·01	2·29	3·00	1·84	1·80
$\chi^2_{[1]}$ (parallelism)	0·08	0·07	0·05	0·02	0·11	0·16
R (mg/mg)	0·334	0·335	0·336	0·337	0·333	0·331
R_L (mg/mg)	0·274	0·274	0·273	0·271	0·275	0·275
R_U (mg/mg)	0·410	0·412	0·415	0·420	0·407	0·401

model (as shown by validity tests) is any less satisfactory than for probits and logits. Moreover the fiducial limits not merely are no wider but in two of the three assays are narrower for these two metameters than for probits and logits. Evidently many assays are unable to discriminate between very different metameters.

17.15 Choice of a transformation

The three commonest metametric transformations are very similar over a wide range of responses, and others could be devised that would also behave like them. Not surprisingly, therefore, they lead to practically indistinguishable results

§ 17.15   QUANTAL RESPONSES AND THE TOLERANCE DISTRIBUTION

in three illustrative examples. Miller (1950) found the same for a fourth set of data, and Berkson (1950) found similar good agreement between probits and logits in ten estimations of ED50s. Armitage and Allen (1950) estimated the ED50 by probit, logit and angle computations in twelve series of data, some of which may be the same as Berkson's. Probits and logits agreed closely, and in no instance did the χ^2 test show markedly poorer agreement with one hypothesis than with the other. The angle transformation was less satisfactory; in one example of 11 doses with about 500 subjects per dose, the angle estimate differed from the probit by only 2 percent but had $\chi^2_{[9]} = 73\cdot 8$, and in four others (with few subjects per dose and many extreme response rates) the estimates were very different from those obtained by probits. Biggers (1952) reported good agreement of calculations based on all three transformations (and also the rectangular) for 17 dose–response lines.

How should a transformation be chosen? The ideal is the 'correct' transformation, i.e. that corresponding to the true tolerance distribution or the true relation between p and x, but this is never known. Where the underlying structure of the dose–response relation is that of a tolerance distribution, and in the absence of evidence to the contrary, the assumption of normality seems reasonable (§ 2.5). As mentioned in § 17.9, the dependence of response on dose does not always derive from variations in individual tolerance, and the logistic assumption may on occasion be preferred. Validity tests (especially that for linearity) will reject a violently wrong assumption, but can seldom discriminate between two as alike as the probit and logit. Narrowness of the fiducial range is in itself irrelevant: the fact that, in Table 17.14.1, the quantities called R_L and R_U are further apart for logits than for probits is no argument against the logit (and would not be, however great the difference), because correctness of the metameter is a necessary condition for the strict interpretation of R_L, R_U as fiducial limits (cf. § 15.6).

Ease of computation, though no criterion for the choice of a 'correct' transformation, has been a consideration of practical importance. If different transformations are going to lead to conclusions as similar as in Tables 17.14.1, 17.14.4 and 17.14.5, greater weight can be given to expediency than to theoretical appropriateness. Berkson's (1944, 1946) suggestion that the logit calculations could be handled more easily than the probit involved substituting the minimization of χ^2 for the maximization of the likelihood as a principle of estimation. This is discussed more fully in § 18.9, where the conclusion is reached that minimum χ^2 has no general theoretical advantages over maximum likelihood, although in some respects it appears to be as good. As Berkson (1949) has shown for the quantal response problem, the two are closely related, and, provided that the number of subjects per dose is not unduly small, minimum χ^2 may be used if found more convenient. An ingenious approximation (Berkson, 1944) makes possible some simplification in the minimization of χ^2 for logits, though it must be used with regard to its inappropriateness at extremes of response (cf. Armitage and Allen, 1950). This does not imply that, in any theoretical sense, minimum χ^2 is the best method for logits, maximum likelihood for probits.

No suggestion has ever been made that the angle transformation has theoretical advantages, but the constancy of w was a convenience when computation was expensive. Thus, in 1947, Knudsen and Curtis considered computational economy a major factor favouring angles. Caution is needed when for any reason extreme response rates occur and place excessive emphasis on discontinuous behaviour of the transformation at the limits of the range, or (Armitage and Allen, 1950) when very large numbers of subjects per dose permit theoretical flaws in the assumed tolerance distribution to show themselves and to disturb estimation.

If an assay must be analyzed on a calculator with no programming facilities, and the data seem so irregular that several cycles of iteration will be needed, labour may be saved by using angles for one or two cycles. Appendix Tables XIV and XV can then be used to convert the fitted regressions approximately to new lines in terms of probits or logits. With good computer facilities, however, full maximum likelihood or minimum χ^2 estimation is not appreciably more costly for probits or logits than for angles. For example, the estimations summarized in Table 17.14.1, conducted on an IBM 370/155, required 4 to 6 seconds each, with no clear advantage for the angles.

17.16 A generalized model

The basic binomial model for responses introduced in § 17.5 is sometimes complicated by additional parameters. Suppose that the population from which subjects are selected has a proportion C of natural responders who will manifest the characteristic response even if they receive zero dose, and a proportion $(1 - D)$ of resistants who will not respond however great the dose. The tolerance distribution then applies only to the remaining proportion $(D - C)$. The total probability that an individual selected at random will respond after receiving a dose x is

$$P^+ = C + (D - C)P, \qquad (17.16.1)$$

where P is as defined by equation (17.4.7). In equation (17.5.1), P^+ must replace P, and the maximum likelihood theory must be generalized in order to estimate C and D as well as α and β (cf. § 16.6).

Most quantal response analyses assume $C = 0$, $D = 1$, as is doubtless justified for many types of response. If a non-zero natural response rate is suspected, a batch of subjects at zero dose can be used to supply empirical information, though estimation ought to be based on all doses. In the less common situation of $D < 1$, a batch of subjects at an excessively high dose will be helpful. A systematic estimation technique similar to that in § 3.13 has been developed (Finney 1944a, 1949a, b, 1971) and has been extensively used. The computations are relatively laborious, and today they are best handled by a computer program for general maximization (§ 18.5). Further details appear in § 20.3.

17.17 Polytomous quantal responses

Occasionally responses are recorded in more than two classes, as when moribund subjects in a toxicity assay are distinguished from 'dead' and 'alive'. Analysis by collapse of classifications into a dichotomy is always permissible, but

§ 17.17   QUANTAL RESPONSES AND THE TOLERANCE DISTRIBUTION

could involve serious sacrifice of information. Evidently one can define a tolerance distribution in respect of the separation between each pair of adjacent classes, and considerations of the internal consistency of the model show that the variances must all be the same. The theory of § 17.5 is readily generalized so as to relate to a series of parallel regression lines for each preparation, of course with the usual requirement of parallelism for different preparations. An iterative procedure for obtaining estimates can easily be devised, though the actual computations are more laborious than for the simple dichotomy.

Aitchison and Silvey (1957) examined this type of problem by maximum likelihood, but without the constraint of equality of variances. White and Graca (1958) used similar methods for situations in which responses were classified according to time-intervals; in this context, papers by Ashford (1959), Ashford, Smith and Brown (1960), and Ashford and Smith (1965) are relevant. Claringbold (1958) described an alternative approach; he kept the classes of response distinct, used no transformation, and applied a canonical procedure to a multivariate analysis of the classes. The objectives of the analysis were not very clear.

More directly in line with the methods of this chapter is a paper by Gurland, Lee and Dahm (1960), though they used minimum χ^2 instead of maximum likelihood. Analysis of their illustrative data (Finney, 1971) showed a perceptible but small gain in precision for the estimated relative potency in consequence of using three classes instead of two. The increase in information will of course always be small in respect of adding a class in which every frequency is small.

$$Y_1 = Y + Q/Z, \qquad (18.1.8)$$

and A is the range,

$$A = 1/Z \qquad (18.1.9)$$

(§ 3.13). Alternatively, working probits may be read directly from Appendix Table VII. Two decimals in y usually suffice; the number should be the same for all doses.

(x) Form columns of products, nwx and nwy.

(xi) For each preparation, sum the nw, nwx, nwy columns and form

$$\bar{x} = \frac{Snwx}{Snw}, \qquad (18.1.10)$$

$$\bar{y} = \frac{Snwy}{Snw}. \qquad (18.1.11)$$

(xii) Calculate adjustments for means in the usual manner. Then for each preparation multiply the nwx, nwy columns by x, y, in turn and obtain

$$S_{xx} = Snwx^2 - \frac{(Snwx)^2}{Snw}, \qquad (18.1.12)$$

$$S_{xy} = Snwxy - \frac{(Snwx)(Snwy)}{Snw}, \qquad (18.1.13)$$

$$S_{yy} = Snwy^2 - \frac{(Snwy)^2}{Snw}. \qquad (18.1.14)$$

(xiii) Use these weighted means and weighted sums of squares and products, just as similar unweighted quantities were used in § 4.11, to give parallel linear regression equations for the two preparations.

(xiv) Evaluate Y_S, Y_T, for each x. If these differ much from the expected probits in (vii) above, repeat the computations with Y_S, Y_T substituted for the original Y column. Repeat steps (viii) to (xiv) until good agreement is obtained. When one place of decimals is used in Y, a good rule is that no value of Y at the end of the last cycle should differ from the corresponding Y at the beginning by as much as 0·1, and that the signs of the differences should not show any obvious association with dose (as would occur if the slope of the regression lines on which the cycle was based were noticeably wrong). In practice, poorer agreement seldom affects potency estimation seriously.

(xv) When agreement has been obtained, use

$$\chi^2_{[f]} = \Sigma S_{yy} - \Sigma \left\{ \frac{(S_{xy})^2}{S_{xx}} \right\} \qquad (18.1.15)$$

as a test of linearity or statistical validity, with f, the number of degrees of freedom, 4 less than the total number of doses tested (cf. the linearity component in Table 4.4.1); the linearity component for quantal responses cannot be subdivided by orthogonal coefficients, even for a symmetric assay design, on account of the differential weighting of the responses. If the χ^2 calculated by equation (18.1.15) is not significantly large, the hypothesis of the normality of the log tolerance distribution is not contradicted, and subsequent calculations will be based on it. In particular, the variance per unit weight will be taken as unity, this being a theoretical instead of an estimated variance: calculations of fiducial limits by Fieller's theorem will use $t = 1 \cdot 960$, the normal deviate for a probability of 0·05. If this χ^2 is significant, the deviations from linearity must be considered in the light of § 4.20. In circumstances which appear to justify retention of the hypothesis of linearity, the variance per unit weight should be taken as

$$s^2 = \frac{\chi^2}{f}, \tag{18.1.16}$$

a quantity known as the *heterogeneity factor* (Finney, 1971, §4.6). All variances must be multiplied by s^2, and must then be regarded as having f degrees of freedom. Assayists must beware of accepting this modification merely because it is convenient.

(xvi) Calculate

$$\chi^2_{[1]} = \Sigma \left\{ \frac{(S_{xy})^2}{S_{xx}} \right\} - \frac{(\Sigma S_{xy})^2}{\Sigma S_{xx}} \tag{18.1.17}$$

as a test of parallelism, and thus of fundamental validity. This is equivalent to a test of equality of variance for the standard and test tolerance distributions (§ 2.6). If a heterogeneity factor has been found necessary, the test must be replaced by a variance ratio test, taking F to be the ratio of the quantity calculated in equation (18.1.17) to s^2; for this test, the degrees of freedom are 1 and f (Appendix Table II).

(xvii) For a valid assay, complete the estimation of potency in the usual manner. As for a parallel line assay, the log potency is given by equation (4.11.5)

$$M = \bar{x}_S - \bar{x}_T - \frac{\bar{y}_S - \bar{y}_T}{b}, \tag{18.1.18}$$

and fiducial limits to M are calculated by Fieller's theorem with

$$\mathrm{Var}(\bar{y}_S - \bar{y}_T) = \Sigma \left(\frac{1}{Snw} \right), \tag{18.1.19}$$

$$\mathrm{Var}(b) = \frac{1}{\Sigma S_{xx}}. \tag{18.1.20}$$

The limits are

$$M_L, M_U = \bar{x}_S - \bar{x}_T + \left[(M - \bar{x}_S + \bar{x}_T)\right.$$
$$\left.\pm \frac{t}{b}\left\{(1-g)\Sigma\left(\frac{1}{Snw}\right) + \frac{(M-\bar{x}_S+\bar{x}_T)^2}{\Sigma S_{xx}}\right\}^{1/2}\right] \Big/ (1-g) \quad (18.1.21)$$

where

$$g = \frac{t^2}{b^2 \Sigma S_{xx}}; \quad (18.1.22)$$

the likeness to equations (4.14.2) and (4.14.3) is apparent. Here t is a normal deviate (1·960 for 0·95 limits), as explained in (xv) above. A heterogeneity factor, if required, must multiply g and also the term within $\{\ \}$ in equation (18.1.21); t then reverts to its usual status as a t-deviate with f degrees of freedom.

This scheme is, in its essentials, the same as that first proposed for biological assay by Gaddum (1933; see § 17.8 above).

Appendix Tables VIII and IX enable logits to be employed exactly as were Tables V and VI in the above instructions. Table X is analogous to Table VII although less detailed; working logits may be read from it with sufficient accuracy for most purposes and more rapidly than by calculation from Table IX. The metameter has been defined to correspond to Wilson's equation (17.9.6), rather than to Berkson's equation (17.9.3), and, as for probits, 5 has been added in order to give values that are almost invariably positive. Perhaps this ought not to be called the logit, but to coin a new name would cause as much confusion as it would save. With this definition, for all but very extreme response rates, the probit and the logit are very similar in numerical value. The transformation is

$$Y = 5 + \tfrac{1}{2} \ln (P/Q), \quad (18.1.23)$$

with a weighting coefficient

$$w = 4PQ, \quad (18.1.24)$$

minimum and maximum working logits,

$$Y_0 = Y - \frac{1}{2Q}, \quad (18.1.25)$$

$$Y_1 = Y + \frac{1}{2P}, \quad (18.1.26)$$

and a range

$$A = \frac{1}{2PQ}. \quad (18.1.27)$$

For the angle transformation, Appendix Tables XI, XII, XIII replace Tables VIII, IX, X. These have been expressed in degrees of arc rather than in radian measure, so that equations (17.12.3) and (17.12.4) are modified slightly. The

§ 18.2 ASSAYS BASED ON QUANTAL RESPONSES 375

constant weighting coefficient is

$$w = \frac{4\pi^2}{(180)^2} = 0.001\,2185;\qquad(18.1.28)$$

the minimum and maximum working angles are

$$Y_0 = Y - 28.6479 \tan Y, \qquad(18.1.29)$$

$$Y_1 = Y + 28.6479 \cot Y, \qquad(18.1.30)$$

and the range is

$$A = 57.2958 \operatorname{cosec} 2Y, \qquad(18.1.31)$$

where Y is measured in degrees throughout.[*] Working logits and working angles are obtainable by using equations (18.1.4)–(18.1.6) with the appropriate definitions of Y_0, Y_1 and A.

18.2 An assay of insulin

One of the earliest published assays for which a complete statistical analysis was attempted was an assay of insulin by the mouse convulsion method (Hemmingsen and Krogh, 1926). At each of 9 doses of S and 5 doses of T, batches of mice were injected with a dose of insulin, and the numbers of mice showing the symptoms of collapse or convulsions were recorded. Unfortunately, Hemmingsen and Krogh published their data only in graphical form. As the assay illustrates several interesting points and is of historical interest, an attempt has been made to reconstruct the data from Hemmingsen and Krogh's diagram, with the aid of an implication in their paper that the number of subjects at each dose was between 30 and 40. Values of n and r have been taken arbitrarily in such a way that the average value of n is about 35 and each r/n agrees with the percentage in the diagram. To this extent, the 'data' analyzed are artificial.

Table 18.2.1 shows the first cycle of iterative calculation. Doses of S were measured in units of 0.001 IU, and doses of T in the same units on the assumption that the potency was 20 IU/mg. The first four columns contain the dose (z), the dose metameter $(x = \log_{10} z)$, the number of mice tested (n), and the number convulsed (r). The empirical response rate, equation (18.1.1), is often shown as a percentage, but its expression as a fraction of unity is more convenient for computation. The empirical probit of p, found from Appendix Table V, is plotted against x in Fig. 18.2.1. Two parallel regression lines have been placed by eye, an easy task when the points lie as close to the lines as do these; allowance was made for a point below the S line at $x = 0.53$. The horizontal distance between the lines, a preliminary indication of potency, is

$$M = -0.175, \qquad(18.2.1)$$

which leads to

---

[*] The factor 57.2958 arises as the number of degrees in 1 radian; 28.6479 is the half of this.

TABLE 18.2.1 First iterative cycle for the assay of insulin

z	x	n	r	p	Empirical probit	Y	nw	y	nwx	nwy	New Y
Standard preparation (S)											
3·4	0·53	33	0	0·00	—	3·5	8·9	2·98	4·717	26·522	3·34
5·2	0·72	32	5	0·16	4·01	4·0	14·0	4·01	10·080	56·140	3·95
7·0	0·85	38	11	0·29	4·45	4·4	21·2	4·45	18·020	94·340	4·36
8·5	0·93	37	14	0·38	4·69	4·7	22·8	4·70	21·204	107·160	4·62
10·5	1·02	40	18	0·45	4·87	4·9	25·4	4·87	25·908	123·698	4·90
13·0	1·11	37	21	0·57	5·18	5·2	23·2	5·18	25·752	120·176	5·19
18·0	1·26	31	23	0·74	5·64	5·7	16·5	5·64	20·790	93·060	5·67
21·0	1·32	37	30	0·81	5·88	5·9	17·4	5·88	22·968	102·312	5·86
28·0	1·45	30	27	0·90	6·28	6·3	10·1	6·28	14·645	63·428	6·28
							159·5		164·084	786·836	
Test preparation (T)											
6·5	0·81	40	2	0·05	3·36	3·8	14·8	3·46	11·988	51·208	3·68
10·0	1·00	30	10	0·33	4·56	4·4	16·7	4·57	16·700	76·319	4·28
14·0	1·15	40	18	0·45	4·87	4·8	25·1	4·88	28·865	122·488	4·76
21·5	1·33	35	21	0·60	5·25	5·4	21·0	5·25	27·930	110·250	5·33
29·0	1·46	37	27	0·73	5·61	5·8	18·6	5·60	27·156	104·160	5·75
							96·2		112·639	464·425	

		S_{xx}	S_{xy}	S_{yy}
$S\begin{cases}\bar{x}=1{\cdot}0287\\ \bar{y}=4{\cdot}9331\end{cases}$		177·5536 168·7997	838·171 809·450	3977·32 3881·57
		8·7539	28·721	95·75
$T\begin{cases}\bar{x}=1{\cdot}1709\\ \bar{y}=4{\cdot}8277\end{cases}$		136·3997 131·8872	557·365 543·788	2285·81 2242·11
		4·5125	13·577	43·70
Totals:		13·2664	42·298	139·45

§ 18.2   ASSAYS BASED ON QUANTAL RESPONSES   377

$$R = 20 \text{ antilog } \bar{1}\cdot 825$$
$$= 13\cdot 4. \qquad (18.2.2)$$

The Y column in Table 18.2.1, read from the lines in Fig. 18.2.1, is used in Appendix Table VI to give the weighting coefficients. For example, for $Y = 3\cdot 5$,

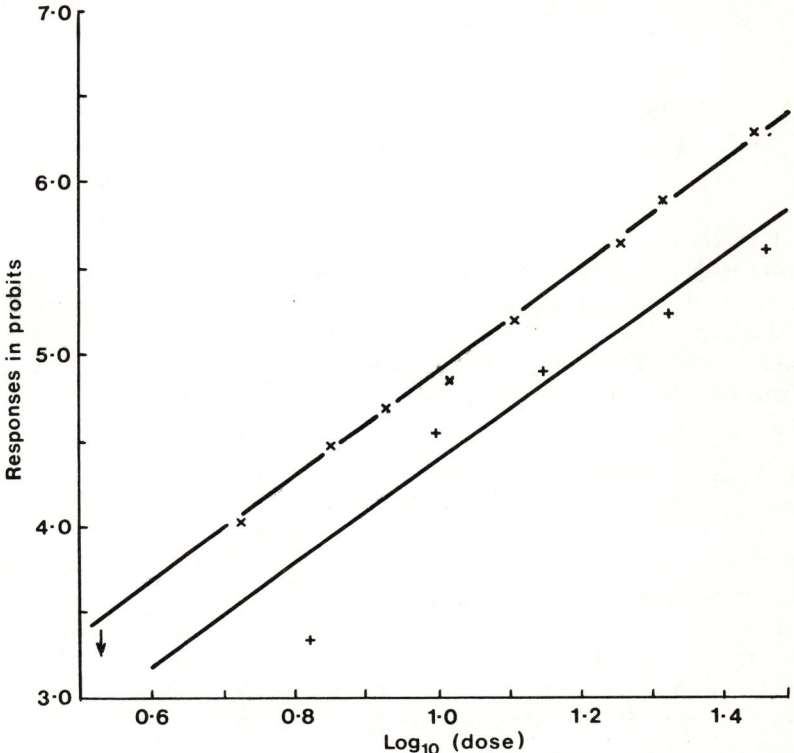

Fig. 18.2.1   Probit regression lines for the insulin assay, Table 18.2.1
×: Empirical probits for standard preparation
+: Empirical probits for test preparation
The arrow indicates a zero response

The lines are those mentioned in § 18.2 as drawn by eye, but the calculated lines (18.2.3) are almost identical with them

$w = 0\cdot 269\,07$ which is multiplied by 33 to give $8\cdot 9$ as the first value of nw. The same table shows $2\cdot 98$ for Y_0, the minimum working probit; since $p = 0$, this is the value of y. For the second dose, $Y = 4\cdot 0$, and, by equation (18.1.2),

$$y = 3\cdot 3443 + 0\cdot 16 \times 4\cdot 1327$$
$$= 4\cdot 01.$$

Alternatively, one of equations (18.1.5), (18.1.6) or Appendix Table VII might have been used. The columns nwx, nwy and the sums of squares and products, S_{xx}, S_{xy}, S_{yy}, are found as described in § 18.1.

The regression coefficient calculated from this cycle, by equation (4.11.1), is

$$b = \frac{42 \cdot 298}{13 \cdot 2664}$$

$$= 3 \cdot 1884.$$

Insertion of means gives the regression lines based on this cycle:

$$\left.\begin{aligned} Y_S &= 4 \cdot 9331 + b(x - 1 \cdot 0287) \\ &= 1 \cdot 653 + 3 \cdot 188x, \\ Y_T &= 4 \cdot 8277 + b(x - 1 \cdot 1709) \\ &= 1 \cdot 094 + 3 \cdot 188x. \end{aligned}\right\} \quad (18.2.3)$$

The final column of Table 18.2.1 shows Y_S, Y_T calculated for each dose. Comparison with the earlier expected probits shows agreement within 0·1 in most instances and no differences exceeding 0·2. A second cycle of calculation could be based on the new Y column, but the consequent change in the conclusions would be small. This question is considered again in § 18.3, but for the present the first cycle will be supposed adequate.

By equation (18.1.13),

$$\chi^2_{[10]} = 139 \cdot 45 - 94 \cdot 23 - 40 \cdot 85$$

$$= 4 \cdot 37,$$

this being calculated exactly as is the sum of squares for deviations from linearity in an ordinary parallel line assay (§ 4.4). Since the χ^2 is well below the significance level (Appendix Table IV), the possibility that the iteration has not been continued far enough for the quantity to behave as a true χ^2, or that small expectations in some classes may have caused serious disturbance, can safely be ignored (cf. § 17.6). Had the value been large, calculation of expected numbers in each class would have been needed. For example, the lowest dose of the standard preparation has $Y_S = 3 \cdot 34$, corresponding to a convulsion rate of 0·05; hence the expected number of responses in this group is only 1·6, and the usual practice with χ^2 tests would require that the data be pooled with those for the next dose before calculation of χ^2 by equation (17.6.2). As a test of parallelism, equation (18.1.17) gives

$$\chi^2_{[1]} = 94 \cdot 23 + 40 \cdot 85 - 134 \cdot 86$$

$$= 0 \cdot 22,$$

and no fundamental invalidity need be feared.

The log potency is obtained from the usual formula, equation (18.1.18), and is

$$M = 1 \cdot 0287 - 1 \cdot 1709 - \frac{4 \cdot 9331 - 4 \cdot 8277}{3 \cdot 1884}$$

$$= -0 \cdot 1422 - 0 \cdot 0331$$

$$= -0{\cdot}1753,$$

very close to the value in equation (18.2.1). Moreover, by equations (18.1.19) and (18.1.20),

$$\mathrm{Var}(\bar{y}_T - \bar{y}_S) = \frac{1}{159{\cdot}5} + \frac{1}{96{\cdot}2},$$

and

$$\mathrm{Var}(b) = \frac{1}{13{\cdot}2664}.$$

Hence, by equation (18.1.22),

$$g = \frac{(1{\cdot}960)^2}{(3{\cdot}1884)^2 \times 13{\cdot}2664}$$

$$= 0{\cdot}0285,$$

the deviate $t = 1{\cdot}960$ being used because all weights are regarded as reciprocals of true variances. Equation (18.1.21) then gives

$$M_L, M_U = -0{\cdot}1422$$

$$- \left[0{\cdot}0331 \pm \frac{1{\cdot}960}{3{\cdot}1884} \left\{ 0{\cdot}9715 \left(\frac{1}{159{\cdot}5} + \frac{1}{96{\cdot}2} \right) + \frac{(0{\cdot}0331)^2}{13{\cdot}2664} \right\}^{1/2} \right] \Big/ 0{\cdot}9715$$

$$= -0{\cdot}1422 - [0{\cdot}0331 \pm 0{\cdot}0784]/0{\cdot}9715$$

$$= -0{\cdot}2570, -0{\cdot}0956.$$

Finally, as for equation (18.2.2),

$$R = 20 \text{ antilog } \bar{1}{\cdot}8247 = 13{\cdot}36,$$

$$R_L = 20 \text{ antilog } \bar{1}{\cdot}7430 = 11{\cdot}07,$$

$$R_U = 20 \text{ antilog } \bar{1}{\cdot}9044 = 16{\cdot}05.$$

The potency of the test preparation is estimated to be 13·36 IU/mg, with fiducial limits at 11·07 IU/mg and 16·05 IU/mg. Hemmingsen and Krogh stated their estimate to be '13·34 ± 0·46' IU/mg. The difference in R is trivial; the figure of 0·46 they stated to be the 'mean error', a phrase no longer interpretable. Their analysis is not fully described, but appears to be related to the characteristic curve method (§ 18.6).

18.3 Speed of convergence

The calculations in § 18.2 used only one cycle of iteration. Theory states that the maximum likelihood estimate is obtained as the limit of the iterative process; it might be expected that several cycles would always be needed in order to approximate to this limit satisfactorily. Experience shows that, unless the irregularity of the data is so great as to make difficult the construction of the provisional regression lines and the first set of expected probits, two cycles

TABLE 18.3.1 Results from three cycles of iteration on the insulin assay in Table 18.2.1

Cycle	I	II	III
b	3·195	3·218	3·219
$\chi^2_{[10]}$ (linearity)	4·203	5·039	5·124
$\chi^2_{[1]}$ (parallelism)	0·266	0·276	0·278
M	$-$ 0·1739	$-$ 0·1735	$-$ 0·1735
g	0·0284	0·0294	0·0295
R (IU/mg)	13·40	13·41	13·41
R_L (IU/mg)	11·11	11·12	11·11
R_U (IU/mg)	16·10	16·12	16·12

almost always suffice and one cycle is often enough. Table 18.3.1 shows results obtained by three iterations for the insulin assay; one extra digit was carried in the p, Y, nw, and y columns of each cycle, in order that small changes from one cycle to the next could show. The regression equations obtained from cycle III would have given a Y column for a fourth cycle identical (to 2 decimals) with that for cycle III, and iteration was therefore stopped.

The potency estimates and their fiducial limits in Table 18.3.1 differ only to a trivial extent. They are a little higher than the results in § 18.2, the extra digits presumably having produced a small and unimportant shift: response rates cannot be satisfactorily estimated to 3 decimals from batches of 40 subjects. (Small differences from Table 17.14.1 may be attributed to the computer program used for that table having automatically evaluated x and p to more digits.) An estimate of potency as 13·4 IU/mg, with fair confidence that the true potency lies between 11·1 IU/mg and 16·1 IU/mg, is in accord with all the analyses. Other quantities have also been included in Table 18.2, notably the two χ^2 values needed for validity tests. The probit method does not necessarily minimize χ^2 (cf. § 18.9). Moreover, unless the maximum likelihood limit has been attained, equations (18.1.15) and (18.1.17) only approximate to what would be obtained by comparison of observed and expected frequencies. Consequently, the quantities termed χ^2 in the system of calculation recommended may sometimes increase in successive cycles of iteration; both increase slightly in Table 18.3.1, but the changes are too small to affect the conclusions.

Similar calculations have been made for other assays. Table 18.3.2 shows results for the much smaller assay reported in Table 17.14.2, in which the responses were less regular; again three iterations sufficed to give good convergence. This assay was used (Finney, 1951) in a study of the success with which scientists inexperienced in probit methods could draw provisional regression lines. In each of 21 independent trials, a single cycle gave R, R_L, R_U near enough to their maximum likelihood values for practical purposes. Nevertheless, when this classical computational system is used, completion of two cycles is a good precaution.

TABLE 18.3.2 Results from three cycles of iteration on the insulin assay in Table 17.14.2

Cycle	I	II	III
b	0·943	0·951	0·951
$\chi^2_{[4]}$ (linearity)	1·504	1·343	1·324
$\chi^2_{[1]}$ (parallelism)	2·995	2·843	2·876
M	0·9151	0·9256	0·9252
g	0·1124	0·1061	0·1070
R (IU/g)	21·74	21·83	21·83
R_L (IU/g)	17·64	17·75	17·74
R_U (IU/g)	28·32	28·27	28·26

18.4 Variants of the method

For many years, great interest attached to finding methods of organizing the arithmetic that would save time. Black (1950) suggested assigning a score Y_0 to each subject that did not respond, Y_1 to each that responded; he tabulated wY_0, wY_1 as a first stage in a scheme equivalent to that of § 18.1. Garwood (1941) showed that basing w on empirical responses secures convergence to the same limits and often requires fewer cycles of iteration, but each cycle involves more computation. Cornfield and Mantel (1950) developed another version of this approach. At the time, these variants had advantages despite obscuring the analogy with linear regression. Modern computing facilities have destroyed the interest that once lay in arguments about the merits of the various possibilities.

18.5 Computer methods

To put the scheme of § 18.1 into a computer program is easy; once this has been done, the time required for analysis of a straightforward assay can be reduced to a few minutes for card punching and a few seconds of computer time. However, there is then no advantage in retaining the analogies with regression, and the maximization of likelihood can be approached as a computing problem free from any preconceptions about method.

Every scientifically oriented computer installation today ought to have available a good routine for maximizing or minimizing an arbitrary but reasonably smooth function. The log-likelihood function, equation (17.5.3), usually changes very smoothly and is relatively flat in the neighbourhood of the maximum; that is to say, for any set of parameter values moderately close to those that maximize, the value of L will be close to its maximum. In such circumstances, almost any sensible routine will converge to the maximum likelihood estimates, to satisfactory accuracy for practical purposes, though the flatness will often prevent more than about 5 digits being determined. Nelder and Mead (1965) proposed a method that has been found very satisfactory, and an adaptation of this has been built into a program specially designed for quantal bioassays.

The method, not described in detail here, involves determining provisional values for the parameters and improving on them by successive trial evaluations of L. Provisional values can be based upon previous knowledge, rough inspection of the data, or approximate computation embodied in the program; the quality of approximation is seldom important with reasonable data, as it has little effect on the time for convergence. In the closing stages, differential coefficients of L are used much as in the classical computations, but these can be approximated from differences. The full set of parameters can be estimated, and a matrix of asymptotic variances and covariances obtained by inverting the matrix of second differential coefficients of L at the maximum. Of course all the remaining arithmetic required for validity tests, potency estimation, and fiducial limits can be incorporated into the program.

Tables 18.5.1–18.5.7 show output from the author's program, as applied to the insulin assay in Table 18.2.1. Most of the output is self-explanatory. The 'activators' control the numbers of iterations of various kinds and optional features of the output. The doses of T were divided by 20, so as to correspond to mg rather than to any assumed potency (cf. § 18.2), and the program then leads directly to the estimated potency. The computations have been done in terms of normal equivalent deviates instead of probits, so that for comparison with § 18.2 the first two parameters need to be increased by 5·0. The 'start' relates to the provisional values, calculated from a very crude rule applied to the data. Equations corresponding to (18.2.3) are taken from the estimated parameters

$$\left.\begin{aligned} Y_S &= 1\cdot6340 + 3\cdot2039x, \\ Y_T &= 5\cdot2420 + 3\cdot2039x. \end{aligned}\right\} \quad (18.5.1)$$

The χ^2 shown is the total of the linearity and parallelism components. In order to get these separately, one should maximize the likelihood for each preparation alone, so obtaining a χ^2 for each:

and
$$\begin{aligned} \chi^2_{[7]} &= 2\cdot245 \quad \text{for } S \\ \chi^2_{[3]} &= 3\cdot165 \quad \text{for } T. \end{aligned}$$

The sum is the required linearity component for the whole assay

$$\chi^2_{[10]} = 5\cdot410,$$

and subtraction from the $\chi^2_{[11]}$ gives

$$\chi^2_{[1]} = 0\cdot293$$

as the parallelism criterion. Though somewhat larger than the values in § 18.2 obtained after a single iteration, they do not suggest any fears of invalidity; they are of course to be regarded as more trustworthy because they correspond to a much more exact maximization of likelihood.

In the program, L was modified by subtraction of the constant L_0, defined by (17.6.4). This would make the maximum precisely zero if the data were such as to fit the parametric model exactly, and it replaces (17.5.3) by

$$\begin{aligned} L = {}& Sr_i \ln P_i + S(n_i - r_i) \ln (1 - P_i) - Sr_i \ln r_i \\ & - S(n_i - r_i) \ln (n_i - r_i) + Sn_i \ln n_i, \end{aligned} \quad (18.5.2)$$

§ 18.5 ASSAYS BASED ON QUANTAL RESPONSES 383

TABLE 18.5.1

```
                                                    THE FOLLOWING ANALYSIS
                                                    USES THE FORTRAN PROGRAM
                                                    +*+ B L I S S  17 +*+

       COMPUTER ANALYSIS OF:
     I N S U L I N   A S S A Y    (HEMMINGSEN AND KROGH, 1926)       ANALYSED ON 18/08/77

NEW CONFIGURATION OF ANALYSIS STARTS HERE

C O N F I G U R A T I O N
=========================
 2 PREPARATIONS
 1 VARIATE
14 DOSE GROUPS, OF WHICH:-
         9 REFER TO THE STANDARD PREPARATION
         5 REFER TO THE TEST PREPARATION
         0 REFER TO UNTREATED SUBJECTS
         0 REFER TO A MAXIMAL DOSE


ANALYSIS USING NORMAL EQUIVALENT DEVIATES   (I.E. PROBIT-5.0)

D A T A   T O   B E   A N A L Y Z E D

PREPARATION   SUBJECTS   RESPONSES      DOSE      DOSE METAMETER

     1           33          0        3.40000        0.53148
                 32          5        5.20000        0.71600
                 38         11        7.00000        0.84510
                 37         14        8.50000        0.92942
                 40         18       10.50000        1.02119
                 37         21       13.00000        1.11394
                 31         23       18.00000        1.25527
                 37         30       21.00000        1.32222
                 30         27       28.00000        1.44716

     2           40          2        0.32500       -0.48812
                 30         10        0.50000       -0.30103
                 40         18        0.70000       -0.15490
                 35         21        1.07500        0.03141
                 37         27        1.45000        0.16137
```

TABLE 18.5.2

HISTORY OF CONVERGENCE (USING MODEL 111)

THE ACTIVATORS ARE: AD = 1,   AH = -8,   AN = -3,   AI = 50,   AE = 0,   AL = 0,   AP = 3,   AT = 1

NELMEAD NO. 1

```
                START  =  -2.149093    0.354897    2.172296
                STEPS  =  -0.644728    0.106469    0.651689
STEPS HAVE BEEN INCREASED TO APPROACH MINIMUM
           NEW STEPS  =  -0.644728   -0.638815    0.651689

SMALLEST VALUE   =  4.12215151, AT  -3.17099317    0.29229083    3.01498358
QUADRATIC PHASE HAS BEEN CALCULATED
FUNCTION NOW     =  3.64532978, AT  -3.35720212    0.24220546    3.19526386
TRUE HESSIAN HAS BEEN EVALUATED
```

---->> ---->> ---->>

NELMEAD NO. 2

```
                START  =  -3.357202    0.242205    3.195264
                STEPS  =   0.009438    0.009940    0.006517

SMALLEST VALUE   =  3.64532978, AT  -3.35720212    0.24220546    3.19526386
QUADRATIC PHASE HAS BEEN CALCULATED
FUNCTION NOW     =  3.64479966, AT  -3.36602860    0.24200840    3.20388705
TRUE HESSIAN HAS BEEN EVALUATED
```

---->> ---->> ---->>

NELMEAD NO. 3

```
                START  =  -3.366029    0.242008    3.203887
                STEPS  =   0.000629    0.000552    0.000652

SMALLEST VALUE   =  3.64479966, AT  -3.36602860    0.24200840    3.20388705
QUADRATIC PHASE HAS BEEN CALCULATED
FUNCTION NOW     =  3.64479966, AT  -3.36601954    0.24200860    3.20387889
TRUE HESSIAN HAS BEEN EVALUATED
```

TABLE 18.5.3

```
SUMMARY  OF  INSTRUCTIONS  AND  CONDITIONS

:+:+ :+:+: +:+:+ :+:+: +:+:+ :+:+: +:+:+ :+:+: +:+:+ :+:+: +:+:+ :+:+:      +*+  B L I S S   1 7  +*+
    COMPUTER ANALYSIS OF:
   I N S U L I N   A S S A Y   (HEMMINGSEN AND KROGH, 1926)                  ANALYSED ON 18/08/77
   ANALYSIS USING NORMAL EQUIVALENT DEVIATES   (I.E. PROBIT-5.0)

THIS ANALYSIS USES MODEL: 111
                             ! NORMAL SIGMOID!                  ::        LOGARITHMS TO BASE 10
NUMBER OF SUBJECTS TESTED: KNOWN    :::   NATURAL RESPONSE RATE: KNOWN    :::   IMMUNITY RATE: KNOWN

:+:+ :+:+: +:+:+ :+:+: +:+:+ :+:+: +:+:+ :+:+: +:+:+ :+:+: +:+:+ :+:+:
                                 S U M M A R Y   O F   R E S U L T S

THE MAXIMUM LOG-LIKELIHOOD =    -3.64479966

P A R A M E T E R   E S T I M A T E S

     I       PARAMETER       STANDARD ERROR       DESCRIPTION
     1       -3.3660195       0.3020756           STANDARD PREPARATION
     2        0.2420086       0.1083863           TEST PREPARATION 1

     3        3.2038789       0.2807099           REGRESSION COEFFICIENT


M A T R I X   O F   V A R I A N C E S,   C O V A R I A N C E S,   A N D   C O R R E L A T I O N S
**  NOTE THAT PARAMETERS ARE IDENTIFIED BY NUMBER; ANY NOT ESTIMATED ARE OMITTED FROM MATRIX

          1                 2                 3
   0.091249652        ( -0.299883)       ( -0.96447)
  -0.009783839         0.011747584       ( 0.30984)
  -0.081782687         0.009426780         0.078798048
```

TABLE 18.5.4

OBSERVED AND EXPECTED FREQUENCIES

DOSE-GROUP	PREPARATION	ALL SUBJECTS	RESPONSES	EXPECTED	DEVIATION	PROBABILITY
1	STANDARD	33	0	1.588	-1.588	0.048134
2		32	5	4.539	0.461	0.141853
3		38	11	9.695	1.305	0.255132
4		37	14	12.910	1.090	0.348907
5		40	18	18.498	-0.498	0.462454
6		37	21	21.475	-0.475	0.580401
7		31	23	23.064	-0.064	0.743998
8		37	30	29.893	0.107	0.807907
9		30	27	26.941	0.059	0.898047
10	TEST 1	40	2	3.724	-1.724	0.093108
11		30	10	7.050	2.950	0.235007
12		40	18	15.986	2.014	0.399640
13		35	21	22.192	-1.192	0.634064
14		37	27	28.715	-1.715	0.776077

FOR HETEROGENEITY OF DEVIATIONS FROM MODEL, KI-SQUARED = 5.7033 WITH 11 DEGREES OF FREEDOM

TABLE 18.5.5

```
:::: ::::: :::: ::::: :::: ::::: :::: ::::: :::: ::::: :::: :::::       +*+ B L I S S   1 7   +*+
         COMPUTER ANALYSIS OF:
   I N S U L I N   A S S A Y   (HEMMINGSEN AND KROGH, 1926)
                                                                        ANALYSED ON 18/08/77
      ANALYSIS USING NORMAL EQUIVALENT DEVIATES   (I.E. PROBIT-5.0)


THIS ANALYSIS USES MODEL:  111
                                 ! NORMAL SIGMOID!                 ::    LOGARITHMS TO BASE 10
NUMBER OF SUBJECTS TESTED: KNOWN    :::   NATURAL RESPONSE RATE: KNOWN   :::   IMMUNITY RATE: KNOWN

:::: ::::: :::: ::::: :::: ::::: :::: ::::: :::: ::::: :::: ::::: :::: ::::: :::: ::::: :::: :::::

   S U M M A R Y   O F   P O T E N C Y   E S T I M A T I O N

INDEX OF REGRESSION SIGNIFICANCE :T**2*V(B)/B**2
      FOR PROBABILITY 0.90, G= 0.020769
      FOR PROBABILITY 0.95, G= 0.029489
      FOR PROBABILITY 0.99, G= 0.050933
             VARIANCE APPROXIMATIONS ARE REASONABLE


              R E L A T I V E   P O T E N C Y

       POTENCY       LIMITS (0.90)        LIMITS (0.95)        LIMITS (0.99)
       ESTIMATE     LOWER     UPPER      LOWER     UPPER       LOWER     UPPER
   1
   2   13.37038    11.42084  15.60482   11.06781  16.08124   10.39048  17.07204


ASYMPTOTIC VARIANCE FOR RELATIVE POTENCY

    1   LOG RHO    VARIANCE    STANDARD ERROR       WEIGHT           RHO
    2   1.12614     0.00166        0.04078         601.45655       13.37038
```

TABLE 18.5.6

```
>>>>THE REMAINDER OF THIS ANALYSIS IS CONCERNED SOLELY WITH EXAMINING DEVIATIONS FROM PARALLELISM<<<<
:+:+ :+:+: +:+:+ :+:+: +:+:+ :+:+: +:+:+ :+:+: +:+:+ :+:+: +:+:+ :+:+: +:+:+ :+:+: +:+:+ :+:+: +:+:+
                                                                          +*+ B L I S S    1 7   +*+
        COMPUTER ANALYSIS OF:
        I N S U L I N   A S S A Y   (HEMMINGSEN AND KROGH, 1926)
                                                                          ANALYSED ON 18/08/77
        ANALYSIS USING NORMAL EQUIVALENT DEVIATES (I.E. PROBIT-5.0)

THIS ANALYSIS USES MODEL: 111

                          ! NORMAL SIGMOID!         ::         LOGARITHMS TO BASE 10
NUMBER OF SUBJECTS TESTED: KNOWN      :::   NATURAL RESPONSE RATE: KNOWN    :::   IMMUNITY RATE: KNOWN

:+:+ :+:+: +:+:+ :+:+: +:+:+ :+:+: +:+:+ :+:+: +:+:+ :+:+: +:+:+ :+:+: +:+:+ :+:+: +:+:+ :+:+: +:+:+

C O N V E R G E N C E   H I S T O R Y   FOR EXTENDED MODEL
                                                                                        NELMEAD NO.  1
  ---->>>--->
         START    =       -3.366020     3.203879     0.242009     3.203879     0.21827512    2.96885348
         STEPS    =       -0.322364     0.325844     0.053235     0.325844

SMALLEST VALUE =  3.50717594, AT     -3.45708721    3.28947282    3.31592299    0.21784348    2.99742669
QUADRATIC PHASE HAS BEEN CALCULATED
FUNCTION NOW   =  3.50158324, AT     -3.48262724    3.31592299
HESSIAN CALCULATED FROM DIFFERENCES
                                                                                        NELMEAD NO.  2
  ---->>>--->
         START    =       -3.482627     3.315923     0.217843     2.997427     0.21784348    2.99742669
         STEPS    =        0.005514     0.003258     0.000532     0.003258

SMALLEST VALUE =  3.50158324, AT     -3.48262724    3.31594657    3.31594657    0.21785577    2.99733975
QUADRATIC PHASE HAS BEEN CALCULATED
FUNCTION NOW   =  3.50158319, AT     -3.48263517    3.31594657
HESSIAN CALCULATED FROM DIFFERENCES
                                                                                        NELMEAD NO.  3
  ---->>>--->
         START    =       -3.482635     3.315947     0.217856     2.997340     0.21785577    2.99739576
         STEPS    =        0.000241     0.000178     0.000009     0.000056

SMALLEST VALUE =  3.50158318, AT     -3.48263517    3.31594657    3.31597436    0.21787085    2.99740862
QUADRATIC PHASE HAS BEEN CALCULATED
FUNCTION NOW   =  3.50158318, AT     -3.48266378    3.31597436
HESSIAN CALCULATED FROM DIFFERENCES
```

§ 18.5 ASSAYS BASED ON QUANTAL RESPONSES

TABLE 18.5.7

SUMMARY OF PARAMETER ESTIMATES

THE MAXIMUM LIKELIHOOD = -3.50158318

PARAMETER ESTIMATES (EXTENDED MODEL)

I	PARAMETER	STANDARD ERROR	DESCRIPTION
1	-3.4826638	0.3801348	STANDARD PREPARATION
2	3.3159744	0.3569402	REGRESSION COEFFICIENT
3	0.21787085	0.11621003	TEST PREPARATION
4	2.9974086	0.4752775	REGRESSION COEFFICIENT

OBSERVED AND EXPECTED FREQUENCIES

DOSE-GROUP	PREPARATION	ALL SUBJECTS	RESPONSES	EXPECTED	DEVIATION	PROBABILITY
1	STANDARD	33	0	1.409	-1.409	0.042690
2		32	5	4.283	0.717	0.133841
3		38	11	9.429	1.571	0.248144
4		37	14	12.739	1.261	0.344308
5		40	18	18.464	-0.464	0.461591
6		37	21	21.594	-0.594	0.583612
7		31	23	23.302	-0.302	0.751681
8		37	30	30.207	-0.207	0.816414
9		30	27	27.178	-0.178	0.905926
10	TEST 1	40	2	4.261	-2.261	0.106527
11		30	10	7.405	2.595	0.246849
12		40	18	16.107	1.893	0.402673
13		35	21	21.787	-0.787	0.622485
14		37	27	28.065	-1.065	0.758522

FOR HETEROGENEITY OF DEVIATIONS FROM MODEL, KI-SQUARED = 5.4101 WITH 10 DEGREES OF FREEDOM

(AFTER 27.67 SECONDS OF CPU TIME)

END OF THIS ANALYSIS

END OF DATA

in which any terms involving the logarithm of zero are to be discarded. At the maximum, $-2L$ is asymptotically distributed as χ^2, corresponding to the whole departure from the fitted model, with degrees of freedom equal to the number of dose groups (14) minus the number of parameters estimated (3). In other words, this approximation corresponds with the total of the linearity and parallelism χ^2, giving now 7·29 instead of the 5·70 calculated by the more usual method. Maximization without the constraint of parallelism gave $-3·50$ as the maximum of L. By the same procedure as before, this leads to

7·00   with 10 d.f.

for linearity, and by subtraction

0·29   with 1 d.f.

for parallelism. Numerically these differ from the values previously obtained, but (as will usually be the case) they give the same conclusions in respect of validity. Indeed, either method of making validity tests could be adopted for quantal assays. For each method, the probability levels tabulated for χ^2 (as in Appendix Table V) are approximations even when the specifications of the model are exactly correct.

The program uses the standard formulae to calculate the estimate of relative potency and its fiducial limits at three different probabilities; values of g are shown, under the name of 'index of regression significance'. A minor change of input enables any of the other five metametric transformations discussed in Chapter 17 to replace the probit. The information for Tables 17.14.1, 17.14.4 and 17.14.5 was obtained from this program.

Such a program destroys all need to worry about the labour of routine statistical analysis of quantal response assays. Preparation of the input for an assay requires only the punching of doses, n_i, r_i on to cards, together with about six cards for headings and other instructions. A typical run time (on the IBM 370/158) is 5 seconds, and even an assay with far more data could be handled in 15 seconds. Moreover, without any change in principle, the program can deal with the generalization described in § 17.16. The log likelihood function need only be written to incorporate either or both the additional parameters C, D, and the maximization routine looks after the rest. Note that C, D are purely nuisance parameters contributing nothing to the estimation of relative potency (cf. § 20.3).

18.6   Approximate methods of analysis

The various arrangements of maximum likelihood estimation all make the same assumptions about the underlying probability structure relating dose to response, except for the alternative formulations embodied in the different metametric transformations (§§ 17.8–17.13). They have the general optimal properties of maximum likelihood (§ 3.12). Whether the classical scheme of computation or one of its variants or a computer program is used, the estimates attained after extensive iteration will be identical. In this sense the group of methods can be regarded as exact.

§ 18.6   ASSAYS BASED ON QUANTAL RESPONSES   391

Ever since the first recognition of the problem of estimating properties of a dose–response relation for quantal responses, alternative methods of analysis have been proposed. Initially this was because no principle pointed to any optimal method; later it was because the classical computations were regarded as intolerably laborious for routine use or for the scientist ill-supplied with mechanical aids to arithmetic. Now that good computation has become quick and relatively cheap, methods whose sole merits were their arithmetic simplicity no longer merit the detailed discussion given in earlier editions; they are briefly described here because of historical interest and because they illustrate a wealth of ingenuity by those who first devised them.

Methods based upon a standard regression curve were no more satisfactory with quantal responses than with quantitative. The example of the standardization of X-rays by their effect on the survival of eggs of *Drosophila melanogaster* was quoted in § 3.14, but this is exceptional. More recently, extensive data from assays of pertussis vaccine by the mouse protection method have been examined in detail (Finney, Holt and Sheffield, 1975), because preliminary inspection suggested that the parameters for the standard preparation changed very little from one occasion to another. In 1970, the Second British Reference Preparation for Pertussis Vaccine (66/84) was used as S in 32 assays, each assay having doses of 0·02, 0·1 and 0·5 IU, with 32 mice per dose and a corresponding arrangement for the various test preparations assayed. Analysis of the records for S alone as though they were for a multiple assay (§ 19.5) raised no doubts about parallelism, but a test of the identity of position of the 32 lines gave

$$\chi^2_{[31]} = 137 \cdot 21.$$

Clearly the probit regression lines differed from occasion to occasion; the sequence of ED50 values gave no indication of a systematic trend. Analysis using logits instead of probits gave essentially the same conclusions. Had the analysis shown no differences between the 32 lines, statutory regulations would still have required that tests of S be included in every assay, but possibly some pooling of results for S would have led to more precise potency estimation. Finney *et al.* presented strong evidence that any pooling so as to estimate the regression for S from the current and some earlier assays would have been disadvantageous: elementary sampling theory showed that any inclusion of data from earlier assays would have lessened the precision of potency estimation. The whole study was repeated with the data for S in 1971, 1972 and 1973: the findings were qualitatively the same, but even less favourable to the pooling of information on S than those for 1970. These analyses reinforce the conclusion that, as for quantitative responses, no method related to that of a standard regression curve should ever be used without strong theoretical or experimental reasons for belief in the constancy of the regression for the standard preparation.

Trevan (1927) suggested that extensive data for S might be used to estimate the percentage change in dose required to produce various responses, as compared with the ED50, the relation between relative dose and response rate being known as the *characteristic curve*. Subsequent single-dose trials of S and T could then be used to give comparable estimates of the ED50 for each, on the assumption that doses for different response rates remain in the same relation to one another as for the original determination of the characteristic curve. Each dose tested is divided by the value from the characteristic curve corresponding to the observed response, and the result is regarded as an estimate of the ED50; the ratio then estimates relative potency. An assay based upon the characteristic curve assumes that it remains unchanged for an indefinite period, and therefore that, although the location may change, the mathematical form and scale parameter of the tolerance distribution may safely be regarded as fixed. The curve is used in practically the same way as was the standard slope method of § 3.15. The method is fairly trustworthy if the two experimental doses show nearly equal response rates: the form of the characteristic curve is then of little importance. On the other hand, if the two response rates are very different, the potency estimate may be seriously in error because the characteristic curve has altered slightly or was never exactly correct. A scheme of computation for fiducial limits could be set up on the analogy of that in § 3.15, though this cannot be regarded as very satisfactory. Unless the investigator is certain of the applicability of a characteristic curve, he should follow the practice recommended for quantitative responses and prefer self-contained assay designs (§ 3.16, Chapter 19).

Here again the findings in the pertussis study are relevant. The parallelism of regressions for S on different occasions suggests that little harm would come from adopting a constant b, or at any rate one based on pooled data from some previous assays. When computation was expensive, this use of a characteristic curve might have been attractive. Today, even if regulations permitted it, the advantages are not great. Some gain in precision could result from being able to concentrate all available mice on a single dose of T, provided that the dose was successfully chosen, but the consequences could be disastrous if ignorance of true potency led to a poor choice. Some gain would result from a smaller $\text{Var}(b)$ in consequence of the pooling, but in good assays (say $g < 0.1$) $\text{Var}(b)$ has too small a part to play in determining R_L, R_U for reduction in it to be very helpful. Against these uncertain gains must be set the loss of any ability to detect the occasional gross anomaly in a test preparation that could produce a very different slope and that might pass undetected if pooling were a routine practice. On balance, continuation of the present scheme of (3, 3) assays, each regarded as self-contained, seems the better policy.

A dose–response diagram, drawn in terms of log dose and response metameter, is often a desirable first step in the statistical analysis of an assay. If two parallel regression lines are drawn by eye, the horizontal distance between them is a first approximation to M. With very little experience, the lines can be so placed that they will commonly give a value that scarcely differs from the maximum likelihood estimate. Not only is this a useful preliminary indication; although never free from the taint of the subjective judgement, it can be sufficiently good in itself without further calculation. Several manufacturers produce specially ruled probability paper – graph paper with a logarithmic scale

§ 18.6 ASSAYS BASED ON QUANTAL RESPONSES 393

horizontally and a probit, logit, or angle scale vertically – which is a great convenience for graphical methods.

Many authors have devised approximations and systems of nomographs to enable the precision of such a graphical estimate to be assessed. Litchfield and Fertig (1941), Miller and Tainter (1944), and de Beer (1945) produced methods that were widely adopted. Curtis, Umberger and Knudsen (1947) and Knudsen and Curtis (1947) gave nomographs for (2, 2) assays using the angle transformation. Haley (1947) suggested a mechanical aid (cf. § 5.7). Perhaps the most ingenious of all graphical procedures is that of Litchfield and Wilcoxon (1949), which works well with reasonably good data and provides approximations to all relevant tests. Finney (1952) claimed that empirical trials showed one cycle of probit iteration to remove a substantial subjective component arising from the placement of the lines. Litchfield and Wilcoxon (1953) showed that, at least for the data under discussion, their recommendation of drawing lines without the constraint of parallelism avoided this weakness because the lines can then be placed more accurately. Though the Litchfield–Wilcoxon method can undoubtedly be regarded as a very effective use of the graphical approach, now that heavy statistical arithmetic has become cheap and quick its practical usefulness may be less. If graphical estimation is wanted at all today, only the simplest is likely to represent a real saving in time and trouble.

Cornfield and Mantel (1950) mentioned a serious weakness in graphical methods that is too often neglected. As they say, 'a variance should be appropriate to the estimator being used rather than to the ideal estimator being approximated'. A variance that attempts to approximate that of the true maximum likelihood estimate makes no allowance for 'the additional errors inherent in the process of graphical estimation'; even supposing that the person who draws the lines is unbiased, he would not obtain exactly the same lines each time he was presented with the same set of data, and appreciable extra error may enter from this source.

None of the above methods introduces a new principle. The aim is merely to simulate the maximum likelihood results. One danger inherent in all approximate methods is that their greatest appeal is to the investigator who has little knowledge of or skill in statistical science and who is therefore least qualified to judge of their adequacy in any one instance. An experienced statistician should recognize when an approximate method may be seriously misleading with particular data, and will reserve its use for data that show no abnormal features.

Many methods of analysis employ special procedures to estimate equivalent doses for S and T separately, the ratio of which then estimates ρ. Usually the ED50 is chosen because this is likely to be more precisely determinable than any other, at least for data well spread over a wide range of doses. Whatever the choice, if m_S, m_T are estimates of equivalent log doses for S and T,

$$M = m_S - m_T \tag{18.6.1}$$

is an estimate of the logarithm of the relative potency. If variances of m_S, m_T are obtained independently, the variance of M will be estimated by

$$\text{Var}(M) = \text{Var}(m_S) + \text{Var}(m_T). \tag{18.6.2}$$

Discussion can therefore be concentrated on one preparation.

One general criticism must be noted. If the condition of similarity is satisfied, the estimate of relative potency and the fiducial limits ought to be independent of any choice of level of response. If the condition of similarity is not satisfied, the relative potency no longer exists as a unique value. A method that uses separate calculations for each preparation will give a value of M that to some extent depends upon whether equation (18.6.1) is based on the ED50, the ED70 or the ED95; it therefore cannot be making the best possible use of the data.

Gaddum (1933) described the method of extreme effective doses, appropriate when only one subject is tested at each dose. Of much greater popularity have been the Dragstedt–Behrens method (Dragstedt and Lang, 1928; Behrens, 1929) and the Reed–Muench method (Reed and Muench, 1938), so similar in form that their names are often confused. These are so simple in structure that they are still often used, yet they have no sound theoretical basis. The late Professor Muench once told the present writer of his surprise that a method devised hastily to meet a particular need should have become accepted as a standard procedure. The two methods have inspired such claims as that of 'making it possible to interpret results in terms of approximately three times as many animals as actually used in the assay' (Barr and Nelson, 1949; cf. Winder, 1947). Even without such extravagence, the methods have been so widely accepted that in some circles they have come to be regarded as standard. The time has come for a change: under the most favourable conditions, neither method is as precise as the Spearman–Kärber (§ 18.7), and often Spearman–Kärber is markedly superior. Dragstedt–Behrens and Reed–Muench estimates are certainly easy to calculate, but no easier than Spearman–Kärber. Except as part of statistical history, both methods should be forgotten. Even as early as 1948, Pittman and Lieberman showed empirical evidence of the inferiority of the Reed–Muench method relative to maximum likelihood.

Bennett (1971) described a method of potency estimation from quantal responses converted to ranks. It is unlikely to be adopted as a standard procedure, but it may perform well with some otherwise troublesome data.

18.7 Spearman–Kärber and moving averages

A particularly simple and easily understood method was first proposed by Spearman (1908) and reintroduced by Kärber (1931). Suppose that, at a dose with metameter x_i, of n_i subjects tested r_i respond, for $i = 1, 2, \ldots, k$. Then

$$p_i = r_i/n_i \tag{18.7.1}$$

is an estimate of the response rate at x_i, and $(p_{i+1} - p_i)$ estimates the proportion of subjects whose tolerances lie between x_i and x_{i+1}. If $p_1 = 0$ and $p_k = 1$, and if intervals between successive x_i are not too wide, the mean of the log tolerance distribution might be estimated by

$$m = S\{(p_{i+1} - p_i)(x_i + x_{i+1})/2\}. \tag{18.7.2}$$

The rule is not exactly equivalent to that for a grouped frequency distribution, because p_i and p_{i+1} are obtained from different groups of subjects and the chances of sampling may happen to make some values of $(p_{i+1} - p_i)$ negative. There is an implicit assumption that doses below x_1, if tested, would have given no responses, and that doses above x_k would have given a response for every subject, so that doses outside the range actually tested would make no contribution to equation (18.7.2). This contradicts the hypothesis of a normal or other unlimited distribution of log tolerances, for which the probability of response is never quite zero or unity at any finite dose, but if the interval between x_1 and x_k is wide relative to the standard deviation of log tolerances, the contradiction will be unimportant.

For equally spaced log doses with (for all i)

$$x_{i+1} - x_i = d, \tag{18.7.3}$$

equation (18.7.2) may be written

$$m = x_k + \tfrac{1}{2}d - d\,\mathrm{S}p_i. \tag{18.7.4}$$

If each dose has n subjects, equation (18.7.4) becomes

$$m = x_k + \tfrac{1}{2}d - \frac{d\,\mathrm{S}r_i}{n}, \tag{18.7.5}$$

the form most suitable for computing. For such a regular set of doses, a rule sometimes stated is that if $p_1 \neq 0$ or $p_k \neq 1$, the next dose in the series, though untested, should be assumed to have given $p_0 = 0$ or $p_{k+1} = 1$; the estimation is then completed as for the longer series. This fabrication of 'data' is obviously without theoretical basis; though often it may do little harm, it could be seriously misleading if applied uncritically. Only equally spaced doses will be considered further here, but the possibility of extending the ideas and formulae will be obvious to the reader.

For equation (18.7.4), the variance is (Irwin and Cheeseman, 1939)

$$\mathrm{Var}(m) = d^2 \mathrm{S}\left(\frac{P_i Q_i}{n_i}\right), \tag{18.7.6}$$

where P_i is the expected response rate at x_i. This may be estimated by

$$\mathrm{Var}(m) = d^2 \mathrm{S}\left(\frac{p_i q_i}{n_i - 1}\right); \tag{18.7.7}$$

the modified divisor removes a bias that can be important if the n_i are small. For constant n_i, a more rapid computation is

$$\mathrm{Var}(m) = \frac{d^2}{n^2(n-1)}\,\mathrm{S}\{r_i(n - r_i)\}. \tag{18.7.8}$$

Equation (18.6.2) gives the variance of M, and fiducial limits are found by an assumption of normality that is not likely to be far wrong unless the n_i are very small.

This account needs a more critical examination. For a specified set of doses, m may be neither an unbiased nor a consistent estimate (§ 3.12) of μ: repeated experimentation need not give values of m whose mean tends to μ, nor need increase in the n_i ensure that m tends to μ. Irwin (1937) and Finney (1950) showed that, when a set of equally-spaced doses is specified, the expectation of m differs from μ by an amount independent of the n_i but dependent upon the location of μ relative to the nearest x_i. If the deviation of μ from the nearest x_i is selected at random between $\frac{1}{2}d$ and $-\frac{1}{2}d$, as is effectively done when the experimenter begins without knowledge of μ, the bias will be removed; with this condition,

$$E(m) = \mu,$$

but the variance of m about μ will be increased by a component due to the location of μ relative to the nearest x_i. In fact, equation (18.7.6) ought to be revised to

$$\operatorname{Var}(m) = d^2 S\left(\frac{P_i Q_i}{n_i}\right) + V_B, \tag{18.7.9}$$

where V_B is independent of the n_i. Finney (1950) examined $\operatorname{Var}(m)$ for the normal tolerance distribution and a perfectly symmetrical arrangement having all the n_i equal; he showed that $\bar{V}(m)$, the value of $\operatorname{Var}(m)$ averaged over all positions of the doses relative to μ, is

$$\bar{V}(m) = \sigma d \left(\frac{0.5642}{n} + B\right), \tag{18.7.10}$$

where

$$B = V_B/\sigma d \tag{18.7.11}$$

and B is a function only of the ratio d/σ. Van der Waerden (1940a, b) showed the factor 0·5642 to be $1/\sqrt{\pi}$. Evidently no increase in n can reduce the variance of m below V_B. Even for d/σ as large as 4·0 (a much higher value than would usually be encountered), B is only 0·0172; for $d/\sigma = 3\cdot0$, B is 0·0019, and for smaller values of d/σ it is negligible. Thus, even for a wide spacing of doses, the contribution of V_B to $\operatorname{Var}(m)$ is scarcely important unless n is 100 or more. The obvious way of avoiding trouble is to use a few subjects at each of many closely spaced doses, rather than the same total number divided into larger groups at more widely spaced doses. The formula

$$\bar{V}(m) \doteq \frac{0.564 \sigma d}{n} \tag{18.7.12}$$

(Gaddum, 1933) is then sufficiently close to the truth for most purposes. The variance for particular doses, equation (18.7.6) with $n_i = n$, will differ from its average $\bar{V}(m)$, but unless d/σ is large the difference is unimportant; the highest and lowest possible values of $S\left(\dfrac{P_i Q_i}{n}\right)$ differ by a factor of 5·6 when $d/\sigma = 4\cdot0$,

§ 18.7   ASSAYS BASED ON QUANTAL RESPONSES   397

a factor of 2·0 when $d/\sigma = 3\cdot 0$, and by less than 10 percent when $d/\sigma \leq 2\cdot 0$. Thus, provided that d is not greater than 2σ, equation (18.7.12) is a close approximation to Var(m) for any set of doses and any reasonable value of n (not unless n exceeded 10 000 would V_B be important). Finney (1953) found similar results for the logistic tolerance distribution.

Epstein and Churchman (1944) advocated the Spearman–Kärber method as a standard in preference to the probit because it does not require an assumption of distributional form. Their reasoning was far from clear and their discussion of estimation and of experimental design seems inadequate. A more important result is that of Cornfield and Mantel (1950), namely that Spearman–Kärber estimation for a logistic tolerance distribution is essentially the same as maximum likelihood estimation, since it is based upon sufficient statistics. This applies with full simplicity only to an assay with equal numbers of subjects per dose, equal logarithmic spacing of doses, and long series, but it helps to explain the frequent empirical finding that Spearman–Kärber performs well relative to probits and logits. Brown (1961, 1966) and Chang and Johnson (1972) have investigated questions relating to optimal planning of assays when Spearman–Kärber analysis is intended. They obtained theoretical results for biases and variances for several tolerance distributions. Though intrinsically interesting, the practical value of this work is limited by the recognition that the optimal design may differ substantially from that which would obtain maximum information; an assayist who knew enough about the relevant response curves to adopt the advice given could probably also follow the principles of §§ 19.1, 19.2.

Thompson (1947) proposed a method that has some relation to the Spearman–Kärber, at least in not relying on a distributional assumption. For as many doses as possible, calculate a moving average

$$p_i^* = (p_i + p_{i+1} + \ldots + p_{i+j-1})/j \tag{18.7.13}$$

for an agreed small integer j, and associate with it

$$x_i^* = (x_i + x_{i+1} + \ldots + x_{i+j-1})/j. \tag{18.7.14}$$

Then seek two consecutive p_i^*, one either side of 0·50, from which a linear interpolation can estimate m as a value of x^* corresponding to $p^* = 0\cdot 50$. (Such interpolation can be grossly wrong for any dose other than the ED50.) Thompson suggested $j = 3$ as a reasonable span for the moving average. As j becomes large, the estimating equation approaches the Spearman–Kärber form, but the moving average method has the advantage of not requiring an unlimited series of doses.

The variance of the estimate can be studied, as was that for Spearman–Kärber, by compounding binomial variances. Finney (1950, 1953) compared variance functions for $j = 1, 2, 3, 4$ with those for Spearman–Kärber. The variance is of the same form as equation (18.7.10), but is appreciably greater when $jd/\sigma < 6\cdot 0$ and increases rapidly when d/σ is small. Thus conditions under which the method is satisfactorily precise, or adequate in extracting the information available in the data, are very restrictive. Thompson and Weil (1952) and Weil (1952) published tables to expedite moving average calculations. Bennett (1952, 1963c) has discussed the optimal deployment of the method, but its inherent weaknesses are scarcely balanced by its computational simplicity in an age when computing is so cheap.

18.8 Comments and comparisons

The methods in §§ 18.6, 18.7 have been advocated primarily on grounds of speed, simplicity and freedom from assumptions. Computers now make the first two of these scarcely relevant. Preparation of data for input to a moving average program would take no less time than for a maximum likelihood program. Computing time might be reduced from 5 seconds to 1 second, a saving negligible relative to the costs of experimentation and the losses in quality of inference. The value of distribution-free techniques in statistical science is undeniable, but the need for them is unconnected with whether or not data are quantal or quantitative. When nothing at all is known about a distribution, their use is mandatory; when a distribution is known to be, say, approximately normal in form, neglect of this knowledge almost certainly reduces the precision of estimation. If normality of distribution, for example, is an acceptable assumption for tolerances that were directly measurable, the mere fact that experimental circumstances permit only quantal records does not destroy this acceptability.

On the other hand, the methods do require assumptions about the symmetry of distributions and the symmetry of doses and subjects relative to the true ED50. The studies already mentioned (Finney, 1950, 1953) showed that precision is almost always higher for Spearman–Kärber than for any of the others. Consequently only that method and perhaps moving averages deserve further consideration; the latter is hopelessly imprecise unless $d > \sigma$, but has advantages in permitting a narrower range of doses to be used.

An ideal Spearman–Kärber experiment would have n subjects at each of an unlimited series of equally spaced doses. With a normal tolerance distribution the average variance for maximum likelihood estimation would be well approximated by

$$\bar{V}(m) = 0.5536\sigma d/n \tag{18.8.1}$$

if $d < 2\sigma$, but it increases markedly if d/σ is larger. At first sight this is little better than equation (18.7.12); if the tolerance distribution were logistic, the two numerical constants would be equal. Some writers have concluded that the general use of maximum likelihood processes has little to recommend it in problems of quantal responses. If d/σ exceeds 2·5, $\bar{V}(m)$ for probits will be greater than the Spearman–Kärber value unless n is large, but even this does not controvert the theoretical maximal efficiency of the maximum likelihood method (§ 3.12). As n is increased, the B component in equation (18.7.10) remains constant and $\bar{V}(m)$ does not tend to zero; the consistency of maximum likelihood estimation ensures that $\bar{V}(m)$ may be made as small as desired by use of a sufficiently large n. In practice, the interval between successive doses is unlikely to exceed 3σ. Unless n is so small as to reduce the effective precision of the probit estimate on account of a large value of g, the difference between the precisions achieved when the two methods of analysis are applied to data for which either is suitable will then be negligible.

A negative virtue of the probit method is not the complete story. The Spearman–Kärber and the moving average methods require a specialized experimental design before they can be safely applied; the underlying theory is more open to objection if the doses are unequally spaced or if the number of subjects per dose is not constant. Moreover, the number of doses needed is usually large, and even though a wide range of doses be adopted, the methods may sometimes break down because an ED50 is much greater or much smaller than was believed when the doses were chosen. The probit method can be applied whatever the doses or the numbers of subjects per dose may be. This is not advocacy of irregular design; circumstances sometimes compel irregularity, and even in a symmetrically designed experiment accidental losses may destroy symmetry. Most important of all, the methods in §§ 18.6 and 18.7 do not provide any validity tests, yet they require a number of doses in excess of the recommendations in § 19.2.

The dependence of Spearman–Kärber upon an unlimited range of doses has been emphasized. Nevertheless, m can be calculated from formulae such as equation (18.7.5) whatever the range of doses, and undoubtedly quantities so calculated have often been used as estimates of the ED50. From one point of view, this assumes that all lower doses would have given $p_i = 0$ and all higher doses would have given $p_i = 1$. Alternatively, the procedure may be regarded as an arbitrary rule for calculating a quantity whose merits as an estimator of m have yet to be investigated. In elaboration of this idea, Bross (1950) compared the Spearman–Kärber, Reed–Muench, and maximum likelihood methods in small samples, by complete enumeration of cases. He based his study on the logistic distribution, but would have obtained qualitatively similar results with the normal distribution. He used four equally spaced doses, chosen to have expected response rates of either 0·10, 0·32, 0·68, 0·90, or 0·59, 0·86, 0·96, 0·99; the first set is symmetric with respect to the ED50, the second is so violently skewed that it does not bracket the ED50, and both have logit differences of about 0·7. He enumerated all possible sets of experimental results with either 2 or 5 subjects per dose. Bross's assessment of the success of different estimation procedures was based upon the relative frequencies with which the absolute value of the difference $(m - \mu)$ lay in various intervals, and so is not directly comparable with the variance criteria used in this chapter. His conclusions were unfavourable to Reed–Muench, but he found the Spearman–Kärber method to give closer approximation to the true ED50 than did the maximum likelihood. The difference was sufficient to suggest a genuine slight superiority when the number of subjects per dose is small. The apparent disagreement with the earlier finding that when d/σ is small, the variances for probit and Spearman–Kärber estimation are almost equal may reflect the inadequacy of the variance as a measure of deviation from the truth when n is very small. To assume that the numerator and denominator of the probit estimator are normally distributed when calculated from an experiment with $n = 2$ at each of four doses is scarcely safe. Bross's results are especially surprising in their indications that the Spearman–Kärber method gives estimates from his skew set of doses of a quality comparable with those from the symmetric doses. They scarcely justify any claim that, even under the computing conditions of 1950, the superiority of Spearman–Kärber estimation is so clear as to make the labour of maximum likelihood unnecessary. Cornfield and Mantel (1951) showed that assumptions about response for the unobserved next lower and next higher doses may have influenced Bross's findings.

The discussion may be summarized as follows:

(i) The extreme effective dose, Reed–Muench, and Dragstedt–Behrens methods ought never to be used. They do not permit assessment of precision from the data of a single assay, they give no validity tests, and they are less efficient than alternatives that are equally simple computationally.

(ii) If the investigator knows nothing about the ED50s for his two preparations before the experiment is performed, he must choose wide dose ranges in order to be sure of bracketing them. If he can give any meaning to the instruction when he is so ignorant about his preparations, he should space his doses fairly closely, but not excessively so! He may use the Spearman–Kärber method. The moving average method, with the largest possible span, is preferable because it avoids flaws at the end of the dose range, but is likely to give almost the same result after more laborious calculation. The probit or other equivalent deviate method will give little or no increase in precision, but it will provide the validity tests that the others lack and the computational cost is no longer a serious objection.

(iii) If existing information makes the investigator fairly sure that each ED50 lies between known limits which are not very far apart (of the order of 4σ, say), his doses should extend over a range rather wider than these limits but not as wide as in (ii). The doses should not be very closely spaced. If enough doses are used, moving average estimation with a span of 3 or greater will be preferable to the Spearman–Kärber method. Whatever the intention at the time of planning the experiment, in the analysis of the results the longest span permitted by the data should be adopted. Probits should be used instead if an unfortunate choice of doses causes the data to be unsuitable for moving averages, and must be used if validity tests are wanted.

(iv) If existing information is more trustworthy, so that both the ED50s and the standard deviation of the tolerance distributions can be guessed before the assay is begun, the design should be planned carefully in accordance with the principles of §§ 19.1 and 19.2. The most economic utilization of subjects will almost certainly demand a design to which neither Spearman–Kärber nor moving average estimation can safely be applied; maximum likelihood or an equivalent is needed in order to extract all available information from the records.

(v) The recommendations in (ii)–(iv) above have assumed that statistical advice is sought at the right time, namely during the planning of an assay. The statistician who is not consulted then, but is later asked to assist in the analysis of data, should of course use the method that appears likely to be most efficient and economical for the assay as actually performed. For example, in the insulin assay of Table 18.2.1 probably little harm would be done by adoption of a modified moving average or Spearman–Kärber estimation. The analgesic data in Table 17.14.3, on the other hand, are quite unsuitable for methods of this type and maximum likelihood estimation (or something like it) must be used.

(vi) Considerations of simplicity and of computing speed are not of negligible importance, but they are today very different from what they were to the originators of some of the methods that have been described. Investigators who still lack good computer facilities may need to act carefully according to (i)–(iv). The more fortunate are probably wise to discard all these other methods and to adopt as a routine maximum likelihood estimation with either the normal or the logistic model. In any event, under the conditions of (iv), any of the other methods represents too high a price for avoiding probits or logits: a good design and maximum likelihood (or similar) estimation are then essential to an economic programme of experimentation (cf. § 6.6).

Finney (1959b) described the application of many different methods of analysis to one set of data.

18.9 Minimum χ^2

Because validity tests for quantal response data are usually made in terms of χ^2, some have advocated estimation of parameters according to a different principle, namely minimization of the heterogeneity χ^2. Indeed, Berkson's proposals for a method of minimum χ^2 have already been mentioned (§ 17.15).

To a mathematical statistician, χ^2 is defined as a sum of squares of independent normal deviates. He would allow use of the name for a statistic calculated from observed and expected frequencies only when the frequencies are so great as to ensure that the probability distribution approximates closely to that of a true χ^2. Here, as is common in applied statistics, χ^2 refers to the quantity that compares observed frequencies with expectations calculated according to a theory under examination in the manner usual for a test of significance of deviations:

$$\chi^2 = S \frac{(\text{observed} - \text{expected})^2}{\text{expected}}. \tag{18.9.1}$$

Thus for quantal responses to a single preparation

$$\chi^2 = S \left\{ \frac{(r-nP)^2}{nP} + \frac{(n-r-nQ)^2}{nQ} \right\}$$

or

$$\chi^2 = S \frac{n(p-P)^2}{PQ}. \tag{18.9.2}$$

Minimization of this χ^2 is a legitimate procedure for estimation; its properties must be considered.

As is well known (Cramér, 1946; Kendall and Stuart, 1973), if the correct probabilistic model is being used for data, estimates of parameters obtained by minimizing χ^2 tend to equality with the maximum likelihood estimates as sample size increases. Thus both methods are consistent and asymptotically efficient, and for each the distribution of the estimates is asymptotically normal. There is no reason to suppose that minimum χ^2 has theoretical advantages over maximum likelihood in large samples; indeed Rao (1961, 1962) has demonstrated certain second-order superiorities of maximum likelihood within a large class of estimation principles that includes minimum χ^2.

Optimal properties in large samples need not apply to experiments of small or moderate size, where little is known about bias and efficiency. Even the words 'large' and 'small' cannot be given exact meaning; on the basis of existing information, it would be unwise to assume that an experiment with 100 subjects at each of 8 doses of a preparation was large enough for the maximum likelihood (or the minimum χ^2) estimate of the ED50 to be for all practical purposes unbiased and of maximum efficiency, and equally unwise to assume that an experiment with 5 subjects at each of 3 doses was too small for this. For all that is known, minimum χ^2 or some third method (Tukey, 1949) might be superior to maximum likelihood in small samples. Berkson (1955a, 1957) presented empirical evidence that minimum χ^2 estimates of the parameters tend to be closer to the truth than maximum likelihood estimates. His simple conclusion was disturbed by Cramer (1962, 1964), who discussed estimation of μ, σ rather than α, β. No clear ruling can be given that one method is generally better than the other in its approach to the true values of the parameters for either normal or logistic models, and indeed it seems unlikely that a consistent superiority of either will ever be demonstrated.

A practical objection to the minimum χ^2 principle is that, when the number of subjects per dose is small, either or both of the expected frequencies (nP and nQ) may be so small as to make χ^2 numerically unstable. This is related to the general result that a loss of information inherent in minimum χ^2 estimation increases without limit as the number of classes increases, even though the total number of observations is large (Kendall and Stuart, 1973). The problem of small class-numbers can be a serious difficulty in bioassay. In his important series of papers, Berkson (1944, 1946, 1949, 1950, 1953, 1955a, b, c, 1957), proposed an arbitrary adjustment for dealing with the most extreme cases and also a modified definition of χ^2 that enables minimization to be accomplished in one stage without iteration. Today these ideas also seem to come into the category of approximations for reducing arithmetic that are no longer particularly relevant.

Maximum likelihood appears to remain the preferable estimation principle, not only because it is less troubled by difficulties when class frequencies are small but because of its general applicability to a wider range of statistical problems including those of continuous variates. Admittedly, when frequencies are very small, estimation can degenerate into indeterminacy, but this is equally true for minimum χ^2 unless Berkson's arbitrary adjustments and modifications are introduced. Finney (1971) discussed these points more fully. Berkson (1949) pointed out that the standard iterative procedure (§ 18.1) can be used for minimizing χ^2 with the one modification that the maximum likelihood weighting coefficients must be multiplied by $\frac{1}{2}\left(\frac{p}{P} + \frac{q}{Q}\right)$. Thus for probits

$$w = \frac{Z^2(qP + pQ)}{2P^2Q^2}, \qquad (18.9.3)$$

for logits

$$w = 2(qP + pQ) \qquad (18.9.4)$$

and for angles

$$w = \frac{0 \cdot 000\ 6092(qP + pQ)}{PQ}, \qquad (18.9.5)$$

with no change in the definition of y, will lead to the required estimates. Obviously only a minor change of programming is involved. As is to be expected, the two weighting coefficients for a metameter tend to equality as n becomes large and p approaches P.

Occasionally the suggestion is made that the parameters for a quantal response assay might be estimated by least squares. Although the principle of least squares is another valuable and widely accepted basis for estimation, it is not appropriate here in any simple sense. Minimization of

$$S\{n(p-P)^2\}$$

has no merits as a method of estimation: it takes no account of the fact that Var(p) depends on P, so that large deviations from P can occur much more easily in the neighbourhood of $P = 0.50$ than at the extremes. If the method is modified and based upon a weighted sum of squares, with the reciprocals of the variances as weights, the quantity to be minimized becomes the χ^2 of equation (18.9.2); the method of least squares is then identical with that of minimum χ^2.

19

Design of assays based on quantal responses

19.1 Principles of good design

The principles of design for bioassays based upon quantal responses are very similar to those for parallel line assays based upon quantitative responses: at this stage §§ 6.1–6.7 should be re-read. The complexities of maximum likelihood estimation prevent as simple a discussion as that in §§ 6.8–6.10, primarily because of the dependence of the weighting coefficient upon the expected value of the response metameter (for every transformation except the angle). Nevertheless, analysis similar to that in § 6.8 leads to conclusions only in part dependent upon the choice of metameter and the behaviour of w.

Once again, the problem is that of economic use of N subjects. In a simple assay involving tests on batches of subjects (selected at random from the available stock) at a series of doses of each preparation, the requirements for high precision may be seen by study of the expression for the quarter-square of the fiducial interval for M; this is analogous to equation (6.8.1), and may be taken from equation (18.1.21) as

$$I = \frac{t^2 s^2}{b^2(1-g)^2}\left[(1-g)\Sigma\left(\frac{1}{Snw}\right) + \frac{(M-\bar{x}_S+\bar{x}_T)^2}{\Sigma S_{xx}}\right]. \qquad (19.1.1)$$

If no heterogeneity factor (§ 18.1) is needed, s^2 will be taken as unity and t read from the normal distribution; for a valid assay that has heterogeneous data, s^2 is the heterogeneity factor and t is read from the t distribution with the appropriate degrees of freedom. The investigator will desire to assign subjects to doses in such a way as to make I small, with the constraint that $\Sigma(Sn) = N$; he will therefore attempt to choose a design for which

(i) the data are homogeneous;
(ii) b is large;
(iii) g is small;
(iv) $\Sigma\left(\frac{1}{Snw}\right)$ is small;
(v) $(M - \bar{x}_S + \bar{x}_T)$ is near to zero;
(vi) ΣS_{xx} is large.

The notes that follow indicate how these ends may be attained (cf. Miller, Bliss and Braun, 1939).

(i) The only sure way of securing homogeneity is by proper randomization. The subjects for each dose should be chosen either completely at random from all the N available or randomly within the framework of suitable block restrictions (§ 19.3). Without randomization a large χ^2 for deviations from

(ii) Selection of subjects for homogeneity in respect of tolerance should reduce the variance of the tolerance distribution. For a normal log tolerance distribution, the regression coefficient is equal to the reciprocal of the standard deviation, and for other distributions it is proportional to that reciprocal. Thus the regression coefficient is a property of the source of subjects and is unaffected by experimental design. Trevan (1927, 1929) recognized its importance to assay precision before methods of equivalent deviate analysis had been developed. Gaddum's claim (1933) that closely inbred Wistar rats showed higher regression coefficients than a mixed group has been challenged by Biggers and Claringbold (1954). Morrell and Allmark (1941) reported only a small, non-significant increase in b as a result of inbreeding their stock used in assaying the trypanocidal activity of neoarsphenamine. As suggested in § 6.8, possibly the best hope of breeding for assay precision lies in using F_1 hybrids of carefully chosen inbred lines.

(iii) As usual, a small value of g is desirable in order that imprecision in b shall be without serious effect on the reliability of R. Ideally, g will be so small that approximate fiducial limits obtained from a formula for $\mathrm{Var}(M)$ are practically the same as those calculated from Fieller's theorem. Reduction of g below 0·05 is of negligible importance, but a value in excess of 0·2 is very undesirable. Measures taken under headings (i), (ii), (iv), and (vi) also benefit g.

(iv) Equal division of N between S and T is generally advantageous for the reduction of $\Sigma\left(\frac{1}{Snw}\right)$. Unlike parallel line assays (Chapter 6), the choice of doses also affects this quantity (except for the angle transformation). For most metameters likely to be used, including the probit and the logit, w has a maximum when the response rate is 0·50; hence, if minimizing $\Sigma\left(\frac{1}{Snw}\right)$ were the only consideration, N would be equally divided between two doses guessed to be about the ED50s of the two preparations. In practice, this would be useless because of the need for estimating the regression coefficient also. If the response rate is kept between 0·05 and 0·95, w for probits will never fall below one-third of its maximum, but if P is allowed to become as extreme as 0·01 or 0·99, w can fall to one-tenth of its maximum. For logits, w decreases more rapidly as the value of P is moved away from 0·50, being one-fifth of its maximum at $P = 0·05$ or 0·95, and only 4 percent at $P = 0·01$ or 0·99. Doses should therefore be chosen, in the light of any pre-existing information on potency, so as to be reasonably sure to give responses between 0·05 and 0·95. Unless special circumstances dictate otherwise, N should be equally divided between all doses.

(v) Exactly as for parallel line assays (§ 6.8), with each dose of S should be associated a dose of T determined by equation (6.8.2):

$$x_T = x_S - \log R_0,$$

where R_0 is a guessed value of ρ. If R_0 is near to the truth, $(M - \bar{x}_S + \bar{x}_T)$ should thus be made small in absolute value.

(vi) The largest contributions to ΣS_{xx} will arise when $w(x - \bar{x})^2$ is a maximum. If doses have been successfully chosen to be symmetric with respect to the ED50 of each preparation and equal numbers of subjects are used at every dose, $\bar{x}$ will be the ED50 and the maximum contribution to S_{xx} will arise when $w(Y - Y_{50})^2$ is a maximum (where Y_{50} is the value of the response metameter at the ED50). For a normal tolerance distribution, the maximum is at probit values of 3·42 and 6·58, or approximately response rates of 0·06 and 0·94. For the logistic distribution, the maximizing logits are 3·80 and 6·20, corresponding to responses of 0·08 and 0·92. For angles, since w is constant, the maximum would be at responses of 0·00 and 1·00. The contribution to S_{xx} declines markedly if more extreme doses are used, catastrophically for angles as the horizontal sections make zero contributions. The consequences are discussed more fully in § 19.2.

A compromise between the optimal requirements under headings (iv) and (vi) is possible without too great a loss on either Snw or S_{xx}. Contributions to Snw and S_{xx} are both about 70 percent of the maximum possible when $P = 0.15$ or $P = 0.85$ for the probit transformation, and correspondingly $P = 0.20$ or $P = 0.80$ for logits. Brown (1966) has described a similar set of rules for guidance in designing an assay using quantal responses to be analyzed by the Spearman–Kärber method (§ 18.7).

19.2 The choice of doses

The complexities of differential weighting make impracticable any comparison of the efficiencies of alternative assay schemes in the detail of Chapter 6. Tables 19.2.1 and 19.2.2 give some information. For an assay using N subjects in a symmetric (k, k) point design with the probit model, Table 19.2.1 shows $Nb^2 \text{Var}(M)$. Here $\text{Var}(M)$ is the asymptotic variance of M (cf. equations (4.12.9) and (6.9.7)), with the assumption that responses are exactly those at which the assayist aimed when he chose his doses; thus by definition

$$Nb^2 \text{Var}(M) = \frac{4k}{Sw}, \tag{19.2.1}$$

where k is the number of doses of each preparation (2, 3, or 4 in the table) and S denotes summation over one subject at each dose of one preparation only. For example, corresponding to values of $-1.80, -0.60, 0.60, 1.80$ for $(Y - Y_{50})$ in probits are weighting coefficients 0·179 94, 0·557 88, 0·557 88, 0·179 94 respectively. The probits represent response rates of 0·04, 0·27, 0·73, 0·96 respectively; hence for the first of the (4, 4) assays in Table 19.2.1,

§ 19.2   DESIGN OF ASSAYS BASED ON QUANTAL RESPONSES

TABLE 19.2.1   Values of g, the variance of M, and the effective variance of M for two sizes of assay, in various designs using a total of N subjects (Probits)

Subjects divided equally between doses giving percentage responses shown below	$Nb^2\text{Var}(M)$	Ng	At probability 0·95 $b^2 V_E(M)$	
			$N = 48$	$N = 240$
4, 96	22·23	6·6	0·537	0·0952
7, 93	14·87	6·3	0·357	0·0636
12, 88	10·80	7·2	0·265	0·0464
18, 82	8·48	10·1	*0·224*	0·0369
27, 73	7·17	19·1	0·248	*0·0325*
33, 67	6·77	32·1	0·425	0·0326
38, 62	6·49	69·3	—	0·0380
44, 56	6·33	270·	—	—
46, 54	6·31	606·	—	—
4, 50, 96	12·04	9·9	0·316	0·0523
7, 50, 93	10·21	9·5	0·265	0·0443
12, 50, 88	8·71	10·8	0·234	0·0380
18, 50, 82	7·60	15·1	*0·231*	0·0338
27, 50, 73	6·85	28·7	0·355	*0·0324*
33, 50, 67	6·60	48·1	—	0·0344
38, 50, 62	6·42	104·	—	0·0472
44, 50, 56	6·32	406·	—	—
46, 50, 54	6·30	909·	—	—
4, 27, 73, 96	10·84	9·8	0·284	0·0471
7, 31, 69, 93	9·41	10·2	0·249	0·0410
12, 34, 66, 88	8·24	12·2	*0·230*	0·0362
18, 38, 62, 82	7·36	17·6	0·242	0·0331
27, 42, 58, 73	6·75	34·0	0·482	*0·0328*
33, 44, 56, 67	6·54	57·4	—	0·0358
38, 46, 54, 62	6·40	124·	—	0·0553

Italics indicate the tabulated value nearest to the minimum; a dash indicates that the fiducial range would be infinite because b would not differ significantly from zero.

$$Nb^2\text{Var}(M) = \frac{16}{2 \times (0 \cdot 179\,94 + 0 \cdot 557\,88)}$$

$$= 10 \cdot 84.$$

Unless N is large, $\text{Var}(M)$ gives an optimistic assessment of precision, and any failure to obtain exactly the planned response rates will increase $\text{Var}(M)$ because $(M - \bar{x}_S + \bar{x}_T)$ is not zero. The table also shows Ng, where

$$Ng = \frac{(1 \cdot 960)^2 \times k}{Sw(Y - Y_{50})^2}; \qquad (19.2.2)$$

again for the first of the (4, 4) assays

$$Ng = \frac{(1 \cdot 960)^2 \times 4}{2 \times 0 \cdot 179\,44 \times (1 \cdot 80)^2 + 2 \times 0 \cdot 557\,88 \times (0 \cdot 60)^2}.$$

All entries in Table 19.2.1 have been calculated for equally spaced values of $(Y - Y_{50})$, and therefore for equally spaced doses, to which the response rates shown correspond approximately. When $(M - \bar{x}_S + \bar{x}_T) = 0$, the effect of g in widening the fiducial limits of M is equivalent to a division of Var(M) by $(1-g)$: if $V_E(M)$, the *effective variance* of M, be defined by

$$V_E(M) = \text{Var}(M)/(1-g), \tag{19.2.3}$$

use of the square root of $V_E(M)$ as though it were a standard error will give the right fiducial limits. Table 19.2.1 shows this effective variance for $N = 48$ and $N = 240$. Table 19.2.2 has been similarly calculated for logits.

TABLE 19.2.2 Values of g, the variance of M, and the effective variance of M for two sizes of assay, in various designs using a total of N subjects (Logits)

Subjects divided equally between doses giving percentage responses shown below	$Nb^2\text{Var}(M)$	At probability 0.95		
		Ng	$bV_E(M)$	
			$N = 48$	$N = 240$
3, 97	38.62	11.4	1.057	0.1690
5, 95	22.13	9.4	0.574	0.0960
8, 92	13.11	8.7	0.334	0.0567
14, 86	8.21	9.7	0.215	0.0357
23, 77	5.62	15.0	*0.170*	0.0250
29, 71	4.87	23.1	0.195	*0.0225*
35, 65	4.37	46.6	3.223	0.0226
43, 57	4.09	175.	–	0.0626
45, 55	4.04	388.	–	–
3, 50, 97	9.94	17.2	0.322	0.0446
5, 50, 95	8.81	14.2	0.261	0.0390
8, 50, 92	7.45	13.1	0.214	0.0328
14, 50, 86	6.08	14.6	*0.182*	0.0270
23, 50, 77	4.95	22.5	0.194	0.0228
29, 50, 71	4.54	34.6	0.339	*0.0221*
35, 50, 65	4.24	70.0	–	0.0249
43, 50, 57	4.06	262.	–	–
45, 50, 55	4.03	582.	–	–
3, 23, 77, 97	9.81	13.0	0.280	0.0432
5, 27, 73, 95	8.27	12.7	0.235	0.0364
8, 31, 69, 92	6.89	13.3	0.199	0.0304
14, 35, 65, 86	5.71	16.1	*0.179*	0.0255
23, 40, 60, 77	4.78	26.1	0.218	0.0224
29, 43, 57, 71	4.44	40.8	0.615	*0.0223*
35, 45, 55, 65	4.20	83.3	–	0.0268

Italics indicate the tabulated value nearest to the minimum; a dash indicates that the fiducial range would be infinite because b would not differ significantly from zero.

The tables illustrate how important allowance for g can be. If in the planning of the assay attention were limited to Var(M), doses near to the ED50 would be chosen (so far as the investigator could guess these), irrespective of the number of subjects to be used. For probits, the column of $Nb^2\text{Var}(M)$ might suggest that any doses giving responses between $P = 0.30$ and $P = 0.70$ would be satisfactory. In fact, even if as many as 240 subjects are available, the optimal design will have its extreme of dose just outside this range, in the neighbourhood of one of the arrangements for which the entry in the last column of Table 19.2.1 is italicized. For fewer subjects, the doses should be spaced more widely. To a rough approximation with which both tables agree, a (4, 4) assay with 240 subjects should be designed with the aim of giving response rates of about 0·25, 0·40, 0·60 and 0·75 for both preparations. The same set of doses with only 48 subjects, however, would give an effective variance of the log potency nearly twice as great as with doses that produce response rates of about 0·15, 0·35, 0·65 and 0·85.

The value of this information in the designing of a particular assay is limited: if the investigator knew the doses that would give the desired responses, he would have no need of an assay. Nevertheless, he will often have some idea of the ED50 of S and perhaps even of that of T; from previous experience of the assay technique, he may also be able to guess b, the reciprocal of the SD of log tolerances. With the aid of these, he can attempt to select doses that will lead to a reliable potency estimate. The objections to a (2, 2) design have been sufficiently stressed earlier; without very strong evidence from past experience, or clear *a priori* knowledge both that T is a dilution of S and that the tolerance distribution is of a particular form, on the basis of which he can assert both fundamental and statistical validity, he should choose a design that permits validity tests. Tables 19.2.1 and 19.2.2 indicate that (3, 3) and (4, 4) designs need not be appreciably less reliable than (2, 2), as a choice of doses near to the optimal makes values of $V_E(M)$ nearly the same for all three. The optimal for any specified number of subjects may easily be found by constructing a column of values of $b^2 V_E(M)$ for the appropriate N, from Table 19.2.1 or Table 19.2.2 and equation (19.2.3). When N is small, a (3, 3) design with doses that give 0·15, 0·50, 0·85 as the response rates, or a (4, 4) design with doses that give 0·15, 0·35, 0·65, 0·85, seems to be about ideal; when the number of subjects is much larger, the doses can with advantage be put closer together. The procedure should be to decide at what response rates to aim, then to use existing information in guessing the corresponding doses, and finally to modify these doses slightly so that they are equally spaced.

A consideration of (2, 2) assays with $N = 48$ illustrates well the use of inadequate information in the choice of doses and the wisdom of erring in the direction of wide spacing. Whether he planned to use probits or logits, an investigator might wish to have responses of 0·25, 0·75 so as to be close to the optimal. If he spaced his doses too narrowly and had response rates of about 0·35, 0·65, the assay could be practically worthless, whereas if his spacing were too wide, even to the extent of giving 0·10, 0·90 as the response rates, the increase in $V_E(M)$ would be much less serious. This discussion assumes perfect symmetry, so that $(M - \bar{x}_S + \bar{x}_T)$ is zero; failure to achieve this will increase all variances, to an extent that depends also upon ΣS_{xx} and is therefore minimized by fairly widely spaced doses (§ 19.1, (vi)). Moreover the effective variance, $V_E(M)$ in equation (19.2.3), depends upon the significance level to be used in the fiducial limits. The numerical values in Tables 19.2.1 and 19.2.2 relate only to 0·95 limits; adoption of 0·99 limits would make wider spacing of doses desirable. The contrast with the recommendations for quantitative responses (§§ 6.8, 6.9), that the widest possible dose range should always be used, is important. If very little is known about the potency of T, however, further doses might be added at either end of the scale with the intention that data might be discarded if they introduced marked non-linearity.

Healy (1950) sought to determine the least number of subjects needed for a specified reliability of estimate, rather than to maximize the reliability obtainable from a specified total number of subjects. His assumptions about the information available for use in planning differ slightly from those adopted here. His recommendations, which he presents in convenient graphical form, are qualitatively similar to those obtained above.

In some circumstances, exact measurement of dose may be impossible until after a subject has been tested. For example, a subject may be allowed to eat food containing approximately the desired dose, but the exact dose received depends upon the amount of food the subject actually consumes. Although the investigator can exercise some control, he will in general find that no two subjects have the same dose. The probit method or one of its analogues can still be applied to the estimation of potency, but validity tests are unsatisfactory unless the number of subjects is large. The standard calculations, with every working probit a maximum or a minimum value, may converge slowly, but with a computer there need be no serious difficulty. Finney (1947c) gave an example (not an assay) from which the procedure should be apparent (cf. Bliss, 1938; Finney, 1971, § 9.3).

19.3   Block restrictions

Constraints that balance inherent differences in the experimental material, or unavoidable differences in the conditions of experimentation, often improve the precision of potency estimation, as has been illustrated earlier (notably in Chapters 9 and 10). Similar constraints of design may be expected also to benefit assays based upon quantal responses. For example, mice from the same litter are likely to vary less in their individual insulin tolerances than are mice from different litters: randomized complete or incomplete block designs, using litters as blocks, might be used in order to balance doses within litters and so to improve the precision of an assay by the mouse convulsion method.

For the analysis of results, the model adopted should generalize the concept of a tolerance distribution exactly as if direct measurement of tolerances were possible, that is to say essentially as in § 2.8. In the simple design blocked according to litters, each litter may be supposed to have its own tolerance distribution such that the means differ but the variance is constant. Thus probit or other metametric regressions could in theory be fitted to litters separately, with the constraint that all are parallel and that the horizontal difference between S and T regressions is the same in every litter (cf. § 14.5). Provided that a set of first approximations to the expected probits can be constructed, the routine of computation can proceed in the ordinary manner. Each cycle uses one line for each preparation in each litter, and the regression coefficient,

$$b = \Sigma S_{xy} / \Sigma S_{xx}, \qquad (19.3.1)$$

is calculated by summations over all litters. Each litter then gives its own value

§ 19.3   DESIGN OF ASSAYS BASED ON QUANTAL RESPONSES   411

of M, and the final estimate of potency is found from a mean of these. For simplicity, the unweighted arithmetic mean might be taken, using essentially a form of calculation described in another context (Finney, 1946b; 1971, § 8.5). This would allow litters of very high or very low mean tolerance to exert undue influence on the final estimate; a more exact procedure would be to weight the values of M inversely in proportion to $\Sigma\left(\dfrac{1}{Snw}\right)$ for each litter, or to apply the methods of Chapter 14.

The same system of calculation could be applied to incomplete block designs, provided that in every litter both S and T are represented by at least one dose. With incomplete blocks the need for using a weighted mean of M increases because of the differing precisions of the separate values (cf. Moore and Bliss, 1942). Fitzhugh *et al.* (1944) described a comparative assay of selenium preparations in respect of their toxicity to rats; they arranged this as a balanced incomplete block design for ten treatments (three doses of each of three preparations plus a control) with litters of six, but ignored the litter classification in the statistical analysis. When the design uses two or more classes of block restriction simultaneously, as in a Latin square, the difficulties are greater. Even for randomized complete block designs, the method of analysis now to be described is indeed the correct one, although that suggested above is likely to be a good enough approximation. Consider the direct assay in § 2.8, but neglect the covariance discussion and confine attention to the tolerances of the 144 cats. The analysis of variance of this Latin square design implicitly assumed that the log tolerance of any cat could be expressed as the sum of five components: a general mean, a deviation attributable to the day of experiment, a deviation attributable to the combination of operator and time of day, a deviation for the particular drug preparation, and a random element for the particular cat. A statement that the log tolerance distribution is normal means that the random elements have a normal distribution. If a Latin square is used for an assay with quantal responses, it is reasonable to make the same assumption that would be made if each quantal response were replaced by a direct measurement of the corresponding tolerance.

TABLE 19.3.1   Design for a (3, 3) assay using six subjects per day balanced over six litters

Litter no.	Dose on day no.					
	1	2	3	4	5	6
I	S_2	T_1	T_3	T_2	S_3	S_2
II	S_2	S_1	T_1	T_3	T_2	S_3
III	T_1	T_2	S_2	S_3	S_1	T_3
IV	S_3	S_2	T_2	T_1	T_3	S_1
V	T_2	T_3	S_3	S_1	S_2	T_1
VI	T_3	S_3	S_1	S_2	T_1	T_2

For example, a (3, 3) assay using six litters of six animals might be arranged as the 6 × 6 Latin square in Table 19.3.1, in order that only six animals would have to be tested each day. Six litters would scarcely give adequate precision, and the assay might be improved by adding further sets of six with other Latin square arrangements, but the scheme shown will serve for illustration here. If the tolerance of each of the 36 subjects can be expressed as the sum of five components and the random elements are normally distributed with constant variance, the general theory of equivalent deviate transformations (§ 17.4) shows that the expected probit for each subject must be expressible as

$$Y = a_i + r_j + c_k + bx, \qquad (19.3.2)$$

where a_i takes one of the two values a_S and a_T (combining the general mean and the preparation components), r_j is the component for a row or litter in Table 19.3.1 ($j =$ I, II, ..., VI), c_k is the component for a column or day ($k = 1, 2, \ldots, 6$), and b is a regression coefficient. The constraints that the r_j sum to zero and the c_k sum to zero are to be imposed. A set of expected probits conforming to this pattern must be guessed. This may not be easy. A reasonable start is to take a_S, a_T, b from a dose–response diagram that ignores rows and columns; r_j might be taken as the deviation of an empirical probit for the subjects of row i (ignoring dose and column) from an empirical probit for the 36 subjects, and similarly for the c_k. Sets of 36 weights and 36 working probits (all maximum or minimum values) are then formed in the usual way. Weighted totals of working probits for each preparation, row and column are equated to the similarly weighted totals of parametric representations of which (19.3.2) is typical, and ΣS_{xy} is equated to ΣS_{xY}. Thus a set of linear equations for the a_i, r_j, c_k and b is obtained, the solutions of which are inserted in equation (19.3.2) to give a revised set of expected probits for the second cycle of iteration. The process is basically the same as for the analysis of variance of non-orthogonal data, the non-orthogonality arising here because of differential weighting. From the final cycle of iteration,

$$M = (a_T - a_S)/b \qquad (19.3.3)$$

is the estimate of $\log \rho$. Inversion of the matrix for the last set of linear equations provides assessments of variances and covariances for all the parameters, and R_L, R_U are found by Fieller's theorem as usual. Irwin and Standfast (1955) described simple approximations for assays of pertussis vaccine using mouse-litters as blocks, but modern computational facilities probably make such approximations unnecessary.

In theory, the problem of the analysis of data from an assay of complex design is solved, for the above method may be adapted to any design. In practice, a complete analysis leading to assessment of the precision of R is laborious unless a general maximum-likelihood computer program is available. If the doses are not too extreme, analysis using angles instead of probits might be tried; this would reduce the labour considerably, because the constancy of w would preserve the orthogonality of the Latin square, and the linear equations for the parameters would be solved by simple evaluation of arithmetic means. After a few cycles of iteration in this way, the expected angles might be replaced by corresponding expected probits or logits with which analysis could continue.

The gain in precision should show itself by a smaller SD of log tolerances (a larger b) than if no restrictions of design were used; in fact the SD within litters and within days is wanted. To analyze responses from the 6×6 Latin square as though they related to a simple unconstrained design with six subjects at each dose would be regrettable. It is equivalent to saying that row and column differences are in reality negligible, and therefore to a denial of the reasons for using the Latin square. The outcome might be approximately correct in the potency estimate, but the limits would be biased and possibly much too wide if inter-litter or inter-day variation were large.

19.4 Cross-over designs

If the quantal response of a subject is some minor reaction of a kind that can be repeated indefinitely without harming or altering the subject, the possibility of using a cross-over design needs to be considered. One of the simplest is the twin cross-over (§ 10.2). Each subject might be tested once or several times with the first dose before passing to the second, so giving either a single quantal record or the proportion of responses in a short series. A more elaborate design would be the 6×6 Latin square in Table 19.3.1, with rows now representing six subjects and columns six occasions of testing (one or several tests on each occasion). As with quantitative responses (Chapter 10), the experimenter might hope to improve the precision of estimation by testing doses of each preparation on every subject and so eliminating inter-subject variation. Unfortunately, the difficulties of analysis encountered in § 19.3 arise again. All designs suggested in Chapters 9, 10 and 11 can also be used for quantal responses; the general method of estimation outlined in § 19.3 is always applicable, but the more complex the design the more laborious will the analysis be and the less safely can any simple approximation be adopted.

Among the few published examples of quantal cross-over assays are Somers and Edge (1947) on anaesthetics using guinea-pigs, Miller *et al.* (1948) on spasmolytic drugs using isolated pieces of tissue, and Blackith (1950) on the paralytic action of insecticides. These authors did not attempt to take account of the cross-over constraints in the analysis, and simply used total response rates at each dose as though each subject had been used only once, although Blackith used an approximate angle analysis to indicate some substantial gain over a design that used each insect only once. Claringbold (1956) suggested analyzing exactly as for quantitative responses with the metameter $y = 1$ for a response, $y = 0$ for a non-response; he claimed that, if all expected responses lie between $P = 0.05$ and $P = 0.95$, this will approximate well to the correct result. The evidence is small, but the idea is at least reasonable and undoubtedly attractive. One interesting example is an assay of relative potency of two vasoconstrictors (McKenzie and Atkinson, 1964), in which responses were assessed in human subjects receiving topical application of ointments.

19.5 Multiple assays

An assay of several test preparations simultaneously against one standard does not of itself present new features for quantal responses. If c test preparations are to be included, the analysis will involve constraining $(c + 1)$ regression lines to be parallel, and the χ^2 for the test of fundamental validity (parallelism) will have c degrees of freedom. The method of calculation for an assay with no block restrictions should be obvious from Chapters 11 and 18, and an example has been given elsewhere (Finney, 1971, §§ 6.2 and 6.6). The program described in § 18.5 includes facilities for simultaneous analysis and estimation for any reasonable value of c. The rule of allocating more subjects to the standard preparation than to each of the test preparations, in the ratio of $c^{1/2}:1$ (§ 11.6), remains applicable, and the choice of dose levels should follow the principles of § 19.2. If the design involves block constraints, the ideas of § 19.3 must be adapted.

19.6 Sequential sampling

An interesting technique for estimating an extreme percentage point was suggested by Bartlett (1946; Finney, 1971, § 10.2). He used a system of inverse sampling to concentrate effort in the neighbourhood of the ED5 or ED99, so as to reduce any biases due to extrapolation. This is not of particular interest for bioassay.

More relevant to the present purpose are various methods of sequential approximation for estimating the ED50. Although these are applicable only when subjects can be tested individually and the response for one subject recorded before the dose for the next is decided, they are attractive in some bioassay situations. Dixon and Mood (1948) take a sequence of equally spaced log doses

$$\ldots, x_{-3}, x_{-2}, x_{-1}, x_0, x_1, x_2, x_3, \ldots,$$

where x_0 is a first guess at the log ED50. They then test subjects one at a time, the first at x_0; if a subject responds, the next subject is tested at a dose one step higher, and if it does not respond the next is tested one step lower. Even if x_0 is not very close to the log ED50, this *staircase rule* forces a concentration of testing in the neighbourhood of the log ED50. The spacing of the doses is important; if it is too narrow, testing may have to continue a considerable time before the log ED50 is approached, whereas if it is too wide, testing is likely simply to oscillate between a high and a low dose. An early recognition of poor spacing can be followed by modification, say to double or half the initial spacing.

Discussion of the merits of the staircase method has been obscured by confusion between the inherent properties of a design and the availability of a simple computational routine. Dixon and Mood proposed to estimate the ED50 by calculations analogous to the Spearman–Kärber method. Brownlee, Hodges and Rosenblatt (1953) claimed much higher efficiency for their version of these calculations than for the probit method. However, the staircase rule for determining doses does not affect the applicability of maximum likelihood estimation, and the usual probit calculations can quite properly be used on responses from a staircase experiment. The gain in precision comes from the improved distribution of subjects over doses consequent upon making each dose dependent upon the information from subjects previously tested. The Dixon and Mood estimator is

$$m = x_0 + d\left(\frac{S(i)}{N} - 0.5\right), \qquad (19.6.1)$$

where d is the dose interval, N the number of subjects tested up to the end of the trial, and $S(i)$ is a summation of the subscripts for the doses x_i at which a subject responds. This has the advantages and some of the flaws of Spearman–Kärber estimation, and probit or other maximum likelihood calculations are preferable as a general procedure. Wetherill (1963) showed a marked dependence of the efficiency of the method on the choice of d.

An analogous method is based upon the Robbins and Monro (1951) procedure for stochastic approximation. A suitable form – there are many variants – is to choose a sequence of doses according to the rule

$$x_{i+1} = x_i - c\left(\frac{r_i}{n_i} - 0.5\right) \bigg/ i, \qquad (19.6.2)$$

where at dose x_i a group of n_i subjects is tested and r_i respond and c is an arbitrary constant; if the tolerance distribution is normal, theory indicates $\sigma\sqrt{2\pi}$ as an ideal value for c. The n_i need not be constant, but $n_i = 1$ is a permissible choice. Again, standard maximum likelihood estimation can be used but, as Wetherill (1963) has shown, the estimator x_{k+1}, determined by equation (19.6.2), is very simple to obtain and is closely related to maximum likelihood.

Cochran and Davis (1963, 1964, 1965) studied this method for totals of 6, 12, 24 and 48 subjects divided equally between doses. They had in mind particularly the needs of bioassay and the advantages of keeping k small so as to expedite completion of the experiment. If x_1 is close to μ, dividing the N subjects between a small number of doses will be satisfactory, but unless the investigator is very confident in his choice of x_1 he will be unwise to take a small k. Cochran and Davis examined many alternatives and suggested variants on the rules for determining successive doses.

With all these methods, estimation of the ED50 must proceed independently for S and T and relative potency is eventually estimated by equation (18.6.1). Precision may not be very readily assessed, unless full maximum likelihood calculations are undertaken and consequently the advantage of remarkably simple arithmetic is lost. The methods are of limited practical value, not least because they do not provide validity tests, but occasionally they are exceptionally suitable. The prospective user would be wise to read carefully the references already quoted and others (Tsutakawa, 1967; Hsi, 1969, Finney, 1971, § 10.3) because of the many special features and the desirability of tailoring a method to the needs of a particular situation.

19.7 Quality control

Control charts were commended in §5.10 as a means of confirming that routine assays are behaving in accordance with statistical theory. Methods may readily be devised for assays based on quantal responses. If a particular experimental technique and assay design is in regular use, b should remain in control, as also should the values of χ^2 for linearity and parallelism; hence each might be made the subject of a control chart. As mentioned in §5.10, the pooling of information on b from several consecutive assays may be considered as an aid to precision; this procedure is not to be encouraged unless there is great confidence that variations in b are due only to the chances of sampling (§18.6).

Throughout this book, the emphasis has been on estimating the potency of a test preparation. When the need is only to inspect the quality of test preparations, in order to ensure that those accepted as satisfactory conform to certain minimal requirements, the statistical problems are different. For example, the permissible level of toxicity of a drug might be defined, either in absolute terms or relative to simultaneous trials with a standard, by the mortality produced in a certain stock of subjects. Perry (1950) discussed the relative merits of different experimental designs, with particular reference to the efficiency with which they discriminate between acceptable and unacceptable preparations. He illustrated his argument by considering a requirement that the toxicity of T relative to S shall not exceed $\rho = 1 \cdot 2$; for various designs, he obtained curves to show (as a function of ρ) the probability that T fails to comply with the criterion.

Perry's approach is closely related to the use of operating characteristics in industrial quality control. Since this book is concerned with assays for the estimation of potency, rather than with routine checks on the quality of drug production, his methods will not be described here.

19.8 Comparison of assay techniques

The foregoing discussion of assay design has been concerned only with statistical aspects of the problem. It has assumed that all comparisons relate to alternative designs to be used with the same population of subjects and the same quantal response. The regression coefficient of equivalent deviate on log dose, however, is dependent upon the variability of individual log tolerances in that population. For a normal tolerance distribution, $1/b$ is an estimate of the SD of log tolerances; for the logistic distribution $\pi/2b\sqrt{3}$ $(= 0 \cdot 9059/b)$ and for the angle distribution $(\pi^2 - 8)^{1/2}/4b$ $(= 19 \cdot 59/b$ for angles measured in degrees) are estimates of the SD (Armitage and Allen, 1950). Any change in population or modification in technique that tends to increase b will improve assay precision.

In a direct assay, whatever the form of the log tolerance distribution, by equation (2.6.2)

$$\text{Var}(M) = s^2 \left(\frac{1}{N_S} + \frac{1}{N_T} \right), \tag{19.8.1}$$

where s is an estimate of the variance per response. For a quantal response assay, if g is small and $(M - \bar{x}_S + \bar{x}_T)$ nearly zero,

§ 19.8 DESIGN OF ASSAYS BASED ON QUANTAL RESPONSES

$$\text{Var}(M) \simeq \frac{1}{b^2} \left\{ \Sigma \frac{1}{Snw} \right\}. \tag{19.8.2}$$

Equation (19.8.2) shows that, for probits, w is the value of the information from a single subject in a good assay relative to its unit value in a direct assay. For logits, w must be multiplied by 0·8225, and for angles by 383·6, in order to give this comparison. A symmetrical direct assay using N subjects will assign $N/2$ to each preparation and give a log potency for which

$$\text{Var}(M) = 4s^2/N. \tag{19.8.3}$$

Consequently the value of $b^2 V_E(M)$ in Table 19.2.1 may be compared with $4/N$ in order to show the loss consequent upon restriction to quantal responses. Even for optimal designs, the effective variance is trebled for $N = 48$ and doubled for $N = 240$; for these numbers of subjects, quantal responses can give only 37 per cent and 51 percent of the information that direct tolerance measurements would contain. As N becomes large, this percentage approaches its limiting value, 63·7, corresponding to the maximum value of w. For logits, $V_E(M)$ in Table 19.2.2 must first be divided by 0·8225 in order to convert $1/b^2$ into a measure of the variance of log tolerance; the efficiencies of optimal designs are 40 percent for $N = 48$, 62 percent for $N = 240$, and 82·2 percent in the limit.

Alternative assay techniques that use quantal response records may be compared in terms of the magnitudes of $1/b$ or its appropriate multiple for equivalent deviates other than the probit. Gaddum (1933) reported 25 assays for which $1/b$ ranged from 0·04 to 0·91. Bliss and Cattell (1943) reported values ranging from 0·01 to 0·76 in 35 unrelated assays analyzed by probits; four methods for digitalis assay gave values between 0·05 and 0·14. The value of $1/b$ is comparable with the inherent standard error per response, s/b, for an assay using quantitative responses (§ 6.11). For example, Bliss and Cattell's five assays of vitamin B_1 based on weight gains of depleted rats gave s/b ranging from 0·06 to 0·24; in three assays using time to recurrence of polyneuritis in rats as a response, s/b ranged from 0·03 to 0·14. They also recorded 0·28 for an assay of this vitamin in which cure of polyneuritis in rats was used as a quantal response, and 0·60 when a similar technique was adopted with pigeons. These findings point to the superiority of assays based on quantitative responses in respect of their efficiency for estimating vitamin B_1. Miller (1944) gave a good elementary account of comparing efficiencies of alternative techniques for digitalis assay.

An indirect quantal assay will always give less information per subject tested than would the corresponding direct assay; it will often, but not inevitably, give less information than an indirect assay based upon quantitative responses. On the other hand, quantal responses may involve simpler, less costly experimental techniques than either of the others, and may sometimes be the only practicable basis for assay. If values of $1/b$ for quantal responses, s for a direct assay, s/b for quantitative responses and relative costs per subject are known, a complete comparison of economic efficiencies is possible.

20
Special problems with quantal responses

20.1 Concomitant variation

In an assay based upon quantal responses, the analysis of covariance can be used in order to estimate the allowance that should be made for the effect of a concomitant variate, such as the initial weights of subjects, on tolerances. All sums of squares and products must be weighted with the weights appropriate to the equivalent deviate transformation adopted. The calculations are likely to be lengthy, but should present no special difficulties to those who have mastered the techniques described in Chapters 12 and 18. In most assays, a batch of subjects will be tested at each dose, and the analysis may then be based upon the mean value of the concomitant for each batch rather than upon values for individual subjects.

20.2 Combination of estimates

As in Chapter 14, when several assays of one test preparation have been performed, a single estimate of potency may be wanted. Methods analogous to those of Chapter 14 are often adequate. If g is small for every assay, the simple weighted mean may be used (§ 14.1; Perry, 1950). If some or all of the values of g are not small, the method of § 14.3 can be tried with reservations noted below. In the notation of that section,

$$\left. \begin{array}{l} v_{11} = \Sigma\left(\dfrac{1}{Snw}\right) + \dfrac{(\bar{x}_T - \bar{x}_S)^2}{\Sigma S_{xx}}, \\ v_{22} = 1/\Sigma S_{xx}, \\ v_{12} = -\dfrac{\bar{x}_T - \bar{x}_S}{\Sigma S_{xx}}, \end{array} \right\} \qquad (20.2.1)$$

where it is to be understood that $1/Snw$ is summed only over the standard and the particular test preparation, whereas S_{xx} is summed over all preparations in an assay. As usual for quantal responses, σ^2 is taken to be 1·0 and normal deviates are used. The remainder of the notation is exactly as in § 14.3.

Although in Chapter 14 the method is an exact maximization of likelihood for the response model applicable, that is not strictly true here. A full solution to the problem of estimation would require that the log likelihood for all the assays be expressed as a function of all the relevant parameters. In the simple instance of a standard and a single test preparation for every assay, this implies parameters $\alpha_{Si}, \alpha_{Ti}, \beta_i$ for each assay, constrained by

$$(\alpha_{Ti} - \alpha_{Si})/\beta_i = \mu \qquad (20.2.2)$$

§ 20.3   SPECIAL PROBLEMS WITH QUANTAL RESPONSES   419

for a constant μ. This is basically similar to what was done in § 14.3, but now, as an extension of equation (17.5.3), L for all the assays has to be written in terms of many binomial components. Maximization with respect to the parameters α_{Si}, α_{Ti} ($i = 1, 2, \ldots, p$) and μ (since β_i can be eliminated) is then required. No further details or numerical example will be given as a general maximum likelihood program should encounter no difficulty other than slow convergence, whereas computations organized for desk calculation would be excessively tedious.

Nothing is known of how the full method compares with the approximation, and empirical study of a few examples could be interesting. If the separate assays agree well in individual estimates of μ, full maximum likelihood should be unnecessary, but any substantial disagreement (whether or not statistically significant) is likely to cause the different weighting implicit in the full method to assume practical numerical importance.

20.3   Natural response rates

In § 17.16, mention was made of a generalization of § 17.5 that takes account of a natural rate of response and also of an ability of some subjects to resist the stimulus, however high the dose. For example, if death is the response, some subjects may die during the course of an experiment from causes quite unconnected with the stimulus. In assays of insecticides, insects kept as untreated controls (or perhaps treated only with the non-toxic carrier used for the insecticide) may die between the time of treatment and the time of examination; in assays of fungicides by tests of spore germination, some spores may fail to germinate even when untreated or treated with very low doses, and evidence from very high doses may indicate that others are effectively immune to the fungicide since they germinate however high the dose. Subjects of either type can themselves give no information on potency: their reactions are independent of dose. Unfortunately, they cannot be discarded before the assay is begun, because they are indistinguishable from the others, nor can they be rejected from the results. The experimenter cannot discover whether a subject that responds does so because of the dose received or would have done so in any case, nor can he be sure that inadequate dose rather than immunity was the reason for non-response. In some circumstances, post-mortem examination of subjects might throw light on these matters; to reject subjects shown by such examination to have died from natural causes or to be immune to the stimulus would be legitimate, provided that *all* subjects were examined, irrespective of their dose or response. For most routine assays this course is impossible, and allowance for the two classes must be made in the statistical analysis. The generalized analysis is described below.

Suppose that a proportion C_1 of the population of subjects will respond whatever dose is given, a proportion C_2 is immune to the stimulus, and the remaining proportion $(1 - C_1 - C_2)$ behaves according to the tolerance distribution. Alternatively, one could express the model in terms of $D = 1 - C_2$, as in § 17.16, but the present notation maintains a greater symmetry. With $P (= 1 - Q)$ defined by equation (17.5.4) as the proportion of subjects in the third segment of the population that responds to the dose of the stimulus, the expected response rate for responses from all sources is

$$P^+ = C_1 + (1 - C_1 - C_2)P. \tag{20.3.1}$$

This equation, previously (17.16.1), assumes natural response and immunity to be entirely independent of the action of the stimulus. Experience has shown the assumption adequate for many sets of data, and Finney (1949a) has defended its use in the absence of evidence favouring some alternative model of the behaviour of the population. If n subjects receive a dose x, the probability that exactly r respond is the binomial probability in equation (17.5.1), except that P^+ replaces P:

$$P(r) = \binom{n}{r}(P^+)^r (Q^+)^{n-r}. \tag{20.3.2}$$

An iteration for estimating the parameters α, β, C_1, C_2 from an experiment that includes several doses of one preparation can be developed, using the principles of §§ 3.13 and 17.5 and ideas from § 3.8. The theory has been described elsewhere (Finney, 1949b, 1971). Successive iterative steps can still be performed as linear regression calculations, but now the regression must use two auxiliary 'independent' variates defined as

$$x' = Q/Z, \tag{20.3.3}$$

$$x'' = -P/Z, \tag{20.3.4}$$

and a modified weighting coefficient

$$w = \frac{Z^2}{\left(P + \dfrac{C_1}{1 - C_1 - C_2}\right)\left(Q + \dfrac{C_2}{1 - C_1 - C_2}\right)}. \tag{20.3.5}$$

Of these, x' and x'' are easily calculated for any of the standard metametric transformations; indeed, equations (18.1.7), (18.1.8) show that they can be read from tables of minimum and maximum working deviates as

$$x' = Y_1 - Y, \tag{20.3.6}$$

$$x'' = Y_0 - Y. \tag{20.3.7}$$

The weighting coefficient could be tabulated as a function of Y, C_1, C_2, though the table would perforce be large; the second form of equation (20.3.5) is perhaps the quicker for calculation. Working equivalent deviates are found exactly as in equations (17.5.10), (18.1.4)–(18.1.6).

For a single preparation, estimation of the four parameters can proceed much as described in Chapters 17 and 18, with modifications for incorporating regression on x' and x''. Since

$$P = \frac{P^+ - C_1}{1 - C_1 - C_2}, \tag{20.3.8}$$

§ 20.3 SPECIAL PROBLEMS WITH QUANTAL RESPONSES

initial guesses c_1, c_2 can be used in association with empirical response rates

$$p^+ = r/n \qquad (20.3.9)$$

to calculate

$$p = \frac{p^+ - c_1}{1 - c_1 - c_2}. \qquad (20.3.10)$$

From values of p, empirical deviates and a provisional regression line can be found in the ordinary way; this in turn gives a set of values of Y as a basis for forming y, nw, x', x''. Iteration then uses

$$\left. \begin{aligned} bS_{xx} + \frac{\delta c_1}{1 - c_1 - c_2} S_{xx'} + \frac{\delta c_2}{1 - c_1 - c_2} S_{xx''} &= S_{xy}, \\ bS_{xx'} + \frac{\delta c_1}{1 - c_1 - c_2} S_{x'x'} + \frac{\delta c_2}{1 - c_1 - c_2} S_{x'x''} &= S_{x'y}, \\ bS_{xx''} + \frac{\delta c_1}{1 - c_1 - c_2} S_{x'x''} + \frac{\delta c_2}{1 - c_1 - c_2} S_{x''x''} &= S_{x''y}, \end{aligned} \right\} \qquad (20.3.11)$$

$$a = \bar{y} - b\bar{x} - \frac{\delta c_1}{1 - c_1 - c_2} \bar{x}' - \frac{\delta c_2}{1 - c_1 - c_2} \bar{x}''. \qquad (20.3.12)$$

From the solutions $\delta c_1/(1 - c_1 - c_2)$, $\delta c_2/(1 - c_1 - c_2)$ are calculated $\delta c_1, \delta c_2$, after which c_1, c_2 are replaced by $(c_1 + \delta c_1), (c_2 + \delta c_2)$ and these with a, b initiate a new cycle of iteration. In the course of convergence, $\delta c_1, \delta c_2$ will become negligibly small.

If some subjects have been tested at dose zero on the absolute scale, so that necessarily $P = 0.0$ and $P^+ = C_1$, the values of $S_{x'x'}$ and $S_{x'y}$ in (20.3.11) must be increased by

$$\frac{n(1 - c_1 - c_2)^2}{c_1(1 - c_1)} \quad \text{and} \quad \frac{n(p^+ - c_1)(1 - c_1 - c_2)}{c_1(1 - c_1)} \qquad (20.3.13)$$

respectively, where p^+ is the empirical response rate for n subjects at the dose. Similarly, if subjects have been tested at a dose so high that $P = 1.0$ and $P^+ = 1 - C_2$ can be asserted, $S_{x''x''}$ and $S_{x''y}$ must be increased by the corresponding values of

$$\frac{n(1 - c_1 - c_2)^2}{c_2(1 - c_2)} \quad \text{and} \quad \frac{n(q^+ - c_2)(1 - c_1 - c_2)}{c_2(1 - c_2)}. \qquad (20.3.14)$$

In the limit, the inverse matrix of the coefficients on the left-hand side of equations (20.3.11) is the variance–covariance matrix. If it is written

$$\begin{pmatrix} v_{11} & v_{12} & v_{13} \\ v_{12} & v_{22} & v_{23} \\ v_{13} & v_{23} & v_{33} \end{pmatrix},$$

then
$$\text{Var}(b) = v_{11}, \tag{20.3.15}$$
$$\text{Var}(c_1) = (1 - c_1 - c_2)^2 v_{22}, \tag{20.3.16}$$
$$\text{Cov}(b, c_2) = (1 - c_1 - c_2) v_{13}, \tag{20.3.17}$$
and so on, and
$$\text{Var}(\bar{y}) = 1/Snw. \tag{20.3.18}$$

Although the estimates of C_1, C_2 may be of some intrinsic interest, for assay purposes chief importance attaches to a, b. Extension of the calculations to two preparations, with all parameters except α the same for both, follows the obvious pattern. Once iteration is complete, estimation of relative potency and calculation of limits proceeds in the ordinary manner, essentially as described in § 18.1.

Two points require to be noted. At any stage of iteration, the current estimate of C_1 may happen to exceed some of the p^+, and the current estimate of C_2 may exceed some values of $(1 - p^+)$. The values of p from (20.3.10) will then be either negative or greater than 1·0. This is purely a chance of sampling and no occasion for alarm. The corresponding expected rates, P, are by definition within the range 0·0 to 1·0, and application of the formulae for working probits enable the observations at these doses to exert their proper influence on the calculations. Secondly, at any stage of iteration, a reasonable approximation to the goodness of fit test is

$$\chi^2 = S_{yy} - bS_{xy} - \frac{\delta c_1}{1 - c_1 - c_2} S_{x'y} - \frac{\delta c_2}{1 - c_1 - c_2} S_{x''y}, \tag{20.3.19}$$

with a deduction of 2 more degrees of freedom for the extra parameters. In the limit, δc_1 and δc_2 are negligible, and only the first two terms on the right are needed, but of course the degrees of freedom are the same. A check by way of comparing observed and expected frequencies is desirable in any critical case.

Unless C_1, C_2 are very small, w in (20.3.5) is likely to be much smaller than Z^2/PQ, so that precision may be seriously reduced. In practice, C_1 seems perhaps more often to need attention than does C_2. If $C_2 = 0·0$ is assumed, the obvious changes are required. In particular, the third of equations (20.3.11) disappears, as do the third terms on the left-hand sides of the other two equations. Finney (1971, Table II) has given an extensive table of w for this case; Fisher and Yates (1963, Table XI$_1$) give a condensed version. Finney also illustrates the calculations in detail for a bioassay.

The classical iteration procedure has been described above, as it may still be required by those who do not have access to good computer facilities. Moreover, it remains an interesting example of iterative regression calculations. A computer program may be better based upon direct appeal to the log likelihood function. That is to say, one returns to (20.3.2), whence the log likelihood is

$$L = \text{constant} + Sr \log P^+ + S(n - r) \log (1 - P^+). \tag{20.3.20}$$

Here P^+ is related to P by (20.3.1), so introducing two parameters, and P in turn is related to dose by one of the standard models introducing the parameters α, β. Any legitimate optimizing routine can then be used to obtain the maximum of L, the corresponding values for the parameters, the variances and covariances, and any additional details to complete potency estimation. The author's program that was mentioned in § 18.5 in fact includes estimation of C_1, C_2, or both, as options requiring no more than a minor variation of input. The additional parameters necessarily make computation a little slower but not to a major extent. The user of the program need scarcely be aware that he is invoking a more complicated model. It should be regarded as an essential feature of a bioassay program for quantal responses that a variant of this kind can be handled naturally and that the quality of output is unimpaired; in particular, the calculations must be taken all the way to the potency estimates and their limits.

20.4 Assays with unknown numbers of subjects

In some quantal assays, to determine the number of subjects tested at each dose is either impossible or impracticable. For example, in assaying an ovicide or larvicide for use against an insect pest in food, the number of individuals in a treated sample that survive and emerge as mature adults can be easily counted, but the number of eggs or larvae present initially may not be determinable without destroying the natural structure of the sample. An independent estimate of the number tested might be available as a count of adults emerging from a parallel untreated sample. The estimate so obtained has one advantage over a direct count, in that it automatically excludes natural mortality and so is truly comparable with the numbers emerging from treated samples. Response rates calculated from these parallel samples must not be analyzed by the methods of Chapter 18, for the denominator of equation (18.1.1) is now only an estimate of the number of subjects treated.

Finney (1949b, 1971) modified Wadley's (1949a) maximum likelihood estimation of the parameters for this kind of experiment, and showed that it came within a general theory that also includes § 17.5. The method requires that the subjects are initially distributed entirely at random within the bulk from which samples are to be drawn; the number of subjects in a randomly selected sample of specified size will then follow a Poisson distribution. If the mean number of subjects per sample is ν, the probability that exactly n occur in a particular sample is

$$\frac{e^{-\nu}\nu^n}{n!}. \qquad (20.4.1)$$

Whether 'response' is regarded as survival or as non-survival is immaterial; all quantal response theory applies with the categories of response and non-response interchanged. A widespread convention is to define response so that increasing dose usually implies an increasing response rate. Here therefore availability of a subject for counting after exposure to a dose will be termed a non-response.

The probability that exactly s non-responses occur in a sample is the probability that the sample contained exactly n subjects, multiplied by the probability that s out of n survive, and summed for all possible values of n:

$$P(s) = \sum_{n=s}^{\infty} e^{-\nu} \left\{ \frac{\nu^n}{n!} \binom{n}{s} P^{n-s} Q^s \right\}$$

$$= \frac{e^{-\nu Q}(\nu Q)^s}{s!}, \tag{20.4.2}$$

and s has a Poisson distribution with mean νQ. As in § 20.3, general theory (Finney, 1949b) leads to computations for estimating the parameters α, β, ν from data on several doses of a single preparation. If the same size of sample is used at each dose, the weight to be attached to the equivalent deviate for that dose is

$$W = \nu w, \tag{20.4.3}$$

where the weighting coefficient is

$$w = Z^2/Q. \tag{20.4.4}$$

If the sample sizes vary, W must be increased in proportion to the size of the particular sample. An auxiliary variate

$$x' = -Q/Z \tag{20.4.5}$$

is defined (cf. equation (20.3.3)). From provisional estimates of the parameters, numerical values of the weight, auxiliary variate, and working equivalent deviate are formed, the last by any of equations (17.5.10), (18.1.4)–(18.1.6); the empirical response rate is

$$p = 1 - q = \frac{\nu - s}{\nu}, \tag{20.4.6}$$

in accord with the suggestions of common sense. The weighted regression of y on x, x' jointly is calculated, the parameter estimates being given by

$$\left. \begin{array}{l} bS_{xx} + \dfrac{\delta\nu}{\nu} S_{xx'} = S_{xy'} \\[1ex] bS_{xx'} + \dfrac{\delta\nu}{\nu} S_{x'x'} = S_{x'y} \end{array} \right\} \tag{20.4.7}$$

and

$$a = \bar{y} - b\bar{x} - \frac{\delta\nu}{\nu} \bar{x}'. \tag{20.4.8}$$

Iteration continues in the usual manner, and the extension to assay calculations with two parallel lines is easy.

If a sample of the standard size is untreated and shows s_c non-responses, this is a direct estimate of ν; in equations (20.4.7) for each cycle of iteration, $S_{x'x'}$ must be increased by the provisional value of ν and $S_{x'y}$ by $(s_0 - \nu)$. If several samples are untreated, each gives its contribution to $S_{x'x'}$ and $S_{x'y}$. Data from untreated subjects are not essential to the calculations. Without them, a value of ν may be guessed by inspection of results for low doses, but the estimates of the parameters may be imprecise because of inadequate information on ν.

Sampling variation in the true numbers of individuals tested at different doses may cause values of s for some low doses to exceed ν; the empirical response rate, (20.4.6), is then negative. These data will play their proper part in increasing the estimate of ν if the calculations are performed exactly as described here, using a negative p in the formation of the working equivalent deviate. Although the computations resemble multiple regression methods, they are a device for solving the non-linear maximum likelihood equations; the anomaly of p being negative is an indication that the analogy is imperfect, not a condemnation of the method.

No restriction has yet been placed upon the form of the tolerance distribution. Wadley's discussion related to the normal tolerance distribution, for which probits are used as the equivalent deviates; Finney (1949b) generalized this along the lines of Chapter 17. For the normal tolerance distribution, a table of w having $Y = 1 \cdot 1(0 \cdot 1)9 \cdot 0$ has been published (Finney, 1971, Table V), and x' is the same as in equation (20.4.5) except for the reversal of sign. Tables for other distributions have not been prepared. No example need be shown here, as the problem is not one that often arises and the method is so similar to that of § 20.3 that familiarity with one is a great aid to the other.

The condition that the distribution of n between samples should be that of (20.4.1) ought to apply to a well-stirred suspension of micro-organisms. It may be far from correct as a representation of an insect population in fruit, and Anscombe (1949) suggested that a negative binomial distribution might then be nearer the truth. Again, numbers of survivors at any one dose will follow the same type of distribution, but two parameters now replace ν. Anscombe indicated a change in experimental design that might help to reduce the complexity of statistical analysis, but today an appeal to a general computer program would perhaps be more appropriate. The models discussed in this Section have not in fact proved very important for bioassay.

Iteration no longer need follow closely the analogy of linear regression. The computer technique outlined in § 18.5 can easily be modified to apply to the Wadley situation. It requires only that a new log likelihood be written into the program, this being based upon (20.4.2) with $(1-P)$ replacing Q and P taking any of the forms discussed in Chapter 17. The model has ν as an additional parameter (analogous to C_1, C_2 in § 20.3), but for the program writer and the program user no new difficulties appear.

20.5 Dilution series

Though generally regarded as a method for population estimation by sampling, one more type of estimation based on quantal responses can justifiably be classed as a biological assay. A point of special interest is that no standard preparation is used: the potency is assessed in absolute units from the behaviour of the test preparation alone. If the density of a bacterial suspension, in terms of the number of bacteria per unit volume, is to be estimated, the obvious procedure is to incubate samples of the suspension (or of a dilution by a known factor) and to count the number of colonies. On the assumption that each colony has developed from one bacterium, the mean number of colonies per sample leads to the estimate required. Colony counting, however, is laborious and may be impracticable if the number of colonies in a sample is large. An alternative, often adopted because of the simplicity with which each observation can be made, is to record only the presence or absence of bacterial growth and to regard this record as a statement of whether a sample is fertile or sterile. Thus the data are quantal. Although the information on bacterial density obtained from a particular set of samples will be less than if accurate colony counts were made, it will be preferable to the uncertain indications of inaccurate counts; moreover, the loss may be compensated by the possibility of using a far greater number of samples.

Whereas colony counts are readily interpretable as estimators of bacterial density, the quantal records require a more complicated analysis before they yield an estimate of density. Two conditions are essential to validity:

(i) The organisms must be distributed entirely at random in the bulk suspension from which small samples are to be removed, so that the frequency distribution of numbers of organisms in replicate samples is Poisson;

(ii) The nature of the culture medium and incubation must be such as to ensure that visible growth will occur in every sample containing one or more organisms.

The first condition is also required if colonies are counted, and implies thorough mixing of the suspension so as to ensure that organisms neither cluster nor repel one another. The second is less stringent than is required for colony counts, where every organism must be certain to develop into a visible and separately identifiable colony.

For convenience of notation, the unit of volume may be taken as the volume of suspension used in each sample. With this convention, suppose that the undiluted suspension has μ organisms per unit volume. Any dilution may be represented by a 'dose' variate z, such that the density in the dilution is μz per unit volume; usually z will be less than unity, but values greater than unity (corresponding to concentrations of the original suspension) can be treated by the same theory. The number of organisms per sample in unit volumes from the dilution z will therefore follow a Poisson distribution with mean μz. If colony counts on n samples from this dilution had a mean value $\bar{y}$, the estimate of μ would be taken as

$$m = \frac{\bar{y}}{z}. \tag{20.5.1}$$

From the properties of a Poisson distribution,

$$\text{Var}(\bar{y}) = \frac{\mu z}{n}, \tag{20.5.2}$$

§ 20.5 SPECIAL PROBLEMS WITH QUANTAL RESPONSES

and therefore

$$\operatorname{Var}(m) = \frac{\mu}{nz}, \tag{20.5.3}$$

a quantity which would be estimated from the data by

$$\operatorname{Var}(m) = \frac{\bar{y}}{nz^2}. \tag{20.5.4}$$

This serves as a basis for comparison later.

Now the probability that a sample is sterile is the probability that it contains no organisms; this is the first term of the Poisson distribution

$$P = e^{-\mu z}. \tag{20.5.5}$$

If an estimate were to be formed from a fertile–sterile classification alone, and if r of the n samples were sterile, the estimate would naturally be taken as m, the solution of the equation

$$e^{-mz} = p, \tag{20.5.6}$$

where

$$p = \frac{r}{n} \tag{20.5.7}$$

is an estimate of P from the data. Hence for this form of estimation

$$m = -\frac{\ln p}{z}. \tag{20.5.8}$$

Since

$$\operatorname{Var}(p) = \frac{PQ}{n}, \tag{20.5.9}$$

it is easily proved that, to a first-order approximation,

$$\operatorname{Var}(m) = \frac{Q}{nPz^2}. \tag{20.5.10}$$

The utility of this formula is limited; when P or n is small, the sampling distribution of r is far from normal, and the formula will not satisfactorily indicate limits of error for m. Still more open to criticism is the variance calculated from the data,

$$\operatorname{Var}(m) = \frac{n-r}{nrz^2}. \tag{20.5.11}$$

Nevertheless, comparison of equations (20.5.3) and (20.5.10) indicates the efficiency of the quantal method by comparison with that of colony counting. This efficiency is the inverse ratio of the two variances:

$$\text{Eff.} = \frac{\mu z P}{Q}$$

$$= \frac{\mu z}{\exp(\mu z) - 1}. \tag{20.5.12}$$

When the density, μz, is small, the efficiency approaches 1·0; this is because the number of organisms per sample is almost certain to be either 0 or 1, and the quantal response is almost as satisfactory as an actual count for determining the number of organisms present. When the density is large, the probability of a sterile sample becomes small and the information from the quantal response is negligible.

The variance in equation (20.5.10) is a minimum when $p \simeq 0.203$, and therefore when $z \simeq 1.59$. Thus the ideal dilution to take in order that the quantal method shall give as much information as possible on μ is one that has about 1·59 organisms per unit volume. Provided that the number of organisms can be kept between 0·9 and 2·5 per unit volume, the relative efficiency remains above 0·85. The variance of the estimate of μ in direct colony counting, equation (20.5.3), decreases steadily as z increases. Table 20.5.1 shows the efficiency of the quantal method for various densities in the sample, reckoned relative to the optimal conditions with $\mu z = 1.5936$ and also relative to colony counts on the same samples; the first is obtained from equation (20.5.10), the second from equation (20.5.12).

The discussion above relates to trials of one dilution only. If the investigator has no idea what the density of organisms is, he will try samples at several dilutions. Even if he intends to count colonies, the impossibility of making satisfactory counts when the density in a sample is high will force him to do this; the

TABLE 20.5.1 Efficiency of fertile–sterile classification of samples for estimating the density of a bacterial suspension

Density of organisms per sample (μz)	Probability of sterile sample (P)	Efficiency relative to optimal μz	Efficiency relative to colony counts at this density
16	0·000 000 113	0·000	0·000
8	0·000 335	0·033	0·003
4	0·018 3	0·461	0·075
2	0·135	0·967	0·313
1·5936	0·203	1·000	0·406
1	0·368	0·899	0·582
1/2	0·607	0·595	0·771
1/4	0·779	0·340	0·880
1/8	0·882	0·181	0·939
1/16	0·939	0·094	0·969
1/32	0·969	0·047	0·984
1/64	0·984	0·024	0·992
1/128	0·992	0·012	0·996
1/256	0·996	0·006	0·998

same policy is desirable for the quantal method in order to ensure that some dilutions have between 4 and $\frac{1}{4}$ organisms per sample and so can give a reasonable amount of information on μ. Information from different dilutions must then be combined.

Suppose that n_i samples (of unit volume) are taken at a dilution z_i, and that $\bar{y}_i$ is their mean colony count. Then, with S indicating summation over k doses,

$$m = \frac{S n_i \bar{y}_i}{S n_i z_i} \qquad (20.5.13)$$

will be taken as the estimate of μ. This generalization of equation (20.5.1) is appropriate because $S n_i \bar{y}_i$ is a Poisson variate with mean $\mu S n_i z_i$; m is the maximum likelihood estimator of μ. Moreover,

$$\text{Var}(m) = \frac{\mu}{S n_i z_i}. \qquad (20.5.14)$$

What is to be done if only the records of fertility and sterility are available? The problem of estimating μ from this type of quantal experiment has long interested bacteriologists and statisticians. Eisenhart and Wilson (1943) surveyed its history. Halvorson and Ziegler (1933a, b) first gave the maximum likelihood equation in a form suitable for calculation, though Fisher (1922) had indicated the method earlier; McCrady (1915) had proposed essentially the same process for estimating what he termed the *most probable number*, and other bacteriologists followed his lead. Cochran (1950) published a useful account of the relation between the two approaches. Various authors, including Halvorson and Ziegler (1933a, b) and Swaroop (1938, 1940, 1941a, b), have given tables from which the maximum likelihood estimate can be read directly for certain assay designs (i.e. particular sets of dilutions and numbers of samples), and have discussed the precision of the estimate. Barkworth and Irwin (1938) showed how to solve the maximum likelihood equation iteratively whatever the arrangement of dilutions and numbers of samples. Ziegler and Halvorson (1935) made an interesting experimental comparison between the precisions obtained in estimation by colony counting and by dilution series for the same suspensions; their findings were in general agreement with theory.

The probability that r out of n plates at dilution z are sterile is

$$P(r) = \binom{n}{r} P^r (1-P)^{n-r}, \qquad (20.5.15)$$

where P is defined by equation (20.5.5). If the metameter Y is defined by

$$P = e^{-Y}, \qquad (20.5.16)$$

evidently

$$Y = \mu z. \qquad (20.5.17)$$

The likeness of equations (20.5.15), (20.5.16) and (20.5.17) to equations

(17.5.1), (17.4.8) and (17.4.9), respectively, suggests that maximum likelihood estimation of μ from a set of dilutions can be based upon an iterative process with Y as an equivalent deviate. The formulae are the same as would be given by application of the methods of Chapter 17 to the tolerance distribution

$$\mu e^{-\mu z} dz \quad (z \geqslant 0; \mu > 0), \tag{20.5.18}$$

although here the tolerance distribution has no real meaning. Only one parameter, μ, has to be estimated, and this is the slope of a regression (of equivalent deviate on z) constrained to pass through $z = 0$, $Y = 0$.

Finney (1947a) developed the equivalent deviate method in the manner shown above, and suggested a further improvement. This was not entirely satisfactory, because the functions required do not lend themselves to convenient tabulation and because the distribution of the estimate is likely to be far from normal. The difficulties can be overcome by estimating $\log \mu$ instead of μ, to which end a transformation proposed by Mather (1949) for a slightly different purpose may be adapted (Finney, 1951c). Analogous computations were proposed by Peto (1953).

Define a new response metameter, the *loglog*, by

$$Y = \ln(-\ln P), \tag{20.5.19}$$

and a dose metameter

$$x = \ln z. \tag{20.5.20}$$

Then equation (20.5.17) may be written

$$Y = \ln \mu + x. \tag{20.5.21}$$

The estimation of $\ln \mu$ may be put into the form of calculating a weighted regression of a working loglog on x, subject to the condition that the regression coefficient is unity. Because of this constancy of the regression coefficient, the arithmetic is made easier by arranging that each iterative cycle gives an adjustment to equation (20.5.21) instead of a new version of the equation. Formally, a response is the occurrence of a sterile plate, and at each dose p_i ($i = 1, 2, \ldots, k$) is defined by equation (20.5.7). The general method of § 17.5 leads to a working loglog

$$y = Y + \frac{p - P}{P \ln P}, \tag{20.5.22}$$

but the recommended alternative is to use a working deviate

$$\eta = \frac{P - p}{P \ln P}. \tag{20.5.23}$$

Minimum and maximum working deviates corresponding to any expected loglog are defined by

$$\eta_0 = -e^{-Y}, \tag{20.5.24}$$

§ 20.5 SPECIAL PROBLEMS WITH QUANTAL RESPONSES

$$\eta_1 = \exp(e^Y - Y) - e^{-Y}, \tag{20.5.25}$$

and the range is

$$A = \exp(e^Y - Y). \tag{20.5.26}$$

The working deviate may be found from any of equations (18.1.4)–(18.1.6) with η, η_0, η_1 replacing y, y_0, y_1; thus

$$\eta = \eta_0 + pA, \tag{20.5.27}$$

$$\eta = \eta_1 - qA, \tag{20.5.28}$$

$$\eta = q\eta_0 + p\eta_1. \tag{20.5.29}$$

The weighting coefficient is now

$$w = \frac{e^{2Y}}{\exp(e^Y) - 1}. \tag{20.5.30}$$

Iteration follows a simple pattern. A value of μ is guessed, and equation (20.5.21) is used to determine expected loglogs at each dilution. Equations (20.5.27)–(20.5.30) give the working deviate and the weight, nw, for each dilution, and the weighted mean of the working deviates is:

$$\bar{\eta} = \frac{Snw\eta}{Snw}. \tag{20.5.31}$$

This $\bar{\eta}$ is *subtracted from* the provisional estimate of $\ln \mu$, to give a revised estimate with which the cycle may be repeated. When satisfactory approximation to the maximum likelihood solution has been attained,

$$\chi^2_{[k-1]} = Snw\eta^2 - \frac{(Snw\eta)^2}{Snw} \tag{20.5.32}$$

is a homogeneity test for the data, equivalent to a χ^2 test on observed and expected numbers of fertile and sterile plates for each dilution. The usual troubles arise with small frequencies; in cases of doubt, appeal must be made to calculations based on equation (17.6.2). For data that are satisfactorily homogeneous, if $\ln m$ is the final estimate,

$$\mathrm{Var}(\ln m) = 1/Snw. \tag{20.5.33}$$

Fiducial limits to μ should be found from limits on the logarithmic scale.

With the aid of tables (Mather, 1949; Finney, 1951c), this process is completed much more rapidly than are probit computations. Empirical loglogs are found from the loglog transformation, Appendix Table XVI. Appendix Table XVII gives the minimum and maximum working deviates, range, and weighting coefficient as functions of the expected loglog. A table of natural logarithms, such as Fisher and Yates's Table XXVI, is also helpful, but logarithms to base 10 can be used by the rule

$$\ln z = 2{\cdot}3026 \log_{10} z. \tag{20.5.34}$$

The χ^2 test, equation (20.5.32), is appropriate for detecting unspecified types of deviation from equation (20.5.5). In some circumstances, it may be possible to specify the pattern of deviation most likely to occur and to choose a more powerful test of this. For example, the essential statistical features of the dilution series appear to be fulfilled in certain techniques of virus assay involving injection of different dilutions of a virus preparation into eggs and subsequent classification of the eggs as sterile or fertile. However, if the eggs themselves differ in their liability to infection, a source of variation additional to the Poisson is introduced; the effect will be to draw out the series of results so as to lengthen the dilution interval between complete sterility and complete fertility. Moran (1954a, b) and Armitage and Spicer (1956) have proposed appropriate significance tests. Moran's appears likely to be the better; it is based on summing $r(n-r)$ over all dilutions and comparing with a theoretical value. Moran (1958) and Stevens (1958) independently proposed a very simple test that unfortunately has never been compared with others. They used as test criterion the *range of transition*, the number of dilutions from the first at which $r \neq 0$ to the last at which $r \neq n$, counted inclusively. Dilutions with $r = 0$ or $r = n$ may occur within the range, which is defined solely by the extreme occurrences of $r \neq 0$ and $r \neq n$. Appendix Table XVIII is an abridged version of Stevens's tabulation of probability levels for the range of transition. For example, with a dilution factor $a = 2$, the probability of the range being 4 or greater is 0·18 for an assay with 1 plate per dilution, 0·42 for $n = 2$, and certainly greater for $n = 5$, although this has not been tabulated.

The Stevens test is strictly correct only for series that extend to infinity in both directions, and is primarily suitable for dilution series in which n is constant over all dilutions; in some instances, approximate allowances for inequalities of n will permit a near certainty about the outcome of a correct test. Its use in an experiment of finite size assumes that the probability of a sterile plate at any higher value of z than those in the assay and the probability of a fertile plate at any lower value of z are both negligible. The test should therefore be used with caution for any assay in which values of r different from $0, n$ occur near the ends of the dilution series, as it may be appreciably biased by a tendency to underestimate the range.

Many related publications discuss dilution series as experimental procedures for studying infectivity. The emphasis is then on the magnitude of effects as determined by doses, rather than on dose as estimated from responses. A useful starting point for this part of the literature is a paper by Armitage and Bartsch (1960), but the biological techniques are not likely to be adopted for assays.

20.6 Density estimation for rope bacillus

From each of 10 dilutions of a suspension of a potato flour, five 1 ml samples were withdrawn and plated. Table 20.6.1 (from Fisher and Yates, 1963) records the numbers of plates that were sterile in respect of spores of 'rope', *Bacillus mesentericus*. The empirical loglogs of each p indicate that the expected loglog is zero near to a dilution of 1/8 g per 100 ml, which may be used as the basis of a

§ 20.6 SPECIAL PROBLEMS WITH QUANTAL RESPONSES

TABLE 20.6.1 Estimation of the density of rope spores in a potato flour, from a dilution series

Dilution (z) in g per 100 ml	n	r	p	Empirical loglog	Y	w	η
4	5	0	0·0	—	3·5	0·000	−0·03
2	5	0	0·0	—	2·8	0·000	−0·06
1	5	0	0·0	—	2·1	0·019	−0·12
1/2	5	0	0·0	—	1·4	0·290	−0·25
1/4	5	1	0·2	0·48	0·7	0·625	0·25
1/8	5	2	0·4	−0·09	0·0	0·582	0·09
1/16	5	3	0·6	−0·67	−0·7	0·383	−0·03
1/32	5	3	0·6	−0·67	−1·4	0·217	−0·94
1/64	5	5	1·0	—	−2·1	0·115	1·06
1/128	5	5	1·0	—	−2·8	0·059	1·03

first cycle. Successive dilutions differ by a factor of 2, and $\ln 2 = 0\cdot69$; the Y column in Table 20.6.1 is easily constructed from equation (20.5.21).

The values of w for each Y are read directly from Table XVII; because n is the same for all dilutions, multiplication by n is unnecessary. Equations (20.5.27) to (20.5.29) enable working deviates to be found from Table XVII. Summations give

$$Sw = 2\cdot290,$$

$$Sw\eta = 0\cdot1010,$$

and therefore

$$\bar{\eta} = 0\cdot1010/2\cdot290$$
$$= 0\cdot0441.$$

The expected loglogs were constructed from the guess that the density was such as to give $Y = 0$ at $z = 1/8$; $\bar{\eta}$ should now be subtracted from each Y and a new column of expected loglogs formed. No new cycle of computation is needed, as the values of Y are scarcely altered (see Finney, 1951c, for the results of a second cycle). Equation (20.5.32) gives

$$\chi^2 = 5 \times \left\{ 0\cdot446 - \frac{(0\cdot1010)^2}{2\cdot290} \right\}$$
$$= 2\cdot21,$$

which might be regarded as having 7 d.f., since only 8 dilutions had non-zero weights and one parameter has been estimated. Here the test is practically worthless, because n is so small; for inferences to be safe in circumstances that make the test of validity insensitive to all but the grossest discrepancies, the *a priori* evidence for validity of the mathematical model must be strong. If the expected frequencies are small only in some classes, expected values of r may be calculated as method requires the dilutions to be spaced regularly in a geometrical progression and the number of samples per dilution to be constant, or rather the method becomes much more troublesome to operate if these restrictions are relaxed. These are usually desirable features of an assay for estimating the density of organisms, but special circumstances or accidents may sometimes intervene; the loglog method, or any other variant of maximum likelihood estimation, has the advantage that it presents no new difficulties.

An investigator who has moderately good advance information on μ should not adopt a wide range of dilutions merely in order that he may be able to estimate his density by a table of the solutions of equation (20.7.3). If N, the total number of samples, is fixed, the ideal allocation would be to use all at the dilution giving 1·59 organisms per sample. Unless the investigator is confident that he can approximate closely to this, he will be wise to use whatever prior knowledge he has in order to choose several dilutions as insurance against a bad guess, and to distribute his N samples between these. Only then does a distinction between maximum likelihood and equation (20.7.3) appear. That equation could indeed be used for estimation even without coverage of the full range of P, but the existing tables are not then applicable and the method has no advantages of simplicity relative to the loglog. For a good experiment, effort should be concentrated at dilutions where the weighting coefficient will be fairly large, say between 5·0 and 0·2 organisms per sample. In the experiment recorded in Table 20.6.1, had the investigator guessed in advance that the density was about 1000 spores per gram, he might have restricted attention to dilutions of 1/2, 1/4, 1/8, 1/16, 1/32 g/100 ml. Had he then assigned 10 plates to each of these, so keeping his total at 50, the value of Snw would have been increased from 11·45 to 20·97; had he been still more confident in his guess and assigned his plates as 5, 10, 20, 10, 5 to the five dilutions, Snw would have been 24·26. These represent great improvements over the design actually used: relative to them, the design of Table 20.6.1 with estimation according to equation (20.7.3) is less than 50 percent efficient. The convenience of the estimation based on T alone, and its high efficiency for data to which it is applicable, must not blind the investigator to the fact that more useful experimental designs may be available.

If μ can be approximately located in advance, concentration of effort in the neighbourhood of 1·6 organisms per sample and maximum likelihood estimation is important; the additional computation time should be compensated by a reduction in the number of samples needed. This is yet one more example of the truth that choice of a satisfactory experimental design depends upon the pre-existing knowledge of the question under investigation. If this knowledge is unreliable, and the investigator fails to admit its unreliability, his experiment may be bad. In the rope spore assay, for example, had the spore density been guessed as 100 per gram when in fact it was 766 per gram, the experiment might have been designed to have 10 plates at each of the dilutions 4, 2, 1, 1/2, 1/4 g/100 ml; Snw would have been 9·34, and the estimate by maximum likelihood would have been less precise than by the use of Fisher's method on the original design. A guess of 10 spores per gram might have had disastrous consequences, unless the investigator had realized that it was untrustworthy and had therefore adopted a wide range of dilutions.

In planning an assay of a density of organisms, the investigator ought first to decide limits between which he is practically certain that his true density lies. If these limits are μ_L and μ_U per unit volume ($\mu_L < \mu_U$), he should choose dilutions to cover a range of 'doses' from $2/\mu_L$ to $1/2\mu_U$, thus ensuring that his first dilution has at least 2 organisms per sample and his last has at most 0·5 per sample. The dilution factor should be as small as is practicable; 2 and 4 are definitely preferable to 10, but the convenience of using a few widely spaced dilutions may be allowed some influence. Thus, if he is sure that the density lies between 100 and 600 organisms per gram, he must have dilutions ranging at least as widely as 1/50 and 1/1200. A suitable set would be 1/40, 1/80, 1/160, 1/320, 1/640, 1/1280 (dilution factor 2), or 1/25, 1/100, 1/400, 1/1600 (dilution factor 4), these being preferable to a design with the same total number of plates distributed between dilutions of 1/30, 1/300, 1/3000 (dilution factor 10).

If he prefers inclusion of extra dilutions to the labour of loglog calculations, the investigator should extend his series so as to be confident that he is for all practical purposes covering the range from $P = 0·0$ to $P = 1·0$; assigning equal numbers of plates to each level, he will plan to base his estimate on T alone. If he can locate μ reasonably closely in advance, he should concentrate his efforts in the centre of the range and use loglogs. Whatever design he adopts, if the outcome of the experiment shows clearly the legitimacy of the method based on T, he may use it with little loss of efficiency, but he must be prepared to use the loglog method if circumstances show the other to be unsuitable. At the stage of design, potential precision in relation to the amount of work put into an experiment should dominate the attention; when the data have been obtained, consideration can be given to whether or not the Fisher and Yates tables for solving equation (20.7.3) can legitimately be applied.

20.9 Information from different types of assay

The reader who has appreciated the problems underlying §§ 14.1–14.3 and 20.2 will realize that no satisfactory general method is available for combining a series of assays of the same preparation when some are parallel line, some slope ratio, and some quantal. If the assays are individually good, with small values of g, the method of weighted means of M (§ 14.1) should be adequate: and nothing better can be proposed at present, even when some g are large. This method has been used in a number of instances of international collaboration for the establishment of a new standard. An early example is the report by Miles and Perry (1950) on a new international digitalis standard, using results from collaborators who had been permitted great freedom in the choice of assay technique. Table 20.9.1 summarizes results from 47 usable assays. There was some indication of heterogeneity of estimates, since $\chi^2_{[46]} = 72·84$, but this is subsidiary to the present discussion. Miles and Perry suggested using the average weight contributed by an animal as a measure of the information per animal in a particular type of assay. They found the frog much less informative than the other animals (to a

TABLE 20.9.1 Combination of results for assays of the Third International Digitalis Standard against the Second

Animal	Assay	No. of assays	Mean potency	Fiducial limits	No. of animals	$\Sigma \omega$	Information per animal
Frog	Quantal	12	1·020	0·977–1·064	874	11 090	13
Cat	Par. line	16	1·087	1·057–1·117	226	25 820	114
Guinea-pig(*)	Par. line	6	1·036	0·997–1·077	152	13 930	92
Guinea-pig(†)	Par. line	6	1·033	0·995–1·072	142	14 650	103
Pigeon	Par. line	7	1·074	1·037–1·113	94	16 540	176

(*) Recommended method (†) Other method

Weighted mean potency: 1·057. Limits: 1·040–1·074

§ 20.9 SPECIAL PROBLEMS WITH QUANTAL RESPONSES

far greater extent than the use of a quantal technique can explain), and the pigeon much better than either cat or guinea-pig. These quantities can be compared with costs per animal in any consideration of economic choice of assay technique.

Among many other published examples may be mentioned a report on the international insulin standard (Bangham and Mussett, 1959), which presented the combined evidence of 268 assays of the Fourth International Standard; these were 147 quantal mouse-convulsion assays, 81 rabbit twin cross-overs, 9 rabbit triple cross-overs, and 31 others. Agreement was good, and the combined estimate of 23·9 IU/mg had limits at 23·6 and 24·2 IU/mg. Rabbits contributed six or seven times as much information as mice per appearance of an animal in the tests.

21

Time as a response

21.1 Time responses

For some assay techniques, the measured response is a time, usually the time that elapses between application of the stimulus to the subject and the occurrence of some reaction; death of the subject and first appearance of certain symptoms are reactions frequently used for this purpose. The measurement is a quantitative response, and in principle the assay may be analyzed by the methods of earlier chapters. If the range of doses is wide, some subjects may respond almost instantaneously and the data may show a considerable increase in variance as the response increases; in the choice of a response metameter, special attention must be paid to the essentially positive nature of the responses and to the homoscedasticity of the dose–response regression (Perry, 1950).

One difficulty peculiar to time-responses is that the assay may end before a response has been measured for every subject. This may happen because bad planning causes the assayist to terminate his observations prematurely. Even in a good assay, however, the mean time at one extreme of dose may be so great as to make continuation until all subjects have reacted impracticable. If a subject is removed from the assay because it suffers damage or death by an accident completely independent of the stimulus, it may be expunged from the records without introducing any bias: the statistical analysis may be made more complicated because of non-orthogonality of the remaining data, but the validity of that analysis is not disturbed. To discard a subject from the analysis merely because experiment showed it to have a long reaction time would bias the estimation of mean reaction time for the corresponding dose; the observations are said to be censored, an idea quite distinct from the missing values in § 4.21.

Three procedures may be considered:—

(i) The data might be converted to quantal form by classifying the subjects as 'reacted' or 'not reacted' at some arbitrary time, and analyzed as in Chapter 18.
(ii) An arbitrary value, independent of dose, might be assigned as the response for all subjects that have not reacted when observation ends.
(iii) A mathematical model of the process of the reaction might be set up, on which could be based a method of estimation (maximum likelihood or other) that would take full account of the incomplete records.

If the number of subjects that have not reacted at the end of the experiment is a high proportion of the total, analysis as in (i) is unlikely to entail much sacrifice of information. If only a few subjects have incomplete records, neglect of the information conveyed by the detailed time-measurements may seriously reduce

§ 21.2 TIME AS A RESPONSE 441

the precision, and (ii) will be preferable. The introduction of an entirely arbitrary value may seem an undesirable feature, but the choice made will usually have little influence on the conclusions because few subjects are concerned. Method (iii), discussed further in § 21.3, would be almost essential to a research programme whose chief object was the study of the reaction, but the computations are unnecessarily complicated for general use in assays. The argument will be clarified by a numerical example.

21.2 An assay of a virus

Gard (1943) reported trials of a standard preparation (No. 21) of the virus of poliomyelitis and of four test preparations. The assay involved inoculation of five male and five female mice at each of five doses of the standard and two doses of each test preparation. Table 21.2.1 shows the number of days elapsing before each mouse became sick. Observation terminated after sixteen days, at which time five subjects were still apparently well. The doses were dilutions of the preparations, at ten-fold intervals for S and hundred-fold for the others. The design thus conforms to the recommendations for multiple assays (Chapter 11) by having more subjects for the standard preparation than for any of the others; the arrangement of the doses permits a wider range to be covered for S than for the others, an insurance against bad guesses if little is known about the potency of the test preparations. On the other hand, if existing knowledge of relative potencies were reasonably good, equal dose ranges for each preparation would have been preferable. The incomplete records naturally occur on the less potent doses.

The methods of analysis mentioned in § 21.1 here become—

(i) Choose some convenient day, perhaps that at which about half of the mice have fallen sick, and use the methods of quantal response analysis on the proportions sick. At day 4, for example, the proportionate responses are 1·0 for both sexes at the first dose of S, 0·6 and 0·4 at the next, and so on. Clearly this will not be very satisfactory, as many doses will show an extreme proportion, even though useful information on the day of sickness is available.

(ii) Assign an arbitrary value, say 20, to the five mice unaffected at day 16; within reasonable limits, the exact choice for this value will have little effect on the results. Inspection of Table 21.2.1 discloses that the variance of the response-time increases as mean response-time increases. A reciprocal transformation is often useful in equalizing variances of time-responses (Box and Cullumbine, 1947; Brownlee and Hamre, 1951); if the animals falling sick on any one day are regarded as grouped at the midpoint of the preceding time-interval, the response metameter may be taken as

$$y = \frac{100}{\text{No. of days} - 0.5} \qquad (21.2.1)$$

This metameter is likely to be more nearly normally distributed than the

TIME AS A RESPONSE

TABLE 21.2.1 Responses of mice to inoculation with poliomyelitis virus

Preparation number	Dilution (as $\log_{10}$)	Sex	Day on which subject fell sick														
			3	4	5	6	7	8	9	10	11	12	13	14	15	16	>16
21 (Standard)	2	M	4	1													
		F	4	1													
	3	M	1	2	1	1											
		F		2	1	1		1									
	4	M		1	3	1											
		F		1		3		1									
	5	M				2	1			1				1			1
		F				3				1					1		
	6	M					1		1	1	1	1					1
		F						1	1	2							1
8	2	M	1	1	2	1											
		F		4	1												
	4	M			1	1	1	1	1								
		F			2	1	1										1
10	2	M	4	1													
		F	5														
	4	M		1		3	1										
		F			1	4											
11	2	M	1	1	3												
		F		5													
	4	M				2	1	1									1
		F				2	1			2							
12	2	M	5	4													
		F		1													
	4	M		2	1	2											
		F		2	3												

time itself, since the transformation will remove the positive skewness usually found in distributions of survival times and like measurements. Hence the metameter is to be recommended from considerations of normality as well as of scedasticity, though more data than are available here would be needed for any adequate confirmation of this. Even an 'infinite' time can be transformed by equation (21.2.1) since its metametric value is zero. To assume that the five unaffected mice would remain unaffected indefinitely, however, is just as arbitrary as to assign them to 20 days, and the latter seems more in keeping with the remainder of the data. Any reasonable value for insertion presumably lies between 16 and 30; these extremes put the five inserted metameters somewhere between 64 and 34, and the difference will have no major impact on potency estimates. Alternatively,

$$y = \log(\text{No. of days} - 0.5). \tag{21.2.2}$$

could be used as a metameter, with much the same conclusions but now without the convenience of accommodating an apparently unlimited response time.

§ 21.2　　　　　　　　TIME AS A RESPONSE　　　　　　　　　　443

(iii)　Method (i) wastes information; method (ii) may appear not quite honest. Any method that is logically more satisfying, however, will be complex and will almost certainly involve more parameters than can be estimated satisfactorily from the data. The distribution of times would have to be formulated mathematically, with allowance for the possibility that mice unaffected by day 16 might be either merely slow to react or immune to the virus; this is scarcely practicable on the evidence of only one experiment.

Method (ii) is less objectionable than it may at first appear. The value assigned to mice that fail to react may be regarded as part of the definition of the response metameter, and in the light of Chapter 15 the only fault in the metameter is its choice to suit a single assay. If a series of assays which stopped at day 16 were all found satisfactorily analyzable in terms of the same metameter, the choice would be as sound as that for other types of quantitative assay. The analysis here will be completed in terms of the inserted value of 20 for the incomplete records; in practice, evidence on the most suitable metametric transformation should be accumulated in a number of assays.

The only evidence on the deviations from linearity of regression comes from S; a preliminary analysis of variance shows these deviations to be significant, not surprisingly because of a flattening of the regression at high doses. The best procedure seems to be to omit the data for $x = 6$; the consequent loss of information must be small, because values of y fall outside the range that is of interest for the test preparations. The analysis of variance for the remaining 120 responses, although easy, requires care because of the lack of symmetry arising from the extra doses of the standard preparation. Table 21.2.2 shows there to be still some suspicion of non-linearity (not enough to cause any serious alarm), but no other indication of invalidity. Some heterogeneity of variance might be expected. The mean squares with 4 d.f. calculated for the 24 groups of five mice range from 0 to 73·00; this is largely attributable to the coarseness of grouping, since, for example, the five female mice at $x = 2$ for preparation 10 are all assigned to $y = 40·0$, but might in reality have had any values between 50·0 and 33·3. There is no indication of any association between the mean and the variance of y, as may be seen by dividing the doses into two sets according to whether the mean value of y is less than or greater than 25·0 and evaluating the mean square for each set; the two mean squares are 25·87 and 33·53, with 52 d.f. and 44 d.f. respectively. Consequently, no ill effects of heteroscedasticity need be feared.

Table 21.2.2 also shows no sign of differential effects for the two sexes. Had there been a significant interaction of sex and regression, separate analyses for males and females would have been needed, and the final potency estimates would have been obtained by methods such as those of Chapter 14. Here, relative potencies may be calculated from all the data at once, with the aid of the regression coefficient obtained by pooling contributions from the five preparations:

TABLE 21.2.2 Analysis of variance of the data in Table 21.2.1, transformed by equation (21.2.1)

Adjustment for mean		79 145·76	
Nature of deviation	d.f.	Sum of squares	Mean square
Preparations	4	1 458·35	364·59
Sex	1	9·08	9·08
Preparations × Sex	4	172·08	43·02
Regression	1	7 814·33	7814·33
Linearity	2	155·53	77·76
Preparations × Regression	4	104·31	26·08
Sex × Regression	1	7·78	7·78
Sex × Linearity	2	70·28	35·14
Preparations × Sex × Regression	4	149·05	37·26
Doses	23	9 940·79	
Error	96	2 820·55	29·381
Total	119	12 761·34	

$$b = -\frac{379 \cdot 3 + 149 \cdot 4 + 195 \cdot 0 + 135 \cdot 2 + 149 \cdot 0}{50 + 20 + 20 + 20 + 20}$$

$$= -\frac{1007 \cdot 9}{130}$$

$$= -7 \cdot 7531. \tag{21.2.3}$$

From totals already used in the calculations for the analysis of variance are obtained the means:

Preparation no.	$\bar{x}$	$\bar{y}$
21	3·5	24·32
8	3·0	23·87
10	3·0	29·11
11	3·0	21·06
12	3·0	31·41

Care is needed in calculating M, because x is a log dilution instead of the more usual log concentration, and b is negative; consideration of Fig. 21.2.1 shows that equation (4.11.5) needs to be altered to

$$M = \bar{x}_T - \bar{x}_S - \frac{\bar{y}_T - \bar{y}_S}{b}$$

$$= -0 \cdot 5 + \frac{\bar{y}_T - 24 \cdot 32}{7 \cdot 7531}. \tag{21.2.4}$$

Hence

§ 21.2   TIME AS A RESPONSE   445

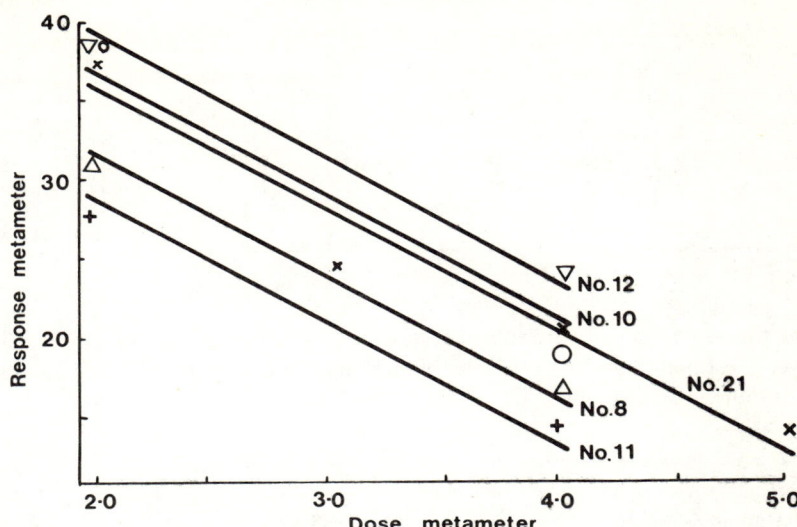

Fig. 21.2.1   Linear regressions for the assay of poliomyelitis virus, Table 21.2.1
              ×: Mean response metameters for standard preparation, no. 21
              △: Mean response metameters for test preparation no. 8
              ○: Mean response metameters for test preparation no. 10
              +: Mean response metameters for test preparation no. 11
              ▽: Mean response metameters for test preparation no. 12

$$M_8 = \bar{1}\cdot 4420,$$
$$M_{10} = 0\cdot 1178,$$
$$M_{11} = \bar{1}\cdot 0795,$$
$$M_{12} = 0\cdot 4145.$$

The ratio of the regression to the error mean square is so large that g is only 0·0148, a value that could be neglected. The full calculation of fiducial limits leads to the summary of potency estimates given in Table 21.2.3. This table also includes a summary based upon the metameter proposed in equation (21.2.2), again with survivors beyond 16 days scored as 20 days and now with the S data for $x = 6$ included; despite the use of all the data, the limits are all somewhat wider.

Comparison with results obtained from analyzing the data by quantal response methods, as suggested in (i) above, is of some interest. If falling sick on or before day 4 is regarded as a quantal response, and the data are analyzed by probits, the standard technique described in § 18.1 gives

$$b = -1\cdot 6357$$
and
$$g = 0\cdot 1379.$$

TABLE 21.2.3 Summary of relative potencies estimated from Table 21.2.1

Preparation no.	Metameter of equation (21.2.1)			Metameter of equation (21.2.2)		
	Potency relative to no. 21	Fiducial limits		Potency relative to no. 21	Fiducial limits	
8	0·277	0·114	0·667	0·255	0·0807	0·835
10	1·31	0·545	3·29	1·39	0·432	4·95
11	0·120	0·0486	0·288	0·119	0·0375	0·383
12	2·60	1·07	6·72	3·00	0·914	11·3

From the regression coefficient, the standard deviation of the tolerance distribution is estimated as 0·611, a value which may be compared with the inherent standard error per response,

$$\frac{s}{b} = \frac{5\cdot420}{7\cdot753} = 0\cdot699$$

in the analysis based on equation (21.2.1). The agreement is good (cf. Perry, 1950). On the other hand, the weight per subject in the quantal analysis is small, the total of all the Snw values being only 34·6 by comparison with an effective total weight of 120 in the other analysis. This explains the marked increase in g and the widening of the fiducial intervals. Table 21.2.4 summarizes the quantal analysis; the estimates (except for preparation no. 8) agree well with those in Table 21.2.3, but the fiducial limits are so widely spaced as to leave no doubt of the advantage that lies in the use of quantitative responses. Had the quantal response been defined in terms of any day other than day 4, the results would have been still less satisfactory.

TABLE 21.2.4 Summary of relative potencies estimated from Table 21.2.1, by analysis of quantal responses

Preparation no.	Potency relative to no. 21	Fiducial limits	
8	0·0838	0·0198	0·385
10	1·16	0·205	5·37
11	0·121	0·0288	0·603
12	4·18	0·914	17·7

Liljestrand (1949) described an assay of neoarsphenamine based on survival time of infected mice. He used a preliminary investigation to establish that

$$y = \log(\text{time in hours} - 48)$$

was a satisfactory metameter and illustrated its use in a (2, 2) assay. He did not try a reciprocal transformation as an alternative to the logarithmic, but probably it would have been equally good.

21.3 Complete analysis

The method in § 21.2 does not take account of the exact form of the regression relation. Methods in earlier chapters are also open to this criticism, but reasons have been given for regarding the dose–response relations used as adequate representations of the exact regressions over a limited range of doses. A response metameter such as equation (21.2.1), with the assumption of a linear regression, is more obviously imperfect, perhaps not so much because it is intrinsically worse than the procedure in other types of assay but because improvements could more easily be suggested.

Rigorous analysis of how time-to-response is related to dose introduces many complications. Even if time, or a simple function of time, is assumed to be normally distributed, estimation of the parameters of the distribution when some records are incomplete involves laborious calculations. Yet this model of the behaviour of the subjects is obviously not sufficiently complex. Of the subjects that do not respond within the period of the experiment, some may have high tolerance for the stimulus, so that they would not respond until a longer time had elapsed; some may be absolutely immune to the stimulus, and therefore would never respond; some may have had a subliminal reaction, after which they have recovered and so have never been observed as responding. All these will be classified together as incomplete records, though their mathematical representations would be different. In addition to those experimentally stimulated, natural responses may occur; for example, if the response were death of the subject and at some doses the survival times were long, natural deaths might be indistinguishable from those due to the treatment. Probabilistic models of these phenomena require considerably more parameters than are usually estimated as part of the statistical analysis of a biological assay. Data such as those of Gard for the assay of poliomyelitis virus (Table 21.2.1) are quite inadequate for the estimation of all the parameters needed in anything approaching a full mathematical model. Despite its theoretical shortcomings, an analysis such as that in § 21.2 is often as good an approximation to the ideal as the data and the knowledge of the experimenter can justify.

Bliss (1937) and Stevens (1937) studied the analysis of time–mortality data for a single treatment. They discussed the artificial truncation of the distribution of times brought about by early conclusion of the experiment and, less completely, the recovery of subjects from the effects of treatment. They did not consider the simultaneous occurrence of both phenomena in one experiment, the possibility of immunity, or the occurrence of natural responses. Boag (1949) investigated some of these problems, with particular reference to cures effected by cancer therapy. The statistical analysis of samples of observations from truncated and censored distributions has been widely discussed; except for a paper by Ipsen (1949), the special application to bioassay has been neglected. For assay purposes, as usual, two or more doses of two preparations must be considered simultaneously, and all the relevant parameters must be estimated. Perhaps nearer in concept to the presentation in this book are papers by Sampford (1952a, b, 1954) on maximum likelihood estimation for time–mortality data. In

§ 4 of the third paper Sampford has outlined the calculations appropriate to assays. Extensive use of time as the response in routine bioassays is unlikely, and no details of Sampford's methods will be presented here. The arithmetic, though heavy, would not tax modern computing facilities. Sampford and Taylor (1959) later extended the methods to censored observations in more highly structured designs; this work would be very relevant to assays in randomized blocks or more complex designs with time as a response. Ipsen's (1941) suggestions on the development of empirical response metameters also have uses with time-responses. Christensen and Finney (1953) have illustrated this in a series of assays of cobra venom, but the situation was so specialized that it scarcely provides much guidance for other circumstances.

References

AITCHISON, J. & SILVEY, S.D. (1957). The generalization of probit analysis to the case of multiple responses. *Biometrika*, **44**, 131–40.
ALLOTT, D.J. & O'NEILL, J.A. (1970). A parallel line assay method for the determination of herbicide soil residues. *Rec. agric. Res. Minist. Agric. N.I.*, **19**, 21–8.
ANON. (1946). Tests and methods of assay for antibiotic drugs – penicillin. *Fed. Register*, **11**, 12 128–36.
ANSCOMBE, F.J. (1949). Note on a problem in probit analysis. *Ann. appl. Biol.*, **36**, 203–5.
ARMITAGE, P. (1970). The combination of assay results. *Biometrika*, **57**, 665–6.
ARMITAGE, P. & ALLEN, I. (1950). Methods of estimating the LD50 in quantal response data. *J. Hyg.*, **48**, 298–322.
ARMITAGE, P., BAILEY, J.M., PETRIE, A., ANNABLE, L. & STACK-DUNNE, M.P. (1974). Studies in the combination of bioassay results. *Biometrics*, **30**, 1–10.
ARMITAGE, P. & BARTSCH, G.E. (1960). The detection of host variability in a dilution series with single observations. *Biometrics*, **16**, 582–92.
ARMITAGE, P. & BENNETT, B.M. (1974). Maximum likelihood solutions for the combination of relative potencies. *J. Hyg.*, **73**, 97–9.
ARMITAGE, P., BENNETT, B.M. & FINNEY, D.J. (1976). Point and interval estimation in the combination of bioassay results. *J. Hyg.*, **76**, 147–62.
ARMITAGE, P. & SPICER, C.C. (1956). The detection of variation in host susceptibility in dilution counting experiments. *J. Hyg.*, **54**, 401–14.
ARRIGUCCI, A., FORTI, G., FIORELLI, G., PAZZAGLI, M. & SERIO, M. (1973). Mathematical analysis of the results of competitive binding methods. *The Endocrine Function of the Human Testis* (edited V.H.T. James, M. Serio & L. Martini), vol. I, 74–90. New York: Academic Press.
ASHFORD, J.R. (1959). An approach to the analysis of data for semi-quantal responses in biological assay. *Biometrics*, **15**, 573–81.
ASHFORD, J.R. & SMITH, C.S. (1965). An analysis of quantal response data in which the measurement of response is subject to error. *Biometrics*, **21**, 811–25.
ASHFORD, J.R., SMITH, C.S. & BROWN, S. (1960). The quantal response analysis of a series of biological assays on the same subjects. *Biometrika*, **47**, 23–32.
ASO, T., GUERRERO, R., CEKAN, S.Z. & DICZFALUSY, E. (1975). A rapid 5 hour radioimmunoassay of progesterone and oestradiol in human plasma. *Clin. Endocr.*, **4**, 173–182.
BACHARACH, A.L. (1945). Biological assay and chemical analysis. *Analyst*, **70**, 394–403.
BACHARACH, A.L., COATES, M.E. & MIDDLETON, T.R. (1942). A biological test for vitamin P activity. *Biochem. J.*, **36**, 407–12.
BANGHAM, D.R. & MUSSETT, M.V. (1959). The fourth international standard for insulin. *Bull. Wld Hlth Org.*, **20**, 1209–20.
BANTING, J.D. (1974). Some statistical considerations in the bioassay of herbicides. Read to *Third int. Congr. Herbicide Chem*.
BARKWORTH, H. & IRWIN, J.O. (1938). Distribution of coliform organisms in milk and the accuracy of the presumptive coliform test. *J. Hyg.*, **38**, 446–57.
BARR, M. & NELSON, J.W. (1949). An accurate and economical method for the biological assay of aconite tincture. *J. Am. pharm. Ass.*, **38**, 518–21.
BARRACLOUGH, C.G. (1955). Statistical analysis of multiple slope ratio assays. *Biometrics*, **11**, 186–200.
BARTLETT, M.S. (1937). Properties of sufficiency and statistical tests. *Proc. R. Soc.*, **A160**, 268–82.
BARTLETT, M.S. (1946). A modified probit technique for small probabilities. *Jl R. statist. Soc. Suppl.*, **8**, 113–17.

REFERENCES

BARTLETT, M.S. (1947). The use of transformations. *Biometrics*, 3, 39–52.
BEHRENS, B. (1929). Zur Auswertung der Digitalisblätter im Froschversuch. *Arch. exp. Path. Pharmakol.*, 140, 237–56.
BENNETT, B.M. (1952). Estimation of LD50 by moving averages. *J. Hyg.*, 50, 157–64.
BENNETT, B.M. (1962). On combining estimates of relative potency in bioassay. *J. Hyg.*, 60, 379–85.
BENNETT, B.M. (1963a). Slope ratio assays and confidence limits. *Metrika*, 7, 117–20.
BENNETT, B.M. (1963b). On combining estimates of a ratio of means. *Jl R. statist. Soc.*, B25, 201–5.
BENNETT, B.M. (1963c). Optimum moving averages for the estimation of median effective dose in bioassay. *J. Hyg.* 61, 401–6.
BENNETT, B.M. (1964). A note on combining correlated estimates of a ratio of multivariate means. *Technometrics*, 6, 463–7.
BENNETT, B.M. (1969). Use of distribution-free methods in bioassay. *Biometr. Z.*, 11, 92–104.
BENNETT, B.M. (1970). Distribution-free methods in bioassay: A review. *Biométr.–Praxim.*, 11, 81–90.
BENNETT, B.M. (1971). Quantal assays using ranks. *Biometr. Z.*, 13, 203–7.
BERKSON, J. (1944). Application of the logistic function to bio-assay. *J. Am. statist. Ass.*, 39, 357–65.
BERKSON, J. (1946). Approximation of chi-square by "probits" and by "logits". *J. Am. statist. Ass.*, 41, 70–4.
BERKSON, J. (1949). Minimum χ^2 and minimum likelihood solution in terms of a linear transformation, with particular reference to bio-assay. *J. Am. statist. Ass.*, 44, 273–8.
BERKSON, J. (1950). Some observations with respect to the error of bio-assay. *Biometrics*, 6, 432–4.
BERKSON, J. (1953). A statistically precise and relatively simple method of estimating the bio-assay with quantal response based on the logistic function. *J. Am. statist. Ass.*, 48, 565–99.
BERKSON, J. (1955a). Maximum likelihood and minimum χ^2 estimates of the logistic function. *J. Am. statist. Ass.*, 50, 130–62.
BERKSON, J. (1955b). Estimate of the integrated normal curve by minimum normit chi-square with particular reference to bio-assay. *J. Am. statist. Ass.*, 50, 529–49.
BERKSON, J. (1955c). Estimation by least squares and by maximum likelihood. *Proc. Third Berkeley Symp. math. Statist. Probab.*, 1, 1–11.
BERKSON, J. (1957). Tables for use in estimating the normal distribution function by normit analysis. *Biometrika*, 44, 411–53.
BIGGERS, J.D. (1952). The calculation of the dose–response line in quantal assays with special reference to oestrogen assays by the Allen–Doisy technique. *J. Endocrin.*, 8, 169–78.
BIGGERS, J.D. & CLARINGBOLD, P.J. (1954). Why use inbred lines? *Nature*, 174, 596–7.
BIRCH, T.W. & HARRIS, L.J. (1934). Bradycardia in the vitamin B_1 deficient rat and its use in vitamin B_1 determinations. *Biochem. J.*, 28, 602–21.
BLACK, A.N. (1950). Weighted probits and their use. *Biometrika*, 37, 158–67.
BLACKITH, R.E. (1950). Bio-assay systems for the pyrethrins. III. Application of the twin cross-over design to crawling insect assays. *Ann. appl. Biol.*, 37, 508–15.
BLISS, C.I. (1934a). The method of probits. *Science*, 79, 38–9.
BLISS, C.I. (1934b). The method of probits – a correction. *Science*, 79, 409–10.
BLISS, C.I. (1935a). The calculation of the dosage–mortality curve. *Ann. appl. Biol.*, 22, 134–67.
BLISS, C.I. (1935b). The comparison of dosage–mortality data. *Ann. appl. Biol.*, 22, 307–333.
BLISS, C.I. (1937). The calculation of the time–mortality curve. *Ann. appl. Biol.*, 24, 815–852.
BLISS, C.I. (1938). The determination of dosage–mortality curves from small numbers. *Q. Jl Pharm. Pharmac.*, 11, 192–216.
BLISS, C.I. (1939). Fly spray testing. *Soap Sanit. Chem.*, 15, no. 4, 103–11.
BLISS, C.I. (1940a). Factorial design and covariance in the biological assay of vitamin D. *J. Am. statist. Ass.*, 35, 498–506.
BLISS, C.I. (1940b). Quantitative aspects of biological assay. *J. Am. pharm. Ass.*, 29, 465–75.

BLISS, C.I. (1944a). The U.S.P. collaborative cat assay for digitalis. *J. Am. pharm. Ass.*, **33**, 225–45.
BLISS, C.I. (1944b). A simplified calculation of the potency of penicillin and other drugs assayed biologically with a graded response. *J. Am. statist. Ass.*, **39**, 479–87.
BLISS, C.I. (1944c). Relative potency as applied to the assay of penicillin. *Science*, **100**, 577–8.
BLISS, C.I. (1945). Confidence limits for biological assays. *Biometrics*, **1**, 57–65.
BLISS, C.I. (1946a). A revised cylinder-plate assay for penicillin. *J. Am. pharm. Ass.*, **35**, 6–12.
BLISS, C.I. (1946b). An experimental design for slope-ratio assays. *Ann. math. Statist.*, **17**, 232–7.
BLISS, C.I. (1946c). Collaborative comparison of three rations for the chick assay of vitamin D. *J. Ass. off. agric. Chem.*, **29**, 396–408.
BLISS, C.I. (1947a). 2 × 2 factorial experiments in incomplete groups for use in biological assays. *Biometrics*, **3**, 69–88.
BLISS, C.I. (1947b). The biological measurement of the depth dose of roentgen rays with lettuce seedlings. *Am. J. Roentg.*, **58**, 222–33.
BLISS, C.I. (1950). The design of biological assays. *Ann. N.Y. Acad. Sci.*, **52**, 877–88.
BLISS, C.I. (1952a). *The Statistics of Bioassay*. New York: Academic Press Inc.
BLISS, C.I. (1952b). Estimation of the error in a clinical assay. *Biometrics*, **8**, 237–45.
BLISS, C.I. (1956). Analysis of biological assays in U.S.P. XV. *Drug Standards*, **24**, 33–68.
BLISS, C.I. & ALLMARK, M.G. (1944). The digitalis cat assay in relation to rate of injection. *J. Pharmac. exp. Ther.*, **81**, 378–89.
BLISS, C.I. & BARTELS, B.L. (1946). The determination of the most efficient response for measuring drug potency. *Proc. Pt II, Fedn Am. Soc. exp. Biol.*, **5**, 167–8.
BLISS, C.I. & CATTELL, McK. (1943). Biological assay. *A. Rev. Physiol.*, **5**, 479–539.
BLISS, C.I. & HANSON, J.C. (1939). Quantitative estimation of the potency of digitalis by the cat method in relation to secular variation. *J. Am. pharm. Ass.*, **28**, 521–30.
BLISS, C.I. & MARKS, H.P. (1939a). The biological assay of insulin. I. Some general considerations directed to increasing the precision of the curve relating dosage and graded response. *Q. Jl Pharm. Pharmac.*, **12**, 82–110.
BLISS, C.I. & MARKS, H.P. (1939b). The biological assay of insulin. II. The estimation of drug potency from a graded response. *Q. Jl Pharm. Pharmac.*, **12**, 182–205.
BLISS, C.I. & PABST, M.L. (1955). Assays for standardizing adrenal cortex extract in production. *Bull. int. statist. Inst.*, **34**(4), 317–38.
BLISS, C.I. & PACKARD, C. (1941). Stability of the standard dosage–effect curve for radiation. *Am. J. Roentg.*, **46**, 400–4.
BLISS, C.I. & ROSE, C.L. (1940). The assay of parathyroid extract from the serum calcium of dogs. *Am. J. Hyg.*, **31**, A 79–98.
BOAG, J.W. (1949). Maximum likelihood estimates of the proportion of patients cured by cancer therapy. *Jl R. statist. Soc.*, **B11**, 15–53.
BORTH, R. (1960). Simplified mathematics for multiple assays. *Acta endocrin.*, **35**, 454–68.
BORTH, R., DICZFALUSY, E. & HEINRICHS, H.D. (1957). Grundlagen der statistischen Auswertung biologischer Bestimmungen. *Arch. Gynäk.*, **188**, 497–538.
BOX, G.E.P. (1953). Non-normality and tests on variances. *Biometrika*, **40**, 318–35.
BOX, G.E.P. & COX, D.R. (1964). An analysis of transformations. *Jl R. statist. Soc.*, **B26**, 211–52.
BOX, G.E.P. & CULLUMBINE, H. (1947). The relationship between survival time and dosage with certain toxic agents. *Br. J. Pharmac. Chemother.*, **2**, 27–37.
BOX, G.E.P. & HAY, W.A. (1953). A statistical design for the efficient removal of trends occurring in a comparative experiment with an application in biological assay. *Biometrics*, **9**, 304–19.
BRAUN, H.A. & SIEGFRIED, A. (1947). The assay of digitalis. V. The guinea-pig method. *J. Am. pharm. Ass.*, **36**, 363–8.
BRITISH PHARMACOPOEIA (1973). London: Her Majesty's Stationery Office.
BRITISH STANDARDS INSTITUTION (1940). *British Standard Method for the Biological Assay of Vitamin D_3 by the Chick Method.* BS 911.
BROSS, I (1950). Estimates of the LD_{50}: A critique. *Biometrics*, **6**, 413–23.
BROWN, B.W. (1961). Some properties of the Spearman estimator in bioassay. *Biometrika*, **48**, 293–302.

BROWN, B.W. (1966). Planning a quantal assay of potency. *Biometrics*, 22, 322–9.
BROWNLEE, K.A., DELVES, C.S., DORMAN, M., GREEN, C.A., GRENFELL, E., JOHNSON, J.D.A. & SMITH, N. (1948). The biological assay of streptomycin by a modified cylinder plate method. *J. gen. Microbiol.*, 2, 40–53.
BROWNLEE, K.A. & HAMRE, D. (1951). Studies on chemotherapy of vaccinia virus. I. An experimental design for testing antiviral agents. *J. Bact.*, 61, 127–34.
BROWNLEE, K.A., HODGES, J.L. & ROSENBLATT, M. (1953). The up-and-down method with small samples. *J. Am. statist. Ass.*, 48, 262–77.
BROWNLEE, K.A. & LAPEDES, D.N. (1951). The effects of design upon the error of a microbiological assay for vitamin B12. *J. Bact.*, 62, 433–4.
BÜLBRING, E. & BURN, J.H. (1935). The estimation of oestrin and of male hormone in oily solution. *J. Physiol. Lond.*, 85, 320–33.
BURGER, H.G., LEE, V.W.K. & RENNIE, G.C. (1973). A generalised computer program for the treatment of data from competitive protein-binding assays including radioimmunoassays. *J. Lab. clin. Med.*, 80, 302–12.
BURN, J.H. (1930). The errors of biological assay. *Physiol. Rev.*, 10, 146–69.
BURN, J.H., FINNEY, D.J. & GOODWIN, L.G. (1950). *Biological Standardization* (2nd edn). London: Oxford University Press.
BUTLER, C.G., FINNEY, D.J. & SCHIELE, P. (1943). Experiments on the poisoning of honeybees by insecticidal and fungicidal sprays used in orchards. *Ann. appl. Biol.*, 30, 143–50.
CARPENTER, K.J., MCDONALD, I. & MILLER, W.S. (1972). Protein quality of feedingstuffs. 5. Collaborative studies on the biological assay of available methionine using chicks. *Br. J. Nutr.*, 27, 7–17.
CHANG, P.C. & JOHNSON, E.A. (1972). Some distribution-free properties of the asymptotic variance of the Spearman estimator in bioassays. *Biometrics*, 28, 882–9.
CHEN, K.K., BLISS, C.I. & ROBBINS, E.B. (1942). The digitalis-like principles of *Calotropis* compared with other cardiac substances. *J. Pharmac. exp. Ther.*, 74, 223–34.
CHRISTENSEN, P.A. & FINNEY, D.J. (1953). Standardization of cobra (*Naja flava*) venom using the graded response method. *J. Immunol.*, 70, 7–20.
CLARINGBOLD, P.J. (1956). The within-animal bioassay with quantal responses. *Jl R. statist. Soc.*, B18, 133–7.
CLARINGBOLD, P.J. (1958). Multivariate quantal analysis. *Jl R. statist. Soc.*, B20, 398–405.
CLARINGBOLD, P.J. (1959). Orthogonal contrasts in slope ratio investigations. *Biometrics*, 15, 307–22.
CLARKE, P.M. (1952). Statistical analysis of symmetrical slope-ratio assays of any number of test preparations. *Biometrics*, 8, 370–9.
COCHRAN, W.G. (1950). Estimation of bacterial densities by means of the "most probable number". *Biometrics*, 6, 105–16.
COCHRAN, W.G. & BLISS, C.I. (1948). Discriminant functions with covariance. *Ann. math. Statist.*, 19, 151–76.
COCHRAN, W.G. & COX, G.M. (1957). *Experimental Designs* (2nd edn). New York: John Wiley and Sons, Inc.
COCHRAN, W.G. & DAVIS, M. (1963). Sequential experiments for estimating the median lethal dose. *Le Plan d'Expériences*, 181–94. Paris: Centre National de la Recherche Scientifique.
COCHRAN, W.G. & DAVIS, M. (1964). Stochastic approximation to the median effective dose in bioassay. *Stochastic Models in Medicine and Biology*, 281–97. Madison: University of Wisconsin Press.
COCHRAN, W.G. & DAVIS, M. (1965). The Robbins–Monro method for estimating the median lethal dose. *Jl R. Statist. Soc.*, B27, 28–44.
COHEN, H., VAN RAMSHORST, J.D. & TASMAN, A. (1959). Consistency in potency assay of tetanus toxoid in mice. *Bull. Wld Hlth Org.*, 20, 1133–50.
COLQUHOUN, D. (1963). Balanced incomplete block designs in biological assay illustrated by the assay of gastrin using a Youden square. *Br. J. Pharmac. Chemother.*, 21, 67–77.
COLQUHOUN, D. (1971). *Lectures in Biostatistics*. Oxford: Clarendon Press.
COOK, B. (1975). Automation and data processing for radioimmunoassays. *Steroid Immunoassay – Proc. Fifth Tenovus Workshop*, 293–310. Cardiff: Alpha Omega Publishing Ltd.

REFERENCES

CORNFIELD, J. (1964). Comparative assays and the role of parallelism. *J. Pharmac. exp. Ther.*, **144**, 143–9.
CORNFIELD, J. (1967). The meaning of bioassay: a comment. *Biometrics*, **23**, 160–2.
CORNFIELD, J. & MANTEL, N. (1950). Some new aspects of the application of maximum likelihood to the calculation of the dosage response curve. *J. Am. statist. Ass.*, **45**, 181–210.
CORNFIELD, J. & MANTEL, N. (1951). Some comments on "Estimates of the LD_{50}: A Critique". *Biometrics*, **7**, 295–8.
COWARD, K.H. (1947). *The Biological Standardisation of the Vitamins* (2nd edn). London: Baillière, Tindall & Cox.
COX, C.P. (1972). On estimating relative potency from quadratic log–dose response relationships. *Biometrics*, **28**, 875–81.
COX, D.R. (1958). *Planning of Experiments*. New York: John Wiley & Sons, Inc.
CRAMER, E.M. (1962). A comparison of three methods of fitting the normal ogive. *Psychometrika*, **27**, 183–92.
CRAMER, E.M. (1964). Some comparisons of methods of fitting the dosage response curve for small samples. *J. Am. statist. Ass.*, **59**, 779–93.
CRAMÉR, H. (1946). *Mathematical Methods of Statistics*. Princeton: University Press.
CURTIS, J.M., UMBERGER, E.J. & KNUDSEN, L.F. (1947). The interpretation of estrogenic assays. *Endocrinology*, **40**, 231–40.
DAGNELIE, P. (1970). *Théorie et Méthodes Statistiques*. Gembloux: Editions J. Duculot, S.A.
DALE, H. (1939). Biological standardisation. *Analyst*, **64**, 554–67.
DAS, M.N. & KULKARNI, G.A. (1966). Incomplete block designs for bio-assays. *Biometrics*, **22**, 706–29.
DE BEER, E.J. (1941). A scale for graphically determining the slopes of dose–response curves. *Science*, **94**, 521–2.
DE BEER, E.J. (1945). The calculation of biological assay results by graphic methods. The all-or-none type of response. *J. Pharmac. exp. Ther.*, **85**, 1–13.
DE BEER, E.J. & SHERWOOD, M.B. (1945). The paper-disc agar-plate method for the assay of antibiotic substances. *J. Bact.*, **50**, 459–67.
DIXON, W.J. & MOOD, A.M. (1948). A method for obtaining and analyzing sensitivity data. *J. Am. statist. Ass.*, **43**, 109–26.
DRAGSTEDT, C.A. & LANG, V.F. (1928). Respiratory stimulants in acute cocaine poisoning in rabbits. *J. Pharmac. exp. Ther.*, **32**, 215–22.
DUFRENOY, J. & GOYAN, F.M. (1947). A graphical calculator for statistical analysis. *J. Am. pharm. Ass.*, **36**, 309–14.
EISENHART, C. & WILSON, P.W. (1943). Statistical method and control in bacteriology. *Bact. Rev.*, **7**, 57–137.
EKINS, R. & NEWMAN, B. (1970). Theoretical aspects of saturation analysis. *Acta endocrin.*, suppl. **147**, 11–36.
ELSTON, R.C. (1965). A simple method of estimating relative potency from two parabolas. *Biometrics*, **21**, 140–9.
EMERY, W.B., LEES, K.A. & TOOTILL, J.P.R. (1951). The assay of vitamin B_{12}. Part IV. The microbiological estimation with *Lactobacillus leichmannii* 313 by the turbidimetric procedure. *Analyst*, **76**, 141–6.
EMMENS, C.W. (1940). The dose/response relation for certain principles of the pituitary gland, and of the serum and urine of pregnancy. *J. Endocrin.*, **2**, 194–225.
EMMENS, C.W. (1948). *Principles of Biological Assay*. London: Chapman & Hall.
EPSTEIN, B. & CHURCHMAN, C.W. (1944). On the statistics of sensitivity data. *Ann. math. Statist.*, **15**, 90–6.
EUROPEAN PHARMACOPOEIA (1971), Vol. II. Paris: Maisonneuve S.A.
FECHNER, G.T. (1860). *Elemente der Psychophysik* (2 vols). Leipzig: Breitkopf und Härtel.
FIELLER, E.C. (1940). The biological standardization of insulin. *Jl R. statist. Soc., Suppl.*, **7**, 1–64.
FIELLER, E.C. (1944). A fundamental formula in the statistics of biological assay, and some applications. *Q. Jl Pharm. Pharmac.*, **17**, 117–23.
FIELLER, E.C. (1947). Some remarks on the statistical background in bio-assay. *Analyst*, **72**, 37–43.

FIELLER, E.C. (1954). Some problems in interval estimation. *Jl R. statist. Soc.*, **B16**, 175–185.
FIELLER, E.C., IRWIN, J.O., MARKS, H.P. & SHRIMPTON, E.A.G. (1939a). The dosage–response relation in the cross-over rabbit test for insulin. Part I. *Q. Jl Pharm. Pharmac.*, **12**, 206–11.
FIELLER, E.C., IRWIN, J.O., MARKS, H.P. & SHRIMPTON, E.A.G. (1939b). The dosage–response relation in the cross-over rabbit test for insulin. Part II. *Q. Jl Pharm. Pharmac.*, **12**, 724–42.
FINNEY, D.J. (1941). The joint distribution of variance ratios based on a common error mean square. *Ann. Eugen.*, **11**, 136–40.
FINNEY, D.J. (1944a). The application of the probit method to toxicity test data adjusted for mortality in the controls. *Ann. appl. Biol.*, **31**, 68–74.
FINNEY, D.J. (1944b). Mathematics of biological assay. *Nature*, **153**, 284.
FINNEY, D.J. (1945a). The microbiological assay of vitamins: The estimate and its precision. *Q. Jl Pharm. Pharmac.*, **18**, 77–82.
FINNEY, D.J. (1945b). Some orthogonal properties of the 4 × 4 and 6 × 6 Latin squares. *Ann. Eugen.*, **12**, 213–19.
FINNEY, D.J. (1946a). Standard errors of yields adjusted for regression on an independent measurement. *Biometrics*, **2**, 53–5.
FINNEY, D.J. (1946b). The analysis of a factorial series of insecticide tests. *Ann. appl. Biol.*, **33**, 160–5.
FINNEY, D.J. (1946c). Orthogonal partitions of the 6 × 6 Latin squares. *Ann. Eugen.*, **13**, 184–96.
FINNEY, D.J. (1946d). The frequency distribution of deviates from means and regression lines in samples from a multivariate normal population. *Ann. math. Statist.*, **17**, 344–9.
FINNEY, D.J. (1947a). The principles of biological assay. *Jl R. statist. Soc., Suppl.*, **9**, 46–91.
FINNEY, D.J. (1947b). The adjustment of biological assay results for variation in concomitant observations. *J. Hyg.*, **45**, 397–406.
FINNEY, D.J. (1947c). The estimation from individual records of the relationship between dose and quantal response. *Biometrika*, **34**, 320–34.
FINNEY, D.J. (1947d). Statistical aspects of microbiological assay. *Biochem. J.*, **41**, v–vii.
FINNEY, D.J. (1949a). The adjustment for a natural response rate in probit analysis. *Ann. appl. Biol.*, **36**, 187–95.
FINNEY, D.J. (1949b). The estimation of the parameters of tolerance distributions. *Biometrika*, **36**, 239–56.
FINNEY, D.J. (1949c). The choice of a response metameter in bio-assay. *Biometrics*, **5**, 261–72.
FINNEY, D.J. (1950). The estimation of the mean of a normal tolerance distribution. *Sankhyā*, **10**, 341–60.
FINNEY, D.J. (1951a). The statistical analysis of slope-ratio assays. *J. gen. Microbiol.*, **5**, 223–30.
FINNEY, D.J. (1951b). Two new uses of the Behrens–Fisher distribution. *Jl R. statist. Soc.*, **B12**, 296–300.
FINNEY, D.J. (1951c). The estimation of bacterial densities from dilution series. *J. Hyg.*, **49**, 26–35.
FINNEY, D.J. (1951d). Subjective judgment in statistical analysis – an experimental study. *Jl R. statist. Soc.*, **B13**, 284–97.
FINNEY, D.J. (1952). Graphical estimation of relative potency from quantal responses. *J. Pharmac. exp. Ther.*, **104**, 440–4.
FINNEY, D.J. (1953). The estimation of the ED50 for a logistic response curve. *Sankhyā*, **12**, 121–36.
FINNEY, D.J. (1955). *Experimental Design and its Statistical Basis.* Chicago: The University Press.
FINNEY, D.J. (1956). Cross-over designs in bioassay. *Proc. R. Soc.*, **B145**, 42–61.
FINNEY, D.J. (1959a). *An Introduction to the Theory of Experimental Design.* Chicago: The University Press.
FINNEY, D.J. (1959b). The design and analysis of an immunological assay. *Acta microbiol. hung.*, **6**, 341–68.
FINNEY, D.J. (1965). The meaning of bioassay. *Biometrics*, **21**, 785–98.

REFERENCES

FINNEY, D.J. (1971). *Probit Analysis* (3rd edn). Cambridge: The University Press.
FINNEY, D.J. (1976a). A computer program for parallel line bioassays. *J. Pharmac. exp. Ther.*, **198**, 497–506.
FINNEY, D.J. (1976b). Radioligand assays. *Biometrics*, **32**, 721–40.
FINNEY, D.J., HOLT, L.B. & SHEFFIELD, F. (1975). Repeated estimations of an immunological response curve. *J. biol. Standard.*, **3**, 1–10.
FINNEY, D.J. & OUTHWAITE, A.D. (1956). Serially balanced sequences in bioassay. *Proc. R. Soc.*, **B145**, 493–507.
FINNEY, D.J. & SCHILD, H.O. (1966). Parallel line assay with successive adjustment of doses. *Br. J. Pharmac. Chemother.*, **28**, 84–92.
FINNEY, D.J. & WOOD, E.C. (1951). Intra-litter replication in biological assays. *Nature*, **167**, 903–4.
FISHER, R.A. (1912). On an absolute criterion for fitting frequency curves. *Mess. Math.*, **41**, 155–60.
FISHER, R.A. (1922). On the mathematical foundations of theoretical statistics. *Phil. Trans. R. Soc.*, **A222**, 309–68.
FISHER, R.A. (1925). Theory of statistical estimation. *Proc. Camb. phil. Soc.*, **22**, 700–25.
FISHER, R.A. (1935). Appendix to Bliss (1935a): The case of zero survivors. *Ann. appl. Biol.*, **22**, 164–5.
FISHER, R.A. (1949). A biological assay of tuberculins. *Biometrics*, **5**, 300–16.
FISHER, R.A. (1956). *Statistical Methods and Scientific Inference*. Edinburgh: Oliver & Boyd.
FISHER, R.A. (1961a). Sampling the reference set. *Sankhyā*, **A23**, 3–8.
FISHER, R.A. (1961b). The weighted mean of two normal samples with unknown variance ratio. *Sankhyā*, **A23**, 103–14.
FISHER, R.A. (1966). *The Design of Experiments* (8th edn). Edinburgh: Oliver & Boyd.
FISHER, R.A. (1970). *Statistical Methods for Research Workers* (14th edn). Edinburgh: Oliver & Boyd.
FISHER, R.A. & YATES, F. (1963). *Statistical Tables for Biological, Agricultural and Medical Research* (6th edn). Edinburgh: Oliver & Boyd.
FITZHUGH, O.G., NELSON, A.N. & BLISS, C.I. (1944). The chronic oral toxicity of selenium. *J. Pharmac. exp. Ther.*, **80**, 289–99.
GADDUM, J.H. (1933). Reports on biological standards. III. Methods of biological assay depending on a quantal response. *Medical Research Council, Special Report Series, no. 183*.
GADDUM, J.H. (1948). *Pharmacology* (3rd edn). London: Oxford University Press.
GADDUM, J.H. (1950). Hormone assay: Introduction. *Analyst*, **75**, 530–3.
GADDUM, J.H. (1953a). Bioassays and mathematics. *Pharmac. Rev.*, **1**, 87–134.
GADDUM, J.H. (1953b). Simplified mathematics for bioassays. *J. Pharm. Pharmac.*, **6**, 345–58.
GARD, S. (1943). *Purification of Poliomyelitis Viruses*. Uppsala: Almqvist & Wiksell.
GARWOOD, F. (1941). The application of maximum likelihood to dosage–mortality curves. *Biometrika*, **32**, 46–58.
GAUTIER, R. (1945). The Health Organization and biological standardisation. *Bull. Hlth Org., League of Nations*, **12**, 1–75.
GEARY, R.C. (1947). Testing for normality. *Biometrika*, **34**, 209–42.
GINSBURG, M. (1951). A method for the assay of antidiuretic activity. *Br. J. Pharmac. Chemother.*, **6**, 411–16.
GOYAN, F.M. & DUFRENOY, J. (1947). A graphical calculator for bio-assays. *J. Am. pharm. Ass.*, **36**, 305–8.
GREWAL, R.S. (1952). A method for testing analgesics in mice. *Br. J. Pharmac. Chemother.*, **7**, 433–7.
GRIDGEMAN, N.T. (1943). The technique of the biological vitamin A assay. *Biochem. J.*, **37**, 127–32.
GRIDGEMAN, N.T. (1944a). Mathematics of biological assay. *Nature*, **153**, 461–2.
GRIDGEMAN, N.T. (1944b). *The estimation of Vitamin A*. London: Lever Brothers & Unilever Ltd.
GRIDGEMAN, N.T. (1951). On the errors of biological assays with graded responses, and their graphical derivation. *Biometrics*, **7**, 201–21.
GUILLEMIN, R. & SAKIZ, E. (1963). Quantitative study of the response to LH after

hypophysectomy in the ovarian ascorbic acid depletion test: effect of prolactin. *Endocrinology*, 72, 813–16.
GULD, J., BENTZON, M.W., BLEIKER, M.A., GRIEP, W.A., MAGNUSSON, M. & WAALER, H. (1958). Standardization of a new batch of purified tuberculin (PPD) intended for international use. *Bull. Wld Hlth Org.*, 19, 845–951.
GURLAND, J., LEE, I. & DAHM, P.A. (1960). Polychotomous quantal response in biological assay. *Biometrics*, 16, 382–98.
GYÖRGY, P. (1951). *Vitamin Methods* (vol. 2). New York: Academic Press Inc.
HALEY, T.J. (1947). An instrument for plotting ED_{50} curves. *Science*, 106, 151.
HALVORSON, H.O. & ZIEGLER, N.R. (1933a). Application of statistics to problems in bacteriology. I. A means of determining bacterial population by the dilution method. *J. Bact.*, 25, 101–21.
HALVORSON, H.O. & ZIEGLER, N.R. (1933b). Application of statistics to problems in bacteriology. III. A consideration of the accuracy of dilution data obtained by using several dilutions. *J. Bact.*, 26, 559–67.
HARDING, B.R., THOMSON, R. & CURTIS, A.R. (1973). A new mathematical model for fitting an HPL radioimmunoassay curve. *J. clin. Path.*, 26, 973–6.
HARRISON, E., LEES, K.A. & WOOD, F. (1951). The assay of vitamin B12. Part VI. Microbiological estimation with a mutant of *Escherichia coli* by the plate method. *Analyst*, 76, 696–705.
HARTE, R.A. (1948). A simple graphical solution for potency calculations of multidose assays. *Science*, 107, 401–2.
HARTLEY, P. (1935). International biological standards. *Pharm. J.*, 81, 625–7.
HARTLEY, P. (1945a). Notes on the international standards for antitoxins and antisera. *Bull. Hlth Org., League of Nations*, 12, 76–97.
HARTLEY, P. (1945b). International biological standards: Prospect and retrospect. *Proc. R. Soc. Med.*, 39, 45–58.
HATCHER, R.A. & BRODY, J.G. (1910). The biological standardization of drugs. *Am. J. Pharm.*, 82, 360–72.
HEALY, M.J.R. (1949). Routine computation of biological assays involving a quantitative response. *Biometrics*, 5, 330–4.
HEALY, M.J.R. (1950). The planning of probit assays. *Biometrics*, 6, 424–31.
HEALY, M.J.R. (1972). Statistical analysis of radioimmunoassay data. *Biochem. J.*, 130, 207–10.
HEMMINGSEN, A.M. (1933). The accuracy of insulin assay on white mice. *Q. Jl Pharm. Pharmac.*, 6, 39–80 & 187–218.
HEMMINGSEN, A.M. & KROGH, A. (1926). The assay of insulin by the convulsive-dose method on white mice. *Publs League of Nations, III, Health*, 7, 40–6.
HSI, B.P. (1969). The multiple sample up-and-down method in bioassay. *J. Am. statist. Ass.*, 64, 147–62.
HUMPHREY, J.H., LIGHTBOWN, J.W., MUSSETT, M.V. & PERRY, W.L.M. (1953). The international standard for aureomycin. *Bull. Wld Hlth Org.*, 9, 851–60.
IPSEN, J. (1941). *Contribution to the Theory of Biological Standardization*. Copenhagen: Nyt Noraisk Forlag (Arnold Busck).
IPSEN, J. (1949). Biometric analysis of graded response with incomplete measurements in assays of analgetic drugs. *Acta pharmac. tox.*, 5, 321–46.
IRWIN, J.O. (1937). Statistical method applied to biological assays. *Jl R. statist. Soc., Suppl.*, 4, 1–60.
IRWIN, J.O. (1950). Biological assays with special reference to biological standards. *J. Hyg.*, 48, 215–38.
IRWIN, J.O. & CHEESEMAN, E.A. (1939). On an approximate method of determining the median effective dose and its error in the case of a quantal response. *J. Hyg.*, 39, 574–80.
IRWIN, J.O. & STANDFAST, A.F.B. (1955). Litter-mate assays of pertussis vaccine. *J. Hyg.*, 53, 106–11.
JERNE, N.K. & PERRY, W.L.M. (1956). The stability of biological standards. *Bull. Wld Hlth Org.*, 14, 167–82.
JERNE, N.K. & WOOD, E.C. (1949). The validity and meaning of the results of biological assays. *Biometrics*, 5, 273–99.

REFERENCES

JOHN, J.A. & QUENOUILLE, M.H. (1977). *Experiments: Design and Analysis* (2nd edn). London & High Wycombe: Charles Griffin & Co. Ltd.

JOHN, P.W.M. (1971). *Statistical Design and Analysis of Experiments*, New York: Macmillan.

JONES, J.I.M. (1945). The biological estimation of vitamin D. *Q. Jl Pharm. Pharmac.* **18**, 92–108.

KAPTEYN, J.C. (1903). *Skew Frequency Curves in Biology and Statistics*. Groningen: P. Noordhoff.

KAPTEYN, J.C. & VAN UVEN, M.J. (1916). *Skew Frequency Curves in Biology and Statistics*. Groningen: Hoitsema Brothers.

KÄRBER, G. (1931). Beitrag zur kollektiven Behandlung pharmakologischer Reihenversuche. *Arch. exp. Path. Pharmak.*, **162**, 480–7.

KEMPTHORNE, O. (1952). *The Design and Analysis of Experiments*. New York: John Wiley & Sons, Inc.

KENDALL, J.W., LIDDLE, G.W., FEDERSPIEL, C.F. & CORNFIELD, J. (1963). Dissociation of corticotropin-suppressing activity from the eosinopenic and hyperglycemic activities of corticosteroid analogues. *J. clin. Invest.*, **42**, 396–403.

KENDALL, Sir Maurice, & STUART, A. *The Advanced Theory of Statistics*, vol. 1 (4th edn, 1977), vol. 2 (3rd edn, 1973). London & High Wycombe: Charles Griffin & Co., Ltd.

KENT-JONES, D.W. & MEIKLEJOHN, M. (1944). Some experiences of microbiological assays of riboflavin, nicotinic acid and other nutrient factors. *Analyst*, **69**, 330–6.

KNUDSEN, L.F. (1945a). Penicillin assay. *Science*, **101**, 46–8.

KNUDSEN, L.F. (1945b). The use of statistics in biological experimentation and assay. *J. Ass. off. agric. Chem.*, **28**, 806–13.

KNUDSEN, L.F. (1950). Statistics in microbiological assay. *Ann. N.Y. Acad. Sci.*, **52**, 889–902.

KNUDSEN, L.F. & CURTIS, J.M. (1947). The use of the angular transformation in biological assays. *J. Am. statist. Ass.*, **42**, 282–96.

KNUDSEN, L.F. & RANDALL, W.A. (1945). Penicillin assay and its control chart analysis. *J. Bact.*, **50**, 187–200.

KNUDSEN, L.F., SMITH, R.B., VOS, B.J. & MCCLOSKY, W.T. (1946). The biological assay of epinephrine. *J. Pharmac. exp. Ther.*, **86**, 339–43.

KOCH, W. (1947). A rapid method for the evaluation of microbiologic tests. *Am. J. clin. Path.*, **17**, 897–903.

KOLB, R.W., CUTCHINS, E.C., JONES, W.P. & AYLOR, H.T. (1961). A comparison of the rabbit scarification technique with titrations in cell cultures for the potency assay of smallpox vaccine. *Bull. Wld Hlth Org.*, **25**, 25–32.

KULSHRESHTHA, A.C. (1969). On the efficiency of modified BIB designs for bio-assays. *Biometrics*, **25**, 591–3.

KULSHRESHTHA, A.C. (1972). A new incomplete block design for slope-ratio assays. *Biometrics*, **28**, 585–7.

LECLERQ, R., TALJEDAL, I.B. & WOLD, S. (1971). Evaluation of radio-isotope data in steroid assays based on competitive protein binding. *Clinica chim. Acta*, **36**, 257–9.

LEECH, F.B. & GRUNDY, P.M. (1953). A nomogram for assays in randomized blocks. *Br. J. Pharmac. Chemother.*, **8**, 281–5.

LEES, K.A. (1949). A semi-automatic calculating machine for the computation of microbiological plate assay results. *Chemy Ind.*, 378.

LEES, K.A. & TOOTILL, J.P.R. (1955a). Microbiological assay on large plates. Part I. General considerations with particular reference to routine assay. *Analyst*, **80**, 95–110.

LEES, K.A. & TOOTILL, J.P.R. (1955b). Microbiological assay on large plates. Part II. Precise assay. *Analyst*, **80**, 110–23.

LEES, K.A. & TOOTILL, J.P.R. (1955c). Microbiological assay on large plates. Part III. High throughput, low precision assays. *Analyst*, **80**, 531–5.

LEPPÄLUOTO, J. (1972). Blood bioassayable thyrotrophin and corticosteroid levels during various physiological and stress conditions in the rabbit. *Acta endocr.*, Supplementum 165.

LIDDLE, G.W., CORNFIELD, J., CASPER, A.G.T. & BARTTER, F.C. (1955). The physiological basis for a method of assaying aldosterone in extracts of human urine. *J. clin. Invest.*, **34**, 1410–16.

LIGHTBOWN, J.W. (1961). Biological standardisation and the analysis. *Analyst*, **86**, 216–30.
LIGHTBOWN, J.W., KOGUT, M. & MUSSETT, M.V. (1965). The international standard for oleandomycin. *Bull. Wld Hlth Org.*, **33**, 227–33.
LILJESTRAND, Å. (1949). Assay of the curative action of neoarsphenamine by time–mortality data. *Pharm. Pharmac.*, **1**, 78–86.
LITCHFIELD, J.T. & FERTIG, J.W. (1941). On a graphic solution of the dosage–effect curve. *Bull. Johns Hopkins Hosp.*, **69**, 276–86.
LITCHFIELD, J.T. & WILCOXON, F. (1949). A simplified method of evaluating dose–effect experiments. *J. Pharmac. exp. Ther.*, **96**, 99–113.
LITCHFIELD, J.T. & WILCOXON, F. (1953). The reliability of graphic estimates of relative potency from dose–per cent effect curves. *J. Pharmac. exp. Ther.*, **108**, 18–25.
LONG, D.A., MILES, A.A. & PERRY, W.L.M. (1954). The assay of tuberculin. *Bull. Wld Hlth Org.*, **10**, 989–1002.
LORD, E. (1947). The use of range in place of standard deviation in the t-test. *Biometrika*, **34**, 41–67.
LUCAS, H.L. (1951). Bias in estimation of error in change-over trials with dairy cattle. *J. agric. Sci.*, **41**, 146–8.
LUCAS, H.L. (1956). Switchback trials for more than two treatments. *J. Dairy Sci.*, **39**, 146–54.
LUCAS, H.L. (1957). Extra-period Latin-square change-over design. *J. Dairy Sci.*, **40**, 225–239.
MCARTHUR, J.W., ULFELDER, H. & FINNEY, D.J. (1966). A flexible computer program for the composite analysis of symmetrical and asymmetrical biologic assays of parallel-line type. *J. Pharmac. exp. Ther.*, **153**, 573–80.
MCCRADY, M.H. (1915). The numerical interpretation of fermentation-tube results. *J. infect. Dis.*, **17**, 183–212.
MCKENZIE, A.W. & ATKINSON, R.M. (1964). Topical activities of betamethasone esters in man. *Archs Derm.*, **89**, 741–6.
MCLAREN, A. & MICHIE, D. (1954). Are inbred strains suitable for bioassay? *Nature*, **173**, 686–7.
MATHER, K. (1949). The analysis of extinction time data in bioassay. *Biometrics*, **5**, 127–143.
MIDGLEY, A.R. (1966). Radioimmunoassay: A method for human chorionic gonadotropin and human luteinizing hormone. *Endocrinology*, **79**, 10–18.
MIDGLEY, A.R., NISWENDER, G.D. & REBAR, R.W. (1969). Principles for the assessment of the reliability of radioimmunoassay methods (precision, accuracy, sensitivity, specificity). *Acta endocr.*, Supplementum 142, 163–84.
MILES, A.A. (1948). Some observations on biological standards. *Analyst*, **73**, 530–8.
MILES, A.A. (1949). The biological unit of activity. *Bull. Wld Hlth Org.*, **2**, 205–13.
MILES, A.A. (1951). Biological standards and the measurement of therapeutic activity. *Br. med. Bull.*, **7**, 283–91.
MILES, A.A. (1952). The concept of biological potency as applied to closely related antibiotics. *Bull. Wld Hlth Org.*, **6**, 131–47.
MILES, A.A. & PERRY, W.L.M. (1950). Third international digitalis standard. *Bull. Wld Hlth Org.*, **2**, 655–72.
MILLER, L.C. (1944). The U.S.P. collaborative digitalis study using frogs (1939–1941). *J. Am. pharmac. Ass.*, **33**, 245–66.
MILLER, L.C. (1950). Biological assays involving quantal responses. *Ann. N.Y. Acad. Sci.*, **52**, 903–19.
MILLER, L.C., BECKER, T.J. & TAINTER, M.L. (1948). The quantitative evaluation of spasmolytic drugs in vitro. *J. Pharmac. exp. Ther.*, **92**, 260–8.
MILLER, L.C., BLISS, C.I. & BRAUN, H.A. (1939). The assay of digitalis. I. Criteria for evaluating various methods using frogs. *J. Am. pharm. Ass.*, **28**, 644–57.
MILLER, L.C. & TAINTER, M.L. (1944). Estimation of the ED_{50} and its error by means of logarithmic-probit graph paper. *Proc. Soc. exp. Biol. Med.*, **57**, 261–4.
MONGAR, J.L. (1959). Use of randomized blocks in local anaesthetic assays. In *Quantitative Methods in Human Pharmacology and Therapeutics*, edited by D.R. Laurence, London: Pergamon Press.
MOORE, W. & BLISS, C.I. (1942). A method for determining insecticidal effectiveness using *Aphis rumicis* and certain organic compounds. *J. econ. Ent.*, **35**, 544–53.

REFERENCES

MORAN, P.A.P. (1954a). The dilution assay of viruses I. *J. Hyg.*, **52**, 189–93.
MORAN, P.A.P. (1954b). The dilution assay of viruses II. *J. Hyg.*, **52**, 444–6.
MORAN, P.A.P. (1958). Another test for heterogeneity of host resistance in dilution assays. *J. Hyg.*, **56**, 319–22.
MORRELL, C.A. & ALLMARK, M.G. (1941). The toxicity and trypanocidal activity of commercial neoarsphenamine. *J. Am. pharm. Ass.*, **30**, 33–8.
MORRELL, C.A., CHAPMAN, C.W. & ALLMARK, M.G. (1938). On the therapeutic assay of neoarsphenamine with *Trypanosoma equiperdum*. *J. Pharmac. exp. Ther.*, **64**, 14–42.
MUNSON, P.L. & SHEPS, M.C. (1958). An improved procedure for the biological assay of androgens by direct application to the combs of baby chicks. *Endrocrinology*, **62**, 173–88.
MURRAY, C.A. (1937). A statistical analysis of fly mortality data. *Soap sanit. Chem.*, **13**, no. 8, 89–105.
MYERSCOUGH, P.R. & SCHILD, H.O. (1958). Quantitative assays of oxytocic drugs on the human postpartum uterus. *Br. J. Pharmac.*, **13**, 207–12.
NELDER, J.A. & MEAD, R. (1965). A simplex method for function minimization. *Comput. J.*, **7**, 308–13.
NOEL, R.H. (1945). The biological assay of epinephrine. *J. Pharmac. exp. Ther.*, **84**, 278–283.
NORTHAM, B.E. & NORRIS, F.W. (1952). A microbiological assay of inositol: its development and statistical analysis. *J. gen. Microbiol.*, **7**, 245–56.
OSGOOD, E.E. (1947). Assay of penicillin, streptomycin, trivalent organic arsenicals, and other bactericidal and bacteriostatic agents. *J. Lab. clin. Med.*, **32**, 444–60.
OSGOOD, E.E. & GRAHAM, S.M. (1947). A simple rapid method for assay of bactericidal and bacteriostatic agents. *Am. J. clin. Path.*, **17**, 93–107.
PATTERSON, H.D. (1950). The analysis of change-over trials. *J. agric. Sci.*, **40**, 375–80.
PATTERSON, H.D. (1951). Change-over trials. *Jl R. statist. Soc.*, **B13**, 256–71.
PATTERSON, H.D. (1952). The construction of balanced designs for experiments involving sequences of treatments. *Biometrika*, **39**, 32–48.
PATTERSON, H.D. & LUCAS, H.L. (1959). Extra-period change-over designs. *Biometrics*, **15**, 116–32.
PEARCE, S.C. (1965). *Biological Statistics*. New York: McGraw-Hill Book Co.
PEARSON, E.S. & HARTLEY, H.O. (1954). *Biometrika Tables for Statisticians*, vol. I. Cambridge: The University Press.
PERRY, W.L.M. (1950). Reports on biological standards. VI. The design of toxicity tests. *Medical Research Council, Special Report Series, no. 270*.
PETO, S. (1953). A dose–response equation for the invasion of micro-organisms. *Biometrics*, **9**, 320–35.
PETRUSZ, P. (1969). Biological and immunological characterization of gonadotrophic profiles. *Acta endocr.*, Supplementum 142, 77–94.
PETRUSZ, P., DICZFALUSY, E. & FINNEY, D.J. (1971a). Bioimmunoassay of gonadotrophins. 1. Theoretical considerations. *Acta endocr.*, **67**, 40–6.
PETRUSZ, P., DICZFALUSY, E. & FINNEY, D.J. (1971b). Bioimmunoassay of gonadotrophins. 2. Practical aspects and tests of additivity. *Acta endocr.*, **67**, 47–62.
PETRUSZ, P., ROBYN, C., DICZFALUSY, E. & FINNEY, D.J. (1970). Bioassay of antigonadotrophic sera. 4. Experimental verification of the principle of additivity. *Acta endocr.*, **63**, 150–60.
PITTMAN, M. & LIEBERMAN, J.E. (1948). An analysis of the Wilson–Worcester method for determining the median effective dose of pertussis vaccine. *Am. J. publ. Hlth*, **38**, 15–21.
PRICE, W.C. (1945). Accuracy of the local-lesion method for measuring virus activity. IV. Southern bean mosaic virus. *Am. J. Bot.*, **32**, 613–19.
PRICE, W.C. (1946). Measurement of virus activity in plants. *Biometrics*, **2**, 81–6.
PRICE, W.C. & SPENCER, E.L. (1943). Accuracy of the local-lesion method for measuring virus activity. III. The standard deviation of the log-ratio of potencies as a measure of the accuracy of measurement. *Am. J. Bot.*, **30**, 720–35.
QAZI, M.H., ROMANI, P. & DICZFALUSY, E. (1974). Discrepancies in plasma LH activities as measured by radioimmunoassay and an "in vitro" assay. *Acta endocr.*, **77**, 672–86.

RAO, C.R. (1954). Estimation of relative potency from multiple response data. *Biometrics*, **10**, 208–20.
RAO, C.R. (1961). Asymptotic efficiency and limiting information. *Proc. Fourth Berkeley Symp. math. Statist. Probab.*, **1**, 531–46.
RAO, C.R. (1962). Efficient estimates and optimum inference procedures in large samples. *Jl R. statist. Soc.*, **B24**, 46–72.
RAO, C.R. (1965). *Linear Statistical Inference and its Applications*. New York: John Wiley & Sons, Inc.
REED, L.J. & MUENCH, H. (1938). A simple method of estimating fifty per cent. endpoints. *Am. J. Hyg.*, **27**, 493–7.
RERUP, C. (1958). The bioassay of corticotrophin A. *Acta endocr.*, Supplementum 42.
RERUP, C. (1959). Advantages of the triplet cross-over design in the intravenous assay of corticotrophin (Sayers test). *Acta endocr.*, **30**, 509–32.
RERUP, C. (1960). On cross-over tests with a graded response: statistical analyses and the concept of the "log-dose difference response–difference line". *Acta pharmac. tox.*, **17**, 390–403.
ROBBINS, H. & MONRO, S. (1951). A stochastic approximation method. *Ann. math. Statist.*, **29**, 400–7.
ROBYN, C. (1969). Biological and immunological characterization of antigonadotrophic profiles. *Acta endocr.*, Supplementum 142, 31–53.
ROBYN, C. & DICZFALUSY, E. (1968a). Bioassay of antigonadotrophic sera. 2. Assay of the human chorionic gonadotrophin (HCG) and luteinising hormone (LH) neutralising potencies. *Acta endocr.*, **59**, 261–76.
ROBYN, C. & DICZFALUSY, E. (1968b). Bioassay of antigonadotrophic sera. 3. Assay of the human follicle stimulating hormone (FSH) neutralising potency. *Acta endocr.*, **59**, 277–97.
ROBYN, C., DICZFALUSY, E. & FINNEY, D.J. (1968). Bioassay of antigonadotrophic sera. 1. Statistical considerations and general principles. *Acta endocr.*, **58**, 593–9.
ROBYN, C., PETRUSZ, P. & DICZFALUSY, E. (1969). Follicle stimulating hormone-like activity in human chorionic gonadotrophin preparations. *Acta endocr.*, **60**, 137–56.
RODBARD, D. (1971). Statistical aspects of radioimmunoassays. *Principles of Competitive Protein Binding Assays* (ed. W.D. Odell & W.H. Daughaday), 204–59. Philadelphia: Lippincott.
RODBARD, D. (1974). Statistical quality control and routine data processing for radioimmunoassays and immunoradiometric assays. *Clin. Chem.*, **20**, 1255–70.
RODBARD, D., BRIDSON, W. & RAYFORD, P.L. (1969). Rapid calculation of radioimmunoassay. *J. Lab. clin. Med.*, **74**, 770–81.
RODBARD, D. & COOPER, J.A. (1970). A model for prediction of confidence limits in radioimmunoassays and competitive protein binding assays. *In Vitro Procedures with Radioisotopes in Medicine*, 659–74. Vienna: International Atomic Energy Agency.
RODBARD, D. & FRAZIER, G.R. (1975). Statistical analysis of radioligand assay data. *Meth. Enzym.*, **37**, 3–32.
RODBARD, D. & HUTT, D.M. (1974). Statistical analysis of radioimmunoassays and immunoradiometric (labelled antibody) assays: a generalized weighted, iterative, least-squares method for logistic curve fitting. *Radioimmunoassay and Related Procedures in Medicine*, vol. I, 165–92, Vienna: International Atomic Energy Agency.
RODBARD, D. & LEWALD, J.E. (1970). Computer analysis of radioligand assay and radioimmunoassay data. *Acta endocr.*, Supplementum 147, 79–103.
ROMANI, P., ROBERTSON, D.M. & DICZFALUSY, E. (1976). Biologically active luteinizing hormone (LH) in plasma. I. Validation of the *in vitro* bioassay when applied to plasma of women. *Acta endocr.*, **83**, 454–65.
ROMANI, P., ROBYN, C., PETRUSZ, P. & DICZFALUSY, E. (1974). Bioassay of antigonadotrophic sera. 5. Further studies on the reliability of the bioassay method for the estimation of human chorionic gonadotrophin (HCG) neutralizing potency. *Acta endocr.*, **76**, 629–44.
SAKIZ, E. & GUILLEMIN, R. (1964). On a method for calculation and analysis of results in the McKenzie assay for thyrotropin (TSH). *Proc. Soc. exp. Biol. Med.*, **115**, 856–60.
SAMPFORD, M.R. (1952a). The estimation of response-time distributions. I. Fundamental concepts and general methods. *Biometrics*, **8**, 13–32.

SAMPFORD, M.R. (1952b). The estimation of response-time distributions. II. Multi-stimulus distributions. *Biometrics*, **8**, 307–69.
SAMPFORD, M.R. (1954). The estimation of response-time distributions. III. Truncation and survival. *Biometrics*, **10**, 531–61.
SAMPFORD, M.R. (1957). Methods of construction and analysis of serially balanced sequences. *Jl R. statist. Soc.*, **B19**, 286–304.
SAMPFORD, M.R. & TAYLOR, J. (1959). Censored observations in randomized block experiments. *Jl. R. statist. Soc.*, **B21**, 214–37.
SCHILD, H.O. (1942). A method of conducting a biological assay on a preparation giving repeated graded responses illustrated by the estimation of histamine. *J. Physiol.*, **101**, 115–30.
SCHILD, H.O. (1950). General approach to biological assays. *Analyst*, **75**, 533–6.
SCHILD, H.O. (1959). The use of incomplete randomized blocks in an oxytocic assay. In *Quantitative Methods in Human Pharmacology and Therapeutics*, ed. by D.R. Laurence. London: Pergamon Press.
SEARLE, S. (1966). *Matrix Algebra for the Biological Sciences*. New York: John Wiley & Sons, Inc.
SEN, P.K. (1963). On the estimation of potency in dilution (-direct) assays by distribution-free methods. *Biometrics*, **19**, 532–52.
SEN, P.K. (1964). Tests for the validity of the fundamental assumption in dilution (-direct) assays. *Biometrics*, **20**, 770–84.
SEN, P.K. (1965). Some further applications of non-parametric methods in dilution (-direct) assays. *Biometrics*, **21**, 799–810.
SEN, P.K. (1971). Robust statistical procedures in problems of linear regression with special reference to quantitative bioassays, I. *Int. statist. Rev.*, **39**, 21–38.
SEN, P.K. (1972). Robust statistical procedures in problems of linear regression with special reference to quantitative bioassays, II. *Int. statist. Rev.*, **40**, 161–72.
SHENTON, L.R. & BOWMAN, K.O. (1977). *Maximum likelihood estimation in small samples*. London & High Wycombe: Charles Griffin & Co., Ltd.
SHEPS, M.C. & HENDRIE, K.H. (1958). A minimum error table for parallel line biological assays. *J. Pharmac. exp. Ther.*, **124**, 94–6.
SHEPS, M.C. & MUNSON, P.I. (1957). The error of replicated potency estimates in a biological assay method of the parallel line type. *Biometrics*, **13**, 131–48.
SHERWOOD, M.B. (1947). Simple formulas for calculating percentage potency in three- and four-dose assay procedures. *Science*, **106**, 152–3.
SHERWOOD, M.B. (1951). A universal line graph for estimating percentage potency in multidose assays. *Science*, **113**, 185–7.
SHERWOOD, M.B., FALCO, E.A. & DE BEER, E.J. (1944). A rapid, quantitative method for the determination of penicillin. *Science*, **99**, 247–8.
SHORACK, G. (1966). Graphical procedures for using distribution-free methods in the estimation of relative potency in dilution (-direct) assays. *Biometrics*, **22**, 610–19.
SMITH, R.B. & VOS, B.J. (1943). The biological assay of posterior pituitary solution. *J. Pharmac. exp. Ther.*, **78**, 72–8.
SMITH, K.W., MARKS, H.P., FIELLER, E.C. & BROOM, W.A. (1944). An extended cross-over design and its use in insulin assay. *Q. Jl Pharm. Pharmac.*, **17**, 108–17.
SNEDECOR, G.W. & COCHRAN, W.G. (1967). *Statistical Methods* (6th edn). Ames, Iowa: Iowa State University Press.
SOKAL, R.R. & ROHLF, F.J. (1973). *Introduction to Biostatistics*, San Francisco: W.H. Freeman & Co.
SOMERS, G.F. (1950). The measurement of thyroidal activity. *Analyst*, **75**, 537–41.
SOMERS, G.F. & EDGE, N.D. (1947). Comparative activities of amethocaine, cinchocaine and procaine as local anaesthetics. *Q. Jl Pharm. Pharmac.*, **20**, 380–7.
SPEARMAN, C. (1908). The method of 'right and wrong cases' ('constant stimuli') without Gauss's formulae. *Br. J. Psychol.*, **2**, 227–42.
STEVENS, W.L. (1937). Appendix to Bliss (1937). The truncated normal distribution. *Ann. appl. Biol.*, **24**, 847–50.
STEVENS, W.L. (1958). Dilution series: a statistical test of technique. *Jl R. statist. Soc.*, **B20**, 205–14.
SWAROOP, S. (1938). Numerical estimation of *B. coli* by dilution method. *Indian J. med. Res.*, **26**, 353–78.

SWAROOP, S. (1940). Error in the estimation of the most probable number of organisms by the dilution method. *Indian J. med. Res.*, **27**, 1129–47.
SWAROOP, S. (1941a). A modification of the routine dilution tests and table showing the most probable number of organisms and the standard error of this number. *Indian J. med. Res.*, **29**, 499–510.
SWAROOP, S. (1941b). A consideration of the accuracy of estimation of the most probable number of organisms by dilution test. *Indian J. med. Res.*, **29**, 511–21.
TAMAOKI, B. (1957). On the analysis of $m \times n$ point assay. *Pharm. Bull.*, **5**, 371–3.
THOMPSON, R.E. (1944). Biological assay of posterior pituitary. *J. Pharmac. exp. Ther.*, **80**, 373–82.
THOMPSON, R.E. (1945). Biological assay of epinephrine. *J. Am. pharm. Ass.*, **34**, 265–9.
THOMPSON, W.R. (1947). Use of moving averages and interpolation to estimate median-effective dose. I. Fundamental formulas, estimation of error, and relation to other methods. *Bact. Rev.*, **11**, 115–45.
THOMPSON, W.R. (1948). On the use of parallel or non-parallel systems of transformed curves in bio-assay: illustrations in the quantitative complement-fixation test. *Biometrics*, **4**, 197–210.
THOMPSON, W.R. & WEIL, C.S. (1952). On the construction of tables for moving-average interpolation. *Biometrics*, **8**, 51–4.
THOMSON, G.H. (1919). A direct deduction of the constant process used in the method of right and wrong cases. *Psychol. Rev.*, **26**, 454–64.
TREVAN, J.W. (1927). The error of determination of toxicity. *Proc. R. Soc.*, **B101**, 483–514.
TREVAN, J.W. (1929). A statistical note on the testing of anti-dysentery sera. *J. Path. Bact.*, **32**, 127–34.
TSUTAKAWA, R.K. (1967). The random walk design in bio-assay. *J. Am. statist. Ass.*, **62**, 842–56.
TUKEY, J.W. (1948). Approximate weights. *Ann. math. Statist.*, **19**, 91–2.
TUKEY, J.W. (1949). Answer to Question 21. *Am. Statistn*, **3**, no. 4, 12.
UNITED STATES PHARMACOPEIA XVIII (1970). Washington: U.S. Pharmacopeial Convention, Inc.
URBAN, F.M. (1909). Die psychophysischen Massmethoden als Grundlagen empirischer Messungen. *Arch. ges. Psychol.*, **15**, 261–355.
URBAN, F.M. (1910). Die psychophysischen Massmethoden als Grundlagen empirischer Messungen (continued). *Arch. ges. Psychol.*, **16**, 168–227.
VAN DER WAERDEN, B.L. (1940a). Biologische Konzentrazionsauswertung. *Ber. Verh. sächs. Akad. Wiss.*, **92**, 41–4.
VAN DER WAERDEN, B.L. (1940b). Wirksamkeits- und Konzentrationsbestimmung durch Tierversuche. *Arch. exp. Path. Pharmak.*, **195**, 389–412.
VAN STRIK, R. (1961). A method of estimating relative potency and its precision in the case of semi-quantitative responses. *Proc. Symp. Quantitative Methods in Pharmacology, Leiden*, 88–100.
VOS, B.J. (1943). Use of the latent period in the assay of ergonovine on the isolated rabbit uterus. *J. Am. pharm. Ass.*, **32**, 138–41.
VOS, B.J. (1950). Statistics in biological assay: An example of the graded response. *Ann. N.Y. Acad. Sci.*, **52**, 920–1.
WADLEY, F.M. (1948). Experimental design in comparison of allergens on cattle. *Biometrics*, **4**, 100–8.
WADLEY, F.M. (1949a). Dosage–mortality correlation with number treated estimated from a parallel sample. *Ann. appl. Biol.*, **36**, 196–202.
WADLEY, F.M. (1949b). The use of biometric methods in comparison of acid-fast allergens. *Am. Rev. Tuberc.*, **60**, 131–9.
WEIL, C.S. (1952). Tables for convenient calculation of median-effective dose (LD_{50} or ED_{50}) and instructions in their use. *Biometrics*, **8**, 249–63.
WETHERILL, G.B. (1963). Sequential estimation of quantal response curves. *Jl R. statist. Soc.*, **B25**, 1–48.
WHITE, R.F. & GRACA, J.G. (1958). Multinomially grouped response times for the quantal response bio-assay. *Biometrics*, **14**, 462–88.
WILLIAMS, D.A. (1973). The estimation of relative potency from two parabolas in symmetric bioassays. *Biometrics*, **29**, 695–700.

WILLIAMS, E.J. (1949). Experimental designs balanced for the estimation of residual effects of treatments. *Aust. J. sci. Res.*, **A2**, 149–68.
WILLIAMS, E.J. (1950). Experimental designs balanced for pairs of residual effects. *Aust. J. sci. Res.*, **A3**, 351–63.
WILLIAMS, E.J. (1959). *Regression Analysis*. New York: John Wiley & Sons, Inc.
WILSON, E.B. & WORCESTER, J. (1943a). The determination of L.D.50 and its sampling error in bio-assay. *Proc. natn. Acad. Sci.*, **29**, 79–85.
WILSON, E.B. & WORCESTER, J. (1943b). The determination of L.D.50 and its sampling error in bio-assay. II. *Proc. natn. Acad. Sci.*, **29**, 114–20.
WILSON, E.B. & WORCESTER, J. (1943c). Bio-assay on a general curve. *Proc. natn. Acad. Sci.*, **29**, 150–4.
WILSON, E.B. & WORCESTER, J. (1943d). The determination of L.D.50 and its sampling error in bio-assay. III. *Proc. natn. Acad. Sci.*, **29**, 257–62.
WINDER, C.V. (1947). Misuse of 'deduced ratios' in the estimation of median effective doses. *Nature*, **159**, 883.
WINDER, C.V. (1950). Some examples of the use of statistics in pharmacology. *Ann. N.Y. Acad. Sci.*, **52**, 838–61.
WINNE, D. (1962). Die Bestimmung von Geraden, die sich in einem Punkt auf der Ordinatenachse schneiden, bei unsymmetrischer Anordnung der Versuchsergebnisse (unsymmetrical multiple slope ratio assay). *Biometr. Z.*, **4**, 217–38.
WOOD, E.C. (1944a). Mathematics of biological assay. *Nature*, **153**, 84–5.
WOOD, E.C. (1944b). Mathematics of biological assay. *Nature*, **153**, 681–2.
WOOD, E.C. (1945). Calculation of the results of microbiological assays. *Nature*, **155**, 632–633.
WOOD, E.C. (1946a). The theory of certain analytical procedures, with particular reference to microbiological assays. *Analyst*, **71**, 1–14.
WOOD, E.C. (1946b). Computation of biological assays. *Nature*, **158**, 835.
WOOD, E.C. (1947a). The computation of microbiological assays of amino-acids and other growth factors. *Analyst*, **72**, 84–90.
WOOD, E.C. (1947b). Short cuts to the estimation of standard errors, particularly in microbiological assays. *Chemy Ind.*, 334–6.
WOOD, E.C. & FINNEY, D.J. (1946). The design and statistical analysis of microbiological assays. *Q. Jl Pharm. Pharmacol.*, **19**, 112–27.
WORCESTER, J. & WILSON, E.B. (1943). A table determining L.D.50 or the fifty per cent. end point. *Proc. natn. Acad. Sci.*, **29**, 207–12.
WORLD HEALTH ORGANIZATION (1971). Twenty-third Report of the Expert Committee on Biological Standardization. *WHO Technical Report Series*. Geneva.
WORLD HEALTH ORGANIZATION (1975). *Biological Substances: International standards and reference preparations*. Geneva: WHO.
YATES, F. (1933). The analysis of replicated experiments when the field results are incomplete. *Emp. J. exp. Agric.*, **1**, 129–42.
YATES, F. (1937). The design and analysis of factorial experiments. *Imperial Bureau of Soil Science, Harpenden, Technical Communication no. 35*.
YATES, F. (1939). An apparent inconsistency arising from tests of significance based on fiducial distributions of unknown parameters. *Proc. Camb. phil. Soc.*, **35**, 579–91.
YOUDEN, W.J. (1937). Use of incomplete block replications in estimating tobacco-mosaic virus. *Contr. Boyce Thompson Inst.*, **9**, 41–8.
YOUDEN, W.J. (1940). Experimental designs to increase accuracy of greenhouse studies. *Contr. Boyce Thompson Inst.*, **11**, 219–28.
YOUNG, D.M. & ROMANS, R.G. (1948). Assays of insulin with one blood sample per rabbit per test day. *Biometrics*, **4**, 122–31.
ZIEGLER, N.R. & HALVORSON, H.O. (1935). Application of statistics to problems in bacteriology. IV. Experimental comparison of the dilution method, the plate count, and the direct count for the determination of bacterial populations. *J. Bact.*, **29**, 609–34.

Appendix tables

Users of bioassay need access to values of standard statistical functions and of elementary mathematical functions. Where statistical analysis is to be undertaken by computer, appropriate subroutines will be built into the programs. Those who do not have good computer facilities, or who are engaged in program development, must use collections of tables. In their *Statistical Tables for Biological, Agricultural and Medical Research*, R.A. Fisher and F. Yates have given both kinds of table to an accuracy sufficient for practical purposes. Because all who apply statistical methods ought to have this or a similar volume, full tables are not included here. Nevertheless, as an aid to the reader who is studying the techniques described in this book, abridged versions of tables of the distributions of t, χ^2, the variance ratio, and the Behrens ratio are included in the following pages. Their accuracy and detail will suffice for many applications; for the numerical examples in this book, tables of greater accuracy have generally been used. Assays based on quantal responses require tables that are not always so readily available. Tables V–XVIII are of the full accuracy generally needed.

I am indebted to Dr Yates, the late Sir Ronald Fisher, and their publishers (Messrs Oliver and Boyd Ltd) for permission to include abridgements of their Tables III, V, V_1, and IV and Table XII in full as Tables I, II, III, IV and XI of this Appendix, to the editors of the *Journal of the Royal Statistical Society*, Series B, for permission to include Table XVIII, and also to the Cambridge University Press for permission to include Tables III and IV from my *Probit Analysis* as Tables VI and VII of this Appendix.

TABLE I The distribution of t

[Abridged from Fisher and Yates (1963, Table III); Merrington (1942)[*] has tabulated for additional probabilities]

Degrees of freedom (f)	Probability			
	0·1	0·05	0·01	0·001
1	6·31	12·7	63·7	637·0
2	2·92	4·30	9·92	31·6
3	2·35	3·18	5·84	12·9
4	2·13	2·78	4·60	8·61
5	2·02	2·57	4·03	6·86
6	1·94	2·45	3·71	5·96
7	1·90	2·36	3·50	5·40
8	1·86	2·31	3·36	5·04
9	1·83	2·26	3·25	4·78
10	1·81	2·23	3·17	4·59
12	1·78	2·18	3·06	4·32
14	1·76	2·14	2·98	4·14
16	1·75	2·12	2·92	4·02
18	1·73	2·10	2·88	3·92
20	1·72	2·09	2·84	3·85
22	1·72	2·07	2·82	3·79
24	1·71	2·06	2·80	3·74
26	1·71	2·06	2·78	3·71
28	1·70	2·05	2·76	3·67
30	1·70	2·04	2·75	3·65
40	1·68	2·02	2·70	3·55
60	1·67	2·00	2·66	3·46
120	1·66	1·98	2·62	3·37
∞ (normal distribution)	1·645	1·960	2·576	3·291

[*] MERRINGTON, M. (1942). Table of percentage points of the t-distribution. *Biometrika*, **32**, 300.

TABLE II The distribution of the variance ratio

[Abridged from Fisher and Yates (1963, Table V); Merrington and Thompson (1943)[*] have tabulated for additional values of f_1 and probabilities]

(a) 0·05 probability

Values of f_2	Values of f_1						
	1	2	3	4	6	12	∞
1	161·	200·	216·	225·	234·	244·	254·
2	18·5	19·0	19·2	19·2	19·3	19·4	19·5
3	10·1	9·6	9·3	9·1	8·9	8·7	8·5
4	7·7	6·9	6·6	6·4	6·2	5·9	5·6
5	6·6	5·8	5·4	5·2	5·0	4·7	4·4
6	6·0	5·1	4·8	4·5	4·3	4·0	3·7
7	5·6	4·7	4·3	4·1	3·9	3·6	3·2
8	5·3	4·5	4·1	3·8	3·6	3·3	2·9
9	5·1	4·3	3·9	3·6	3·4	3·1	2·7
10	5·0	4·1	3·7	3·5	3·2	2·9	2·5
15	4·5	3·7	3·3	3·1	2·8	2·5	2·1
20	4·4	3·5	3·1	2·9	2·6	2·3	1·8
30	4·2	3·3	2·9	2·7	2·4	2·1	1·6
60	4·0	3·2	2·8	2·5	2·3	1·9	1·4
∞	3·8	3·0	2·6	2·4	2·1	1·8	1·0

(b) 0·01 probability

Values of f_2	Values of f_1						
	1	2	3	4	6	12	∞
1	4052·	4999·	5403·	5625·	5859·	6106·	6366·
2	98·5	99·0	99·2	99·2	99·3	99·4	99·5
3	34·1	30·8	29·5	28·7	27·9	27·1	26·1
4	21·2	18·0	16·7	16·0	15·2	14·4	13·5
5	16·3	13·3	12·1	11·4	10·7	9·9	9·0
6	13·7	10·9	9·8	9·1	8·5	7·7	6·9
7	12·2	9·5	8·5	7·8	7·2	6·5	5·6
8	11·3	8·6	7·6	7·0	6·4	5·7	4·9
9	10·6	8·0	7·0	6·4	5·8	5·1	4·3
10	10·0	7·6	6·6	6·0	5·4	4·7	3·9
15	8·7	6·4	5·4	4·9	4·3	3·7	2·9
20	8·1	5·8	4·9	4·4	3·9	3·2	2·4
30	7·6	5·4	4·5	4·0	3·5	2·8	2·0
60	7·1	5·0	4·1	3·6	3·1	2·5	1·6
∞	6·6	4·6	3·8	3·3	2·8	2·2	1·0

[*] MERRINGTON, M. & THOMPSON, C.M. (1943). Tables of percentage points of the inverted beta (F) distribution. *Biometrika*, 33, 73–88.

TABLE III The Behrens distribution

[Abridged from Fisher and Yates (1963, Table V_1)]

(a) 0·05 probability

f_2	Values of θ	Values of f_1 6	8	12	24	∞
6	0°	2·45	2·45	2·45	2·45	2·45
	15°	2·44	2·43	2·42	2·42	2·41
	30°	2·44	2·40	2·37	2·34	2·32
	45°	2·44	2·36	2·30	2·25	2·20
8	0°	2·31	2·31	2·31	2·31	2·31
	15°	2·31	2·30	2·29	2·29	2·28
	30°	2·33	2·29	2·26	2·24	2·22
	45°	2·36	2·29	2·23	2·18	2·13
12	0°	2·18	2·18	2·18	2·18	2·18
	15°	2·19	2·18	2·18	2·17	2·16
	30°	2·24	2·20	2·17	2·14	2·12
	45°	2·30	2·23	2·17	2·11	2·06
24	0°	2·06	2·06	2·06	2·06	2·06
	15°	2·09	2·08	2·07	2·06	2·06
	30°	2·16	2·12	2·08	2·06	2·04
	45°	2·25	2·18	2·11	2·06	2·01
∞	0°	1·96	1·96	1·96	1·96	1·96
	15°	1·99	1·98	1·97	1·97	1·96
	30°	2·08	2·04	2·01	1·98	1·96
	45°	2·20	2·13	2·06	2·01	1·96

(b) 0·01 probability

f_2	Values of θ	Values of f_1 6	8	12	24	∞
6	0°	3·71	3·71	3·71	3·71	3·71
	15°	3·65	3·64	3·64	3·63	3·63
	30°	3·56	3·50	3·45	3·42	3·40
	45°	3·51	3·36	3·25	3·16	3·09
8	0°	3·36	3·36	3·36	3·36	3·36
	15°	3·33	3·32	3·31	3·30	3·30
	30°	3·31	3·24	3·19	3·16	3·13
	45°	3·36	3·21	3·08	2·99	2·92
12	0°	3·06	3·06	3·06	3·06	3·06
	15°	3·05	3·04	3·03	3·02	3·01
	30°	3·10	3·03	2·98	2·94	2·91
	45°	3·25	3·08	2·95	2·85	2·78
24	0°	2·80	2·80	2·80	2·80	2·80
	15°	2·82	2·80	2·79	2·78	2·78
	30°	2·94	2·86	2·80	2·76	2·73
	45°	3·16	2·99	2·85	2·75	2·66
∞	0°	2·58	2·58	2·58	2·58	2·58
	15°	2·63	2·61	2·60	2·58	2·58
	30°	2·80	2·72	2·66	2·61	2·58
	45°	3·09	2·92	2·78	2·66	2·58

For $\theta > 45°$, note that the d-deviate for (f_1, f_2, θ) is identical with that for $(f_2, f_1, 90°-\theta)$.

TABLE IV The distribution of χ^2

[Abridged from Fisher and Yates (1963, Table IV); Thompson (1941)[*] has tabulated for additional values of f and probabilities]

Degrees of freedom (f)	Probability			
	0·1	0·05	0·01	0·001
1	2·7	3·8	6·6	10·8
2	4·6	6·0	9·2	13·8
3	6·3	7·8	11·3	16·3
4	7·8	9·5	13·3	18·5
5	9·2	11·1	15·1	20·5
6	10·6	12·6	16·8	22·5
7	12·0	14·1	18·5	24·3
8	13·4	15·5	20·1	26·1
9	14·7	16·9	21·7	27·9
10	16·0	18·3	23·2	29·6
12	18·5	21·0	26·2	32·9
14	21·1	23·7	29·1	36·1
16	23·5	26·3	32·0	39·3
18	26·0	28·9	34·8	42·3
20	28·4	31·4	37·6	45·3
22	30·8	33·9	40·3	48·3
24	33·2	36·4	43·0	51·2
26	35·6	38·9	45·6	54·1
28	37·9	41·3	48·3	56·9
30	40·3	43·8	50·9	59·7

When χ^2 is based on more than 30 degrees of freedom the tabular value may be taken as approximately $\{\sqrt{(f-\frac{1}{2})} + B\}^2$ where $B = 0.91, 1.16, 1.64,$ and 2.19 for the 0·1, 0·05, 0·01 and 0·001 probabilities respectively.

[*] THOMPSON, C.M. (1941). Table of percentage points of the χ^2 distribution. *Biometrika*, **32**, 187–91.

TABLE V The probit transformation

[Fisher and Yates (1963) and Finney (1971) have tabulated the probit at intervals of 0·001 in the response rate]

Response rate	0·00	0·01	0·02	0·03	0·04	0·05	0·06	0·07	0·08	0·09
0·00	–	2·67	2·95	3·12	3·25	3·36	3·45	3·52	3·59	3·66
0·10	3·72	3·77	3·82	3·87	3·92	3·96	4·01	4·05	4·08	4·12
0·20	4·16	4·19	4·23	4·26	4·29	4·33	4·36	4·39	4·42	4·45
0·30	4·48	4·50	4·53	4·56	4·59	4·61	4·64	4·67	4·69	4·72
0·40	4·75	4·77	4·80	4·82	4·85	4·87	4·90	4·92	4·95	4·97
0·50	5·00	5·03	5·05	5·08	5·10	5·13	5·15	5·18	5·20	5·23
0·60	5·25	5·28	5·31	5·33	5·36	5·39	5·41	5·44	5·47	5·50
0·70	5·52	5·55	5·58	5·61	5·64	5·67	5·71	5·74	5·77	5·81
0·80	5·84	5·88	5·92	5·95	5·99	6·04	6·08	6·13	6·18	6·23
0·90	6·28	6·34	6·41	6·48	6·55	6·64	6·75	6·88	7·05	7·33

Response rate	0·000	0·001	0·002	0·003	0·004	0·005	0·006	0·007	0·008	0·009
0·97	6·88	6·90	6·91	6·93	6·94	6·96	6·98	7·00	7·01	7·03
0·98	7·05	7·07	7·10	7·12	7·14	7·17	7·20	7·23	7·26	7·29
0·99	7·33	7·37	7·41	7·46	7·51	7·58	7·65	7·75	7·88	8·09

TABLE VI Minimum and maximum working probits, ranges, and weighting coefficients

Minimum working probit		Range	Maximum working probit		Weighting coefficient
Expected probit Y	Y_0	A	Y_1	Expected probit Y	w
1·1	0·8579	5034·	9·1421	8·9	·00082
1·2	0·9522	3425·	9·0478	8·8	·00118
1·3	1·0462	2354·	8·9538	8·7	·00167
1·4	1·1400	1634·	8·8600	8·6	·00235
1·5	1·2334	1146·	8·7666	8·5	·00327
1·6	1·3266	811·5	8·6734	8·4	·00451
1·7	1·4194	580·5	8·5806	8·3	·00614
1·8	1·5118	419·4	8·4882	8·2	·00828
1·9	1·6038	306·1	8·3962	8·1	·01104
2·0	1·6954	225·6	8·3046	8·0	·01457
2·1	1·7866	168·00	8·2134	7·9	·01903
2·2	1·8772	126·34	8·1228	7·8	·02458
2·3	1·9673	95·96	8·0327	7·7	·03143
2·4	2·0568	73·62	7·9432	7·6	·03977
2·5	2·1457	57·05	7·8543	7·5	·04979
2·6	2·2339	44·654	7·7661	7·4	·06168
2·7	2·3214	35·302	7·6786	7·3	·07564
2·8	2·4081	28·189	7·5919	7·2	·09179
2·9	2·4938	22·736	7·5062	7·1	·11026
3·0	2·5786	18·522	7·4214	7·0	·13112
3·1	2·6624	15·2402	7·3376	6·9	·15436
3·2	2·7449	12·6662	7·2551	6·8	·17994
3·3	2·8261	10·6327	7·1739	6·7	·20774
3·4	2·9060	9·0154	7·0940	6·6	·23753
3·5	2·9842	7·7210	7·0158	6·5	·26907
3·6	3·0606	6·6788	6·9394	6·4	·30199
3·7	3·1351	5·8354	6·8649	6·3	·33589
3·8	3·2074	5·1497	6·7926	6·2	·37031
3·9	3·2773	4·5903	6·7227	6·1	·40474
4·0	3·3443	4·1327	6·6557	6·0	·43863
4·1	3·4083	3·7582	6·5917	5·9	·47144
4·2	3·4687	3·4519	6·5313	5·8	·50260
4·3	3·5251	3·2025	6·4749	5·7	·53159
4·4	3·5770	3·0010	6·4230	5·6	·55788
4·5	3·6236	2·8404	6·3764	5·5	·58099
4·6	3·6643	2·7154	6·3357	5·4	·60052
4·7	3·6982	2·6220	6·3018	5·3	·61609
4·8	3·7241	2·5573	6·2759	5·2	·62742
4·9	3·7407	2·5192	6·2593	5·1	·63431
5·0	3·7467	2·5066	6·2533	5·0	·63662
5·1	3·7401	2·5192	6·2599	4·9	·63431
5·2	3·7187	2·5573	6·2813	4·8	·62742
5·3	3·6798	2·6220	6·3202	4·7	·61609
5·4	3·6203	2·7154	6·3797	4·6	·60052
5·5	3·5360	2·8404	6·4640	4·5	·58099
5·6	3·4220	3·0010	6·5780	4·4	·55788
5·7	3·2724	3·2025	6·7276	4·3	·53159
5·8	3·0794	3·4519	6·9206	4·2	·50260
5·9	2·8335	3·7582	7·1665	4·1	·47144
6·0	2·5229	4·1327	7·4771	4·0	·43863
6·1	2·1325	4·5903	7·8675	3·9	·40474
6·2	1·6429	5·1497	8·3571	3·8	·37031
6·3	1·0295	5·8354	8·9705	3·7	·33589
6·4	0·2606	6·6788	9·7394	3·6	·30199
6·5	−0·7051	7·7210	10·7051	3·5	·26907

TABLE VII Working probits
($Y = 2 \cdot 0 – 2 \cdot 9$; $p = 0 \cdot 00 – 0 \cdot 33$)

Response rate, p	Expected probit, Y									
	2·0	2·1	2·2	2·3	2·4	2·5	2·6	2·7	2·8	2·9
·00	1·695	1·787	1·877	1·967	2·057	2·146	2·234	2·321	2·408	2·494
·01	3·951	3·467	3·141	2·927	2·793	2·716	2·681	2·674	2·690	2·721
·02	6·207	5·147	4·404	3·886	3·529	3·287	3·127	3·027	·972	·949
·03	8·463	6·827	5·667	4·846	4·265	·857	·574	·380	3·254	3·176
·04		8·507	6·931	5·806	5·002	4·428	4·020	·733	·536	·403
·05			8·194	6·765	·738	·998	·467	4·086	·818	·631
·06			9·458	7·725	6·474	5·569	4·913	4·440	4·099	3·858
·07				8·684	7·210	6·139	5·360	·793	·381	4·085
·08				9·644	·946	·710	·806	5·146	·663	·313
·09					8·683	7·280	6·253	·499	·945	·540
·10					9·419	·851	·699	·852	5·227	·767
·11						8·421	7·146	6·205	5·509	4·995
·12						·992	·592	·558	·791	5·222
·13						9·562	8·039	·911	6·073	·449
·14							·486	7·264	·355	·677
·15							·932	·617	·636	·904
·16							9·379	7·970	6·918	6·132
·17							·825	8·323	7·200	·359
·18								·676	·482	·586
·19								9·029	·764	·814
·20								·382	8·046	7·041
·21								9·735	8·328	7·268
·22									·610	·496
·23									·892	·723
·24									9·173	·950
·25									·455	8·178
·26									9·737	8·405
·27										·633
·28										·860
·29										9·087
·30										·315
·31										9·542
·32										·769
·33										·997

TABLE VII (continued)   Working probits
($Y = 3.0$–3.9; $p = 0.00$–0.50)

Response rate, p	\multicolumn{10}{c}{Expected probit, Y}									
	3·0	3·1	3·2	3·3	3·4	3·5	3·6	3·7	3·8	3·9
·00	2·579	2·662	2·745	2·826	2·906	2·984	3·061	3·135	3·207	3·277
·01	2·764	2·815	2·872	2·932	2·996	3·061	3·127	3·193	3·259	3·323
·02	·949	·967	·998	3·039	3·086	·139	·194	·252	·310	·369
·03	3·134	3·120	3·125	·145	·176	·216	·261	·310	·362	·415
·04	·319	·272	·252	·251	·267	·293	·328	·369	·413	·461
·05	·505	·424	·378	·358	·357	·370	·395	·427	·465	·507
·06	3·690	3·577	3·505	3·464	3·447	3·447	3·461	3·485	3·516	3·553
·07	·875	·729	·632	·570	·537	·525	·528	·544	·568	·599
·08	4·060	·882	·758	·677	·627	·602	·595	·602	·619	·645
·09	·246	4·034	·885	·783	·717	·679	·662	·660	·671	·690
·10	·431	·186	4·012	·889	·808	·756	·728	·719	·722	·736
·11	4·616	4·339	4·138	3·996	3·898	3·834	3·795	3·777	3·774	3·782
·12	·801	·491	·265	4·102	·988	·911	·862	·835	·825	·828
·13	·986	·644	·391	·208	4·078	·988	·929	·894	·877	·874
·14	5·172	·796	·518	·315	·168	4·065	·996	·952	·928	·920
·15	·357	·948	·645	·421	·258	·142	4·062	4·010	·980	·966
·16	5·542	5·101	4·771	4·527	4·348	4·220	4·129	4·069	4·031	4·012
·17	·727	·253	·898	·634	·439	·297	·196	·127	·083	·058
·18	·913	·406	5·025	·740	·529	·374	·263	·185	·134	·104
·19	6·098	·558	·151	·846	·619	·451	·330	·244	·186	·149
·20	·283	·710	·278	·953	·709	·528	·396	·302	·237	·195
·21	6·468	5·863	5·405	5·059	4·799	4·606	4·463	4·361	4·289	4·241
·22	·653	6·015	·531	·165	·889	·683	·530	·419	·340	·287
·23	·839	·168	·658	·272	·979	·760	·597	·477	·392	·333
·24	7·024	·320	·785	·378	5·070	·837	·664	·536	·443	·379
·25	·209	·472	·911	·484	·160	·914	·730	·594	·495	·425
·26	7·394	6·625	6·038	5·591	5·250	4·992	4·797	4·652	4·546	4·471
·27	·580	·777	·165	·697	·340	5·069	·864	·711	·598	·517
·28	·765	·930	·291	·803	·430	·146	·931	·769	·649	·563
·29	·950	7·082	·418	·910	·520	·223	·997	·827	·701	·608
·30	8·135	·234	·545	6·016	·610	·300	5·064	·886	·752	·654
·31	8·320	7·387	6·671	6·122	5·701	5·378	5·131	4·944	4·804	4·700
·32	·506	·539	·798	·229	·791	·455	·198	5·002	·855	·746
·33	·691	·692	·925	·335	·881	·532	·265	·061	·907	·792
·34	·876	·844	7·051	·441	·971	·609	·331	·119	·958	·838
·35	9·061	·996	·178	·548	6·061	·687	·398	·177	5·010	·884
·36	9·247	8·149	7·305	6·654	6·151	5·764	5·465	5·236	5·061	4·930
·37	·432	·301	·431	·760	·242	·841	·532	·294	·113	·976
·38	·617	·454	·558	·867	·332	·918	·599	·353	·164	5·022
·39	·802	·606	·685	·973	·422	·995	·665	·411	·216	·068
·40	·987	·758	·811	7·079	·512	6·073	·732	·469	·267	·113
·41		8·911	7·938	7·186	6·602	6·150	5·799	5·528	5·319	5·159
·42		9·063	8·065	·292	·692	·227	·866	·586	·370	·205
·43		·216	·191	·398	·782	·304	·932	·644	·422	·251
·44		·368	·318	·505	·873	·381	·999	·703	·473	·297
·45		·520	·445	·611	·963	·459	6·066	·761	·525	·343
·46		9·673	8·571	7·717	7·053	6·536	6·133	5·819	5·576	5·389
·47		·825	·698	·824	·143	·613	·200	·878	·628	·435
·48		·978	·825	·930	·233	·690	·266	·936	·679	·481
·49			·951	8·036	·323	·767	·333	·994	·731	·527
·50			9·078	·143	·414	·845	·400	6·053	·782	·572

TABLE VII (continued) Working probits
($Y = 3 \cdot 0 – 3 \cdot 9$; $p = 0 \cdot 51 – 1 \cdot 00$)

Response rate, p	Expected probit, Y									
	3·0	3·1	3·2	3·3	3·4	3·5	3·6	3·7	3·8	3·9
·51			9·205	8·249	7·504	6·922	6·467	6·111	5·834	5·618
·52			·331	·355	·594	·999	·534	·170	·885	·664
·53			·458	·462	·684	7·076	·600	·228	·937	·710
·54			·585	·568	·774	·154	·667	·286	·988	·756
·55			·711	·674	·864	·231	·734	·345	6·040	·802
·56			9·838	8·781	7·954	7·308	6·801	6·403	6·091	5·848
·57			·965	·887	8·045	·385	·868	·461	·143	·894
·58				·993	·135	·462	·934	·520	·194	·940
·59				9·100	·225	·540	7·001	·578	·246	·986
·60				·206	·315	·617	·068	·636	·297	6·031
·61				9·312	8·405	7·694	7·135	6·695	6·349	6·077
·62				·419	·495	·771	·201	·753	·400	·123
·63				·525	·585	·848	·268	·811	·452	·169
·64				·631	·676	·926	·335	·870	·503	·215
·65				·738	·766	8·003	·402	·928	·555	·261
·66				9·844	8·856	8·080	7·469	6·986	6·606	6·307
·67				·950	·946	·157	·535	7·045	·658	·353
·68					9·036	·234	·602	·103	·709	·399
·69					·126	·312	·669	·162	·761	·445
·70					·216	·389	·736	·220	·812	·491
·71					9·307	8·466	7·803	7·278	6·864	6·536
·72					·397	·543	·869	·337	·915	·582
·73					·487	·621	·936	·395	·967	·628
·74					·577	·698	8·003	·453	7·018	·674
·75					·667	·775	·070	·512	·070	·720
·76					9·757	8·852	8·136	7·570	7·121	6·766
·77					·848	·929	·203	·628	·173	·812
·78					·938	9·007	·270	·687	·224	·858
·79						·084	·337	·745	·276	·904
·80						·161	·404	·803	·327	·950
·81						9·238	8·470	7·862	7·379	6·995
·82						·315	·537	·920	·430	7·041
·83						·393	·604	·978	·482	·087
·84						·470	·671	8·037	·533	·133
·85						·547	·738	·095	·585	·179
·86						9·624	8·804	8·154	7·636	7·225
·87						·701	·871	·212	·688	·271
·88						·779	·938	·270	·739	·317
·89						·856	9·005	·329	·791	·363
·90						·933	·072	·387	·842	·409
·91							9·138	8·445	7·894	7·454
·92							·205	·504	·945	·500
·93							·272	·562	·997	·546
·94							·339	·620	8·048	·592
·95							·405	·679	·100	·638
·96							9·472	8·737	8·151	7·684
·97							·539	·795	·203	·730
·98							·606	·854	·254	·776
·99							·673	·912	·306	·822
1·00							·739	·970	·357	·868

TABLE VII (continued) Working probits
($Y = 4.0\text{–}4.9$; $p = 0.00\text{–}0.50$)

Response rate, p	Expected probit, Y									
	4·0	4·1	4·2	4·3	4·4	4·5	4·6	4·7	4·8	4·9
·00	3·344	3·408	3·469	3·525	3·577	3·624	3·664	3·698	3·724	3·741
·01	3·386	3·446	3·503	3·557	3·607	3·652	3·691	3·724	3·750	3·766
·02	·427	·487	·538	·589	·637	·680	·719	·751	·775	·791
·03	·468	·521	·572	·621	·667	·709	·746	·777	·801	·816
·04	·510	·559	·607	·653	·697	·737	·773	·803	·826	·841
·05	·551	·596	·641	·685	·727	·766	·800	·829	·852	·867
·06	3·592	3·634	3·676	3·717	3·757	3·794	3·827	3·856	3·878	3·892
·07	·634	·671	·710	·749	·787	·822	·854	·882	·903	·917
·08	·675	·709	·745	·781	·817	·851	·882	·908	·929	·942
·09	·716	·747	·779	·813	·847	·879	·909	·934	·954	·967
·10	·758	·784	·814	·845	·877	·908	·936	·960	·980	·993
·11	3·799	3·822	3·848	3·877	3·907	3·936	3·963	3·987	4·005	4·018
·12	·840	·859	·883	·909	·937	·964	·990	4·013	·031	·043
·13	·882	·897	·917	·941	·967	·993	4·017	·039	·057	·068
·14	·923	·934	·952	·973	·997	4·021	·044	·065	·082	·093
·15	·964	·972	·986	4·005	4·027	·050	·072	·092	·108	·119
·16	4·006	4·010	4·021	4·038	4·057	4·078	4·099	4·118	4·133	4·144
·17	·047	·047	·056	·070	·087	·106	·126	·144	·159	·169
·18	·088	·085	·090	·102	·117	·135	·153	·170	·184	·194
·19	·130	·122	·125	·134	·147	·163	·180	·196	·210	·219
·20	·171	·160	·159	·166	·177	·192	·207	·223	·236	·245
·21	4·212	4·198	4·194	4·198	4·207	4·220	4·235	4·249	4·261	4·270
·22	·253	·235	·228	·230	·237	·248	·262	·275	·287	·295
·23	·295	·273	·263	·262	·267	·277	·289	·301	·312	·320
·24	·336	·310	·297	·294	·297	·305	·316	·327	·338	·345
·25	·377	·348	·332	·326	·327	·334	·343	·354	·363	·370
·26	4·419	4·385	4·366	4·358	4·357	4·362	4·370	4·380	4·389	4·396
·27	·460	·423	·401	·390	·387	·391	·397	·406	·415	·421
·28	·501	·461	·435	·422	·417	·419	·425	·432	·440	·446
·29	·543	·498	·470	·454	·447	·447	·452	·459	·466	·471
·30	·584	·536	·504	·486	·477	·476	·479	·485	·491	·496
·31	4·625	4·573	4·539	4·518	4·507	4·504	4·506	4·511	4·517	4·522
·32	·667	·611	·573	·550	·537	·533	·533	·537	·542	·547
·33	·708	·649	·608	·582	·567	·561	·560	·563	·568	·572
·34	·749	·686	·642	·614	·597	·589	·588	·590	·594	·597
·35	·791	·724	·677	·646	·627	·618	·615	·616	·619	·622
·36	4·832	4·761	4·711	4·678	4·657	4·646	4·642	4·642	4·645	4·648
·37	·873	·799	·746	·710	·687	·675	·669	·668	·670	·673
·38	·915	·836	·780	·742	·717	·703	·696	·695	·696	·698
·39	·956	·874	·815	·774	·747	·731	·723	·721	·721	·723
·40	·997	·912	·849	·806	·777	·760	·750	·747	·747	·748
·41	5·039	4·949	4·884	4·838	4·807	4·788	4·778	4·773	4·773	4·774
·42	·080	·987	·918	·870	·837	·817	·805	·799	·798	·799
·43	·121	5·024	·953	·902	·867	·845	·832	·826	·824	·824
·44	·163	·062	·988	·934	·897	·873	·859	·852	·849	·849
·45	·204	·099	5·022	·966	·927	·902	·886	·878	·875	·874
·46	5·245	5·137	5·057	4·998	4·957	4·930	4·913	4·904	4·900	4·900
·47	·287	·175	·091	5·030	·987	·959	·941	·931	·926	·925
·48	·328	·212	·126	·062	5·017	·987	·968	·957	·952	·950
·49	·369	·250	·160	·094	·047	5·015	·995	·983	·977	·975
·50	·411	·287	·195	·126	·078	·044	5·022	5·009	5·003	5·000

APPENDIX TABLES

TABLE VII (continued)   Working probits
($Y = 4.0$–4.9; $p = 0.51$–1.00)

Response rate, p	Expected probit, Y									
	4·0	4·1	4·2	4·3	4·4	4·5	4·6	4·7	4·8	4·9
·51	5·452	5·325	5·229	5·158	5·108	5·072	5·049	5·035	5·028	5·025
·52	·493	·363	·264	·190	·138	·101	·076	·062	·054	·051
·53	·535	·400	·298	·222	·168	·129	·103	·088	·079	·076
·54	·576	·438	·333	·254	·198	·157	·131	·114	·105	·101
·55	·617	·475	·367	·286	·228	·186	·158	·140	·131	·126
·56	5·659	5·513	5·402	5·318	5·258	5·214	5·185	5·167	5·156	5·151
·57	·700	·550	·436	·351	·288	·243	·212	·193	·182	·177
·58	·741	·588	·471	·383	·318	·271	·239	·219	·207	·202
·59	·783	·626	·505	·415	·348	·299	·266	·245	·233	·227
·60	·824	·663	·540	·447	·378	·328	·294	·271	·258	·252
·61	5·865	5·701	5·574	5·479	5·408	5·356	5·321	5·298	5·284	5·277
·62	·907	·738	·609	·511	·438	·385	·348	·324	·310	·303
·63	·948	·776	·643	·543	·468	·413	·375	·350	·335	·328
·64	·989	·814	·678	·575	·498	·441	·402	·376	·361	·353
·65	6·031	·851	·712	·607	·528	·470	·429	·402	·386	·378
·66	6·072	5·889	5·747	5·639	5·558	5·498	5·456	5·429	5·412	5·403
·67	·113	·926	·781	·671	·588	·527	·484	·455	·437	·429
·68	·155	·964	·816	·703	·618	·555	·511	·481	·463	·454
·69	·196	6·001	·851	·735	·648	·583	·538	·507	·489	·479
·70	·237	·039	·885	·767	·678	·612	·565	·534	·514	·504
·71	6·279	6·077	5·920	5·799	5·708	5·640	5·592	5·560	5·540	5·529
·72	·320	·114	·954	·831	·738	·669	·619	·586	·565	·555
·73	·361	·152	·989	·863	·768	·697	·647	·612	·591	·580
·74	·402	·189	6·023	·895	·798	·725	·674	·638	·617	·605
·75	·444	·227	·058	·927	·828	·754	·701	·665	·642	·630
·76	6·485	6·265	6·092	5·959	5·858	5·782	5·728	5·691	5·668	5·655
·77	·526	·302	·127	·991	·888	·811	·755	·717	·693	·680
·78	·568	·340	·161	6·023	·918	·839	·782	·743	·719	·706
·79	·609	·377	·196	·055	·948	·868	·809	·770	·744	·731
·80	·650	·415	·230	·087	·978	·896	·837	·796	·770	·756
·81	6·692	6·452	6·265	6·119	6·008	5·924	5·864	5·822	5·796	5·781
·82	·733	·490	·299	·151	·038	·953	·891	·848	·821	·806
·83	·774	·528	·334	·183	·068	·981	·918	·874	·847	·832
·84	·816	·565	·368	·215	·098	6·010	·945	·901	·872	·857
·85	·857	·603	·403	·247	·128	·038	·972	·927	·898	·882
·86	6·898	6·640	6·437	6·279	6·158	6·066	6·000	5·953	5·923	5·907
·87	·940	·678	·472	·311	·188	·095	·027	·979	·949	·932
·88	·981	·716	·506	·343	·218	·123	·054	6·006	·975	·958
·89	7·022	·753	·541	·375	·248	·152	·081	·032	6·000	·983
·90	·064	·791	·575	·407	·278	·180	·108	·058	·026	6·008
·91	7·105	6·828	6·610	6·439	6·308	6·208	6·135	6·084	6·051	6·033
·92	·146	·866	·644	·471	·338	·237	·162	·110	·077	·058
·93	·188	·903	·679	·503	·368	·265	·190	·137	·102	·084
·94	·229	·941	·713	·535	·398	·294	·217	·163	·128	·109
·95	·270	·979	·748	·567	·428	·322	·244	·189	·154	·134
·96	7·312	7·016	6·783	6·600	6·458	6·350	6·271	6·215	6·179	6·159
·97	·353	·054	·817	·632	·488	·379	·298	·242	·205	·184
·98	·394	·091	·852	·664	·518	·407	·325	·268	·230	·210
·99	·436	·129	·886	·696	·548	·436	·353	·294	·256	·235
1·00	·477	·166	·921	·728	·578	·464	·380	·320	·281	·260

TABLE VII (continued) Working probits
($Y = 5.0$–5.9; $p = 0.00$–0.50)

Response rate, p	Expected probit, Y									
	5·0	5·1	5·2	5·3	5·4	5·5	5·6	5·7	5·8	5·9
·00	3·747	3·740	3·719	3·680	3·620	3·536	3·422	3·272	3·079	2·834
·01	3·772	3·765	3·744	3·706	3·647	3·564	3·452	3·304	3·114	2·871
·02	·797	·790	·770	·732	·675	·593	·482	·336	·148	·909
·03	·822	·816	·795	·758	·702	·621	·512	·368	·183	·946
·04	·847	·841	·821	·785	·729	·650	·542	·400	·217	·984
·05	·872	·866	·846	·811	·756	·678	·572	·433	·252	3·021
·06	3·897	3·891	3·872	3·837	3·783	3·706	3·602	3·465	3·287	3·059
·07	·922	·916	·898	·863	·810	·735	·632	·497	·321	·097
·08	·947	·942	·923	·890	·838	·763	·662	·529	·356	·134
·09	·972	·967	·949	·916	·865	·792	·692	·561	·390	·172
·10	·997	·992	·974	·942	·892	·820	·722	·593	·425	·209
·11	4·022	4·017	4·000	3·968	3·919	3·848	3·752	3·625	3·459	3·247
·12	·047	·042	·025	·994	·946	·877	·782	·657	·494	·284
·13	·073	·068	·051	4·021	·973	·905	·812	·689	·528	·322
·14	·098	·093	·077	·047	4·000	·934	·842	·721	·563	·360
·15	·123	·118	·102	·073	·028	·962	·872	·753	·597	·397
·16	4·148	4·143	4·128	4·099	4·055	3·990	3·902	3·785	3·632	3·435
·17	·173	·168	·153	·126	·082	4·019	·932	·817	·666	·472
·18	·198	·194	·179	·152	·109	·047	·962	·849	·701	·510
·19	·223	·219	·204	·178	·136	·076	·992	·881	·735	·548
·20	·248	·244	·230	·204	·163	·104	4·022	·913	·770	·585
·21	4·273	4·269	4·256	4·230	4·191	4·132	4·052	3·945	3·804	3·623
·22	·298	·294	·281	·257	·218	·161	·082	·977	·839	·660
·23	·323	·320	·307	·283	·245	·189	·112	4·009	·873	·698
·24	·348	·345	·332	·309	·272	·218	·142	·041	·908	·735
·25	·373	·370	·358	·335	·299	·246	·172	·073	·942	·773
·26	4·398	4·395	4·383	4·362	4·326	4·275	4·202	4·105	3·977	3·811
·27	·423	·420	·409	·388	·353	·303	·232	·137	4·011	·848
·28	·449	·445	·435	·414	·381	·331	·262	·169	·046	·886
·29	·474	·471	·460	·440	·408	·360	·292	·201	·080	·923
·30	·499	·496	·486	·466	·435	·388	·322	·233	·115	·961
·31	4·524	4·521	4·511	4·493	4·462	4·417	4·352	4·265	4·149	3·999
·32	·549	·546	·537	·519	·489	·445	·382	·297	·184	4·036
·33	·574	·571	·563	·545	·516	·473	·412	·329	·219	·074
·34	·599	·597	·588	·571	·544	·502	·442	·361	·253	·111
·35	·624	·622	·614	·598	·571	·530	·472	·393	·288	·149
·36	4·649	4·647	4·639	4·624	4·598	4·559	4·502	4·425	4·322	4·186
·37	·674	·672	·665	·650	·625	·587	·532	·457	·357	·224
·38	·699	·697	·690	·676	·652	·615	·562	·489	·391	·262
·39	·724	·723	·716	·702	·679	·644	·592	·521	·426	·299
·40	·749	·748	·742	·729	·706	·672	·622	·553	·460	·337
·41	4·774	4·773	4·767	4·755	4·734	4·701	4·652	4·585	4·495	4·374
·42	·799	·798	·793	·781	·761	·729	·682	·617	·529	·412
·43	·825	·823	·818	·807	·788	·757	·712	·649	·564	·450
·44	·850	·849	·844	·833	·815	·786	·742	·682	·598	·487
·45	·875	·874	·869	·860	·842	·814	·772	·714	·633	·525
·46	4·900	4·899	4·895	4·886	4·869	4·843	4·802	4·746	4·667	4·562
·47	·925	·924	·921	·912	·897	·871	·832	·778	·702	·600
·48	·950	·949	·946	·938	·924	·899	·862	·810	·736	·637
·49	·975	·975	·972	·965	·951	·928	·892	·842	·771	·675
·50	5·000	5·000	·997	·991	·978	·956	·922	·874	·805	·713

APPENDIX TABLES

TABLE VII (continued) Working probits
($Y = 5{\cdot}0\text{–}5{\cdot}9$; $p = 0{\cdot}51\text{–}1{\cdot}00$)

Response rate, p	Expected probit, Y									
	5·0	5·1	5·2	5·3	5·4	5·5	5·6	5·7	5·8	5·9
·51	5·025	5·025	5·023	5·017	5·005	4·985	4·953	4·906	4·840	4·750
·52	·050	·050	·048	·043	·032	5·013	·983	·938	·874	·788
·53	·075	·075	·074	·069	·059	·041	5·013	·970	·909	·825
·54	·100	·100	·100	·096	·087	·070	·043	5·002	·943	·863
·55	·125	·126	·125	·122	·114	·098	·073	·034	·978	·901
·56	5·150	5·151	5·151	5·148	5·141	5·127	5·103	5·066	5·012	4·938
·57	·175	·176	·176	·174	·168	·155	·133	·098	·047	·976
·58	·201	·201	·202	·201	·195	·183	·163	·130	·082	5·013
·59	·226	·226	·227	·227	·222	·212	·193	·162	·116	·051
·60	·251	·252	·253	·253	·250	·240	·223	·194	·151	·088
·61	5·276	5·277	5·279	5·279	5·277	5·269	5·253	5·226	5·185	5·126
·62	·301	·302	·304	·305	·304	·297	·283	·258	·220	·164
·63	·326	·327	·330	·332	·331	·325	·313	·290	·254	·201
·64	·351	·352	·355	·358	·358	·354	·343	·322	·289	·239
·65	·376	·378	·381	·384	·385	·382	·373	·354	·323	·276
·66	5·401	5·403	5·406	5·410	5·412	5·411	5·403	5·386	5·358	5·314
·67	·426	·428	·432	·437	·440	·439	·433	·418	·392	·351
·68	·451	·453	·458	·463	·467	·467	·463	·450	·427	·389
·69	·476	·478	·483	·489	·494	·496	·493	·482	·461	·427
·70	·501	·504	·509	·515	·521	·524	·523	·514	·496	·464
·71	5·526	5·529	5·534	5·541	5·548	5·553	5·553	5·546	5·530	5·502
·72	·551	·554	·560	·568	·575	·581	·583	·578	·565	·539
·73	·577	·579	·585	·594	·603	·609	·613	·610	·599	·577
·74	·602	·604	·611	·620	·630	·638	·643	·642	·634	·615
·75	·627	·630	·637	·646	·657	·666	·673	·674	·668	·652
·76	5·652	5·655	5·662	5·673	5·684	5·695	5·703	5·706	5·703	5·690
·77	·677	·680	·688	·699	·711	·723	·733	·738	·737	·727
·78	·702	·705	·713	·725	·738	·752	·763	·770	·772	·765
·79	·727	·730	·739	·751	·765	·780	·793	·802	·806	·802
·80	·752	·755	·764	·777	·793	·808	·823	·834	·841	·840
·81	5·777	5·781	5·790	5·804	5·820	5·837	5·853	5·866	5·875	5·878
·82	·802	·806	·816	·830	·847	·865	·883	·898	·910	·915
·83	·827	·831	·841	·856	·874	·894	·913	·930	·944	·953
·84	·852	·856	·867	·882	·901	·922	·943	·962	·979	·990
·85	·877	·881	·892	·908	·928	·950	·973	·995	6·014	6·028
·86	5·902	5·907	5·918	5·935	5·956	5·979	6·003	6·027	6·048	6·066
·87	·927	·932	·943	·961	·983	6·007	·033	·059	·083	·103
·88	·953	·957	·969	·987	6·010	·036	·063	·091	·117	·141
·89	·978	·982	·995	6·013	·037	·064	·093	·123	·152	·178
·90	6·003	6·007	6·020	·040	·064	·092	·123	·155	·186	·216
·91	6·028	6·033	6·046	6·066	6·091	6·121	6·153	6·187	6·221	6·253
·92	·053	·058	·071	·092	·118	·149	·183	·219	·255	·291
·93	·078	·083	·097	·118	·146	·178	·213	·251	·290	·329
·94	·103	·108	·122	·144	·173	·206	·243	·283	·324	·366
·95	·128	·133	·148	·171	·200	·234	·273	·315	·359	·404
·96	6·153	6·159	6·174	6·197	6·227	6·263	6·303	6·347	6·393	6·441
·97	·178	·184	·199	·223	·254	·291	·333	·379	·428	·479
·98	·203	·209	·225	·249	·281	·320	·363	·411	·462	·517
·99	·228	·234	·250	·276	·309	·348	·393	·443	·497	·554
1·00	·253	·259	·276	·302	·336	·376	·423	·475	·531	·592

TABLE VII (continued)   Working probits
($Y = 6\cdot0$–$6\cdot9$; $p = 0\cdot00$–$0\cdot50$)

Response rate, p	\multicolumn{10}{c}{Expected probit, Y}									
	6·0	6·1	6·2	6·3	6·4	6·5	6·6	6·7	6·8	6·9
·00	2·523	2·132	1·643	1·030	0·261					
·01	2·564	2·178	1·694	1·088	0·327					
·02	·606	·224	·746	·146	·394					
·03	·647	·270	·797	·205	·461					
·04	·688	·316	·849	·263	·528					
·05	·730	·362	·900	·321	·595					
·06	2·771	2·408	1·952	1·380	0·661					
·07	·812	·454	2·003	·438	·728					
·08	·854	·500	·055	·496	·795					
·09	·895	·546	·106	·555	·862					
·10	·936	·591	·158	·613	·928	0·067				
·11	2·978	2·637	2·209	1·671	0·995	0·144				
·12	3·019	·683	·261	·730	1·062	·221				
·13	·060	·729	·312	·788	·129	·299				
·14	·102	·775	·364	·846	·196	·376				
·15	·143	·821	·415	·905	·262	·453				
·16	3·184	2·867	2·467	1·963	1·329	0·530				
·17	·226	·913	·518	2·022	·396	·607				
·18	·267	·959	·570	·080	·463	·685				
·19	·308	3·005	·621	·138	·530	·762				
·20	·350	·050	·673	·197	·596	·839				
·21	3·391	3·096	2·724	2·255	1·663	0·916				
·22	·432	·142	·776	·313	·730	·993	0·062			
·23	·474	·188	·827	·372	·797	1·071	·152			
·24	·515	·234	·879	·430	·864	·148	·243			
·25	·556	·280	·930	·488	·930	·225	·333			
·26	3·598	3·326	2·982	2·547	1·997	1·302	0·423			
·27	·639	·372	3·033	·605	2·064	·379	·513			
·28	·680	·418	·085	·663	·131	·457	·603			
·29	·721	·464	·136	·722	·197	·534	·693			
·30	·763	·509	·188	·780	·264	·611	·784			
·31	3·804	3·555	3·239	2·838	2·331	1·688	0·874			
·32	·845	·601	·291	·897	·398	·766	·964			
·33	·887	·647	·342	·955	·465	·843	1·054	0·050		
·34	·928	·693	·394	3·014	·531	·920	·144	·156		
·35	·969	·739	·445	·072	·598	·997	·234	·262		
·36	4·011	3·785	3·497	3·130	2·665	2·074	1·324	0·369		
·37	·052	·831	·548	·189	·732	·152	·415	·475		
·38	·093	·877	·600	·247	·799	·229	·505	·581		
·39	·135	·923	·651	·305	·865	·306	·595	·688		
·40	·176	·969	·703	·364	·932	·383	·685	·794		
·41	4·217	4·014	3·754	3·422	2·999	2·460	1·775	0·900		
·42	·259	·060	·806	·480	3·066	·538	·865	1·007		
·43	·300	·106	·857	·539	·132	·615	·955	·113	0·035	
·44	·341	·152	·909	·597	·199	·692	2·046	·219	·162	
·45	·383	·198	·960	·655	·266	·769	·136	·326	·289	
·46	4·424	4·244	4·012	3·714	3·333	2·846	2·226	1·432	0·415	
·47	·465	·290	·063	·772	·400	·924	·316	·538	·542	
·48	·507	·336	·115	·830	·466	3·001	·406	·645	·669	
·49	·548	·382	·166	·889	·533	·078	·496	·751	·795	
·50	·589	·428	·218	·947	·600	·155	·586	·857	·922	

TABLE VII (continued) Working probits
($Y = 6 \cdot 0 – 6 \cdot 9$; $p = 0 \cdot 51 – 1 \cdot 00$)

Response rate, p	Expected probit, Y									
	6·0	6·1	6·2	6·3	6·4	6·5	6·6	6·7	6·8	6·9
·51	4·631	4·473	4·269	4·006	3·667	3·233	2·677	1·964	1·049	
·52	·672	·519	·321	·064	·734	·310	·767	2·070	·175	0·022
·53	·713	·565	·372	·122	·800	·387	·857	·176	·302	·175
·54	·755	·611	·424	·181	·867	·464	·947	·283	·429	·327
·55	·796	·657	·475	·239	·934	·541	3·037	·389	·555	·480
·56	4·837	4·703	4·527	4·297	4·001	3·619	3·127	2·495	1·682	0·632
·57	·879	·749	·578	·356	·068	·696	·218	·602	·809	·784
·58	·920	·795	·630	·414	·134	·773	·308	·708	·935	·937
·59	·961	·841	·681	·472	·201	·850	·398	·814	2·062	1·089
·60	5·003	·887	·733	·531	·268	·927	·488	·921	·189	·242
·61	5·044	4·932	4·784	4·589	4·335	4·005	3·578	3·027	2·315	1·394
·62	·085	·978	·836	·647	·401	·082	·668	·133	·442	·546
·63	·127	5·024	·887	·706	·468	·159	·758	·240	·569	·699
·64	·168	·070	·939	·764	·535	·236	·849	·346	·695	·851
·65	·209	·116	·990	·823	·602	·313	·939	·452	·822	2·004
·66	5·251	5·162	5·042	4·881	4·669	4·391	4·029	3·559	2·949	2·156
·67	·292	·208	·093	·939	·735	·468	·119	·665	3·075	·308
·68	·333	·254	·145	·998	·802	·545	·209	·771	·202	·461
·69	·375	·300	·196	5·056	·869	·622	·299	·878	·329	·613
·70	·416	·346	·248	·114	·936	·700	·390	·984	·455	·766
·71	5·457	5·392	5·299	5·173	5·003	4·777	4·480	4·090	3·582	2·918
·72	·499	·437	·351	·231	·069	·854	·570	·197	·709	3·070
·73	·540	·483	·402	·289	·136	·931	·660	·303	·835	·223
·74	·581	·529	·454	·348	·203	5·008	·750	·409	·962	·375
·75	·623	·575	·505	·406	·270	·086	·840	·516	4·089	·528
·76	5·664	5·621	5·557	5·464	5·336	5·163	4·930	4·622	4·215	3·680
·77	·705	·667	·608	·523	·403	·240	5·021	·728	·342	·832
·78	·747	·713	·660	·581	·470	·317	·111	·835	·469	·985
·79	·788	·759	·711	·639	·537	·394	·201	·941	·595	4·137
·80	·829	·805	·763	·698	·604	·472	·291	5·047	·722	·290
·81	5·870	5·851	5·814	5·756	5·670	5·549	5·381	5·154	4·849	4·442
·82	·912	·896	·866	·815	·737	·626	·471	·260	·975	·594
·83	·953	·942	·917	·873	·804	·703	·561	·366	5·102	·747
·84	·994	·988	·969	·931	·871	·780	·652	·473	·229	·899
·85	6·036	6·034	6·020	·990	·938	·858	·742	·579	·355	5·052
·86	6·077	6·080	6·072	6·048	6·004	5·935	5·832	5·685	5·482	5·204
·87	·118	·126	·123	·106	·071	6·012	·922	·792	·609	·356
·88	·160	·172	·175	·165	·138	·089	6·012	·898	·735	·509
·89	·201	·218	·226	·223	·205	·166	·102	6·004	·862	·661
·90	·242	·264	·278	·281	·272	·244	·192	·111	·988	·814
·91	6·284	6·310	6·329	6·340	6·338	6·321	6·283	6·217	6·115	5·966
·92	·325	·355	·381	·398	·405	·398	·373	·323	·242	6·118
·93	·366	·401	·432	·456	·472	·475	·463	·430	·368	·271
·94	·408	·447	·484	·515	·539	·553	·553	·536	·495	·423
·95	·449	·493	·535	·573	·605	·630	·643	·642	·622	·576
·96	6·490	6·539	6·587	6·631	6·672	6·707	6·733	6·749	6·748	6·728
·97	·532	·585	·638	·690	·739	·784	·824	·855	·875	·880
·98	·573	·631	·690	·748	·806	·861	·914	·961	7·002	7·033
·99	·614	·677	·741	·807	·873	·939	7·004	7·068	·128	·185
1·00	·656	·723	·793	·865	·939	7·016	·094	·174	·255	·338

TABLE VII (continued)   Working probits
$(Y = 7{\cdot}0\text{–}7{\cdot}9;\ p = 0{\cdot}60\text{–}1{\cdot}00)$

Response rate, p	Expected probit, Y									
	7·0	7·1	7·2	7·3	7·4	7·5	7·6	7·7	7·8	7·9
·60	0·013									
·61	0·198									
·62	·383									
·63	·568									
·64	·753									
·65	·939									
·66	1·124									
·67	·309	0·003								
·68	·494	·231								
·69	·680	·458								
·70	·865	·685								
·71	2·050	0·913								
·72	·235	1·140								
·73	·420	·367								
·74	·606	·595	0·263							
·75	·791	·822	·545							
·76	2·976	2·050	0·827							
·77	3·161	·277	1·108							
·78	·347	·504	·390							
·79	·532	·732	·672	0·265						
·80	·717	·939	·954	·618						
·81	3·902	3·186	2·236	0·971						
·82	4·087	·414	·518	1·324						
·83	·273	·641	·800	·677	0·175					
·84	·458	·868	3·082	2·030	·621					
·85	·643	4·096	·364	·383	1·068					
·86	4·828	4·323	3·645	2·736	1·514					
·87	5·014	·551	·927	3·089	·961	0·438				
·88	·199	·778	4·209	·442	2·408	1·008				
·89	·384	5·005	·491	·795	·854	·579				
·90	·569	·233	·773	4·148	3·301	2·149	0·581			
·91	5·754	5·460	5·055	4·501	3·747	2·720	1·317			
·92	·940	·687	·337	·854	4·194	3·290	2·054	0·356		
·93	6·125	·915	·619	5·207	·640	·861	·790	1·316		
·94	·310	6·142	·901	·560	5·087	4·431	3·526	2·275	0·542	
·95	·495	·369	6·182	·914	·533	5·002	4·262	3·235	1·806	
·96	6·681	6·597	6·464	6·267	5·980	5·572	4·998	4·194	3·069	1·493
·97	·866	·824	·746	·620	6·426	6·143	5·735	5·154	4·333	3·173
·98	7·051	7·051	7·028	·973	·873	·713	6·471	6·114	5·596	4·853
·99	·236	·279	·310	7·326	7·319	7·284	7·207	7·073	6·859	6·533
1·00	·421	·506	·592	·679	·766	·854	·943	8·033	8·123	8·213

TABLE VIII The logit transformation

Response rate	0·00	0·01	0·02	0·03	0·04	0·05	0·06	0·07	0·08	0·09
0·00	–	2·70	3·05	3·26	3·41	3·53	3·62	3·71	3·78	3·84
0·10	3·90	3·95	4·00	4·05	4·09	4·13	4·17	4·21	4·24	4·27
0·20	4·31	4·34	4·37	4·40	4·42	4·45	4·48	4·50	4·53	4·55
0·30	4·58	4·60	4·62	4·65	4·67	4·69	4·71	4·73	4·76	4·78
0·40	4·80	4·82	4·84	4·86	4·88	4·90	4·92	4·94	4·96	4·98
0·50	5·00	5·02	5·04	5·06	5·08	5·10	5·12	5·14	5·16	5·18
0·60	5·20	5·22	5·24	5·27	5·29	5·31	5·33	5·35	5·38	5·40
0·70	5·42	5·45	5·47	5·50	5·52	5·55	5·58	5·60	5·63	5·66
0·80	5·69	5·73	5·76	5·79	5·83	5·87	5·91	5·95	6·00	6·05
0·90	6·10	6·16	6·22	6·29	6·38	6·47	6·59	6·74	6·95	7·30

Response rate	0·000	0·001	0·002	0·003	0·004	0·005	0·006	0·007	0·008	0·009
0·97	6·74	6·76	6·77	6·79	6·81	6·83	6·85	6·87	6·90	6·92
0·98	6·95	6·97	7·00	7·03	7·06	7·09	7·13	7·16	7·21	7·25
0·99	7·30	7·35	7·41	7·48	7·55	7·65	7·76	7·90	8·11	8·45

482 APPENDIX TABLES

TABLE IX Minimum and maximum working logits, ranges, and weighting coefficients

Expected logit Y	Minimum working logit Y_0	Range A	Maximum working logit Y_1	Expected logit Y	Weighting coefficient w
1·1	0·5998	1221·3	9·4002	8·9	·00164
1·2	0·6997	1000·1	9·3003	8·8	·00200
1·3	0·7997	819·0	9·2003	8·7	·00244
1·4	0·8996	670·7	9·1004	8·6	·00298
1·5	0·9995	549·3	9·0005	8·5	·00364
1·6	1·0994	449·92	8·9006	8·4	·00445
1·7	1·1993	368·55	8·8007	8·3	·00543
1·8	1·2992	301·92	8·7008	8·2	·00662
1·9	1·3990	247·38	8·6010	8·1	·00808
2·0	1·4988	202·72	8·5012	8·0	·00987
2·1	1·5985	166·15	8·4015	7·9	·01204
2·2	1·6982	136·22	8·3018	7·8	·01468
2·3	1·7977	111·71	8·2023	7·7	·01790
2·4	1·8972	91·64	8·1028	7·6	·02182
2·5	1·9966	75·21	8·0034	7·5	·02659
2·6	2·0959	61·759	7·9041	7·4	·03238
2·7	2·1950	50·747	7·8050	7·3	·03941
2·8	2·2939	41·732	7·7061	7·2	·04793
2·9	2·3925	34·351	7·6075	7·1	·05822
3·0	2·4908	28·308	7·5092	7·0	·07065
3·1	2·5888	23·362	7·4112	6·9	·08561
3·2	2·6863	19·313	7·3137	6·8	·10356
3·3	2·7833	15·999	7·2167	6·7	·12501
3·4	2·8796	13·287	7·1204	6·6	·15053
3·5	2·9751	11·068	7·0249	6·5	·18071
3·6	3·0696	9·2527	6·9304	6·4	·21615
3·7	3·1629	7·7690	6·8371	6·3	·25743
3·8	3·2546	6·5569	6·7454	6·2	·30502
3·9	3·3446	5·5679	6·6554	6·1	·35920
4·0	3·4323	4·7622	6·5677	6·0	·41997
4·1	3·5174	4·1075	6·4826	5·9	·48692
4·2	3·5991	3·5775	6·4009	5·8	·55906
4·3	3·6767	3·1509	6·3233	5·7	·63474
4·4	3·7494	2·8107	6·2506	5·6	·71158
4·5	3·8161	2·5431	6·1839	5·5	·78645
4·6	3·8753	2·3374	6·1247	5·4	·85564
4·7	3·9256	2·1855	6·0744	5·3	·91514
4·8	3·9648	2·0811	6·0352	5·2	·96104
4·9	3·9906	2·0201	6·0094	5·1	·99007
5·0	4·0000	2·0000	6·0000	5·0	1·00000
5·1	3·9893	2·0201	6·0107	4·9	·99007
5·2	3·9541	2·0811	6·0459	4·8	·96104
5·3	3·8889	2·1855	6·1111	4·7	·91514
5·4	3·7873	2·3374	6·2127	4·6	·85564
5·5	3·6408	2·5431	6·3592	4·5	·78645
5·6	3·4399	2·8107	6·5601	4·4	·71158
5·7	3·1724	3·1509	6·8276	4·3	·63474
5·8	2·8234	3·5775	7·1766	4·2	·55906
5·9	2·3751	4·1075	7·6249	4·1	·48692
6·0	1·8055	4·7622	8·1945	4·0	·41997
6·1	1·0875	5·5679	8·9125	3·9	·35920
6·2	0·1885	6·5569	9·8115	3·8	·30502
6·3	−0·9319	7·7690	10·9319	3·7	·25743
6·4	−2·3223	9·2527	12·3223	3·6	·21615
6·5	−4·0428	11·0677	14·0428	3·5	·18071

APPENDIX TABLES

TABLE X Working logits

Y	\multicolumn{11}{c}{Empirical response rate}

Y	0·00	0·05	0·10	0·15	0·20	0·25	0·30	0·35	0·40	0·45	0·50
2·0	1·50										
2·1	1·60	9·91									
2·2	1·70	8·51									
2·3	1·80	7·38									
2·4	1·90	6·48									
2·5	2·00	5·76	9·52								
2·6	2·10	5·18	8·27								
2·7	2·20	4·73	7·27	9·81							
2·8	2·29	4·38	6·47	8·55							
2·9	2·39	4·11	5·83	7·55	9·26						
3·0	2·49	3·91	5·32	6·74	8·15	9·57					
3·2	2·69	3·65	4·62	5·58	6·55	7·51	8·48	9·45			
3·4	2·88	3·54	4·21	4·87	5·54	6·20	6·87	7·53	8·19	8·86	9·52
3·6	3·07	3·53	3·99	4·46	4·92	5·38	5·85	6·31	6·77	7·23	7·70
3·8	3·25	3·58	3·91	4·24	4·57	4·89	5·22	5·55	5·88	6·21	6·53
4·0	3·43	3·67	3·91	4·15	4·38	4·62	4·86	5·10	5·34	5·58	5·81
4·2	3·60	3·78	3·96	4·14	4·31	4·49	4·67	4·85	5·03	5·21	5·39
4·4	3·75	3·89	4·03	4·17	4·31	4·45	4·59	4·73	4·87	5·01	5·15
4·6	3·88	3·99	4·11	4·23	4·34	4·46	4·58	4·69	4·81	4·93	5·04
4·8	3·96	4·07	4·17	4·28	4·38	4·49	4·59	4·69	4·80	4·90	5·01
5·0	4·00	4·10	4·20	4·30	4·40	4·50	4·60	4·70	4·80	4·90	5·00
5·2	3·95	4·06	4·16	4·27	4·37	4·47	4·58	4·68	4·79	4·89	4·99
5·4	3·79	3·90	4·02	4·14	4·25	4·37	4·49	4·61	4·72	4·84	4·96
5·6	3·44	3·58	3·72	3·86	4·00	4·14	4·28	4·42	4·56	4·70	4·85
5·8	2·82	3·00	3·18	3·36	3·54	3·72	3·90	4·08	4·25	4·43	4·61
6·0	1·81	2·04	2·28	2·52	2·76	3·00	3·23	3·47	3·71	3·95	4·19
6·2	0·19	0·52	0·84	1·17	1·50	1·83	2·16	2·48	2·81	3·14	3·47
6·4							0·45	0·92	1·38	1·84	2·30
6·6											0·48

TABLE X (continued)   Working logits

Y	\multicolumn{11}{c}{Empirical response rate}

Y	0·50	0·55	0·60	0·65	0·70	0·75	0·80	0·85	0·90	0·95	1·00
3·4	9·52										
3·6	7·70	8·16	8·62	9·08	9·55						
3·8	6·53	6·86	7·19	7·52	7·84	8·17	8·50	8·83	9·16	9·48	9·81
4·0	5·81	6·05	6·29	6·53	6·77	7·00	7·24	7·48	7·72	7·96	8·19
4·2	5·39	5·57	5·75	5·92	6·10	6·28	6·46	6·64	6·82	7·00	7·18
4·4	5·15	5·30	5·44	5·58	5·72	5·86	6·00	6·14	6·28	6·42	6·56
4·6	5·04	5·16	5·28	5·39	5·51	5·63	5·75	5·86	5·98	6·10	6·21
4·8	5·01	5·11	5·21	5·32	5·42	5·53	5·63	5·73	5·84	5·94	6·05
5·0	5·00	5·10	5·20	5·30	5·40	5·50	5·60	5·70	5·80	5·90	6·00
5·2	4·99	5·10	5·20	5·31	5·41	5·51	5·62	5·72	5·83	5·93	6·04
5·4	4·96	5·07	5·19	5·31	5·42	5·54	5·66	5·77	5·89	6·01	6·12
5·6	4·85	4·99	5·13	5·27	5·41	5·55	5·69	5·83	5·97	6·11	6·25
5·8	4·61	4·79	4·97	5·15	5·33	5·51	5·69	5·86	6·04	6·22	6·40
6·0	4·19	4·42	4·66	4·90	5·14	5·38	5·62	5·85	6·09	6·33	6·57
6·2	3·47	3·79	4·12	4·45	4·78	5·11	5·43	5·76	6·09	6·42	6·75
6·4	2·30	2·77	3·23	3·69	4·15	4·62	5·08	5·54	6·01	6·47	6·93
6·6	0·48	1·14	1·81	2·47	3·13	3·80	4·46	5·13	5·79	6·46	7·12
6·8				0·55	1·52	2·49	3·45	4·42	5·38	6·35	7·31
7·0						0·43	1·85	3·26	4·68	6·09	7·51
7·1							0·74	2·45	4·17	5·89	7·61
7·2								1·45	3·53	5·62	7·71
7·3								0·19	2·73	5·27	7·80
7·4									1·73	4·82	7·90
7·5									0·48	4·24	8·00
7·6										3·52	8·10
7·7										2·62	8·20
7·8										1·49	8·30
7·9										0·09	8·40
8·0											8·50

TABLE XI The angle transformation
[Fisher and Yates, 1963, Table XII]

Response rate	0·00	0·01	0·02	0·03	0·04	0·05	0·06	0·07	0·08	0·09
0·00	0	5·7	8·1	10·0	11·5	12·9	14·2	15·3	16·4	17·5
0·10	18·4	19·4	20·3	21·1	22·0	22·8	23·6	24·4	25·1	25·8
0·20	26·6	27·3	28·0	28·7	29·3	30·0	30·7	31·3	31·9	32·6
0·30	33·2	33·8	34·4	35·1	35·7	36·3	36·9	37·5	38·1	38·6
0·40	39·2	39·8	40·4	41·0	41·6	42·1	42·7	43·3	43·9	44·4
0·50	45·0	45·6	46·1	46·7	47·3	47·9	48·4	49·0	49·6	50·2
0·60	50·8	51·4	51·9	52·5	53·1	53·7	54·3	54·9	55·6	56·2
0·70	56·8	57·4	58·1	58·7	59·3	60·0	60·7	61·3	62·0	62·7
0·80	63·4	64·2	64·9	65·6	66·4	67·2	68·0	68·9	69·7	70·6
0·90	71·6	72·5	73·6	74·7	75·8	77·1	78·5	80·0	81·9	84·3

Response rate	0·000	0·001	0·002	0·003	0·004	0·005	0·006	0·007	0·008	0·009
0·97	80·0	80·2	80·4	80·5	80·7	80·9	81·1	81·3	81·5	81·7
0·98	81·9	82·1	82·3	82·5	82·7	83·0	83·2	83·5	83·7	84·0
0·99	84·3	84·6	84·9	85·2	85·6	85·9	86·4	86·9	87·4	88·2

486   APPENDIX TABLES

TABLE XII   Minimum and maximum working angles and ranges

Minimum working angle		Range	Maximum working angle	
Expected angle Y	Y_0	A	Y_1	Expected angle Y
1	0·5	1641·7	89·5	89
2	1·0	821·4	89·0	88
3	1·5	548·1	88·5	87
4	2·0	411·7	88·0	86
5	2·5	330·0	87·5	85
6	3·0	275·6	87·0	84
7	3·5	236·8	86·5	83
8	4·0	207·9	86·0	82
9	4·5	185·4	85·5	81
10	4·9	167·5	85·1	80
11	5·4	152·9	84·6	79
12	5·9	140·9	84·1	78
13	6·4	130·7	83·6	77
14	6·9	122·0	83·1	76
15	7·3	114·6	82·7	75
16	7·8	108·1	82·2	74
17	8·2	102·5	81·8	73
18	8·7	97·5	81·3	72
19	9·1	93·1	80·9	71
20	9·6	89·1	80·4	70
21	10·0	85·6	80·0	69
22	10·4	82·5	79·6	68
23	10·8	79·7	79·2	67
24	11·2	77·1	78·8	66
25	11·6	74·8	78·4	65
26	12·0	72·7	78·0	64
27	12·4	70·8	77·6	63
28	12·8	69·1	77·2	62
29	13·1	67·6	76·9	61
30	13·5	66·2	76·5	60
31	13·8	64·9	76·2	59
32	14·1	63·7	75·9	58
33	14·4	62·7	75·6	57
34	14·7	61·8	75·3	56
35	14·9	61·0	75·1	55
36	15·2	60·2	74·8	54
37	15·4	59·6	74·6	53
38	15·6	59·0	74·4	52
39	15·8	58·6	74·2	51
40	16·0	58·3	74·0	50

TABLE XII (*cont.*)   Minimum and maximum working angles and ranges

Minimum working angle		Range	Maximum working angle	
Expected angle Y	Y_0	A	Y_1	Expected angle Y
41	16·1	57·9	73·9	49
42	16·2	57·6	73·8	48
43	16·3	57·4	73·7	47
44	16·3	57·3	73·7	46
45	16·4	57·3	73·6	45
46	16·4	57·3	73·6	44
47	16·3	57·4	73·7	43
48	16·2	57·6	73·8	42
49	16·0	57·9	74·0	41
50	15·7	58·3	74·3	40
51	15·6	58·6	74·4	39
52	15·4	59·0	74·6	38
53	15·0	59·6	75·0	37
54	14·6	60·2	75·4	36
55	14·1	61·0	75·9	35
56	13·5	61·8	76·5	34
57	12·9	62·7	77·1	33
58	12·2	63·7	77·8	32
59	11·3	64·9	78·7	31
60	10·3	66·2	79·7	30
61	9·3	67·6	80·7	29
62	8·1	69·1	81·9	28
63	6·8	70·8	83·2	27
64	5·3	72·7	84·7	26
65	3·6	74·8	86·4	25

APPENDIX TABLES

TABLE XIII Working angles

Y	Empirical response rate										
	0·00	0·05	0·10	0·15	0·20	0·25	0·30	0·35	0·40	0·45	0·50
1	0·5	82·6									
2	1·0	42·1	83·1								
3	1·5	28·9	56·3	83·7							
4	2·0	22·6	43·2	63·8	84·3						
5	2·5	19·0	35·5	52·0	68·5	85·0					
6	3·0	16·8	30·5	44·3	58·1	71·9	85·7				
7	3·5	15·3	27·2	39·0	50·9	62·7	74·5	86·4			
8	4·0	14·4	24·8	35·2	45·5	55·9	66·3	76·7	87·1		
9	4·5	13·7	23·0	32·3	41·5	50·8	60·1	69·4	78·6	87·9	
10	5·0	13·3	21·7	30·1	38·5	46·8	55·2	63·6	72·0	80·3	88·7
12	5·9	13·0	20·0	27·0	34·1	41·1	48·2	55·2	62·3	69·3	76·3
14	6·9	13·0	19·1	25·2	31·3	37·4	43·5	49·6	55·7	61·8	67·9
16	7·8	13·2	18·6	24·0	29·4	34·8	40·2	45·6	51·0	56·4	61·8
18	8·7	13·6	18·4	23·3	28·2	33·1	37·9	42·8	47·7	52·6	57·4
20	9·6	14·0	18·5	22·9	27·4	31·9	36·3	40·8	45·2	49·7	54·1
22	10·4	14·6	18·7	22·8	26·9	31·0	35·2	39·3	43·4	47·5	51·7
24	11·2	15·1	19·0	22·8	26·7	30·5	34·4	38·2	42·1	45·9	49·8
26	12·0	15·7	19·3	22·9	26·6	30·2	33·8	37·5	41·1	44·7	48·4
28	12·8	16·2	19·7	23·1	26·6	30·0	33·5	37·0	40·4	43·9	47·3
30	13·5	16·8	20·1	23·4	26·7	30·0	33·3	36·6	39·9	43·2	46·5
32	14·1	17·3	20·5	23·7	26·8	30·0	33·2	36·4	39·6	42·8	46·0
34	14·7	17·8	20·9	24·0	27·0	30·1	33·2	36·3	39·4	42·5	45·6
36	15·2	18·2	21·2	24·2	27·2	30·2	33·3	36·3	39·3	42·3	45·3
38	15·6	18·6	21·5	24·5	27·4	30·4	33·3	36·3	39·2	42·2	45·1
40	16·0	18·9	21·8	24·7	27·6	30·5	33·4	36·3	39·2	42·1	45·0
42	16·2	19·1	22·0	24·9	27·7	30·6	33·5	36·4	39·3	42·1	45·0
44	16·3	19·2	22·1	24·9	27·8	30·7	33·5	36·4	39·3	42·1	45·0
46	16·3	19·2	22·1	24·9	27·8	30·7	33·5	36·4	39·3	42·1	45·0
48	16·2	19·1	21·9	24·8	27·7	30·6	33·5	36·3	39·2	42·1	45·0
50	15·9	18·8	21·7	24·6	27·5	30·4	33·3	36·2	39·1	42·0	45·0
52	15·3	18·3	21·2	24·2	27·1	30·1	33·0	36·0	39·0	41·9	44·9
54	14·6	17·6	20·6	23·6	26·6	29·6	32·6	35·7	38·7	41·7	44·7
56	13·5	16·6	19·7	22·8	25·9	29·0	32·1	35·2	38·2	41·3	44·4
58	12·2	15·3	18·5	21·7	24·9	28·1	31·3	34·5	37·6	40·8	44·0
60	10·4	13·7	17·0	20·3	23·6	26·9	30·2	33·5	36·8	40·2	43·5
62	8·1	11·6	15·0	18·5	21·9	25·4	28·9	32·3	35·8	39·2	42·7
64	5·3	8·9	12·5	16·2	19·8	23·4	27·1	30·7	34·3	38·0	41·6
66	1·6	5·5	9·4	13·2	17·1	20·9	24·8	28·6	32·5	36·3	40·2
68		1·2	5·3	9·5	13·6	17·7	21·8	26·0	30·1	34·2	38·3
70			0·2	4·7	9·1	13·6	18·0	22·5	26·9	31·4	35·9
72					3·3	8·2	13·1	17·9	22·8	27·7	32·6
74						1·1	6·5	11·9	17·3	22·7	28·2
76								3·8	9·9	16·0	22·1
78										6·6	13·7
80											1·3

APPENDIX TABLES 489

TABLE XIII (*cont.*) Working angles

Y	Empirical response rate										
	0·50	0·55	0·60	0·65	0·70	0·75	0·80	0·85	0·90	0·95	1·00
10	88·7										
12	76·3	83·4									
14	67·9	74·0	80·1	86·2							
16	61·8	67·3	72·7	78·1	83·5	88·9					
18	57·4	62·3	67·2	72·1	76·9	81·8	86·7				
20	54·1	58·6	63·1	67·5	72·0	76·4	80·9	85·3	89·8		
22	51·7	55·8	59·9	64·0	68·2	72·3	76·4	80·5	84·7	88·8	
24	49·8	53·7	57·5	61·4	65·2	69·1	72·9	76·8	80·6	84·5	88·4
26	48·4	52·0	55·7	59·3	62·9	66·6	70·2	73·8	77·5	81·1	84·7
28	47·3	50·8	54·2	57·7	61·1	64·6	68·1	71·5	75·0	78·4	81·9
30	46·5	49·8	53·2	56·5	59·8	63·1	66·4	69·7	73·0	76·3	79·6
32	46·0	49·2	52·4	55·5	58·7	61·9	65·1	68·3	71·5	74·7	77·8
34	45·6	48·7	51·8	54·8	57·9	61·0	64·1	67·2	70·3	73·4	76·5
36	45·3	48·3	51·3	54·3	57·4	60·4	63·4	66·4	69·4	72·4	75·4
38	45·1	48·1	51·0	54·0	57·0	59·9	62·9	65·8	68·8	71·7	74·7
40	45·0	48·0	50·9	53·8	56·7	59·6	62·5	65·4	68·3	71·2	74·1
42	45·0	47·9	50·8	53·7	56·5	59·4	62·3	65·2	68·1	70·9	73·8
44	45·0	47·9	50·7	53·6	56·5	59·3	62·2	65·1	67·9	70·8	73·7
46	45·0	47·9	50·7	53·6	56·5	59·3	62·2	65·1	67·9	70·8	73·7
48	45·0	47·9	50·7	53·6	56·5	59·4	62·3	65·1	68·0	70·9	73·8
50	45·0	47·9	50·8	53·7	56·6	59·5	62·4	65·3	68·2	71·1	74·0
52	44·9	47·8	50·8	53·7	56·7	59·6	62·6	65·5	68·5	71·4	74·4
54	44·7	47·7	50·7	53·7	56·7	59·8	62·8	65·8	68·8	71·8	74·8
56	44·4	47·5	50·6	53·7	56·8	59·9	63·0	66·0	69·1	72·2	75·3
58	44·0	47·2	50·4	53·6	56·8	60·0	63·2	66·3	69·5	72·7	75·9
60	43·5	46·8	50·1	53·4	56·7	60·0	63·3	66·6	69·9	73·2	76·5
62	42·7	46·1	49·6	53·0	56·5	60·0	63·4	66·9	70·3	73·8	77·2
64	41·6	45·3	48·9	52·5	56·2	59·8	63·4	67·1	70·7	74·3	78·0
66	40·2	44·1	47·9	51·8	55·6	59·5	63·3	67·2	71·0	74·9	78·8
68	38·3	42·5	46·6	50·7	54·8	59·0	63·1	67·2	71·3	75·4	79·6
70	35·9	40·3	44·8	49·2	53·7	58·1	62·6	67·1	71·5	76·0	80·4
72	32·6	37·4	42·3	47·2	52·1	56·9	61·8	66·7	71·6	76·4	81·3
74	28·2	33·6	39·0	44·4	49·8	55·2	60·6	66·0	71·4	76·8	82·2
76	22·1	28·2	34·3	40·4	46·5	52·6	58·7	64·8	70·9	77·0	83·1
78	13·7	20·7	27·7	34·8	41·8	48·9	55·9	63·0	70·0	77·0	84·1
80	1·3	9·7	18·0	26·4	34·8	43·2	51·5	59·9	68·3	76·7	85·0
81		2·1	11·4	20·6	29·9	39·2	48·5	57·7	67·0	76·3	85·5
82			2·9	13·3	23·7	34·1	44·5	54·8	65·2	75·6	86·0
83				3·6	15·5	27·3	39·1	51·0	62·8	74·7	86·5
84					4·3	18·1	31·9	45·7	59·5	73·2	87·0
85						5·0	21·5	38·0	54·5	71·0	87·5
86							5·7	26·2	46·8	67·4	88·0
87								6·3	33·7	61·1	88·5
88									6·9	47·9	89·0
89										7·4	89·5

TABLE XIV Transformation of angles to probits

Angle	0	1	2	3	4	5	6	7	8	9
0	—	1·57	1·97	2·22	2·41	2·57	2·71	2·83	2·93	3·03
10	3·12	3·21	3·29	3·36	3·43	3·50	3·57	3·63	3·69	3·75
20	3·81	3·87	3·92	3·98	4·03	4·08	4·13	4·18	4·23	4·28
30	4·33	4·37	4·42	4·47	4·51	4·56	4·60	4·65	4·69	4·74
40	4·78	4·82	4·87	4·91	4·96	5·00	5·04	5·09	5·13	5·18
50	5·22	5·26	5·31	5·35	5·40	5·44	5·49	5·53	5·58	5·63
60	5·67	5·72	5·77	5·82	5·87	5·92	5·97	6·02	6·08	6·13
70	6·19	6·25	6·31	6·37	6·43	6·50	6·57	6·64	6·71	6·79
80	6·88	6·97	7·07	7·17	7·29	7·43	7·59	7·78	8·03	8·43

TABLE XV Transformation of angles to logits

Angle	0	1	2	3	4	5	6	7	8	9
0	—	0·95	1·65	2·05	2·34	2·56	2·75	2·90	3·04	3·16
10	3·26	3·36	3·45	3·53	3·61	3·68	3·75	3·81	3·88	3·93
20	3·99	4·04	4·09	4·14	4·19	4·24	4·28	4·33	4·37	4·41
30	4·45	4·49	4·53	4·57	4·61	4·64	4·68	4·72	4·75	4·79
40	4·82	4·86	4·90	4·93	4·97	5·00	5·03	5·07	5·10	5·14
50	5·18	5·21	5·25	5·28	5·32	5·36	5·39	5·43	5·47	5·51
60	5·55	5·59	5·63	5·67	5·72	5·76	5·81	5·86	5·91	5·96
70	6·01	6·07	6·12	6·19	6·25	6·32	6·39	6·47	6·55	6·64
80	6·74	6·84	6·96	7·10	7·25	7·44	7·66	7·95	8·35	9·05

TABLE XVI   The loglog transformation

Proportion sterile	·000	·001	·002	·003	·004	·005	·006	·007	·008	·009
·00	—	1·93	1·83	1·76	1·71	1·67	1·63	1·60	1·57	1·55
·00	·00	·01	·02	·03	·04	·05	·06	·07	·08	·09
·0	—	1·53	1·36	1·25	1·17	1·10	1·03	0·98	0·93	0·88
·1	0·83	0·79	0·75	0·71	0·68	0·64	0·61	0·57	0·54	0·51
·2	0·48	0·45	0·41	0·39	0·36	0·33	0·30	0·27	0·24	0·21
·3	0·19	0·16	0·13	0·10	0·08	0·05	0·02	−0·01	−0·03	−0·06
·4	−0·09	−0·11	−0·14	−0·17	−0·20	−0·23	−0·25	−0·28	−0·31	−0·34
·5	−0·37	−0·40	−0·42	−0·45	−0·48	−0·51	−0·55	−0·58	−0·61	−0·64
·6	−0·67	−0·70	−0·74	−0·77	−0·81	−0·84	−0·88	−0·92	−0·95	−0·99
·7	−1·03	−1·07	−1·11	−1·16	−1·20	−1·25	−1·29	−1·34	−1·39	−1·45
·8	−1·50	−1·56	−1·62	−1·68	−1·75	−1·82	−1·89	−1·97	−2·06	−2·15
·9	−2·25	−2·36	−2·48	−2·62	−2·78	−2·97	−3·20	−3·49	−3·90	−4·60
·000	·000	·001	·002	·003	·004	·005	·006	·007	·008	·009
·97	−3·49	−3·53	−3·56	−3·60	−3·64	−3·68	−3·72	−3·76	−3·81	−3·85
·98	−3·90	−3·95	−4·01	−4·07	−4·13	−4·19	−4·26	−4·34	−4·42	−4·50
·99	−4·60	−4·71	−4·82	−4·96	−5·11	−5·30	−5·52	−5·81	−6·21	−6·91

TABLE XVII Minimum and maximum working loglog deviates, ranges, and weighting coefficients

Expected loglog Y	Minimum working deviate η_0	Range A	Maximum working deviate η_1	Weighting coefficient w
2·5	−0·0821	16034·	−	0·00076
2·4	−0·0907	5559·0	−	0·00198
2·3	−0·1003	2152·1	−	0·00463
2·2	−0·1108	920·59	−	0·00980
2·1	−0·1225	431·03	−	0·01895
2·0	−0·1353	219·00	−	0·03376
1·9	−0·1496	119·81	−	0·05587
1·8	−0·1653	70·080	69·915	0·08653
1·7	−0·1827	43·552	43·369	0·12622
1·6	−0·2019	28·589	28·387	0·17448
1·5	−0·2231	19·721	19·498	0·22985
1·4	−0·2466	14·228	13·981	0·29005
1·3	−0·2725	10·689	10·417	0·35223
1·2	−0·3012	8·3321	8·0309	0·41342
1·1	−0·3329	6·7138	6·3809	0·47080
1·0	−0·3679	5·5750	5·2071	0·52204
0·9	−0·4066	4·7163	4·3097	0·57071
0·8	−0·4493	4·1601	3·7108	0·59975
0·7	−0·4966	3·7201	3·2235	0·62471
0·6	−0·5488	3·3944	2·8456	0·64034
0·5	−0·6065	3·1541	2·5476	0·64716
0·4	−0·6703	2·9797	2·3094	0·64598
0·3	−0·7408	2·8572	2·1164	0·63780
0·2	−0·8187	2·7771	1·9584	0·62369
0·1	−0·9048	2·7323	1·8275	0·60473
0·0	−1·0000	2·7183	1·7183	0·58198
−0·1	−1·1052	2·7315	1·6263	0·55638
−0·2	−1·2214	2·7697	1·5483	0·52880
−0·3	−1·3499	2·8316	1·4817	0·49999
−0·4	−1·4918	2·9163	1·4245	0·47057
−0·5	−1·6487	3·0238	1·3751	0·44107
−0·6	−1·8221	3·1544	1·3323	0·41192
−0·7	−2·0138	3·3088	1·2950	0·38345
−0·8	−2·2255	3·4880	1·2625	0·35592
−0·9	−2·4596	3·6935	1·2339	0·32951
−1·0	−2·7183	3·9270	1·2087	0·30435
−1·1	−3·0042	4·1907	1·1865	0·28054
−1·2	−3·3201	4·4870	1·1669	0·25811
−1·3	−3·6693	4·8188	1·1495	0·23708
−1·4	−4·0552	5·1893	1·1341	0·21744
−1·5	−4·4817	5·6020	1·1203	0·19916

APPENDIX TABLES

TABLE XVII *(cont.)* Minimum and maximum working loglog deviates, ranges, and weighting coefficients

Expected loglog Y	Minimum working deviate η_0	Range A	Maximum working deviate η_1	Weighting coefficient w
−1·6	−4·9530	6·0611	1·1081	0·18220
−1·7	−5·4739	6·5711	1·0972	0·16650
−1·8	−6·0496	7·1370	1·0874	0·15201
−1·9	−6·6859	7·7646	1·0787	0·13866
−2·0	−7·3891	8·4599	1·0708	0·12638
−2·1	−8·1662	9·2300	1·0638	0·11511
−2·2	−9·0250	10·083	1·0575	0·10478
−2·3	−9·9742	11·026	1·0518	0·09532
−2·4	−11·023	12·070	1·0468	0·08667
−2·5	−12·182	13·224	1·0422	0·07876
−2·6	−13·464	14·502	1·0381	0·07155
−2·7	−14·880	15·914	1·0344	0·06497
−2·8	−16·445	17·476	1·0310	0·05898
−2·9	−18·174	19·202	1·0280	0·05352
−3·0	−20·086	21·111	1·0253	0·04856
−3·1	−22·198	23·221	1·0229	0·04404
−3·2	−24·533	25·554	1·0207	0·03994
−3·3	−27·113	28·132	1·0187	0·03621
−3·4	−29·964	30·981	1·0169	0·03282
−3·5	−33·115	34·130	1·0153	0·02974
−3·6	−36·598	37·612	1·0138	0·02695
−3·7	−40·447	41·459	1·0125	0·02442
−3·8	−44·701	45·712	1·0113	0·02212
−3·9	−49·402	50·412	1·0102	0·02004
−4·0	−54·598	55·607	1·0092	0·01815
−4·1	−60·340	61·348	1·0083	0·01644
−4·2	−66·686	67·694	1·0075	0·01488
−4·3	−73·700	74·707	1·0068	0·01348
−4·4	−81·451	82·457	1·0062	0·01220
−4·5	−90·017	91·023	1·0056	0·01105
−4·6	−99·484	100·49	1·0050	0·01000
−4·7	—	110·95	1·0046	0·00905
−4·8	—	122·51	1·0041	0·00820
−4·9	—	135·29	1·0037	0·00742
−5·0	—	149·41	1·0034	0·00672
−5·1	—	165·02	1·0031	0·00608
−5·2	—	182·27	1·0028	0·00550
−5·3	—	201·34	1·0025	0·00498
−5·4	—	222·41	1·0023	0·00451
−5·5	—	245·69	1·0020	0·00408
−5·6	—	271·43	1·0018	0·00369
−5·7	—	299·87	1·0017	0·00334
−5·8	—	331·30	1·0015	0·00302
−5·9	—	366·04	1·0014	0·00274
−6·0	—	404·43	1·0012	0·00248

TABLE XVII (*cont.*)   Minimum and maximum working loglog deviates, ranges, and weighting coefficients

Expected loglog Y	Minimum working deviate η_0	Range A	Maximum working deviate η_1	Weighting coefficient w
−6·1	−	446·86	1·0011	0·00224
−6·2	−	493·75	1·0010	0·00203
−6·3	−	545·57	1·0009	0·00183
−6·4	−	602·85	1·0008	0·00166
−6·5	−	666·14	1·0008	0·00150
−6·6	−	736·10	1·0007	0·00136
−6·7	−	813·41	1·0006	0·00123
−6·8	−	898·85	1·0006	0·00111
−6·9	−	993·28	1·0005	0·00101
−7·0	−	1097·6	1·0005	0·00091
−7·1	−	1213·0	1·0004	0·00082
−7·2	−	1340·0	1·0004	0·00075
−7·3	−	1481·3	1·0003	0·00068
−7·4	−	1637·0	1·0003	0·00061
−7·5	−	1809·0	1·0003	0·00055
−7·6	−	1999·2	1·0003	0·00050
−7·7	−	2209·3	1·0002	0·00045
−7·8	−	2441·6	1·0002	0·00041
−7·9	−	2698·3	1·0002	0·00037
−8·0	−	2982·0	1·0002	0·00034
−8·1	−	3295·5	1·0002	0·00030
−8·2	−	3641·9	1·0001	0·00027
−8·3	−	4024·9	1·0001	0·00025
−8·4	−	4448·1	1·0001	0·00022
−8·5	−	4915·8	1·0001	0·00020
−8·6	−	5432·7	1·0001	0·00018
−8·7	−	6003·9	1·0001	0·00017
−8·8	−	6635·2	1·0001	0·00015
−8·9	−	7333·0	1·0001	0·00014
−9·0	−	8104·1	1·0001	0·00012

TABLE XVIII Distribution of range of transition in dilution series

[Table of probability that the range exceeds a stated value, abridged from Stevens (1958)]

Dilution ratio (*a*)	2	2	4	4	4	4
Plates per dilution (*n*)	1	2	1	2	3	4
Range of transition						
1	·500	·930	·158	·717	·890	·955
2	·500	·820	·158	·373	·511	·682
3	·317	·625	·045	·123	·209	·294
4	·183	·415	·012	·034	·060	·088
5	·099	·246	·003	·009	·015	·023
6	·052	·135	·001	·002	·004	·006
7	·026	·071	—	·001	·001	·001
8	·013	·037	—	—	—	—
9	·007	·019	—	—	—	—
10	·003	·009	—	—	—	—
11	·002	·005	—	—	—	—
12	·001	·002	—	—	—	—
13	—	·001	—	—	—	—
14	—	·001	—	—	—	—

Dilution ratio (*a*)	10	10	10	10	10	10	10
Plates per dilution (*n*)	1	2	3	4	5	6	8
Range of transition							
1	·041	·525	·731	·838	·899	·936	·973
2	·041	·114	·193	·271	·340	·404	·512
3	·004	·013	·023	·035	·047	·060	·087
4	—	·001	·002	·004	·005	·006	·009
5	—	—	—	—	—	·001	·001

Index of authors

Aitchison   369, 449
Allen   367–8, 416, 449
Allmark   38, 117, 405, 451, 459
Allott   236, 449
Annable   273, 449
Anscombe   425, 449
Armitage   273, 276–9, 281, 367–8, 416, 432, 449
Arrigucci   330, 449
Ashford   369, 449
Aso   340, 449
Atkinson   413, 458
Aylor   457

Bacharach   2, 45, 449
Bailey   273, 449
Bangham   439, 449
Banting   339, 449
Barkworth   429, 449
Barr   394, 449
Barraclough   235, 449
Bartels   260, 451
Bartlett   13, 25, 55, 57, 74, 127–9, 152, 162, 270, 273, 414, 449–50
Bartsch   432, 449
Bartter   457
Becker   458
Behrens   82–6, 189, 191, 270, 272, 394, 400, 435, 450
Bennett   120, 266, 273, 281, 290, 394, 397, 449–50
Bentzon   456
Berkson   358–60, 367, 374, 401–3, 450
Biggers   140, 367, 405, 450
Birch   148–9, 450
Black   381, 450
Blackith   413, 450
Bleiker   456
Bliss   3–4, 16, 18, 23, 38, 39, 43, 65, 80, 91, 95, 110, 117–19, 121, 140, 146, 153, 192, 210, 215, 235, 242, 246–9, 260, 268, 285, 357–8, 404, 410–11, 417, 447, 450, 452, 455, 458
Boag   447, 451
Borth   228, 451
Box   43, 57, 222, 308, 441, 451
Braun   27, 404, 451, 458
Bridson   460
Brody   17, 27, 456

Broom   461
Bross   399–400, 451
Brown, B.W.   397, 406, 451–2
Brown, S.   369, 449
Brownlee   235–7, 414, 441, 452
Bülbring   88, 452
Burger   330, 452
Burn   3, 17–18, 55, 88, 148, 203, 452
Butler   47, 452

Carpenter   149, 452
Casper   457
Cattell   3, 43, 146, 417, 451
Cauchy   360–6, 371
Cekan   339, 449
Chang   397, 452
Chapman   117, 459
Cheeseman   395, 456
Chen   27, 33, 36, 452
Christensen   448, 452
Churchman   397, 453
Claringbold   140, 166, 369, 405, 413, 450, 452
Clarke   235, 452
Coates   449
Cochran   6, 37, 55, 177, 183, 191–2, 195, 209, 235, 268, 415, 429, 435, 452, 461
Cohen   121, 452
Colquhoun   6, 24, 211, 452
Cook   335, 346, 452
Cooper   330, 334, 460
Cornfield   297, 381, 393, 397, 400, 453, 457
Coward   3, 453
Cox, C.P.   122, 125, 453
Cox, D.R.   6, 37, 43, 177, 308, 451–3
Cox, G.M.   6, 38, 177, 183, 191–2, 195, 209, 235, 452
Cramer, E.M.   403, 453
Cramér, H.   22, 133, 303, 357, 402, 453
Cullumbine   441, 451
Curtis, A.R.   456
Curtis, J.M.   361, 368, 393, 453, 457
Cutchins   457
Cuthbertson   316

Dagnelie   6, 453
Dahm   369, 456
Dale   3, 5, 453

498 INDEX

Das 201, 453
Davis 415, 452
De Beer 112, 117, 119, 139, 393, 453, 461
Delves 452
Diczfalusy 138, 290, 292–4, 296, 449, 451, 459–60
Dixon 414–15, 453
Dorman 452
Dragstedt 394, 400, 435, 453
Dufrenoy 119, 453, 455

Edge 413, 461
Ehrlich 3, 5
Einstein 24
Eisenhart 428, 453
Ekins 330, 453
Elston 122–3, 453
Emery 47, 57, 105, 127, 453
Emmens 3, 46, 55, 94, 138, 149, 453
Epstein 397, 453
Euclid 305

Falco 117, 461
Fechner 357, 453
Federspiel 457
Fertig 393, 458
Fieller 22–3, 57, 59, 80–7, 92, 99, 104, 118, 125, 139, 149, 155–6, 181, 187–9, 195, 197, 203, 222–3, 225, 237, 241–2, 246, 255–6, 266, 273, 276–9, 284, 289, 304, 320, 325, 337, 341, 346, 356, 373, 405, 412, 453–4, 461
Finney 3, 6, 17, 34, 37, 40, 44, 47, 59, 63, 84, 88, 108, 117, 126, 129, 144, 149, 153, 160, 177, 204, 211, 220–222, 232, 242–3, 262, 273, 285, 290, 293, 296–7, 308, 331–2, 335, 339, 344, 350, 354, 356–8, 368–70, 373, 380, 391, 393, 396–8, 401, 403, 410–11, 414–15, 420, 422–5, 430–1, 433, 448, 449, 452, 454–5, 458–60, 463, 469
Fiorelli 449
Fisher 6, 10, 21, 37, 51, 55–6, 58, 63, 83–4, 107–8, 116, 137–8, 177, 181–3, 189, 191–2, 209–10, 261, 269, 355, 359, 370, 422, 431–2, 434–7, 455, 465–9, 485
Fitzhugh 411, 455
Forti 449
Frazier 329, 460

Gaddum 3, 38, 42, 297, 353, 357, 374, 394, 396, 405, 417, 455
Gard 441, 447, 455
Garwood 63, 381, 455
Gautier 3, 5, 455
Geary 303, 455

Ginsburg 261, 455
Goodwin 3, 17, 452
Gosset 148
Goyan 119, 453, 455
Graca 369, 462
Graham 118, 459
Green 452
Grenfell 452
Grewal 366, 455
Gridgeman 117–19, 121, 123, 179, 182, 308, 455
Griep 456
Grundy 118–19, 457
Guerrero 449
Guillemin 242–3, 246–7, 258, 455, 460
Guld 235, 456
Gurland 369, 456
György 3, 456

Haley 393, 456
Halvorson 429, 456, 463
Hamre 441, 452
Hanson 23, 451
Harding 330, 456
Harris 148–9, 450
Harrison 236, 456
Harte 118, 456
Hartley, H.O. 108, 459
Hartley, P. 3, 5, 42, 456
Hatcher 17, 27, 456
Hay 222, 451
Healy 118–19, 334–5, 347, 410, 456
Heinrichs 451
Hemmingsen 365, 375, 379, 383–9, 456
Hendrie 139, 461
Hodges 414, 452
Holt 332, 391, 455
Hsi 415, 456
Hudson 148
Humphrey 96, 456
Hutt 329, 334–5, 460

Ipsen 46, 447–8, 456
Irwin 3, 395–6, 412, 429, 449, 454, 456

James 242
Jerne 3, 5, 259, 297–305, 313, 456
John, J.A. 6, 177, 457
John, P.W.M. 177, 457
Johnson, E.A. 397, 452
Johnson, J.D.A. 452
Jones, J.I.M. 88, 121, 146, 457
Jones, W.P. 457

Kapteyn 46, 308, 457
Kärber 394–401, 406, 414–15, 434–5, 457
Kempthorne 6, 177, 183, 457
Kendall, J.W. 297, 457
Kendall, M.G. 21–2, 133, 303, 357, 402, 457

INDEX

Kent-Jones 150, 457
Knight 276
Knudsen 117–19, 121, 139, 285, 361, 368, 393, 453, 457
Koch 118, 457
Kogut 458
Kolb 41, 457
Krogh 375, 379, 383–9, 456
Kulkarni 201, 453
Kulshreshtha 201, 457

Lang 394, 453
Lapedes 236–7, 452
Leclerq 330, 457
Lee, I. 369, 456
Lee, V.W.K. 452
Leech 118–19, 457
Lees 47, 119, 127, 231, 235–6, 453, 456–7
Leppäluoto 242, 457
Lewald 334, 460
Liddle 268, 457
Lieberman 394, 459
Lightbown 3–4, 42, 271, 456, 458
Liljestrand 446, 458
Litchfield 393, 458
Long 177, 458
Lord 120, 458
Lucas 211, 458–9

McArthur 126, 458
McClosky 118, 457
McCrady 429, 458
McDonald 149, 452
McKenzie 413, 458
McLaren 140, 458
Maclaurin 61
Magnusson 456
Mantel 381, 393, 397, 400, 453
Marks 91, 117–18, 242, 451, 454, 461
Mather 430–1, 458
Mead 336, 381, 459
Meiklejohn 150, 457
Merrington 466–7
Michie 140, 458
Middleton 449
Midgley 330, 338, 458
Miles 3–5, 38, 42, 177, 297, 437, 458
Miller, L.C. 367, 393, 404, 413, 417, 458
Miller, W.S. 149, 452
Mongar 147, 458
Monro 415, 460
Mood 414–15, 453
Moore 411, 458
Moran 432, 459
Morrell 117, 405, 459
Muench 394, 399–400, 435, 460
Munson 88, 268, 459, 461
Murray 95, 459
Mussett 439, 449, 456, 458
Myerscough 147, 211, 459

Nelder 336, 381, 459

Nelson, A.N. 455
Nelson, J.W. 394, 449
Newman 330, 453
Newton 24, 348
Niswender 330, 458
Noah 2–3
Noel 121, 220, 459
Norris 46, 459
Northam 46, 459

O'Neill 236, 449
Osgood 118, 459
Outhwaite 220, 455

Pabst 121, 451
Packard 65, 358, 451
Patterson 211, 459
Pazzagli 449
Pearce 6, 459
Pearson 108, 459
Perry 5, 38, 177, 416, 418, 437, 440, 446, 456, 458–9
Peto 430, 459
Petrie 273, 449
Petrusz 293–4, 296, 459–60
Phillips 339
Pittman 394, 459
Poisson 334, 344, 423–4, 426–8, 432, 434
Price 197–8, 459
Ptolemy 24

Qazi 296, 459
Quenouille 6, 177, 457

Randall 117–18, 121, 139, 285, 457
Rao 58, 265–6, 403, 460
Raphson 348
Rayford 460
Rebar 330, 458
Reed 394, 399–400, 435, 460
Rennie 452
Rerup 204, 207, 460
Robbins 415, 452, 460
Robertson 138, 460
Robyn 290, 292–4, 459–60
Rodbard 329–35, 337, 346, 460
Rohlf 6, 461
Romani 138, 293, 459–60
Romans 88, 463
Rose 210, 215, 242, 247–9, 451
Rosenblatt 414, 452

Sakiz 242–3, 246–7, 258, 455, 460
Sampford 220, 447–8, 460–1
Schiele 47, 452
Schild 117, 147, 211, 220–2, 297, 455, 459, 461
Searle 102, 461
Sen 24, 120, 461
Serio 449
Sheffield 332, 391, 455
Sheps 88, 139, 268, 459, 461

Sherwood   112, 117–18, 139, 453, 461
Shorack   24, 461
Shrimpton   454
Siegfried   27, 451
Silvey   369, 449
Smith, C.S.   369, 449
Smith, K.W.   204, 279, 461
Smith, N.   452
Smith, R.B.   118, 220, 457, 461
Snedecor   6, 55, 461
Sokal   6, 461
Somers   146, 413, 461
Spearman   394–401, 406, 414–15, 434–5, 461
Spencer   198, 459
Spicer   432, 449
Stack-Dunne   273, 449
Standfast   412, 456
Stevens   432, 434, 447, 461, 495
Stuart   21–2, 133, 303, 357, 402, 457
Sukhatme   12, 15, 83
Swaroop   429, 461–2

Tainter   393, 458
Taljedal   457
Tamaoki   228, 462
Tasman   452
Taylor, B.   61
Taylor, J.   448, 461
Thompson, C.M.   467, 469
Thompson, R.E.   221, 462
Thompson, W.R.   22, 76, 397, 462
Thomson, G.H.   357, 462
Thomson, R.   456
Tootill   47, 127, 231, 235, 453, 457
Trevan   350, 391, 405, 462
Tsutakawa   415, 462
Tukey   305, 403, 462

Ulfelder   126, 458
Umberger   393, 453
Urban   360–6, 371, 462

Van der Waerden   396, 462
Van Ramshorst   452
Van Strik   120, 462
Van Uven   46, 457
Vos   118, 220–1, 457, 461–2

Waaler   456
Wadley   228, 230, 423, 425, 462
Weil   397, 462
Wetherill   415, 462
White   369, 462
Wilcoxon   393, 458
Williams, D.A.   122, 462
Williams, E.J.   6, 166, 211, 462–3
Wilson, E.B.   359–66, 371, 374, 463
Wilson, P.W.   428, 453
Winder   314, 394, 463
Winne   228, 463
Wold   457
Wood, E.C.   1–3, 40, 42–3, 119, 123, 134, 138, 149, 153–4, 160–1, 165, 259, 285, 297–305, 313, 316, 326, 455–6, 463
Wood, F.   236, 456
Worcester   359–66, 371, 463

Yates   10, 21, 51, 55–6, 83–4, 99, 107–8, 116, 137, 177, 181–3, 189, 191–2, 209–10, 249, 355, 359, 370, 422, 431–2, 435, 437, 455, 463, 465–9, 485
Youden   209, 211–15, 236, 463
Young   88, 463

Ziegler   429, 456, 463

Index of subjects

Additivity   290–6
Aldosterone   268
Allocation of subjects   6, 26, 134–7, 141, 145, 169, 173, 175, 179, 191, 209, 215, 235–9, 326, 404–10, 414, 436
Amino-acid   316
Analgesics   366
Analytic dilution assay   22, 40–1, 76, 133, 147, 154, 181, 269–70, 297, 299
Angle, expected   412
  sigmoid   55, 361–6, 371, 374, 393, 403, 404–6, 412–13, 416–17
  , working   375
Animal units   3, 18
Antibody   5, 329
Antigen   5, 328
Antigonadotrophic serum   292
Antisera   290–6
Antiserum neutralization   294–6
Ark   2
Asymptotic efficiency   58–9, 402
  normality   58, 402
Autocatalytic function   359
Autoregression   203
Auxiliary variate   420, 424

Bacillus mesentericus   432–5
  subtilis   113
Bacterial density   426–37
Balanced incomplete blocks   136, 183–95, 211, 213–15, 230, 235, 247–9, 285
Bartlett's test   13, 25, 55–7, 74–5, 127–9, 152, 162, 270, 273
Behrens distribution   82–6, 189, 270, 272
Bias   58, 60, 77, 91, 155, 211, 279, 314, 316, 329, 396–7, 402, 440
Binomial distribution   353, 420
Blanks   14, 153–5, 158, 163–4, 175, 238, 313
BLISS   382–90
Block size   177–8, 192–5, 197–202
Body weight   18, 26–7, 37, 88–9, 240–1, 247–58, 314–15
Bound count   329–31
Box and Cox transformation   308–13
Breeding of subjects   67, 140, 168, 405
British Pharmacopoeia   4–5, 88
British Standards Institution   69

Cage variation   93–4
Cat   17–18, 23, 26–8, 39, 146, 438–9
  method   17, 349
  unit   3, 18
Cauchy–Urban sigmoid   360–6, 371
Censoring   440, 447–8
Central limit theorem   22, 24, 60, 121, 300, 303, 306, 357
Characteristic curve   379, 391–2
  response   39, 349, 352
Check calculation   9, 102, 286, 370
Chemical analysis   2, 4, 22
χ^2 distribution   10, 270, 276, 355–6, 390, 404
  test   56, 294, 355–6, 372–3, 378, 380, 382, 386, 389, 404, 414, 416, 431–4, 437
Choice of design   134, 137, 141, 201,
  dilutions   435–7
  doses   72, 75, 91, 109, 136, 141, 144, 173, 179, 221, 235, 298, 326–7, 338, 405–10
  metameter   43–6, 51–5, 135, 145, 240–2, 297–315, 366–8, 404, 413
  subjects   146–7, 175, 437–9
Coding   69
Cod-liver oil   69–70, 72, 79, 87–8
Coefficient of variation   20
Colony counting   426–8
Colour matching   155
Combination of estimates   8, 122–5, 183, 219, 227, 255–7, 260, 265–8, 269–96, 418–19
  measurements   147, 240–57, 260–8
Comparative dilution assay   22, 41, 154, 270, 297, 299
Computer   8–9, 27, 54, 69, 82, 99, 102, 117–20, 125–32, 165, 201, 258, 277, 279–80, 327–9, 334–8, 340–2, 346–8, 361–2, 368, 370, 380–1, 390–1, 398, 401, 412, 422, 425, 448
Concomitant variate   27, 36–8, 240–59, 268, 314, 418
Confidence interval   21
Confounding   93, 145, 178–201, 204–5, 211, 215, 231–2, 235, 252, 285
  , partial   178–9, 183, 191
Consistency   58, 133, 396, 398, 402
Constant standard design   221

Constraints of design   93–5, 123, 126,
    145, 168, 177, 232, 238, 300, 404,
    413
Contrast   13–14, 90–2, 96, 99, 106–17,
    122, 139, 158–66, 178–9, 183–8,
    191, 196–200, 204–7, 215–18,
    221–4, 229, 249–51, 254, 262–5,
    281–2, 309
Control chart   121–2, 165, 307, 348, 416
Correlation, serial   203, 221
Corticotrophin   1, 204–8, 223–6
Costs   94, 117, 125, 132, 134, 136–7,
    147, 258–9, 351, 391–3, 398, 417
Covariance, analysis of   6–7, 27–36, 89,
    100, 241–59, 261–5, 314–15, 418
Cow   177, 269
Cross-over design   192, 203–26, 235–6,
    242, 247, 249, 279–81, 413, 439
Curvature   14, 48, 51, 77, 95, 107, 116,
    122–5, 129–30, 153–5, 163–4,
    168, 181, 186, 228, 245, 266,
    309–11, 316, 319, 364
Cylinder plate assay   114, 121, 138, 199,
    230, 285

Design   1, 6–8, 17, 27, 37–8, 86, 93–6,
    105, 134–7, 157–65, 167, 169–75,
    177–202, 203–26, 227, 235–9,
    281–6, 303, 306, 326–7, 338, 345,
    351, 397–400, 404–17, 425, 429,
    435–7
Design   (*see also under* Balanced incomplete block *and other specific names*)
    , choice of   134, 137, 141, 201
    constraints   93–5, 123, 126, 145, 168,
    177, 232, 238, 300, 404, 413
    , (2, 2)-point   90, 112–15, 117, 123,
    141–5, 192, 197–8, 204, 209–211, 215, 219–20, 233, 242, 247,
    261, 277, 281, 299, 393, 409,
    446
    , (3, 3)-point   105, 115–17, 141–5,
    192–3, 198–200, 204–5, 209,
    211–12, 220, 242, 262, 271,
    282–3, 292, 409, 411
    , (4, 4)-point   116–17, 122, 145,
    193–4, 200–1, 212–13, 406,
    409
    , (k, k)-point   105–12, 122, 136,
    140–5, 197, 406,
    , (0, 2, 2)-point   164–5, 173–5
    , (0, 3, 3)-point   164–5
    , (0, 4, 4)-point   164–5
    , (0, k, k)-point   159, 164–5, 173–5,
    201
    , (1, 1, 1)-point   159, 170–5, 194, 213
    , (1, 2, 2)-point   159–63, 173, 194,
    214
    , (1, 3, 3)-point   162–63, 193, 195,
    214
    , (1, 4, 4)-point   163, 173, 194–5,
    201, 214–15
    , (1, k, k)-point   157–64, 169–75,
    194, 201, 214
Desk calculator   8–9, 69, 102, 117, 125,
    127, 132, 165, 321–2, 346, 361,
    368, 419
Digitalis   3, 17, 27, 39, 349, 357, 417,
    437–8
Dilution assay   7, 22–4, 37, 40–1, 76,
    133, 135, 147, 154, 181, 269–270, 297, 299, 409
    series   8, 425–37
Dilutions, choice of   435–7
Diphtheria antitoxin   3, 5
Direct assays   7, 17–38, 39, 258, 350,
    357, 416–17
Discriminant analysis   7, 147, 240, 260–8
Distribution   (*see under* Behrens *and other specific names*)
Distribution-free methods   24, 120–1,
    300, 307, 398
Dog   216–19, 247–57
Dose   1–4, 8, 10, 17, 40, 113, 426
Dose errors   300–1
    metameter   13, 45–8, 53, 57, 60,
    69–70, 105, 109–10, 127–8,
    148–9, 229, 261–2, 265, 300, 306,
    316, 351–3, 430
    range   75–9, 123, 135, 141, 145, 167,
    326, 395, 400, 405, 440–1
Dose–response diagram   70–2, 75–9, 89,
    95, 119, 128–31, 322, 392, 412,
    447–8
    relation   1–2, 8, 10, 17, 39–68,
    95, 105–6, 129, 134, 147, 183,
    249, 297–8, 316, 357, 361, 367,
    447
Doses, choice of   72, 75, 91, 109, 136,
    141, 144, 173, 179, 221, 235,
    298, 326–7, 338, 405–10
    , number of   72, 142–5, 173, 235,
    327, 345, 406–10
    , order of   138–9, 208–9, 219–23,
    231
    , range of   75–9, 123, 135, 141, 145,
    167, 326, 395, 400, 405, 440–1
    , spacing of   69, 72, 157, 235, 327,
    395–6, 400, 407, 409, 436
Dragstedt–Behrens method   394, 400,
    435
Drosophila melanogaster   65, 358, 391
Drug standardization   5–6, 42, 65
d-statistic   12, 15, 83–6, 181–2, 190,
    256, 272–3

INDEX

Dummy variate   99–100, 153, 183, 223, 247

Economics   37, 117, 134–5, 175, 227–8, 237, 258–9, 268, 401, 404, 417, 435, 439
Effective constituent   19, 22, 41–2, 78, 146
   variance   407–9, 417
Efficiency   36–8, 58–9, 135, 149, 157, 162, 167–76, 208–9, 236–8, 304, 402, 414–17, 428, 434, 437
Empirical logit   332
   probit   370, 375–7, 412
   response   64, 354, 370, 375, 381, 421, 424–5, 432
   response metameter   62, 448
Equivalent deviate   351–4, 405–6, 412, 417, 424, 430
   deviate, working   354, 359–62, 420, 424, 430, 433
Ergometrine   211
Estimates, combination of   8, 122–5, 183, 219, 227, 255–7, 260, 265–8, 269–96, 418–19
   , homogeneity of   270–1, 276, 281, 294, 437–9
Estimation   6–7, 57–68, 79–81, 119, 122, 133, 155–6, 165, 187, 269–96, 297–8, 305, 307, 329, 336, 351, 353–4, 370, 418, 435
   by eye   46, 60, 71, 80, 89, 150–1, 375, 392
   , graphical   44, 68, 71, 80, 89, 118–120, 148, 150, 329, 392–3
European Pharmacopoeia   4
Expectation   81, 83, 282–4
Expected angle   412
   logit   413
   loglog   430
   probit   371, 379, 412–13
   response   39–40, 64, 128, 148, 282, 298, 420
Experimental conditions   42, 65–7, 94, 121, 135, 139, 165, 195, 410
Extreme effective doses method   394
Eye estimation   46, 60, 71, 80, 89, 150–151, 375, 392

Factorial design   38, 145
Fiducial interval   11, 14, 21–3, 26, 35–7, 67, 76, 80–8, 93, 103–4, 111, 118, 125, 128–9, 139–40, 145, 155–60, 167–9, 181–2, 187–8, 191, 197, 208, 223, 226, 230, 242, 252–7, 266, 269, 284–5, 295, 303, 305, 307–14, 316, 321, 326–7, 337, 341–3, 356, 365–7, 373, 379–82, 390, 394–5, 404–5, 408–9, 431, 434, 445–6

Fieller's theorem   22–3, 80–7, 92, 99, 104, 118, 125, 139, 149, 155–6, 181, 187–9, 195, 197, 222, 225, 246, 255, 266, 273, 276–9, 284, 289, 320, 325, 337, 341, 346, 356, 373, 405, 412
Follicle stimulating hormone (FSH)   293, 339
Food and Drug Administration, United States   121
Free count   329
Fundamental validity   25, 76–7, 134, 153, 175, 258, 297–9, 306, 326–7, 373, 378, 409, 414
Fungicides   3, 349, 419

g 81, 84, 86–7, 104, 111, 118, 128–9, 140–2, 146, 158–60, 168–75, 181–2, 187, 191, 252, 254, 269–70, 290, 309–13, 343, 374, 379–81, 390, 404–8, 418, 437
Gaussian sigmoid   331
GENSTAT   126
Glossary of symbols   12–16
Gonadotrophin   124, 292–3, 296
Graphical estimation   44, 68, 71, 80, 89, 118–20, 148, 150, 329, 392–3
Guinea pig   177, 228, 235, 269, 413, 438–9

Haphazard selection   137
Herbicides   236, 339
Heterogeneity factor   96, 373–4, 404–5
Heteroscedasticity   54–7, 69, 74–5, 78, 92, 127–8, 152, 162, 177, 304, 309–10, 313, 316–17, 443
Homogeneity of estimates   270–1, 276, 281, 294, 437–9
   subjects   140, 300, 355, 405
   variance   25, 54–7, 74–5, 107, 128–9, 155, 203, 211, 270, 272, 274, 280–6, 311, 404–5, 443
Homoscedasticity   54–7, 59, 69, 74–5, 88, 149, 304–6, 311, 316, 440
Human chorionic gonadotrophin (HCG)   292–3, 296
Human subjects   147, 211, 413

Immunity   419–20, 443, 447
Immunoradiometric assay (IRMA)   329, 331, 347
Incomplete blocks   132, 136, 145, 177–202, 209, 220, 235, 242, 249, 285, 410–13
Index of regression significance (g)   81, 84, 86–7, 104, 111, 118, 128–9, 140–2, 146, 158–60, 168–75, 181–2, 187, 191, 252, 254, 269–70, 290, 309–13, 343, 374, 379–81, 390, 404–8, 418, 437

INDEX

Indirect assay 7, 17, 38, 39, 203, 417
Information 58–9, 84, 99, 122, 125, 188, 204, 228, 240–59, 284, 369, 397, 400, 417, 428, 437–9, 440, 443
Insecticides 22, 39, 95, 138, 349, 357, 413, 419
Insulin 3–4, 88, 203, 240–2, 279–81, 349, 365, 375–89, 410, 439
Interaction 179, 205–7, 233, 281–6
Inter-block variation 89, 179–202, 249–57, 285, 413
International standard 5–6, 42, 437–9
  unit 4–5, 42, 208, 311, 365–6, 375, 379, 391, 439
Intersection 14, 46, 60, 67, 153–4, 158, 163–4, 175, 313
Inter-subject variation 203–26, 249–57, 413
Intra-block variation 179–202, 281–9
Intra-subject variation 203–26, 249–57, 413
Inverse matrix 102, 155, 342, 347–8, 382, 412, 421
Iteration 60–5, 67, 84, 97, 132, 275–6, 322–3, 328–30, 334, 336, 345, 347, 352, 354–5, 365, 368, 372, 375, 378–90, 403, 410–13, 420–5, 430

Lactobacillus arabinosus 150, 316
  helveticus 154, 161
  leichmannii 10, 47-9, 127
Latin square 28, 36, 137, 177, 209–20, 232, 236, 411–13
League of Nations 5, 42
Least squares 59, 61, 97, 99, 223, 293, 296, 327–9, 334–47, 403
Likelihood, maximum 9, 16, 47, 57–65, 125, 274–6, 279, 289, 316, 329, 336, 339, 344–5, 353–5, 357, 361–2, 365–8, 370, 379–90, 393, 397–403, 404, 412, 415, 418–19, 425, 428–31, 434–6, 447
Limits, fiducial 11, 14, 21–3, 26, 35–7, 67, 76, 80–8, 93, 103–4, 111, 118, 125, 128–9, 139–40, 145, 155–60, 167–9, 181–2, 187–8, 191, 197, 208, 223, 226, 230, 242, 252–7, 266, 269, 284–5, 295, 303, 305, 307–14, 316, 321, 326–7, 337, 341–3, 356, 365–7, 373, 379–82, 390, 394–5, 404–5, 408–9, 431, 434, 445–6
Linearity 50, 66, 68, 72, 75, 88, 91, 95–6, 107–8, 110, 115, 123, 145, 153, 168, 175, 233, 266, 285, 299, 304, 306, 318–19, 324–7, 334, 356–7, 365–6, 372–3, 378, 380–2, 390, 410, 416, 443

Linearity, range of 51, 66, 72, 75–9, 113, 141, 146, 162, 164, 168–9, 175, 298–9, 326–7, 345
Linearization 43, 45–6, 59–60, 64, 75, 300, 316, 326, 334
Litter-mate control 88–91, 93, 111, 123, 137, 140, 178–80, 202, 208, 281–6, 300, 410–13
Litter size 121, 178, 201–2, 281
Logarithmic dose metameter 45, 53–4, 69, 105, 109–10, 302, 330, 430
  transformation 23, 25, 46, 177, 316–17
Logistic sigmoid 46–7, 298, 330–48, 358–60, 362–6, 371, 374, 397, 399, 406, 416, 435
Logit 330, 332–4, 340, 348, 359–60, 367, 373, 403, 405–6, 408–9, 417
  , empirical 332
  , expected 413
  , working 334, 374–5
Loglog 430–7
  , expected 430
  , working 430
Lognormal distribution 23–4
Luteinizing hormone (LH) 138, 243–6, 292, 339

Matrix inversion 102, 155, 342, 347–8, 382, 412, 421
Maximum likelihood 9, 16, 47, 57–65, 125, 274–6, 279, 289, 316, 329, 336, 339, 344–5, 353–5, 357, 361–2, 365–8, 370, 379–90, 393, 397–403, 404, 412, 415, 418–19, 425, 428–31, 434–6, 447
Maximum working angle 375
  logit 374
  probit 371
  response 420, 430–1
MAXLIKE 336
Median effective dose (ED50) 350–1, 391–400, 405, 408–9, 414–15
Metameter, choice of 43–6, 51–5, 135, 145, 240–2, 297–315, 366–8, 404, 413
  , dose 13, 45–8, 53, 57, 60, 69–70, 105, 109–10, 127–8, 148–9, 229, 261–2, 265, 300, 306, 316, 351–3, 430
  , response 13, 45–6, 51, 56–7, 60–2, 69, 105, 127–8, 177, 240–2, 260–8, 300–1, 304, 306, 308–15, 316, 340, 344, 351, 392, 404, 413, 430, 440–3, 447–8
Metametric transformation 45, 47, 51, 57, 63, 70, 95, 126, 177, 300–3, 307–15, 316–17, 340, 344–5, 350, 352, 358–68, 390, 420, 442–3

INDEX

Microbiological assay   93, 121, 127, 138, 149, 156, 168, 171, 230, 313, 316
Midrange (ID50)   331, 333
Minimal effective dose   350
Minimum χ^2 59, 367–8, 401–3
Minimum working angle   375
   logit   374
   probit   371, 377
   response   63, 420, 430–1
MINISQUARE   328, 336, 340, 347–8
Missing values   8, 96–104, 204, 223–6, 280, 440
Monotony   43–4, 46, 75
Most probable number   429
Mouse   124, 138, 375, 391, 410, 439, 441–3
  convulsion method   349, 365, 375, 439
  unit   3
Moving average method   394, 397–400, 435
Multiple assay   38, 132, 141, 227–39, 293, 296, 391, 414, 441
  regression   99–101, 110, 149, 153, 183, 201, 223, 261, 425

National Institute for Medical Research, London   5
Natural response   368, 419–23, 447
Negative binomial distribution   425
Neoarsphenamine   405, 446
Nicotinic acid   150–1
Nomograph   9, 118–20, 393
Non-linear regression   46–7, 65, 69, 95, 110, 328–48, 350
Non-parallel regressions   76
Non-parametric estimation   24
Normal distribution   22–3, 57–8, 60, 82, 88, 119–20, 139, 261, 267, 273–4, 282, 300, 303, 306, 334, 344, 350–352, 354, 356–7, 396, 398–9, 404–407, 411, 416, 425, 441, 447
  equivalent deviate (N.E.D.)   269, 353, 357–8, 360, 382–9
  sigmoid   357–8, 362–6, 435
Number of doses   72, 142–5, 173, 235, 327, 345, 406–10
  subjects   135, 142, 167, 175, 227, 236, 258, 268, 350, 404–10, 415, 417

Objective function   61, 327–8, 336
Oestradiol   339–46, 349
Oestrone   88–9, 97–9, 118, 136, 145, 178, 202, 285
Oleandomycin   271–2
Operating characteristic   416
Optimization   336, 347–8, 381, 423
Order of doses   138–9, 208–9, 219–23, 231

Orthogonal coefficients   51, 90, 106–7, 116–18, 158–66, 217, 249–51
  partition   232
Orthogonality   90–1, 94, 96, 99–100, 103, 107, 113, 132, 158, 178–9, 183–5, 190, 196, 217, 246, 412, 440
Ouabain   18–23, 25–7, 35
Output from computer   126–31, 346, 383–9

Parallelism   46, 60, 67, 69, 76–9, 91, 107, 115, 134, 179, 181, 198, 204, 221, 232–3, 249, 257–8, 261, 285, 288, 298–9, 311, 319, 324–7, 353, 373, 378, 380–2, 390, 392, 414, 416
Parallel line assay   7, 14, 69–104, 105–132, 135, 139–47, 178, 204, 242, 261, 269–71, 274, 282–5, 303, 336–338, 345, 369, 373, 404, 437
Parathyroid extract   210, 215–19, 242, 247–58
PARLIN   126–32
Partial balance   195
  confounding   178–9, 183, 191
Penicillin   4, 42, 112–15, 139, 144, 178, 285
Pertussis vaccine   391–2, 412
Pharmacopoeias   4–5, 88, 205, 314
Pilot investigation   135–6, 139, 338
Pocket calculator   8, 125
Poisson distribution   334, 344, 423–8
Poliomyelitis virus   441–7
Polytomous response   368–9
Potency   2–4, 6–7, 17–22, 26–7, 36, 38, 39–41, 65, 72, 79–80, 88, 92–3, 99, 104, 107, 115, 128–9, 133, 155–6, 165, 167, 179, 230, 235, 243–6, 253–7, 268, 295, 297, 305, 308–13, 316, 350–1, 365–6, 380–2, 415, 425
Power   134, 178, 191, 200–1, 204, 208, 230, 249, 306
  dose metameter   45, 148–9, 298
Precision   19–22, 27, 37, 39, 69, 74, 89, 91–3, 119–22, 133, 136, 143–4, 147, 167, 170–4, 178, 201, 203, 230, 235, 238, 241, 253, 260, 271, 304, 326–7, 351, 398, 400, 404–5, 413, 415, 422, 429
Preliminary regression investigation   43, 46–55, 59–68, 148, 306, 316
Probit   8, 321, 348, 350, 357–8, 363, 367, 370–1, 380, 398–400, 403, 405–7, 409–10, 412, 414–17, 425, 445
  , empirical   370, 375–7, 412
  , expected   371, 379, 412–13
  , working   371–2, 410, 412, 422

Programming   8, 27, 82, 118, 125–32, 165, 201, 280, 327–8, 334–8, 340–1, 346–8, 368, 370, 380–90, 403, 412, 414, 419, 422–5
Prolactin   243, 262, 310–12
Provisional regression line   61, 67, 379, 421

Quadratic curvature   107, 122–5, 181, 186, 228, 266
  regression   47, 51, 122–5
Qualitative assay   1
Quality control   79, 121, 165, 307, 348, 416
Quantal response   2, 7–8, 10, 12, 17, 39, 63–5, 135, 304, 321, 334, 336, 345, 348, 349–69, 440, 445–6
Quantitative response   1, 7, 17, 39–68, 135, 240, 304, 350–1, 356, 391, 404, 409, 413, 417, 440, 443
Quasi-Latin square   231–2, 235

Rabbit   203, 240, 242, 279, 439
RADIMM   348
Radioimmunoassay (RIA)   47, 328–49
Radioligand assay (RLA)   8, 237, 316, 328–48
Randomization   27–8, 76, 94–6, 137–9, 154, 162, 168, 177, 182–3, 191, 215, 221, 231, 233, 240, 257–8, 285, 301, 319, 324, 327, 404
Randomized blocks   38, 86, 89, 91, 97, 113, 119, 126, 129, 136–7, 157, 168, 177–8, 208, 220, 228, 235, 410–13, 448
Range   13, 63, 372, 374–5, 431
  estimation   119–20, 136, 139, 165
  of doses   75–9, 123, 135, 141, 145, 167, 326, 395, 400, 405, 440–1
  of linearity   51, 66, 72, 75–9, 113, 141, 146, 162, 164, 168–9, 175, 298–9, 326–7, 345
  of transition   432, 434
Rank estimation   394
Rat   10, 88–9, 97, 121, 146, 148, 179, 196, 204–5, 208, 223, 242–4, 247, 262, 286, 349, 417
Reciprocal transformation   45–6, 441
Rectangular distribution   362–6
Reed–Muench method   394, 399–400, 435
Regression   6, 17, 27, 29, 40, 43, 46–55, 59–65, 76, 91, 100, 201, 244–7, 253–4, 261, 298, 300–1
  , multiple   99–101, 110, 149, 153, 183, 201, 223, 261, 425
  , non-linear   46–7, 65, 69, 95, 110, 328–48, 350
  , provisional   61, 67, 379, 421

Reliability   143–5, 159, 167–75, 182, 240, 405, 410
Residuals   210–11
Response   1–4, 10, 22, 26, 39, 42, 147, 176, 240, 260, 349, 419, 423
  , composite   147, 240–57, 260–8
  , empirical   64, 354, 370, 375, 381, 421, 424–5, 432
  , expected   39–40, 64, 128, 148, 282, 298, 420
  metameter   13, 45–6, 51, 56–7, 60–62, 69, 105, 127–8, 177, 240–242, 260–8, 300–1, 304, 306, 308–15, 316, 340, 344, 351, 392, 404, 413, 430, 440–3, 447–8
  , empirical   62, 448
  , natural   368, 419–23, 447
  , quantal   2, 7–8, 10, 12, 17, 39, 63–65, 135, 304, 321, 334, 336, 345, 348, 349–69, 440, 445–446
  , quantitative   1, 7, 17, 39–68, 135, 240, 304, 350–1, 356, 391, 404, 409, 413, 417, 440, 443
  , working   62–3, 321–2, 354, 359–362, 420, 424, 430–1, 433
Responsiveness   12, 67, 75, 140, 220–1, 230, 242, 258
Riboflavin   134, 154, 161, 171, 311–13
Rope bacillus   432–5
Routine assays   8–9, 121–2, 128, 136, 165, 237, 306, 338, 348, 448

Scedasticity transformation   54, 59–60, 62, 64, 304, 442
Sensitivity   12
Sequential sampling   135, 414–15
Serial correlation   203, 221
Serially balanced sequence   220
Sigmoid   59, 330–4, 346, 349–69
  , angle   55, 361–6, 371, 374, 393, 403, 404–6, 412–13, 416–17
  , Cauchy–Urban   360–6, 371
  , logistic   46–7, 298, 330–48, 358–360, 362–6, 371, 374, 397, 399, 406, 416, 435
  , normal   357–8, 362–6, 435
  , Wilson–Worcester   360–6, 371
Significance tests   7, 11, 74–9, 96, 108, 116, 120, 178, 261, 263–6, 270, 276, 298, 303, 309, 318, 355, 372–3, 378
Similarity   41–3, 45, 69, 75–6, 134, 153, 221, 297–9, 304, 306, 325, 337, 350, 353, 394
Simultaneous trial estimation   67–8
Single subject assay   219–23
Size of block   177–8, 192–5, 197–202
Skewness   307, 337, 442

Slope ratio assay   7, 12, 14, 134–5, 148–166, 167–76, 178, 234–5, 238, 261, 269, 285, 290, 303, 311, 437
Spacing of doses   69, 72, 157, 235, 327, 395–6, 400, 407, 409, 436
Spearman–Kärber method   394–401, 406, 414–15, 434–5
Square-root transformation   45–6, 177
Staircase method   135, 414–15
Standard curve estimation   65–8, 150, 340, 391
   preparation   4–5, 7, 10, 17, 22, 39–41, 69, 128, 238, 297
   slope estimation   66–8, 80, 203, 392
State Serum Institute, Copenhagen   5
Statistical inference   57–9, 137
   validity   68, 75–9, 134, 144–5, 153–154, 159, 175, 233, 288, 299–307, 310, 356, 372, 409
Stimulus   2–3, 10, 17, 22, 38, 39, 146, 260–1, 299, 315, 349, 419–20, 447
Stochastic approximation   415
Streamlined computation   10, 117–18, 136
Strophanthus   18–23, 25–6
Subject   2–3, 10, 17, 26, 38, 39, 146–7, 197, 211, 219–23, 240, 260, 349, 410, 413, 419, 440
Subjects, allocation of   6, 26, 134–7, 141, 145, 169, 173, 175, 179, 191, 209, 215, 235–9, 326, 404–10, 414, 436
  , breeding of   67, 140, 168, 405
  , choice of   146–7, 175, 437–9
  , homogeneity of   140, 300, 355, 405
  , number of   135, 142, 167, 175, 227, 236, 258, 268, 350, 404–10, 415, 417
Sufficiency   59, 397
Super-complete blocks   195, 211
Symbols   12–16
Symmetry   69, 77, 88–91, 96, 105–32, 136, 141–4, 157–65, 169–75, 177–8, 183, 191, 202, 235–6, 247, 274, 298, 304, 316, 327, 399, 409

Taylor–Maclaurin expansion   61
t distribution   10, 21–2, 26, 81–3, 271–272, 303, 357, 404–5
Test preparation   4, 7, 10, 17, 22, 39–41, 69, 128, 133, 227, 234–9, 297, 346, 416, 418, 426
Time-lag   38, 39
Time response   8, 440–8
Tolerance   18–19, 22–3, 26, 28, 33, 39, 349, 418, 447
   distribution   15, 22–4, 349–69, 373, 392, 394, 397–8, 405, 409–11, 415–16, 419, 425, 430, 446

Transformation, angle   55, 361–6, 371, 374, 393, 403, 404–6, 412–13, 416–17
  , Box and Cox   308–13
  , Cauchy–Urban   360–6, 371
  , equivalent deviate   351–4, 405–6, 412, 417, 424, 430
  , logarithmic   23, 25, 46, 177, 316–17
  , logit   46–7, 298, 330–48, 358–60, 362–7, 371, 373–4, 397, 399, 403, 405–6, 408–9, 416–17, 435
  , loglog   430–7
  , metametric   45, 47, 51, 57, 63, 70, 95, 126, 177, 300–3, 307–15, 316–17, 340, 344–5, 350, 352, 358–68, 390, 420, 442–3
  , probit   8, 321, 348, 350, 357–8, 362–7, 370–1, 380, 398–400, 403, 405–7, 409–10, 412, 414–417, 425, 435, 445
  , reciprocal   45–6, 441
  , scedasticity   54, 59–60, 62, 64, 304, 442
  , square-root   45–6, 177
  , Wilson–Worcester   360–6, 371
Transition range   432, 434
Triple cross-over   204–8, 223–6, 439
Truncation   447
Tryptophan   316–27
Tuberculin   177, 228–30, 269
Twin cross-over   204, 249, 279, 413, 439

United States Food and Drug Administration   121, 205
United States Pharmacopeia   4, 205, 314
Unknown number of subjects   423–5
Unreduced design   192–4

Validity   7–8, 38, 41, 69, 88, 107, 119, 133–4, 146, 150, 153–4, 159, 165, 182, 187, 201, 207, 222, 228, 233, 235, 244–5, 276, 293, 297–315, 325–6, 390, 405, 426, 435, 443
  , fundamental   25, 76–7, 134, 153, 175, 258, 297–9, 306, 326–7, 373, 378, 409, 414
  , statistical   68, 75–9, 134, 144–5, 153–4, 159, 175, 233, 288, 299–307, 310, 356, 372, 409
   tests   74–6, 91–2, 99, 103, 105, 107–8, 110–11, 118, 122–3, 133–6, 139, 141–2, 152–5, 162, 169, 175, 178, 181, 186–8, 195, 200, 204, 221, 230, 235, 244–7, 254, 257–9, 273, 284, 306–13, 347, 354, 365–7, 370, 380–2, 399–401, 409–10, 415, 433–4

INDEX

Variance, analysis of 6, 28, 32, 48–52, 72–4, 81, 90–1, 94, 97–8, 100–3, 105, 107, 110–13, 115, 124, 128–30, 150, 152, 154, 157–8, 160, 178–80, 183–91, 196, 203–7, 216–18, 228–9, 232–3, 249, 261, 281–3, 286–8, 318–20, 324–5, 412, 443–4
— , effective 407–9, 417
— function 13–14, 132, 304–5, 316, 326, 334–5, 339, 344–5, 347
— homogeneity 25, 54–7, 74–5, 107, 128–9, 155, 203, 211, 270, 272, 274, 280–6, 311, 404–5, 443
— ratio distribution 10, 266–7, 276, 286, 309
— test 25, 76, 108, 111, 276, 373
Variation, coefficient of 20
— , inter-block 89, 179–202, 249–57, 285, 413
— , inter-subject 203–26, 249–57, 413
— , intra-block 179–202, 281–9
— , intra-subject 203–26, 249–57
Vasopressin 262
Virus assay 197–8, 432, 441–7
Vitamin A 179–91, 195–7
— B_1 148, 417

Vitamin
— B_1 148, 417
— B_{12} 10, 47–54, 65–7, 105–9, 111, 116, 118, 126–31, 144, 230–4, 236
— D 88, 146, 242, 308–10
— D_3 69–71, 86–8, 286–9

Weighted mean 83–5, 191, 222, 269–73, 322, 418, 437
Weighting coefficient 13, 61, 63, 321, 354, 359–62, 371, 374–5, 377, 404, 420, 424, 431, 435–6
Wilson–Worcester sigmoid 360–6, 371
Working angle 375
— equivalent deviate 354, 359–62, 420, 424, 430, 433
— logit 334, 374–5
— loglog 430
— probit 371–2, 410, 412, 422
— response 62–3, 321–22, 354, 359–362, 420, 424, 430–1, 433
World Health Organization 5, 42, 271

X-rays 65, 88, 146, 358, 391

Youden square 209, 211–15, 236